Handbook of Structural and Construction Engineering

Handbook of Structural and Construction Engineering

Edited by **Ray Anton**

Willford Press
New York

Published by Willford Press,
118-35 Queens Blvd., Suite 400,
Forest Hills, NY 11375, USA
www.willfordpress.com

Handbook of Structural and Construction Engineering
Edited by Ray Anton

International Standard Book Number: 978-1-68285-253-8 (Hardback)

Printed in China.

Contents

Preface

Every book is a source of knowledge and this one is no exception. The idea that led to the conceptualization of this book was the fact that the world is advancing rapidly; which makes it crucial to document the progress in every field. I am aware that a lot of data is already available, yet, there is a lot more to learn. Hence, I accepted the responsibility of editing this book and contributing my knowledge to the community.

Infrastructure is a key factor contributing towards the development of any nation. Advances in technology have revolutionized the way we look at buildings today. Design methods, repairs, strengthening and maintenance of structures, failure investigation, structural safety, etc. are some of the topics that have been discussed within this text. The various studies that are constantly contributing towards advancing technologies and evolution of structural and construction engineering are examined in detail.

While editing this book, I had multiple visions for it. Then I finally narrowed down to make every chapter a sole standing text explaining a particular topic, so that they can be used independently. However, the umbrella subject sinews them into a common theme. This makes the book a unique platform of knowledge.

I would like to give the major credit of this book to the experts from every corner of the world, who took the time to share their expertise with us. Also, I owe the completion of this book to the never-ending support of my family, who supported me throughout the project.

Editor

1

Vibration of wavy single-walled carbon nanotubes based on nonlocal Euler Bernoulli and Timoshenko models

Payam Soltani[1,2*], Amir Kassaei[1], Mohammad Mehdi Taherian[3] and Anooshiravan Farshidianfar[2]

Abstract

The transverse vibration of a single-walled carbon nanotube (SWCNT) with light waviness along its axis is modeled by the nonlocal Euler-Bernoulli and Timoshenko beam theory. Unlike the Euler-Bernoulli beam model (EBM), the effects of transverse shear deformation and rotary inertia are considered within the framework of the Timoshenko beam model (TBM). The surrounding elastic medium is described as both Winkler-type and Pasternak-type foundation models. The governing equations are derived using Hamilton's principle, and the Galerkin method is applied to solve these equations. According to this study, the results indicate that the frequency calculated by TBM is lower than that obtained by EBM. Detailed results show that the importance of transverse shear deformation and rotary inertia become more significant for stocky SWCNTs with clamped-clamped boundary conditions. Moreover, the influences of the amplitude of waviness, nonlocal parameter, medium constants, boundary conditions and aspect ratio are analyzed and discussed. It is shown that waviness in the curved SWCNT causes an obvious increase in the natural frequency in comparison with the straight SWCNT, especially for a compliant medium, pinned-pinned boundary condition, short SWCNT and large nonlocal coefficient.

Keywords: Curved single-walled carbon nanotubes, Vibration, Timoshenko beam model, Nonlocal elasticity theory

Introduction

The discovery of carbon nanotubes (CNTs) by Ijima (Iijima 1991) in the past decade has stimulated research communities in nanotechnology and the development of nano-scale functional devices designed for, and devoted to, carbon nanostructures. The superior mechanical, thermal and electrical properties of CNTs have been investigated in extensive research. For instant, the results show that in its mechanical properties, CNTs have exhibited excellent mechanical stiffness and strength (Dresselhaus et al. 2004; Treacy et al. 1996; Walgraef 2007). These properties have caused CNTs to become one of the most promising materials for nano-electronics, nano-devices and nano-composites (Chen et al. 2004; Hafner et al. 1999; Nishide et al. 2003; Tsukagoshi et al. 2002).

Recently, the mechanical behavior has become the main topic of interest. Therefore, several studies have been done to investigate the vibration of CNTs, since it is essential to understand their dynamical behavior. There are two main categories for simulating the mechanical and physical properties of CNTs: the first is atomic modeling, which includes such techniques as classical molecular dynamics (MD). This method is time consuming, complex, expensive in computational cost and limited to maximum system sizes of about 10^9 atoms (Lau et al. 2004; Liew et al. 2004). The second type is continuum-based modeling, which includes elastic beam models (Yoon et al. 2005; Zhang et al. 2006; Wang et al. 2008; Kuang et al. 2009) and elastic shell theories (Yakobson et al. 1997; Nguyen et al. 2009; Yan et al. 2009; Liew & Wang 2007). For the beam models, the Euler-Bernoulli beam model (EBM) (Fu et al. 2006; Chaterjee & Pohit 2009; Chowdhury et al. 2009) as well as the Timoshenko beam model (TBM) (Hsu et al. 2008; Wang et al. 2006; Lee & Chang 2009; Mahdavi et al. 2011) have been widely employed. The EBM ignores the

* Correspondence: payam.soltani@gmail.com
[1]Department of Mechanical Engineering, Semnan Branch, Islamic Azad University, Semnam, Iran
[2]Department of Mechanical Engineering, Ferdowsi University of Mashhad, Mashhad, Iran
Full list of author information is available at the end of the article

effect of rotary inertia and transverse shear deformation and is applicable to analysis of vibrational characteristics of CNTs. Therefore, the TBM model offers a more precise model compared with the Euler-Bernoulli theory (Rao 2007). The above mentioned literature cannot consider the nano-scale effect on the formulation. The small-scale effect in elasticity is described by assuming that the stress at a point is a function, not only of the strain at that point, but the stress at all reference points and is a function of strain field at every point in the body, as proposed in Eringen's nonlocal elasticity theory (Eringen 1983). Nowadays, many studies have been devoted to analysis of the vibration of CNTs, based on this non-classical theory (Murmu & Pradhan 2009a; Şimşek 2011; Mehdipour et al. 2011; Wang & Ni 2008; Yang et al. 2010; Murmu & Pradhan 2009b).

Most previous theoretical studies are for straight nanotubes, but theoretic and experimental investigations show that the CNTs are not straight in the environment. There are few investigations performed to examine the vibration of curved CNTs. For example, Joshi et al. (Joshi et al. 2010) investigated the vibrational response analysis of carbon nanotubes with their waviness treated as a thin shell. Furthermore, Mayoof and Hawwa (Hawwa & Mayoof 2009) used classical EBM to study the dynamics of the CNT when it acted as the first-mode resonator, with a focus on the chaotic behavior of a curved carbon nanotube under harmonic excitation.

In the present study, both nonlocal Euler-Bernoulli and Timoshenko beam theories are used to simulate the linear transverse vibration of an embedded SWCNT with waviness. To the best of our knowledge, no previous work has been done concerning the vibration behavior of curved CNTs based on TBM. The surrounding elastic medium is described as a Winkler and Pasternak model. The equations of motion are derived by Hamilton's principle, and they are solved using the Galerkin method to obtain fundamental frequencies. The importance of rotary inertia and transverse shear deformation has been explored by comparing the natural frequencies for both EBM and TBM theories. In addition, the effects of amplitude of curvature on the fundamental frequency are discussed. Moreover, the variation of frequency has been considered based on the different parameters such as the surrounding elastic medium, the boundary conditions, the aspect ratio of the SWCNT and the nonlocal coefficient.

Modeling

The system under consideration in Figure 1 is a doubly simply supported ends SWCNT with immovable ends, length L, initial waviness amplitude H, Young's modulus E, shear modulus G, cross-sectional area A, cross-sectional moment I, mass moment inertia J and mass per unit length m. The surrounding elastic medium is simulated by both Pasternak-type and Winkler-type variables. The Winkler-type elastic foundation is assumed to act as a vertical linear spring while the Pasternak-type explains the transverse shear stress due to interaction between SWCNT and shear deformation of its medium (Murmu & Pradhan 2009b). This medium is defined by K_W and K_G to represent the Winkler modulus and Pasternak shear modulus of the elastic medium, respectively.

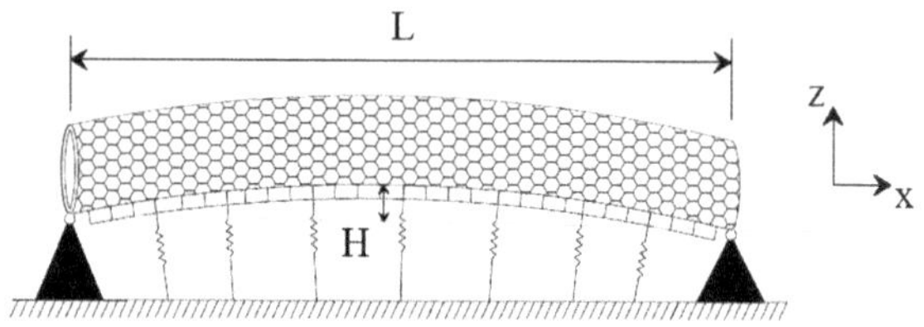

Figure 1 **Configuration of the doubly simply support single-walled carbon nanotube resonator having waviness defined by the amplitude *H*.**

Nonlocal Timoshenko Beam Theory

Because it considers the transverse shear deformation and rotary inertia, TBM provides more reliable results compared to other elastic beam theories. Thus, based on this theory, the displacement field in *x-z* plane and at any point in the body is as

$$\tilde{U}(x,z,t) = U(x,t) + z\psi(x,t),\ \tilde{W}(x,z,t) = W(x,t) \tag{1}$$

Where x is the longitude coordinate, z the coordinate measured from the mid-plane of the beam and t is time. Moreover, $U(x,t)$ and $W(x,t)$ are axial and transverse displacement components in the mid-plane, respectively, and ψ is the rotation of the cross- section.

According to Eq. (1), the linear and nonzero strains are expressed as

$$\varepsilon_{xx} = \frac{\partial U}{\partial x} + \frac{\partial R}{\partial x}\frac{\partial W}{\partial x} + z\left(\frac{\partial \psi}{\partial x}\right),\ \varepsilon_{zz} = 0,\ \varepsilon_{xz} = \frac{\partial W}{\partial x} + \psi \tag{2}$$

Where, ε_{xx} is the axial strain, ε_{xz} is the shear strain and $R(x)$ represents the small rise function that is described as $R(x) = H\sin(\pi x/L)$.

The potential energy of strain V_s is given by

$$V_s = \frac{1}{2}\int_0^L \int_A (\sigma_{xx}\varepsilon_{xx} + \sigma_{xz}\varepsilon_{xz})dAdx \tag{3}$$

Where, σ_{xx} and σ_{xz} are normal stress and shear stress, respectively. By substituting Eq. (2) into Eq. (3), the strain energy V_s can be written as

$$\begin{aligned} V_s &= \frac{1}{2}\int_0^L \int_A \left\{ \sigma_{xx}\left[\frac{\partial U}{\partial x} + \frac{\partial \phi}{\partial x}\frac{\partial W}{\partial x} + z\left(\frac{\partial \psi}{\partial x}\right)\right] + \sigma_{xz}\left(\frac{\partial W}{\partial x} + \psi\right)\right\} dAdx \\ &= \frac{1}{2}\int_0^L \left\{ N_x\left[\frac{\partial U}{\partial x} + \frac{\partial \phi}{\partial x}\frac{\partial W}{\partial x}\right] + M_x\left(\frac{\partial \psi}{\partial x}\right) + Q_x\left(\frac{\partial W}{\partial x} + \psi\right)\right\} dx \end{aligned} \tag{4}$$

The axial resultant force N_x, bending moment M_x and transverse resultant force Q_x can be obtained from

$$N_x = \int_A \sigma_{xx} dA, M_x = \int_A \sigma_{xx} z dA, Q_x = \int_A \sigma_{xz} dA. \tag{5}$$

The kinetic energy T is calculated from

$$T = \frac{1}{2} m \int_0^L \left(\frac{\partial W}{\partial t}\right)^2 dx + \int_0^L J\frac{\partial \psi}{\partial t} dx. \tag{6}$$

The work V_e done by external load can be expressed as

$$V_e = \int_0^L f_e W dx. \tag{7}$$

f_e indicates the external load and corresponds to

$$f_e = -K_W W + K_G \frac{\partial^2 W}{\partial x^2}. \tag{8}$$

According to the nonlocal elasticity theory of Eringen, shear force and bending moment in the nonlocal beam theory are different from those in the classical TBM due to the nonlocal constitutive equations (Eringen 2002), The one-dimensional form for nonlocal constitutive relations can be approximated as

$$\sigma_{xx} - (e_0 a)^2 \frac{d^2 \sigma_{xx}}{dx^2} = E\varepsilon_{xx} \tag{9 – a}$$

$$\sigma_{xz} - (e_0 a)^2 \frac{d^2 \sigma_{xz}}{dx^2} = G\varepsilon_{xz}. \tag{9 – b}$$

It should be noted that by setting the nonlocal parameter $e_0 a = 0$ the constitutive relations in classical elasticity theories can be recovered.

Substituting Eqs. (4), (6) and (7) into Hamilton's principle $\int_0^L (\delta T - \delta V_s + \delta V_e) dt = 0$, integrating by parts and setting the coefficients of δW, δU and $\delta \psi$ to zero leads to the differential equations of motion as

$$\delta U : \frac{\partial N_x}{\partial x} = 0 \tag{10 – a}$$

$$\delta W : \frac{\partial^2 M_x}{\partial x^2} = m\left(\frac{\partial^2 W}{\partial t^2}\right) - \frac{\partial}{\partial x}\left(N_x \frac{\partial W}{\partial x}\right) - \frac{\partial}{\partial x}\left(N_x \frac{\partial Z}{\partial x}\right) - \frac{\partial Q_x}{\partial x} + f_e \tag{10 – b}$$

$$\delta \psi : \frac{\partial Q_x}{\partial x} = J\frac{\partial^2 \psi}{\partial t^2} - \frac{\partial M}{\partial x} + Q_x \tag{10 – c}$$

and corresponding boundary conditions at beam ends ($x = 0, L$) entail

$$U = 0 or N_x = 0, W = 0 or Q_x = 0, \psi = 0 or M_x = 0. \tag{11}$$

By using Eqs. (2), (5) and (9), the normal resultant force, bending moment and shear force can be expressed as

$$N_x - (e_0 a)^2 \frac{\partial^2 N_x}{\partial x^2} = EA\left(\frac{\partial U}{\partial x} + \frac{\partial R}{\partial x}\frac{\partial W}{\partial x}\right) \tag{12 – a}$$

$$M_x - (e_0 a)^2 \frac{\partial^2 M_x}{\partial x^2} = EI\frac{\partial \psi}{\partial x} \tag{12 – b}$$

$$Q_x - (e_0 a)^2 \frac{\partial^2 Q_x}{\partial x^2} = kAG\left(\frac{\partial W}{\partial x} + \psi\right). \tag{12 – c}$$

Where k is the shear correction factor, depending on the shape of the cross-section of the beam. By sequence substituting Eq. (10-a), (10-b) and (10-c) into Eqs. (12-a), (12-b) and (12-c), the explicit expressions of nonlocal normal resultant force N_x, bending moment M_x and shear force Q_x can be obtained as

$$N_x = EA\left(\frac{\partial U}{\partial x} + \frac{\partial R}{\partial x}\frac{\partial W}{\partial x}\right) \tag{13 – a}$$

$$M_x = EI\frac{\partial \psi}{\partial x} + (e_0 a)^2 \left(m\frac{\partial^2 W}{\partial t^2} - \frac{\partial}{\partial x}\left(N_x \frac{\partial R}{\partial x}\right) + J\frac{\partial^3 \psi}{\partial t^2 \partial x} - f_e\right) \tag{13 – b}$$

$$Q_x = kAG\left(\frac{\partial W}{\partial x} + \psi\right) + (e_0a)^2 \left(m\frac{\partial^3 W}{\partial x \partial t^2} - \frac{\partial^2}{\partial x^2}\left(N_x\frac{\partial R}{\partial x}\right) - \frac{\partial f_e}{\partial x}\right). \quad (13-c)$$

The governing equation for nonlocal SWCNTs can be derived by inserting Eqs. (13) into Eqs. (10)

$$EA\left(\frac{\partial^2 U}{\partial x^2} + \frac{\partial^2 R}{\partial x^2}\frac{\partial W}{\partial x} + \frac{\partial R}{\partial x}\frac{\partial^2 W}{\partial x^2}\right) = 0 \quad (14-a)$$

$$EA\left[(e_0a)^2\Sigma_1 - \Sigma_2\right] - (KAG + K_G)\frac{\partial^2 W}{\partial x^2} + m\frac{\partial^2 W}{\partial t^2} - kAG\frac{\partial \psi}{\partial x} + K_W W - m\frac{\partial^4 W}{\partial x^2 \partial t^2} - K_W\frac{\partial^2 W}{\partial x^2} + K_G\frac{\partial^4 W}{\partial x^4} = 0 \quad (14-b)$$

$$J\left[(e_0a)^2\frac{\partial^4 \psi}{\partial x^2 \partial t^2} - \frac{\partial^2 \psi}{\partial t^2}\right] - KAG\left(\frac{\partial W}{\partial x} + \psi\right) + EI\frac{\partial^2 \psi}{\partial x^2} = 0 \quad (14-c)$$

Where

$$\Sigma_1 = \frac{\partial R}{\partial x}\left(\frac{\partial^4 U}{\partial x^4} + 2\frac{\partial^4 R}{\partial x^4}\frac{\partial W}{\partial x} + 6\frac{\partial^3 R}{\partial x^3}\frac{\partial^2 W}{\partial x^2} + 6\frac{\partial^2 R}{\partial x^2}\frac{\partial^3 W}{\partial x^3} + \frac{\partial R}{\partial x}\frac{\partial^4 W}{\partial x^4}\right) + \frac{\partial^2 R}{\partial x^2}\left(3\frac{\partial^3 U}{\partial x^3} + 6\frac{\partial^3 R}{\partial x^3}\frac{\partial W}{\partial x} + 6\frac{\partial^2 R}{\partial x^2}\frac{\partial^2 W}{\partial x^2}\right) + 3\frac{\partial^3 R}{\partial x^3}\frac{\partial^2 U}{\partial x^2} + \frac{\partial^4 R}{\partial x^4}\frac{\partial U}{\partial x} \quad (15-a)$$

$$\Sigma_2 = \frac{\partial R}{\partial x}\left(\frac{\partial^2 U}{\partial x^2} + 2\frac{\partial^2 R}{\partial x^2}\frac{\partial W}{\partial x} + \frac{\partial R}{\partial x}\frac{\partial^2 W}{\partial x^2}\right) - \frac{\partial^2 R}{\partial x^2}\frac{\partial U}{\partial x}. \quad (15-b)$$

By integrating over the domain [0-L] from Eq. (14-a), displacement along x axes can be obtained as

$$U = -\int_0^x \left[\frac{1}{2}\left(\frac{\partial W}{\partial x}\right)^2 + \frac{\partial R}{\partial x}\frac{\partial W}{\partial x}\right]dx + x\frac{1}{L}\int_0^L \left[\frac{1}{2}\left(\frac{\partial W}{\partial x}\right)^2 + \frac{\partial R}{\partial x}\frac{\partial W}{\partial x}\right]dx. \quad (16)$$

Eqs. (14-a), (14-b) and 14-c) are consistent basic equations of the curved TBM. By using Eq. (16) and eliminating ψ, Eqs. (14) are summarized as an uncoupled equation as

$$\frac{(e_0a)^4}{kAG}\left(-\frac{EJA}{L}\frac{\partial^6 R}{\partial x^6}\int_0^L \frac{\partial R}{\partial x}\frac{\partial W}{\partial x}dx + JK_W\frac{\partial^6 W}{\partial x^4 \partial t^2} - K_G\frac{\partial^8 W}{\partial x^6 \partial t^2} + Jm\frac{\partial^8 W}{\partial x^4 \partial t^4}\right)\frac{-b \pm \sqrt{b^2 - 4ac}}{2a}$$
$$+ \frac{(e_0a)^2}{kAG}\left[-\frac{2JEA}{L}\frac{\partial^4 R}{\partial x^4}\int_0^L \frac{\partial R}{\partial x}\frac{\partial^3 W}{\partial x \partial t^2}dx + \left(EI\frac{\partial^6 R}{\partial x^6} - kAG\frac{\partial^4 R}{\partial x^4}\right)\frac{EA}{L}\int_0^L \frac{\partial R}{\partial x}\frac{\partial W}{\partial x}dx + 2JK_W\frac{\partial^4 W}{\partial x^2 \partial t^2} + Jm\frac{\partial^6 W}{\partial x^2 \partial t^4} - mEI\frac{\partial^6 W}{\partial x^4 \partial t^2} - JK_G\frac{\partial^6 W}{\partial x^4 \partial t^2} + EI\left(K_G\frac{\partial^6 W}{\partial x^6} - K_W\frac{\partial^4 W}{\partial x^4}\right)\right]$$
$$+ (e_0a)^2\left(J\frac{\partial^6 W}{\partial x^4 \partial t^2} + m\frac{\partial^4 W}{\partial x^2 \partial t^2} - K_G\frac{\partial^4 W}{\partial x^4} + K_W\frac{\partial^2 W}{\partial x^2}\right)$$
$$+ \frac{1}{KAG}\left[\left(-EI\frac{\partial^4 R}{\partial x^4} + kAG\frac{\partial^2 R}{\partial x^2}\right)\frac{EA}{L}\int_0^L \frac{\partial R}{\partial x}\frac{\partial W}{\partial x}dx + \frac{JEA}{L}\frac{\partial^2 R}{\partial x^2}\int_0^L \frac{\partial R}{\partial x}\frac{\partial^3 W}{\partial x \partial t^2}dx + (JK_G + mEI)\frac{\partial^4 W}{\partial x^2 \partial t^2} - (EIK_G + Jm)\frac{\partial^4 W}{\partial x^4} + (EI - J)K_W\frac{\partial^4 W}{\partial x^4}\right]$$
$$+ J\frac{\partial^4 W}{\partial x^2 \partial t^2} + EI\frac{\partial^4 W}{\partial x^4} - m\frac{\partial^2 W}{\partial x^2} + K_G\frac{\partial^2 W}{\partial x^2} + K_W W = 0. \quad (17)$$

Nonlocal Euler-Bernoulli Beam Model

The governing equation of EBM contains only the deflection of the beam because the effects of transverse shear deformation and rotary inertia are ignored. Therefore, the linear strain–displacement relations for curved Euler beam are expressed as

$$\varepsilon_{xx} = \frac{\partial U}{\partial x} + \frac{\partial R}{\partial x}\frac{\partial W}{\partial x} - z\left(\frac{\partial^2 W}{\partial x^2}\right), \varepsilon_{zz} = \varepsilon_{xz} = 0. \quad (18)$$

By performing a similar approach to that presented for the TBM, the differential equation that governs the vibration of the curved SWCNT based on the EBM is obtained as

$$
\begin{aligned}
&EI\frac{\partial^4 W}{\partial x^4} + m\left(\frac{\partial^2 W}{\partial t^2} - (e_0 a)^2\frac{\partial^4 W}{\partial x^2 \partial t^2}\right) \\
&\quad + K_G\left((e_0 a)^2\frac{\partial^4 W}{\partial x^4} - \frac{\partial^2 W}{\partial x^2}\right) \\
&\quad + K_W\left(W - (e_0 a)^2\frac{\partial^2 W}{\partial x^2}\right) \\
&\quad = EA\left(\frac{\partial^2 R}{\partial x^2} - (e_0 a)^2\frac{\partial^4 R}{\partial x^4}\right)\int_0^x\left(\frac{\partial R}{\partial x}\frac{\partial W}{\partial x}\right)dx.
\end{aligned} \tag{19}
$$

The boundary conditions for SWCNTs with immovable ends can be considered as

$$
W = \psi = 0 x = 0, L. \tag{20}
$$

For clamped ends and

$$
\begin{aligned}
&W = 0, \\
&M_x = EI\frac{\partial \psi}{\partial x} + (e_0 a)^2 \quad m\frac{\partial^2 W}{\partial t^2} - \frac{\partial}{\partial x}\left(N_x\frac{\partial R}{\partial x}\right) \\
&+J\frac{\partial^3 \psi}{\partial t^2 \partial x} + K_W W - K_G\frac{\partial^2 W}{\partial x^2}\Bigg) = 0 \qquad x = 0, L
\end{aligned} \tag{21}
$$

For pinned ends.

Solution Method

In this study, the free vibration equation of the curved SWCNT has been investigated by using the nonlocal TBM and EBM. The influences of transverse shear deformation and rotary inertia on the vibration frequencies are investigated by comparing the TBM results with those based on the EBM.

For rewriting the governing equations in a non-dimensional form, introducing the following dimensionless quantities is important

$$
\begin{aligned}
&\tau = t\omega, \omega = \sqrt{\frac{EI}{mL^4}}, \Omega = \frac{\omega}{\omega_0}\xi = \frac{x}{L}, \\
&w = \frac{W}{L}, b = \frac{J}{mL^2} r = \frac{R}{L}, s = \frac{AL^2}{I}, \\
&k_W = \frac{K_W L^4}{EI}, h = \frac{H}{L} k_G = \frac{K_G L^2}{EI}, e_n = \frac{e_0 a}{L}, \\
&p = \sqrt{\frac{EI}{kAGL^2}}, k = \frac{2(1+\upsilon)}{4+3\upsilon}
\end{aligned} \tag{22}
$$

Where ω and υ are the natural frequency andpoisson ratio respectively.

By utilizing Eqs. (22), the TBM equation of motion (17) can be rewritten in dimensionless form

$$
\begin{aligned}
&{}_{en}{}^4 sbk_G\frac{\partial^8 w}{\partial \zeta^6 \partial \tau^2} - e_n{}^4 sb\frac{\partial^8 w}{\partial \zeta^4 \partial \tau^4} \quad e_n{}^2 + 2e_n{}^2 k_W sb + b + s + sbk_G\Big) \\
&\qquad \frac{\partial^4 w}{\partial \zeta^2 \partial \tau^2} \\
&-e_n{}^2(b + s + e_n{}^2 k_W sb + 2k_W sb)\frac{\partial^6 w}{\partial \zeta^4 \partial \tau^2} + 2e_n{}^2 sb\frac{\partial^6 w}{\partial \zeta^2 \partial \tau^4} \\
&\qquad + e_n{}^2 k_G s\frac{\partial^6 w}{\partial \zeta^6} \\
&-(sk_W e_n{}^2 + e_n{}^2 k_G + sk_G + 1)\frac{\partial^4 w}{\partial \zeta^4} - sb\frac{\partial^4 w}{\partial \tau^4} \\
&\qquad + \left[k_W \quad s - e_n{}^2\Big) + k_G\right]\frac{\partial^2 w}{\partial \zeta^2} \\
&-(sbk_W + 1)\frac{\partial^2 w}{\partial \tau^2} + p\frac{\partial^2 r}{\partial \zeta^2} + pse_n{}^2\frac{\partial^6 r}{\partial \zeta^6} - \frac{\partial^4 r}{\partial \zeta^4}(pe_n{}^2 + sp) \\
&\qquad \int_0^1 \frac{\partial r}{\partial \zeta}\frac{\partial w}{\partial \zeta}d\zeta \\
&+psb\left(e_n{}^4\frac{\partial^6 r}{\partial \zeta^6} - 2e_n{}^2\frac{\partial^4 r}{\partial \zeta^4} + 1\right)\int_0^1 \frac{\partial r}{\partial \zeta}\frac{\partial^3 w}{\partial \zeta \partial \tau^2}dx - k_W w
\end{aligned} \tag{23}
$$

Moreover, the non-dimensional EBM governing equation is obtained from Eq. (19) as

$$
\begin{aligned}
&\frac{\partial^4 w}{\partial \xi^4} + \frac{\partial^2 w}{\partial \tau^2} - e_n{}^2\frac{\partial^4 w}{\partial \xi^2 \partial \tau^2} + k_G\left(\frac{\partial^4 w}{\partial \xi^4} - e_n{}^2\frac{\partial^2 w}{\partial \xi^2}\right) \\
&+k_W\left(w - e_n{}^2\frac{\partial^2 w}{\partial \xi^2}\right) = s\left(\frac{\partial^2 r}{\partial \xi^2} - e_n{}^2\frac{\partial^4 r}{\partial \xi^4}\right)\int_0^1\left(\frac{\partial r}{\partial \xi}\frac{\partial w}{\partial \xi}\right)d\xi
\end{aligned} \tag{24}
$$

The Galerkin method is a powerful solution technique to solve the differential equations. The Galerkin method of decomposition is used to obtain the governing ordinary differential equation (ODE) from a partial counterpart. For one-term approximation the deflection of the beam $w(\xi,\tau)$ separates as

$$
w(\xi, \tau) = \varphi(\xi).e^{i\Omega\tau} \tag{25}
$$

Where $\phi(\xi)$ is the fundamental mode shape for boundary conditions of beam and Ω is the dimensionless natural frequency.

The non-dimensional form of boundary conditions from Eqs. (20) and (21) can be written as

$$\phi = \psi = 0\xi = 0, 1 \tag{26}$$

For clamped ends and

$$\begin{aligned} &\phi = 0, \\ &\frac{\partial \psi}{\partial \xi} + e_n{}^2 \quad \Omega^2 \phi - \frac{\partial}{\partial \xi}\left(n_x \frac{\partial r}{\partial \xi}\right. \\ &\left. + b\Omega^2 \frac{\partial \psi}{\partial \xi} + k_W \phi - k_G \frac{\partial^2 \phi}{\partial \xi^2}\right) = 0 \quad \xi = 0, 1 \end{aligned} \tag{27}$$

For pinned ends. Here n_x is the dimensionless axial resultant force expressed as $n_x = N_x L^2/EI$.

It is interesting to note that there isn't any discrepancy in boundary equations between classical and nonlocal beam theories for simply supported boundary condition (Murmu & Pradhan 2009b). This is in view of $\phi = 0$ at the boundaries. Here the dependency of ϕ and ψ is assumed. Therefore, the nonlocal effects are ignored in the Eq. (27). By using the assumption that $\psi = \mathrm{d}\phi/\mathrm{d}\xi$, the boundary conditions can be simplified as

$$\phi = \frac{d\phi}{d\xi} = 0\xi = 0, 1. \tag{28}$$

For clamped ends and

$$\phi = \frac{d^2\phi}{d\xi^2} = 0\xi = 0, 1. \tag{29}$$

In this study three boundary value problems with immovable ends (one for simply supported beam, the second for double clamped and another for clamped-pinned beam) are considered.

Consequently, ϕ (ξ) from the boundary conditions will be obtained as follows

1) A beam clamped at both ends or the clamped–clamped boundary condition (C–C):
$$\begin{aligned}\phi(\xi) = {} & \cosh(4.73\xi) - 0.9825\sinh(4.73\xi) \\ & - \cos(4.73\xi) + 0.9825\sin(4.73\xi)\end{aligned} \tag{30}$$

2) A beam clamped at one end and simply supported at the other end, i.e. clamped–pinned condition (C–P):
$$\begin{aligned}\phi(\xi) = {} & \cosh(3.92\xi) - 1.0008\sinh(3.92\xi) \\ & - \cos(3.92\xi) + 1.0008\sin(3.92\xi)\end{aligned} \tag{31}$$

3) And a simply supported beam at both ends or pinned–pinned condition (P–P):
$$\phi(\xi) = \sin(3.14\xi) \tag{32}$$

Applying Eqs. (30–32) into partial equations of motion SWCNT and multiplying of the these equations by $\phi(\xi)$ and then integrating over the length of SWCNT, the ODE of motion can be obtained.

Therefore, the following linear fourth-order ODE is obtained from TBM as

$$\left(A\Omega^4 - B\Omega^2 + C\right)q = 0 \tag{33}$$

Where A, B and C are the coefficients of general equation of motion (33) that represented in appendix A. The dimensionless linear fundamental frequency represented as

$$\Omega_{TBM} = \frac{1}{2}\left(B - \sqrt{B^2 - 4AC}\right). \tag{34}$$

Similarly, by applying the EBM for the free vibration of a curved SWCNT, the governing equation can be simplified to the following dimensionless equation

$$\left(M_{eq}\Omega^2 + K_{eq}\right)q = 0. \tag{35}$$

Where M_{eq} and K_{eq} are the equivalent of mass and stiffness of the vibrational system that are described in appendix B, and the resonant frequency becomes

$$\Omega_{EBM} = \sqrt{\frac{K_{eq}}{M_{eq}}}. \tag{36}$$

Results and discussion

In this case study, the diameter, aspect ratio, thickness of SWCNT and Young's modulus of the nanotube are assumed to be $d_e = 3.19$ nm, $L/d_e = 6.052$ nm, $t_c = 0.137$ nm, and $E = 2.407$ TPa, respectively (Gupta et al. 2010). The mass density of the SWCNT is 2300 kg/m^3 with nonlocal parameter $e_0 a$ of 2 nm. Also, the Winkler modulus and Pasternak modulus are estimated at the values of $K_W = 1$ MPa and $K_G = 5$nN, in that order (Murmu & Pradhan 2009b). Moreover, the amplitude of curve H is 1 nm (Mayoof & Hawwa 2009).

In order to explore the effects of rotary inertia and transverse shear deformation on the frequency of the curved SWCNT, one makes a comparison between the results obtained from the EBM and TBM as mentioned in Table 1. The values in the brackets are the relative errors between the EBM and TBM. It is found from this table that the resonant frequencies of the TBM are smaller than those predicted by EBM, which overestimates the resonant frequencies for a linear vibration. It is noted that the difference between the values of the resonant frequencies predicted by different beam models in this study is more significant for the stiffer boundary conditions and stocky SWCNTs. In other words, for long and slender SWCNTs with a high aspect ratio, the impacts of transverse shear deformation and rotary

Table 1 Comparison of the dimensionless linear fundamental frequency by TBM and EBM for different boundary conditions and length ($d_e = 3.19$ nm, $H = 1$ nm, $e_0a = 2$ nm, $K_W = 1$ MPa, $K_G = 5$nN)

boundary condition	L = 20 nm		L = 40 nm		L = 60 nm	
	EBM	TBM	EBM	TBM	EBM	TBM
C-C	17.93	15.19 (15.2%)	21.65	20.94 (3.2%)	22.86	22.58 (1.2%)
C-P	13.24	11.96 (9.7%)	15.71	15.34 (2.3%)	16.65	16.49 (0.9%)
P-P	10.02	9.51(5%)	11.41	11.23 (1.5%)	12.1	12(0.6%)

inertia on the vibration of curved SWCNTs can be ignored, especially for P-P boundary condition.

As mentioned before, another main purpose of this paper is to show the significance of curvature in modeling of the SWCNT vibration. Hence, Figure 2 illustrates the fundamental frequencies of TBM $\mathbf{\Omega_{TBM}}$ against the amplitude of curvature H for various boundary conditions. It has been shown from this figure that with increasing the amplitude of waviness H, the frequency increases. Furthermore, the fundamental frequencies are completely dependent on the boundary conditions and they are increased while the bending stiffness of the SWCNT rises from P-P to C-C, in particular for low curvature amplitude.

Moreover, to see the effects of curvature clearly, the *difference percent* (DP) is defined as a parameter that shows the percent increment of frequency for a curved SWCNT ($H \neq 0$ nm) compared with a straight nanotube ($H = 0$ nm).

$$DifferencePercent(DP) = \frac{\left(\Omega_{TBM}^{H=Hnm} - \Omega_{TBM}^{H=0nm}\right) \times 100}{\Omega_{TBM}^{H=0nm}} \tag{37}$$

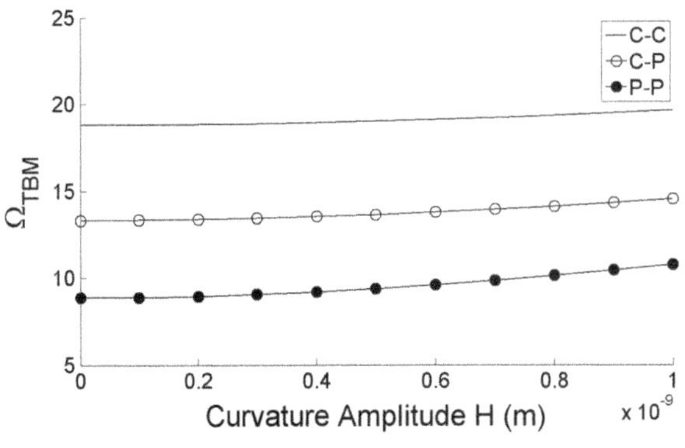

Figure 2 **The fundamental frequency Ω_{TBM} against the curvature amplitude H for three typical boundary conditions ($d_e = 3.19$ nm, $e_0a = 2$ nm, $L/d_e = 6.052$ nm, $K_W = 1$ MPa, $K_G =$ 5nN).**

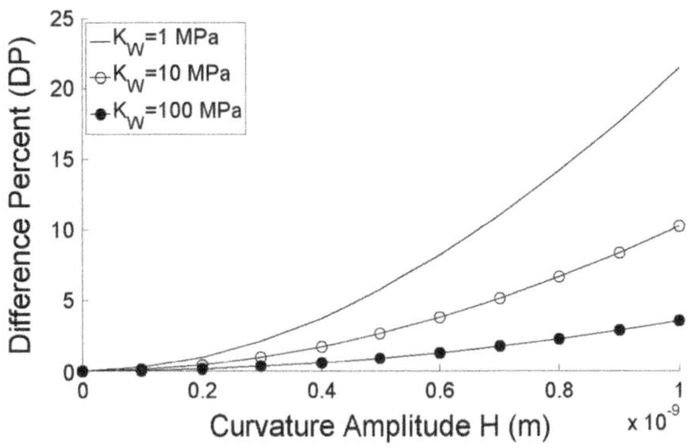

Figure 3 **The difference percent DP against the curvature amplitude H for a clamped-clamped SWCNT with different values of the Winkler modulus K_W ($d_e = 3.19$ nm, $e_0a = 2$ nm, $L/d_e = 6.052$ nm, $K_G = 5$nN).**

Certainly, DP gives a better illustration for the pure effects of the amplitude of curvature H. Figures 3, 4, 5, 6, 7 represent the difference percent DP as a function of the waviness amplitude H, while the effects of a certain parameter such as the stiffness of model, the length of SWCNT and the nonlocal parameter have been evaluated in each figure. Obviously, the variation of fundamental frequencies is increased when the waviness is amplified, in all the figures.

The surrounding elastic medium can be described by Winkler and Pasternak modulus that are determined by the material properties of the elastic medium (Yoon et al. 2003; Lanir & Fung 1972). The DP is very responsive to the stiffness of the model, due to foundation and boundary conditions. As the Winkler modulus K_W and Pasternak modulus K_G of the surrounding elastic medium increase, the shift of the DP is significant (Figures 3 and 4). Therefore, in a soft elastic medium, the effects of the curvature

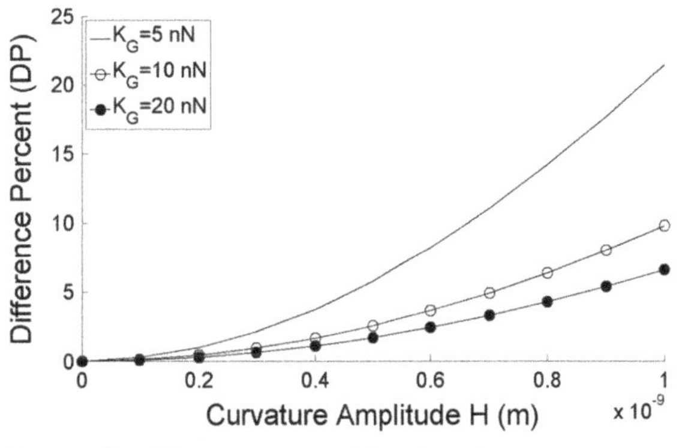

Figure 4 **The difference percent DP against the curvature amplitude H for a clamped-clamped SWCNT with different values of the Pasternak modulus K_G ($d_e = 3.19$ nm, $e_0a = 2$ nm, $L/d_e = 6.052$ nm, $K_W = 1$ MPa).**

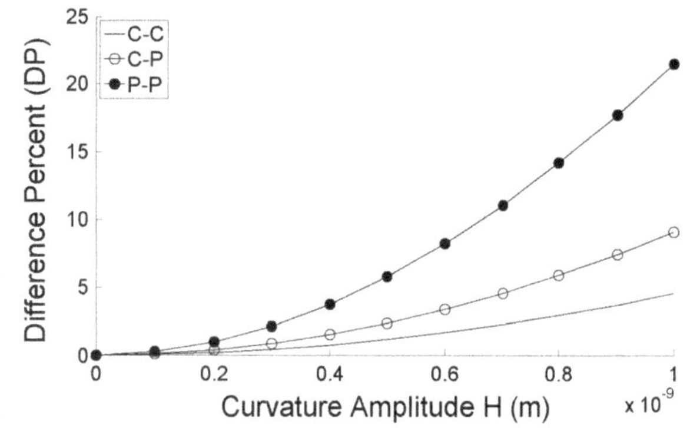

Figure 5 **The difference percent DP against the curvature amplitude *H* with different types of boundary conditions (d_e = 3.19 nm, e_0a = 2 nm, L/d_e = 6.052 nm, K_W = 1 MPa, K_G = 5nN).**

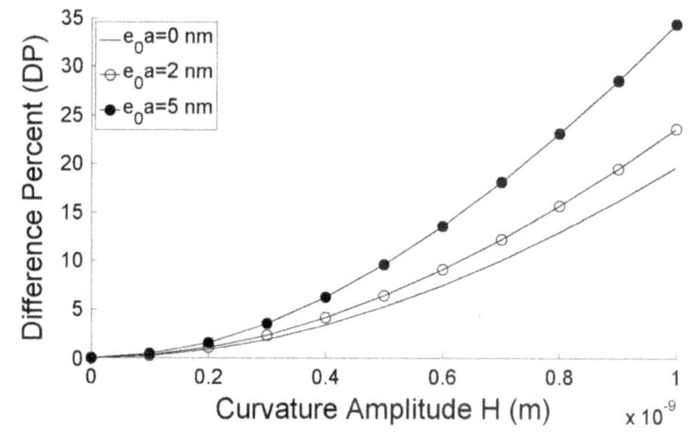

Figure 7 **The difference percent DP against the curvature amplitude *H* for a clamped-clamped SWCNT with different values of the nonlocal parameter e_0a (d_e = 3.19 nm, L/d_e = 6.052 nm, K_G = 5nN, K_W = 1 MPa).**

on the vibration of SWCNT increase. In addition, stiffness of boundaries has a similar behavior as the stiffness of the medium. As mentioned previously, to show the effects of boundary condition, this model solved for P-P, C-P and P-P boundary conditions. Figure 5 indicates that the boundary conditions have considerable effects on the DP. The illustration demonstrates that with decreasing the stiffness of the SWCNT from C-C to P-P, the effects of waviness on vibrational frequency increase.

Figure 6 is presented in order to see the role of the SWCNT dimension. It shows the importance of the SWCNT length in the natural frequency and associated difference percent. Results show that an increase in the slenderness ratio causes the DP to considerably decrease. According to this figure, the rise of frequency with curvature amplitude is found to be significantly dependent on the aspect ratio. It means for a long nanotube, the difference of frequency between the straight SWCNT and curved SWCNT is reduced.

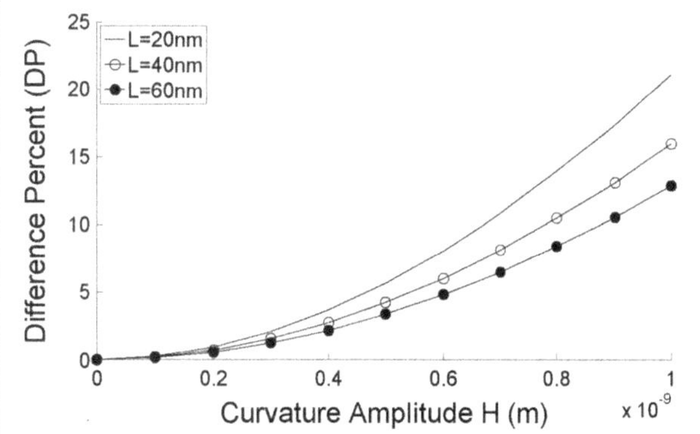

Figure 6 **The difference percent DP against the curvature amplitude *H* for a clamped-clamped SWCNT with different values of the length *L* (d_e = 3.19 nm, e_0a = 2 nm, K_G = 5nN, K_W = 1 MPa).**

Finally, to investigate the effects of nonlocal theory, the DP is plotted as a function of the curvature amplitude H and the nonlocal parameter e_0a, in Figure 7. Nonlocal curves have been plotted for three different e_0a values. The e_0a values taken are 0 nm, 2 nm and 5 nm. The nonlocal elasticity theory makes nano-devices more flexible and reduces the stiffness of the SWCNTs. As the figure shows, with an increase in model stiffness, by decreasing the nonlocal coefficient, the DP decreases consequently. Therefore, the larger DP occurred at higher nonlocal values and the nonlocal theory provides a higher sensitivity of curved amplitude for the waviness SWCNT.

Conclusions

The nonlocal Timoshenko beam model (TBM) and Euler-Bernoulli beam model (EBM) have been introduced to analyze the effect of waviness on the curved single-walled carbon nanotube (SWCNT). The surrounding elastic medium is simulated using Winkler and Pasternak models. Comparing the results predicted by these two beam models, it is found that the EBM may provide as accurate results as the TBM for SWCNTs with larger length-to-diameter ratio. However, the Timoshenko beam model is highly recommended for stiffer boundary conditions and stocky SWCNTs due to the effects of shear deformation and rotary inertia. In addition, the results show that the waviness has a significant effect on the vibrational behavior of the SWCNT. Detailed results indicate the influence of the curvature, the stiffness of medium around the SWCNT, the boundary conditions, the length of SWCNT and the nonlocal parameter on the natural frequency. Our investigation demonstrates that increasing the amplitude of curvature causes the fundamental frequency to increase.

Furthermore, with an increase in the stiffness of the model, due to the boundary conditions and the foundation, the length of SWCNT and the nonlocal constant, a higher frequency value is obtained for a curved SWCNT in comparison to a straight SWCNT.

Appendix A
The coefficients of Eq. (33)
C-P condition

$$A = 2 + 46.05{e_n}^4 + 475.34{e_n}^2$$

$$B = \frac{475.34{e_n}^2}{s} + \frac{475.34{e_n}^2}{b} + \frac{23.02{e_n}^2}{sb} + 5.99{e_n}^2\pi^5 ph^2$$
$$+23.02k_G + 5473.78k_G{e_n}^4 + 475.34k_W{e_n}^4 + 2\pi^4 k_W{e_n}^2$$
$$+ 2.99{e_n}^4\pi^7 ph^2 + 2k_W + 996.73k_W{e_n}^2 + 2.99e^2\pi^3 p$$
$$+\frac{23.02}{s} + \frac{23.02}{b} + \frac{2}{sb}$$

$$C = \frac{1}{b}\left(23.02k_W + 475.34k_G + 475.34{e_n}^2 k_W + 916.65ph^2 + 9047.04{e_n}^2 ph^2 + 5473.78k_G{e_n}^2\right)$$
$$+\frac{1}{sb}\left(23.02k_G + 475.34 + 2k_W + 475.34{e_n}^2 k_G + 92.87ph^2 + 916.65{e_n}^2 ph^2 + 23.02k_W{e_n}^2\right)$$

C-C condition

$$A = 2 + 49.2{e_n}^4 + 1001.15{e_n}^2$$

$$B = \frac{1001.15{e_n}^2}{s} + \frac{1001.15{e_n}^2}{b} + \frac{24.6{e_n}^2}{sb} + 1865.17{e_n}^2 ph^2$$
$$+24.6k_G + 12315.84k_G{e_n}^4 + 1001.15k_W{e_n}^4$$
$$+ 2051.51k_W{e_n}^2 + 9194.93{e_n}^4 ph^2 + 2k_W + 996.73k_W{e_n}^2$$
$$+ 94.58h^2 p + \frac{24.6}{s} + \frac{24.6}{b} + \frac{2}{sb}$$

$$C = \frac{1}{b}\left(24.6k_W + 1001.15k_G + 1001.15{e_n}^2 k_W + 932.58ph^2 + 9194.93{e_n}^2 ph^2 + 12315.84k_G{e_n}^2\right)$$
$$+\frac{1}{sb}\left(24.6k_G + 1001.15 + 2k_W + 1001.15{e_n}^2 k_G + 94.58ph^2 + 932.58{e_n}^2 ph^2 + 24.6k_W{e_n}^2\right)$$

P-P condition

$$A = 2 + 39.47{e_n}^4 + 194.81{e_n}^2$$

$$B = \frac{194.81{e_n}^2}{s} + \frac{194.81{e_n}^2}{b} + \frac{19.73{e_n}^2}{sb} + 1922.73{e_n}^2 ph^2$$
$$+19.73k_G + 1922.73k_G{e_n}^4 + 194.81k_W{e_n}^4 + 429.1k_W{e_n}^2$$
$$+ 9488.33{e_n}^4 ph^2 + 2k_W + 996.73k_W{e_n}^2 + 97.4h^2 p$$
$$+\frac{19.73}{s} + \frac{24.6}{b} + \frac{2}{sb}$$

$$C = \frac{1}{b}\left(19.73k_W + 194.81k_G + 194.81{e_n}^2 k_W + 961.36ph^2 + 9488.33{e_n}^2 ph^2 + 1922.73k_G{e_n}^2\right)$$
$$+\frac{1}{sb}\left(19.73k_G + 194.81 + 2k_W + 194.81{e_n}^2 k_G + 97.4ph^2 + 961.36{e_n}^2 ph^2 + 19.73k_W{e_n}^2\right)$$

Appendix B
The coefficients of Eq. (36)
C-C condition

$$K_{eq} = 1 + 12.3{e_n}^2$$
$$M_{eq} = 12.3k_W{e_n}^2 + 500.56k_G{e_n}^2 + 12.3k_G + 467.53ph^2{e_n}^2 + k_W + 47.37ph^2 + 500.56$$

C-P

$$K_{eq} = 11.51{e_n}^2 + 1$$
$$M_{eq} = 237.7{e_n}^2 k_G + 11.5{e_n}^2 k_W + 11.5k_G + k_W + 458.4ph^2{e_n}^2 + 46.4ph^2 + 237.7$$

P-P

$$K_{eq} = 9.8{e_n}^2 + 1$$
$$M_{eq} = 97.4{e_n}^2 k_G + 9.8{e_n}^2 k_W + 9.8k_G + k_W + 480.6ph^2{e_n}^2 + 48.7ph^2 + 97.4$$

Competing interests
The authors declare that they have no competing interests.

Authors' contributions
The main idea of this paper was suggested, modified, and developed by PS; the modeling procedure, and analytic solution were done by AK. The discussion section was prepared by MMT. Reviewing process and verifying the used procedure were done by AF. All authors read and approved the final manuscript.

Author details
[1]Department of Mechanical Engineering, Semnan Branch, Islamic Azad University, Semnam, Iran. [2]Department of Mechanical Engineering, Ferdowsi

University of Mashhad, Mashhad, Iran. [3]Young Researchers Club, Semnan Branch, Islamic Azad University, Semnam, Iran.

References

Iijima S (1991) Helical microtubes of graphitic carbon. Nature 354:56–58

Dresselhaus MS et al (2004) *Electronic, thermal and mechanical properties of carbon nanotubes*. Philosophical Transactions of the Royal Society of London. Ser A Math Phys Eng Sci 362(1823):2065–2098

Treacy MMJ, Ebbesen TW, Gibson JM (1996) Exceptionally high Young's modulus observed for individual carbon nanotubes. Nature 381(6584):678–680

Walgraef D (2007) On the mechanics of deformation instabilities in carbon nanotubes. Eur Phys J Spec Top 146(1):443–457

Chen L et al (2004) Single-Walled Carbon Nanotube AFM Probes: Optimal Imaging Resolution of Nanoclusters and Biomolecules in Ambient and Fluid Environments. Nano Lett 4(9):1725–1731

Hafner JH, Cheung CL, Lieber CM (1999) Growth of nanotubes for probe microscopy tips. Nature 398(6730):761–762

Nishide D et al (2003) High-yield production of single-wall carbon nanotubes in nitrogen gas. Chem Phys Lett 372(1–2):45–50

Tsukagoshi K et al (2002) Carbon nanotube devices for nanoelectronics. Physica B 323(1–4):107

Lau K-T et al (2004) On the effective elastic moduli of carbon nanotubes for nanocomposite structures. Compos Part B Eng 35(2):95–101

Liew KM, He XQ, Wong CH (2004) On the study of elastic and plastic properties of multi-walled carbon nanotubes under axial tension using molecular dynamics simulation. Acta Mater 52(9):2521–2527

Yoon J, Ru CQ, Mioduchowski A (2005) Vibration and instability of carbon nanotubes conveying fluid. Compos Sci Technol 65(9 SPEC. ISS):1326–1336

Zhang YY, Wang CM, Tan VBC (2006) Effect of omitting terms involving tube radii difference in shell models on buckling solutions of DWNTs. Comput Mater Sci 37(4):578–581

Wang L et al (2008) The thermal effect on vibration and instability of carbon nanotubes conveying fluid. Phys E 40(10):3179–3182

Kuang YD et al (2009) Analysis of nonlinear vibrations of double-walled carbon nanotubes conveying fluid. Comput Mater Sci 45(4):875–880

Yakobson BI et al (1997) High strain rate fracture and C-chain unraveling in carbon nanotubes. Comput Mater Sci 8(4):341–348

Nguyen HLT, Elishakoff I, Nguyen VT (2009) Buckling under the external pressure of cylindrical shells with variable thickness. Int J Solid Struct 46(24):4163–4168

Yan Y, Wang WQ, Zhang LX (2009) Dynamical behaviors of fluid-conveyed multi-walled carbon nanotubes. Appl Math Model 33(3):1430–1440

Liew KM, Wang Q (2007) Analysis of wave propagation in carbon nanotubes via elastic shell theories. Int J Eng Sci 45(2–8):227–241

Fu YM, Hong JW, Wang XQ (2006) Analysis of nonlinear vibration for embedded carbon nanotubes. J Sound Vib 296(4–5):746–756

Chaterjee S, Pohit G (2009) A large deflection model for the pull-in analysis of electrostatically actuated microcantilever beams. J Sound Vib 322(4–5): 969–986

Chowdhury R, Adhikari S, Mitchell J (2009) Vibrating carbon nanotube based bio-sensors. Phys E 42(2):104–109

Hsu J-C, Chang R-P, Chang W-J (2008) Resonance frequency of chiral single-walled carbon nanotubes using Timoshenko beam theory. Phys Lett A 372 (16):2757–2759

Wang CM, Tan VBC, Zhang YY (2006) Timoshenko beam model for vibration analysis of multi-walled carbon nanotubes. J Sound Vib 294(4–5):1060–1072

Lee H-L, Chang W-J (2009) A closed-form solution for critical buckling temperature of a single-walled carbon nanotube. Phys E 41(8):1492–1494

Mahdavi MH, Jiang LY, Sun X (2011) Nonlinear vibration of a double-walled carbon nanotube embedded in a polymer matrix. Phys E 43(10):1813–1819

Rao S (2007) Vibration of continuous systems. Wiley, Hoboken, New Jersey

Eringen AC (1983) On differential equations of nonlocal elasticity and solutions of screw dislocation and surface waves. J Appl Phys 54(9):4703–4710

Murmu T, Pradhan SC (2009a) Thermo-mechanical vibration of a single-walled carbon nanotube embedded in an elastic medium based on nonlocal elasticity theory. Comput Mater Sci 46(4):854–859

Şimşek M (2011) Vibration analysis of a single-walled carbon nanotube under action of a moving harmonic load based on nonlocal elasticity theory. Phys E 50:2112–2123

Mehdipour I, Barari A, Domairry G (2011) Application of a cantilevered SWCNT with mass at the tip as a nanomechanical sensor. Comput Mater Sci 50: 1830–1833

Wang L, Ni Q (2008) On vibration and instability of carbon nanotubes conveying fluid. Comput Mater Sci 43(2):399–402

Yang J, Ke LL, Kitipornchai S (2010) Nonlinear free vibration of single-walled carbon nanotubes using nonlocal Timoshenko beam theory. Phys E 42 (5):1727–1735

Murmu T, Pradhan SC (2009b) Buckling analysis of a single-walled carbon nanotube embedded in an elastic medium based on nonlocal elasticity and Timoshenko beam theory and using DQM. Phys E 41(7):1232–1239

Joshi AY, Sharma SC, Harsha SP (2010) Dynamic Analysis of a Clamped Wavy Single Walled Carbon Nanotube Based Nanomechanical Sensors. J Nanotechnol Eng Med 1(3):031007–7

Hawwa MA, Mayoof FN (2009) Nonlinear oscillations of a carbon nanotube resonator. International Symposium on Mechatronics and its Applications, Sharjah

Eringen AC (2002) Nonlocal continuum field theories. Springer Verlag

Gupta SS, Bosco FG, Batra RC (2010) Wall thickness and elastic moduli of single-walled carbon nanotubes from frequencies of axial, torsional and inextensional modes of vibration. Comput Mater Sci 47(4):1049–1059

Mayoof FN, Hawwa MA (2009) Chaotic behavior of a curved carbon nanotube under harmonic excitation. Chaos Solitons Fractals 42(3):1860–1867

Yoon J, Ru CQ, Mioduchowski A (2003) Vibration of an embedded multiwall carbon nanotube. Compos Sci Technol 63(11):1533–1542

Lanir Y, Fung YCB (1972) Errata on "Fiber Composite Columns Under Compression" J. COMPOSITE MATERIALS, Vol. 6 (July 1972), pp. 387–401. J Compos Mater 6(4):533–535

2

Behaviour of reinforced concrete column under biaxial cyclic loading—state of the art

Hugo Rodrigues[1,2,4*], Humberto Varum[1], Antonio Arêde[3] and Aníbal G Costa[1]

Abstract

The cyclic behaviour of reinforced concrete columns has been object of many experimental studies in the last years, mostly focused on the unidirectional loading of columns under constant axial load conditions. In this research work, the existing test on reinforced concrete (RC) columns under biaxial load has been reviewed, underlying their main findings. In general, the experimental results show that the RC columns' response is highly dependent on the loading pattern, and the biaxial loading induces a decrease in the maximum strength and anticipates each damage state. Thus, in columns where demands are expected with large moments in both directions, specific detailing should be provided in their critical regions in order to improve the columns' performance and avoid premature failure.

Keywords: RC column, Seismic behaviour, Cyclic behaviour, Biaxial testing

Review

Introduction

Columns are key structural elements for the seismic performance of buildings. Therefore, special attention should be given to their structural response under load reversals. Moreover, earthquake effects generally require the inclusion of two horizontal component loads that are recognised to be more damaging than single direction actions. The interest in the inelastic response of axially loaded members under biaxial bending moment histories is relatively recent, and the available experimental results are limited. This is possibly due in part to the uncertainty of combining histories of bending moments in the two orthogonal directions, adding considerable complications to the problem.

The practical result is that our present-day knowledge of the inelastic behaviour of reinforced concrete (RC) columns under biaxial cyclic moments is much behind than our understanding of their behaviour under uniaxial cyclic bending with axial load. In fact, besides the fibre-based models, the existing simplified analytical models are not mature enough to be incorporated into code standards, by contrast with uniaxial simplified global models which are already accepted in international codes Rodrigues et al. (2012b). This paper aims at providing a brief review of past experimental studies in the field of biaxial behaviour of reinforced concrete RC columns and to identify the existing open problems.

Typical causes and consequences of column failure during earthquakes

The earthquake performance of RC buildings has been well documented from the observation of past seismic events. The collapse of a RC building is caused, in the majority of cases, by the failure of the vertical members. The seismic behaviour deficiencies of RC buildings can be associated with design, detailing and construction deficiencies, as well as deterioration and structural modifications. Varum (2003) points out the ten most common causes of failure or damage in RC buildings: (1) lack of stirrups/hoops, confinement and ductility; (2) bond/anchorage/lap-splices slipping and bond splitting; (3) inadequate shear capacity; (4) inadequate flexural capacity; (5) inadequate shear strength of the joints; (6) influence of the infill masonry on the seismic behaviour of frames; (7) vertical and horizontal irregularities, abrupt change in structural and/or element properties; (8) higher modes' effects; (9) strong-beam weak-column mechanisms; and finally, (10) structural deficiencies due to architectural requirements.

Some of these categories are in fact related to design, detailing and construction deficiencies, for which the

* Correspondence: hrodrigues@ua.pt
[1]Civil Engineering Department, University of Aveiro, Aveiro, Portugal
[2]Faculty of Natural Sciences, Engineering and Technology, Oporto Lusophone University, Oporto, Portugal
Full list of author information is available at the end of the article

extensive damage or collapse of structures can occur through the contribution of many of these causes. Columns properly designed for earthquake loading are able to prevent brittle failure and to have a ductile behaviour. In fact adequate reinforcement and detailing is very important in providing the ductile behaviour of RC elements EASY (1997). Moreover, the flexural behaviour is conditioned by the axial force and by the amount of reinforcement in the plastic hinge region. Figure 1 shows examples of RC columns with failures associated with inadequate flexural capacity after an earthquake event, although in these examples, shear and confinement deficiencies are also evident (transversal reinforcement with large spacing at the columns extremities).

Deficient flexural behaviour can be more evident in exterior corner columns where varying levels of axial force can be expected during earthquakes, leading to high levels of axial force. In some cases, it may be difficult to differentiate flexural compression and shear compression failure, as both take place in or near the column ends and involve crushing Otani (2002).

Inadequate transversal reinforcement in terms of size, spacing and detailing is the principal cause of shear failure of RC columns. This problem can be increased in exterior corner columns in buildings with in-plan irregularities Varum (2003). Columns with shear failure commonly exhibit a diagonal fracture in the column mid-zone, as illustrated in Figure 2a,b. Another typical cause of shear failure is related with insufficient area of transversal reinforcing steel, with wide spacing and deficiently anchored to the concrete core, where ties should have enough length or proper bar bents to promote adequate splice anchorage (see Figure 2c).

Beam-column connections play also an important role in the seismic behaviour of RC buildings; poor behaviour of beam-column joints can lead to the collapse or severe damage of buildings during earthquakes. In some cases, the beam longitudinal reinforcement is not properly anchored in the beam-column joint Otani (2002; see Figure 3a) or there is no adequate transverse reinforcement in the joints (see Figure 3b,c).

The contribution of non-structural elements to the structural response is typically not considered in the design of new buildings and in the assessment of existing ones Rodrigues et al. (2010b). In the case of the infill masonry panels, two principal mechanisms have been often observed. The first is associated with cases where masonry infill walls leave a short portion of clear column, creating the so-called short column; this situation is caused by openings in the infill walls, for doors or windows, or for landing slabs of staircases. If this effect is not considered in the design, a short column is actually developed with increased stiffness which becomes subjected to shear force level that can lead to shear failure of the column, as illustrated Figure 4a. The second is related to the absence of the infill masonry panels in one storey, frequently in the ground floor storey due to its use for car parking or commercial areas, which induces an abrupt change of the storey stiffness, leading to a potential global soft-storey mechanism Vicente et al. (2010, 2012). Moreover, the asymmetric distribution of the infill masonry panels can drive torsion phenomena not predicted in the design, which can induce additional unconsidered forces, especially in the concrete columns of the outer frames Fardis (2006).

Experimental studies on the seismic behaviour of RC columns

Introduction

The experimental study of RC elements under earthquake loads is very important since it permits the observation and measurement of the particular performance of the element. Regarding the behaviour of RC elements under lateral loads, a large number of various types of studies have been carried out in past years which, apart from the specific subject of each study, have generally

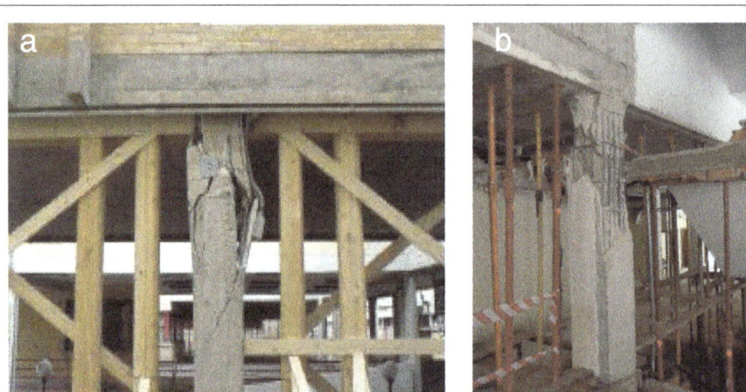

Figure 1 Flexural deficient behaviour in RC columns. (**a**) In the San Salvatore Hospital, in the 2009 earthquake in L'Aquila, Italy; (**b**) in a residential building, in the 2011 earthquake in Lorca, Spain.

Figure 2 Shear deficient behaviour in RC columns in residential buildings in the 2009 L'Aquila earthquake. (**a**) Square column (Rodrigues et al. 2010a); (**b**) circular column (Ricci et al. 2011); failure is related to the transversal reinforcing steel; (**c**) deficient flexural/shear behaviour in a corner column in a residential building in the 2011 earthquake in Lorca, Spain.

covered elements with or without axial loads, as well as uniaxial and biaxial bending. In the case of zero axial force, these studies are typically associated with beams; therefore, most of them focus only on the uniaxial behaviour of these elements. Columns are studied with non-zero axial load, constant or variable, under uniaxial or biaxial bending. In the case of slender elements, the behaviour is governed by flexure, but depending on the cross-section geometry and the shear span ratio, the shear force can also represent an important subject for the study. Studies of RC elements can, therefore, also be divided into elements where the global behaviour is governed essentially by flexure or by shear, or by both mechanisms.

In view of the present study, it is important to perform a retrospective review of experimental tests performed on rectangular columns under horizontal loads, especially with biaxial behaviour. The response of RC members subjected to axial loads together with biaxial bending moment reversals is recognised as a very important research topic for building structures in earthquake-prone regions. On one hand, studies of the response of RC building columns to earthquake actions deal in general with its three-dimensional response, due to the random characteristics of the earthquake direction and to the actual building irregularities. On the other hand, the biaxial features of bending moment histories applied to a given RC column section tend to reduce its actual capacity and to accelerate the strength and stiffness deterioration process during successive load reversals. In addition, the 3D response of frame structures to actual earthquake motions generally does not induce the same type of increased deterioration in beams because they behave essentially in only one direction (vertical), i.e., the potential development plastic hinges in beams are not aggravated by that fact. This means that both the biaxial loading effects in columns and the 3D features of the general structure response positively contribute to inelasticity and damage concentration in the columns rather than in the beams, which is essentially the opposite of what present-day design code requirements created to avoid collapse of RC frame structures under lateral load reversals (plastic hinges in the beams rather than in the columns) CEB (1996).

Experimental research work on the inelastic response of RC members under compression axial force and biaxial lateral cyclic bending loading conditions is currently very limited. Uncertainties concerning the relationship and combination of the two orthogonal horizontal loading paths, associated to the complexity of the experimental set-up, certainly justify this gap.

As a consequence, current knowledge concerning the inelastic response of RC columns under biaxial cyclic moments is very much less than that of the uniaxial cyclic bending behaviour with compressive axial load (CEB 1996; Marante and Flórez-López 2002; Paulay and Priestley 1992).

The available test results for biaxial bending under constant axial load (CAL) are not so extensive when compared to those on uniaxial bending, although they have been delivered over a period of almost 30 years. On the basis of an extensive analysis of international experimental databases and from a literature review of studies covering tests on rectangular RC columns subjected to cyclic loading, a statistical analysis was performed based on the data collected for 453 column tests (see Figure 5). From the data analysis, it was verified that (1) only 27 of these columns (6%) were tested with variable axial load (VAL) and (2) only 56 (12%) were tested in biaxial loading conditions. These statistical figures emphasise the insufficient experimental results for the comprehensive characterization of the biaxial cyclic behaviour of RC columns.

Figure 3 Inadequate joint behaviour in residential buildings. (**a**) Inappropriate anchorage of beam longitudinal reinforcement; (**b**) deficient detailing of a beam-column joint in a residential building (2009 earthquake in L'Aquila, Italy); (**c**) RC building collapse in the 2011 earthquake in Lorca, Spain.

Due to testing difficulties and because there are still open questions regarding the cyclic behaviour both in biaxial bending with constant axial force and in uniaxial bending with simultaneously varying axial load, very few experimental studies have, as yet, tackled the more general problem of biaxial bending with varying axial force (CEB 1996; Coelho 1992), as can be seen in Figure 5. From the literature review, only the tests performed by Low and Moehle (1987; two columns) and by Chang (2010a; two columns) refer to rectangular cross-sections with different dimensions and amount of longitudinal reinforcement in the two principal directions, leading to different characteristics in terms of stiffness and strength.

In recent years, a very small number of tests have also been performed in beam-column sub-assemblages where the columns were subjected to biaxial loading (Suzuki et al. 1984; Monti and Nuti 1992; Li et al. 2008; Akguzel and Pampanin 2010).

The next sections present a summary of the previous works involving experimental studies performed with

Figure 4 Column failure due to the presence of non-structural elements. Short column effect caused by window openings in residential building. (**a**) 2009 earthquake in L'Aquila, Italy and (**b**) 2011 earthquake in Lorca, Spain.

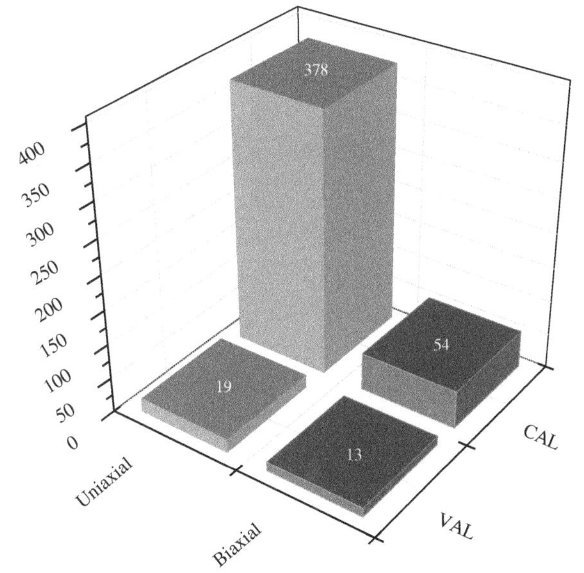

Figure 5 Experimental tests on RC rectangular columns according to function of the loading conditions. Uniaxial or biaxial, and CAL or VAL. Statistics are based on a review of the literature and test databases (Rodrigues et al. 2013).

biaxial bending. The analyses focus on the main findings obtained from tests on columns subjected to biaxial lateral loading (with constant and varying axial force) and under different test conditions; biaxial tests performed on global bare frame structures are also analysed. All relevant experimental works performed before 1996 are very well summarised in CEB (1996), and so the present section will focus on the main findings of the experimental studies and work performed after 1996.

Specimen geometries

Two types of column specimens were used in the cyclic biaxial tests described in the literature. The prototype test considering the column specimens fixed against rotation on both ends (Figure 6a) was used only in one study Takiguchi et al. (1980), while all the other studies report the use of the cantilever type (Figure 6b). Other test configurations can be used, namely, the double-ended cantilever with flexible-base and the hammerhead configurations Berry et al. (2004).

Double curvature specimens with rigid zones on top and bottom can be assumed as the best models to simulate a typical building column. However, due to many factors, most of the authors use the cantilever-type specimen. There are no known studies on the differences between these specimen configuration types.

The cantilever-type model assumes that the inflection point of a column is located at mid-height of the column and takes no damage. This can be considered to be true in columns governed by flexural behaviour, where the damage is concentrated in the so-called plastic hinge zones located at the column ends.

Regarding the column height and cross-section dimensions, many different sections have been tested as described in Table 1, where the main geometric characteristics of the columns are summarised. As previously mentioned, in most of the cases, the columns have a square cross-section, although there is no clear justification for a square section; mainly, it can be used due to the difficulties in developing the test set-up, the need to perform two uniaxial tests (one in each principal direction) and the new variable to analyse the differences in the stiffness and strength in each direction.

Displacement patterns

The behaviour of RC columns under cyclic behaviour is especially influenced by the geometric and mechanical characteristics of the column cross-sections and the *f* level of axial load. However, the displacement pattern can also influence the global behaviour, and in terms of biaxial bending forces, the displacement history and the way the biaxial forces are combined in two orthogonal

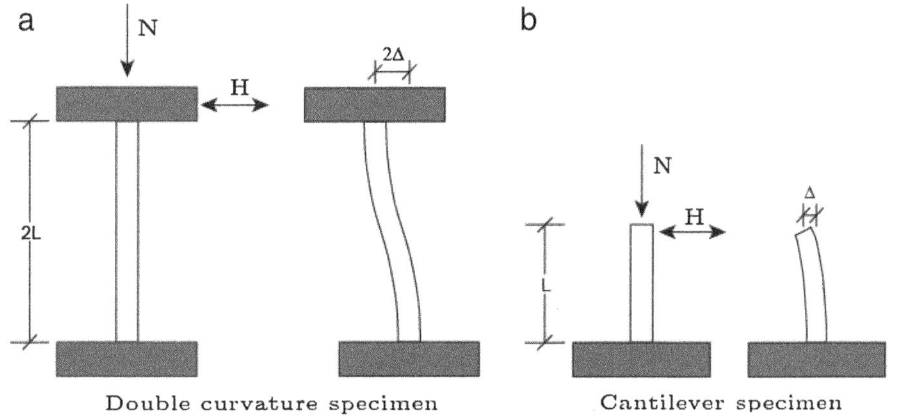

Figure 6 Column test configurations.

Table 1 Geometric characteristics of rectangular RC columns tested under biaxial loading (adapted and updated from CEB (1996))

Reference	Number of specimens	Length (mm)	Cross-section dimensions (b × h; mm)	Axial load
Takizawa and Aoyama (1976)	4	600	200 × 200	C
Otani et al. (1980)	4	1,372	305 × 305	C
Takiguchi et al. (1980)	5	250	150 × 150	C
Park et al. (1984)	1	1,600	400 × 400	C
Low and Moehle (1987)	2	546	127 × 165	C
	2	546	127 × 165	V
Li et al. (1988)	1	600	200 × 200	C
	4	600	200 × 200	V
Saatcioglou and Ozcebe (1989)	2	1,000	350 × 350	C
	1	1,000	350 × 350	V
Zahn et al. (1989)	2	1,600	400 × 400	C
Bousias et al. (1992)	9	1,500	250 × 250	C
	1	1,500	250 × 250	V
Kim and Lee (2000)	2	1,200	100 × 100	C
	4	1,200	200 × 100	C
Qiu et al. (2002)	6	900	200 × 200	C
Tsuno and Park (2004)	2	2,750	550 × 550	C
Bechtoula et al. (2005)	2	625	250 × 250	C
	1	1,200	560 × 560	C
	2	1,200	560 × 560	V
	2	625	242 × 242	V
	1	1,200	600 × 600	V
Kawashima et al. (2006)	6	1,750	400 × 400	C
Chang (2010b)	1	355	750 × 600	C
Rodrigues et al. (2013)	2	1,700	200 × 400	C
	5	1,700	300 × 300	C
	5	1,700	300 × 400	C
	5	1,700	300 × 500	C

C, constant axial load; V, variable axial load.

directions is a very important subject. This section will, therefore, review the displacement patterns used in published studies and will briefly analyse their influence on the column behaviour.

From the literature review, it was possible to describe several types of laws considered in displacement control biaxial tests. Bixial displacement patterns can be applied only along the x and y axes, illustrated in Figure 7 as L1 (unidirectional) and L2 (cruciform), creating a cruciform displacement pattern. Other configurations of displacement patterns have been used by different authors, namely: the diagonal cruciform (L3), the rhombus or diamond (L4), expanding square (L5), the square in each quadrant in the form of an eight (L6) and the circular (L7) or elliptical (L8). Some variations of the presented displacement patterns can be found in the literature.

The use of these (or other) displacement patterns is based on the main objectives of the performed biaxial tests: on one hand, the comparison between uniaxial and biaxial load paths, and on the other hand, the evaluation of the effect of the different biaxial displacement patterns in the column response. For example, in the study of Saatcioglou and Ozcebe (1989), similar hysteretic behaviour was found between uniaxial and diagonal load paths; however, for the elliptically shaped displacement pattern, an earlier progressive degradation of strength and stiffness was observed. In the studies of Li et al. (1988) and of Kawashima et al. (2006), based on the displacement path with the eight shape, a characteristic drop of the restoring force at zero displacement was observed. The biaxial loading history had a clear effect on the column response and can induce particular

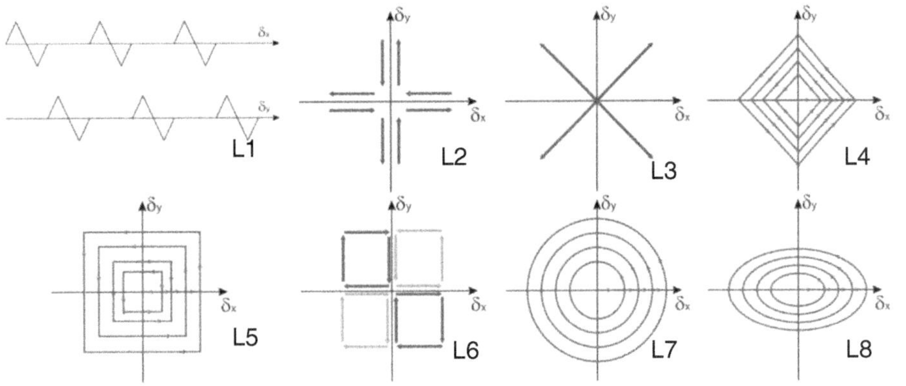

Figure 7 Load paths used by different authors.

effects as a result of the coupling between the two orthogonal directions.

Additionally, the experimental results are intended to be used in the development and calibration of mathematical models that represent the column behaviour, including the biaxial interaction. From the observation of the results, it is clear that the column behaviour is path-dependent and particularly the failure mode Taylor et al. (1996). This fact is even more relevant in the case of the biaxial loading, where the coupling effect between the axial and the two transversal directions is important. Similar tests tested with different loading paths present a very distinct damage evolution Rodrigues et al. (2013). Additional research on this topic should be conducted in order to correlate the load path with the failure mode. The presented displacement paths cannot be considered as representative of an earthquake displacement path; however, the intention of calibrating numerical models imposes the definition of simplified displacement paths with smooth increase of the peak displacements and with the repetition of each displacement level in order to capture the strength and stiffness degradation along the tests.

Behaviour of RC columns under biaxial bending with constant axial load

As stated before, the available test results for biaxial bending under constant axial load are not so extensive when compared to those for uniaxial bending, although they have been delivered over a period of almost 30 years. Contributions are summarised in Table 1. In general, most research findings agree that, additionally to the expected significant influence of axial loads on the hysteretic response of columns, the biaxial transversal load cycles are responsible for increasing the strength and stiffness degradation when compared to the uniaxial response. In addition, the failure mechanism of RC columns is found to be very dependent on the loading path/history and strongly affects both the ductile and energy dissipation capacity of the columns. On the other hand, there is some experimental evidence that plastic hinge zone lengths tend to be stable at around theoretical values and are not strongly affected by biaxial loading.

The CEB Report CEB (1996) includes the major findings of the experimental studies about biaxial flexure with constant axial load of RC columns until 1992. The major findings are summarised next. Observing the force path measured with the square displacement path, several authors (Takizawa and Aoyama 1976; Otani et al. 1980) have reported the expected elastoplastic behaviour of the RC columns, for which after yielding, the square force paths show a tendency to cluster into a single square.

Several authors (Takiguchi et al. 1980; Takizawa and Aoyama 1976; Otani et al. 1980; Bousias et al. 1992) reported a rotation of the measured force paths by about 10° to 20° with respect to the applied displacement path. This rotation occurs when the displacement in one direction is changed, while the displacement in the opposite direction is maintained at a practically constant level; the force required in this lateral direction consequently drops. The effect of this coupling on the hysteretic curves induces an almost vertical unloading branch, which increases the level of dissipated energy in each loop CEB (1996).

Saatcioglou and Ozcebe (1989) reported on the use of elliptical displacement paths with increasing ellipse amplitude, for which the force-displacement curves show an even more rounded shape characteristic of the phase lag between the resultant force and the imposed displacement vector. This fact is associated with the effect already reported for the square load paths. The increase in roundness of the force-displacement curves means that they are much wider when compared with the uniaxial response, thus increasing the energy dissipation due to the coupling between the two directions. For the circular load path, Bousias et al. (1992) have found similar results, concluding that the phase lag is almost constant at each deflection level, but increases with the deflection amplitude and at the test end, when failure is appearing.

Low and Moehle (1987) performed the first biaxial tests on columns with rectangular cross-sections. In the third test, besides the abrupt drop of the force at the beginning of the unloading branch, as also reported by other authors, a similar force drop under constant zero displacement was observed. This phenomenon is associated with the type of displacement path similar to L6 in Figure 7. This fact was also reported by Li et al. (1998).

Bousias et al. (1992) focused on the effect of load path. The examined load paths did not duplicate those of earlier researchers, but covered different transverse displacement-controlled paths. The study emphasised the significant coupling between the three loading directions. The strong coupling between the two transverse directions produced an apparent reduction in strength and stiffness in each of the two transverse directions when compared with the loading separately in each direction, but also increased the hysteretic energy dissipation. Despite the adverse effect on the structural response, some effects can be considered beneficial to the structural response, such as the increase of hysteretic energy dissipation.

Kawashima et al. (2006) studied the effect of bilateral excitation on the seismic performance of RC bridge columns with 1.35 m tall and 400 × 400 mm cross-section, using cyclic and hybrid loading tests. In the cyclic loading test, the columns were loaded unilaterally or bilaterally using diagonal, square, circular and elliptic displacement paths. For the hybrid loading cases, the ground accelerations of the Kobe and Northridge earthquakes were used. In the cyclic tests, the loading and unloading hysteresis curves are round near the peak displacements under the bilateral loading, which results from the interaction of restoring forces in two directions. Only the hysteresis at the first excursions, in each loading step, has a feature similar to that under the unilateral loading. This is because, under the circular-path loading, the column was first loaded only in the N direction in the first excursions at each loading step. Comparing to the strength of the column under bilateral hybrid loading, the strength of the column under bilateral cyclic loading is 14% to 23% and 15% to 35% smaller in NS and EW directions, respectively. Furthermore, the deterioration of the lateral restoring force is significant after 3% of drift, while deterioration did not occur until 6.3% peak drift under the bilateral hybrid loading. Flexural strength and ductility capacity of RC bridge columns with a square section significantly deteriorate under the bilateral excitation in comparison with unilateral excitation in both cyclic and hybrid loading tests. Failure of the columns under the cyclic loading test, for which the loading displacement was increased stepwise with three loading excursions at each loading step, is more extensive than the failure developed in the hybrid loading test. In fact, loading protocol is important, and it has to be carefully determined in the cyclic loading tests.

Qiu et al. (2002) tested seven specimens of RC column subjected to biaxial loading with different load paths. The interactions of biaxial deformation, under biaxial loads, were found to weaken the biaxial strength and the hysteretic energy dissipation capacity. According to the test result, although the accumulative hysteresis dissipation energy relates closely to the loading position and path length, the energy parameter of a specimen under biaxial load is apparently larger than that under unidirectional loading. The specimen damage under biaxial loading is greater than that under uniaxial loading, which agrees with the reduction of the plastic deformation capacity.

Tsuno and Park (2004) also performed cyclic bidirectional tests on two RC columns 2.75 m tall and with 550 × 550 mm cross-section. In this study, it was possible to conclude that the plastic hinge zone length tends towards theoretical values after some cyclic loadings and is not affected by bi-directional loading. The ultimate displacement of a column in a bi-directional cyclic loading is smaller than that of the same column subjected to the standard unidirectional loading pattern. Tsuno and Park, in agreement with Ohno and Nishioka (1984), state that the total dissipated energy until the ultimate state is achieved is approximately the same for biaxial and uniaxial conditions, meaning, the biaxial loading does not affect the total energy dissipation capacity of a column that reaches failure.

Chang (2010b) pseudo-dynamically tested two RC bridge columns at 2/5 reduced scale, and one identical column was also tested under biaxial cyclic horizontal load with constant axial load. The results showed that the distinct characteristics of biaxial hysteretic loops are rounded at corners and negative stiffness values. In addition, scatter plots of pseudo-dynamic results reveal that there is no bias in the orientations of the resultant flexural moments. Therefore, it seems appropriate to make a seismic design based on biaxial bending without bias in any specific direction. The complicated biaxial hysteretic loops confirm the difficulty of developing suitable load-displacement models for non-linear dynamic analysis. These pseudo-dynamic outputs, such as displacement and hysteretic responses, can be treated as references for analytical models proposed to simulate the load-displacement relationship. After comparing biaxial with uniaxial hysteretic loops for both cyclic loading and pseudo-dynamic tests, it is important to note that biaxial hysteretic loops show greater stiffness degradation and pinching during unloading. This reveals that damage caused in one direction weakens the seismic resistance in the other direction.

Bechtoula et al. (2005) tested eight large-scale and eight small-scale cantilever RC columns under various vertical and horizontal loading patterns. Square and circular, as well as unidirectional, horizontal displacement patterns were considered. The axial load was considered both as constant (moderate load and high load) and variable during the test. From the tests observation, it was clear that the bi-directional loading had a significant influence on the envelope curves as well as on the damage progress. The observed damage to large-scale columns was much more severe than that observed for the small-scale columns.

Finally, Rodrigues et al. (2013) and Rodrigues (2012) tested 17 RC rectangular columns with four types of full-scale quadrangular building columns tested for different loading histories. The horizontal loading patterns considered were cruciform, diamond, expanding quadrangular and circular. In this study, the comparison of the biaxial results is performed with similar columns under uniaxial load. Based on the obtained results, it was verified that (1) the initial column stiffness in both directions is not significantly affected by the biaxial load path, (2) when comparing the maximum strength in one specific direction of the columns for each biaxial test against the corresponding uniaxial test, lower values were obtained for all biaxial tests than uniaxial ones. The biaxial loading induces a 20% to 30% reduction of the maximum strength of the columns in their weak direction, *Y*, while reductions from 8% to 15% for the stronger direction, *X*; (3) the ultimate ductility is significantly reduced in columns subjected to biaxial load paths; (4) the strength degradation is practically zero, in the first loading cycles, increasing after displacement ductility demands of about 3. From the strength degradation analysis, more pronounced strength degradation was observed for biaxial tests when compared with corresponding uniaxial tests; (5) the biaxial loading can introduce higher energy dissipation (circular, rhombus and cruciform load paths) than uniaxial loading, as previously recognised by other authors. It was confirmed that the energy dissipation also depends on the column's geometry Rodrigues et al. (2012a). (6) The viscous damping highly depends on the biaxial load path. The repetition of cycles, for the same maximum displacement level, has practically no influence on the equivalent damping.

Behaviour of RC columns under biaxial bending with varying axial load

As widely known, from the axial load-bending interaction diagrams, it is possible to observe that both yielding and ultimate moments increase with the level of axial load until the balance point is achieved. Variation in the axial load during an earthquake can change the strength, stiffness and ultimate displacement capacity, as well as all the hysteretic properties of a RC section. Such variations can occur due to the vertical component of the seismic load, or in the external columns of the bottom storeys of RC frames, due to the overturning moments which increase the axial load on one side and decrease it on the opposite side. In fact, the variation in the axial force during the response cycle may significantly affect the inelastic response of the columns CEB (1996).

Due to testing difficulties, the number of RC column tested under bidirectional displacement with varying axial load is very reduced, and it is always associated with tests with constant axial load, as summarised in Table 1; thus, the few available results do not allow for solid conclusions to be drawn about coupling behaviour between biaxial bending and the varying axial force.

The early results of both Li et al. (1988) and Low and Moehle (1987) evidence similar effects of axial load variations on uniaxial and biaxial flexure. In particular, it was found that the axial load variation simultaneously, with the transverse forces and deformations, leads to stiffness and strength increase, while the strength degradation is larger for higher axial load, and also decreases when the axial load decreases.

The results obtained by Bousias et al. (1992) present a strong coupling between the axial and transverse directions. For the relatively low levels of compressive axial loads considered therein, deflections induce an axial extension which has a magnitude roughly proportional to their resultant vector. For lower compressive loads, the cycling of transverse forces or deflections causes a gradual shortening in the axial direction under the axial load alone, a ratcheting extension that rapidly turns into shortening when failure is imminent. As a result of axial/lateral forces coupling, the cycling of the axial force below the balance load causes a ratcheting increase in the deflection under constant transverse force.

Bechtoula et al. (2005) found that the axial load intensity had little effect on the envelope curve of the second cycle of the load-displacement plot for specimens under unidirectional horizontal load with constant or variable axial load. Equivalent viscous damping increased with the increase in axial load, and columns under variable axial load showed equivalent viscous damping values similar to those of columns under constant/moderate and constant/high axial loads.

Biaxial tests on global bare frame structures

The experimental study of the seismic response of RC frame structures is very important to understand the behaviour of RC buildings, since this inelastic response is dependent on each element and on the distribution and participation of each element in the global response. The available tests in singular elements or structural sub-assemblages give an important source of information for understanding the global structural behaviour; however, the available tests on concrete frame structures are very limited. Most of the tests performed on frame structures only focused on planar structures, or three-dimensional structures loaded along one single direction. However, it has been earlier recognised that three-dimensional behaviour may play an important role in the global behaviour of frame systems, in particular when the torsional response assumes significant importance due to in-plan irregular distribution of the strength and stiffness or mass. Therefore, only 3D model testing allows capturing the failures associated with the non-linear bidirectional flexure response CEB (1996).

Oliva and Clough (1987) tested a 7/10 scale model of a frame structure with two storeys and four rectangular columns by mounting it on a shaking table with its longitudinal principal axis at a 25°-angle relative to the axis of shaking table horizontal motion. The specimen was subjected to the N69W component of the 1952 Taft earthquake record, with increasing intensity motions.

From the frame behaviour observation, it was possible to find that the existence of simultaneous biaxial moments in the columns resulted in yielding, even though the moments along either of the column's principal axes were lower than uniaxial yield moments. The non-symmetrical structural stiffness distribution, relative to the loading direction, led to a global torsion in the frame response. The effect of the global torsional response on the frame corner columns and the varying axial load caused by the overturning moment were different from column to column. Thus, yielding of each column is reached for different steps, which influences the evolution of the overall lateral stiffness of the frame system, increasing the biaxial flexure demands in the corner columns.

From the analysis of column behaviour, it was observed that the response along the strong column axis, parallel to the frame longitudinal axis, exhibited readily identifiable yielding, and the column flexural response in the weak (or transverse) direction was significantly different from the cyclic flexural response normally seen in RC members without an obvious yield point even though the deformations were greater than those along the columns' strong axis. The yielding and damage induced by the strong axis motion resulted in a considerable decrease in the weak axis stiffness, with large deformations and low energy dissipation.

The SPEAR structure is a three-storey building representative of old constructions in southern European countries, without specific provisions for earthquake resistance Mola and Negro (2005). The structure is regular in elevation; the plan configuration is non-symmetric in two directions, with two-bay frames spanning from 3 to 6 m. There is a balcony, shifting the centre of the stiffness away from the centre of the mass. Eight columns have a square 250 × 250 mm cross-section, while one has a cross-section of 250 × 750 mm, which makes it much stiffer and stronger than the others along the *Y* direction. The test programme consisted of three different sets of pseudo-dynamic tests on the specimen in the original and in two different retrofitted configurations. Each set consisted of three tests at different PGA levels: 0.02, 0.15 and 0.20 g Molina et al. (2004). The *Y* direction, parallel to the strong column direction, was globally stronger. Thus, as expected, the levels of interstorey shear reached in this direction were larger, in agreement with the global value of the absorbed energy; on the contrary, much larger displacements were reached in the *X* direction on the second floor. These findings were confirmed by the time histories of the absorbed energy at each floor. The most affected level was the second level. The interstorey drifts at each storey were different for each of the nine columns of the specimen, due to torsional effects. The tests highlighted the strong effects of torsional irregularity on the column drifts, even for a limited level of plan eccentricity and relatively low levels of excitation Mola and Negro (2005).

Finally, it is worth referring that most of the testing campaigns available in the literature were developed, aiming to study the buildings response to three-dimensional loadings, so they were not particularly focused in the study of the biaxial behaviour of columns. So, the conclusions from those research projects regarding the analysis of the column response at local level are limited.

Conclusions

The response of RC columns is recognised as a very important topic that should be taken into consideration for building structures in earthquake-prone regions. The response of RC building columns to earthquake actions deals in general with three-dimensional responses, due to the random characteristics of the earthquake direction and to the building irregularities.

Experimental research on the inelastic response of RC members under compression axial force and biaxial lateral cyclic bending loading conditions is currently very limited. Uncertainties concerning the relationship and combination of the two orthogonal horizontal loading paths, associated with the complexity of the experimental set-up, certainly justify this lacuna.

From the analysis of the available experimental work, it was possible to verify that, besides the expected significant influence of axial loads on the hysteretic response of columns, the bidirectional transversal load cycles are responsible for an increasing in the stiffness and strength degradation, when compared to the uniaxial response. In addition, the failure mechanism of RC columns is found to be very dependent on the loading path, which strongly affects both the ductility and energy dissipation capacity of the columns. On the other hand, there is some experimental evidence that plastic hinge lengths tend to be stable at around theoretical values and are not strongly affected by bidirectional loading. Based on the revision of the work done in the last years, many questions are still open in the field of the biaxial behaviour of RC columns, especially those related with the response dependency of the load paths. Thus, additional experimental tests are needed to be performed by imposing displacement paths representative of the displacements actually imposed by earthquakes.

Competing interest
The authors declare that they have no competing interests.

Authors' contributions
HR participated in the review of the experimental tests on RC column and structures. HV participated in the review behaviour of RC columns, in particular, the damages during recent earthquakes. AA participated in the review and discussions of the experimental studies presented. AC participated in the global revision and discussion of the review performed. All authors read and approved the final manuscript.

Authors' information
HR is assistant professor at the Faculty of Natural Sciences, Engineering and Technology, Oporto Lusophone University and is a researcher at the University of Aveiro, Portugal. He received his Ph.D. degree in Civil Engineering from Aveiro. His research interests include seismic engineering, structural analysis methods, and repair and strengthening of building structures, including experimental and numerical activities on RC structures. HV is an associate professor at the Civil Engineering Department of the University of Aveiro, Portugal. His work comprises experimental testing and nonlinear analytical modelling of structural systems. In his teaching, he has specialised in the dynamic of structures, strength of materials and rehabilitation of structures. His current research interests include assessment, strengthening and repair of existing structures, structural testing and modelling, reliability of structures, and earth construction. AA is an associate professor at the Civil Engineering Department of the Faculty of Engineering, University of Porto, Portugal, having taught since 1990 several topics such as structural analysis (static and dynamic), earthquake engineering, reinforced concrete, and structural rehabilitation. His major research interests are earthquake engineering and structural rehabilitation, including experimental and numerical activities on both ancient stone masonry and contemporary RC structures. He is the executive director and co-founder of the Laboratory for Earthquake and Structural Engineering (LESE) of FEUP since 2002. AC is currently a full professor at the Civil Engineering Department of the University of Aveiro, Portugal. His work comprises structural rehabilitation and strengthening and seismic engineering. He is the founder of the Conservation and Rehabilitation of Buildings and Patrimony Group, which is integrated in the Structure's Division of the Faculty of Engineering, University of Porto. AC is the vice president of the Portuguese Society of Seismic Engineering.

Acknowledgements
This paper reports research developed under the financial support provided by Fundação para a Ciência e Tecnologia (FCT), Portugal, namely through the research project PTDC/ECM/102221/2008.

Author details
[1]Civil Engineering Department, University of Aveiro, Aveiro, Portugal. [2]Faculty of Natural Sciences, Engineering and Technology, Oporto Lusophone University, Oporto, Portugal. [3]Civil Engineering Department, Faculty of Engineering, University of Porto, Oporto, Portugal. [4]Civil Engineering Department, University of Aveiro, Campus Universitário de Santiago, Aveiro 3810-193, Portugal.

References
Akguzel U, Pampanin S (2010) Effects of variation of axial load and bidirectional loading on seismic performance of GFRP retrofitted reinforced concrete exterior beam-column joints. J Compos Constr 14:94–104

Bechtoula H, Kono S, Watanabe F (2005) Experimental and analytical investigations of seismic performance of cantilever reinforced concrete columns under varying transverse and axial loads. J Asian Archit Build Eng 4:467–474

Berry M, Parrish M, Eberhard M (2004) PEER structural performance database—user's manual. University of California, Berkeley

Bousias SN, Verzelleti G, Fardis MN, Magonette G (1992) RC columns in cyclic biaxial bending and axial load. In: Paper presented at the 10th world conference on earthquake engineering., Madrid, Spain, 19–24 July 1992

CEB (1996) RC frames under earthquake loading. Telford, London

Chang SY (2010) Experimental studies of reinforced concrete bridge columns under axial load plus biaxial bending. IJASE 136:12–25

Coelho E (1992) Comportamento sísmico de estruturas em pórtico de betão armado - avaliação da resposta não-linear histerética. Thesis, IST, Lisbon

EASY (1997) Earthquake engineering slide information system., http://www.ikpir.com/easy/html/ang/INDEX.HTM. Accessed 07 February 2013

Fardis MN (2006) Seismic design issues for masonry-infilled RC frames. Paper presented at the proceedings of the first European conference on earthquake engineering and seismology, Geneva, Switzerland, 3–8 September 2006

Kawashima K, Ogimoto H, Hayakawa R, Watanabe G (2006) Effect of bilateral excitation on the seismic performance of reinforced concrete bridge columns. Paper presented at the 8th U.S. National conference on earthquake engineering, San Francisco, CA, 18–22 April 2006

Kim JK, Lee SS (2000) The behaviour of reinforced concrete columns subjected to axial force and biaxial bending. Eng Struct 23:1518–1528

Li K-N, Aoyana H, Otani S (1988) Reinforced concrete columns under varying axial load and bi-directional lateral load reversals. Paper presented at the proceedings of the 9th world conference on earthquake engineering, Tokyo-Kyoto, Japan, 2–9 August 1988

Li KN, Aoyama H, Otani S (1998) Reinforced concrete columns under varying axial load and bidirectional horizontal load reversals. Proceedings of the 9th world conference on earthquake engineering, Tokyo-Kyoto, Japan, 2–9 August 1988

Li L, Mander JB, Dhakal RP (2008) Bidirectional cyclic loading experiment on a 3D beam-column joint designed for damage avoidance. J Struct Eng 134:1733–1742

Low S, Moehle JP (1987) Experimental study of reinforced concrete columns subject to multiaxial cyclic loading. University of California, Berkeley

Marante ME, Flórez-López J (2002) Model of damage for RC elements subjected to biaxial bending. Eng Struct 24:1141–1152

Mola E, Negro P (2005) Full-scale PSD testing of the torsionally unbalanced spear structure in the 'as-built' and retrofitted configurations. Paper presented at the proceeding of SPEAR International Workshop, Ispra, Italy, 4–5 April 2005

Molina J, Buchet P, Magonette GE, Hubert O, Negro P (2004) Bidirectional pseudodynamic technique for testing a three-storey reinforced concrete building. Paper presented at the 13th world conference on earthquake engineering, Vancouver, B.C., Canada, 1–6 August 2004

Monti G, Nuti C (1992) Nonlinear cyclic behavior of reinforcing bars including buckling. J Struct Eng-ASCE 118:3268–3284

Ohno T, Nishioka T (1984) An experimental study on energy absorption capacity of columns in reinforced concrete structures. SE/EE 1(2):137–147

Oliva MG, Clough RW (1987) Biaxial seismic response of R/C frames. J Struct Eng 113:1264–1281

Otani S (2002) Rose School short course on nonlinear analysis of reinforced concrete buildings, Winter 2002. Earthquake Engineering, Rose School, Italy, Specialization

Otani S, Cheung VWT, Lai SS (1980) Reinforced concrete columns subjected to biaxial lateral load reversals. Paper presented at the 7th World Conference on Earthquake Engineering, Istanbul, Turkey, 9–13 September 1980

Park R, Zahn FA, Falconer TJ (1984) Strength and ductility of reinforced and prestressed concrete columns and piles under seismic loading. Paper presented at the world conference on earthquake engineering, San Francisco, USA, 21–28 July 1984

Paulay T, Priestley MJN (1992) Seismic design of RC and masonry buildings. Wiley, New York

Qiu F, Li W, Pan P, Qian J (2002) Experimental tests on RC columns under biaxial quasi-static loading. Engrg Structures 24:419–428

Ricci P, De Luca F, Verderame G (2011) 6th April 2009 L'Aquila earthquake, Italy: reinforced concrete building performance. B Earthq Eng 9:285–305

Rodrigues H (2012) Biaxial seismic behaviour of reinforced concrete columnns. Thesis, University of Aveiro, Aveiro

Rodrigues H, Romão X, Costa AG, Arêde A, Varum H, Guedes J, Vicente R, Costa AA, Paupério E (2010a) Sismo de L'Aquila de 6 de Abril de 2009. Ensinamentos para Portugal. Paper presented at the 8th congresso nacional de sismologia e engenharia sísmica - SISMICA 2010, Aveiro, 20–23 October 2010

Rodrigues H, Varum H, Costa A (2010b) Simplified macro-model for infill masonry panels. J Earthquake Eng 14:390–416

Rodrigues H, Varum H, Arêde A, Costa A (2012a) A comparative analysis of energy dissipation and equivalent viscous damping of RC columns subjected to uniaxial and biaxial loading. Eng Struct 35:149–164

Rodrigues H, Varum H, Arêde A, Costa A (2012b) Comparative efficiency analysis of different nonlinear modelling strategies to simulate the biaxial response of RC columns. Earthq Eng Eng Vibrat 11:553–566

Rodrigues H, Arêde A, Varum H, Costa AG (2013) Experimental evaluation of rectangular reinforced concrete column behaviour under biaxial cyclic loading. Earthquake Eng Struc 42(2):239–259

Saatcioglu M, Ozcebe G (1989) Response of reinforced concrete columns to simulated seismic loading. ACI Struct J 86:3–12

Suzuki N, Otani S, Kobayashi Y (1984) Three-dimensional beam-column sub-assemblages under bi-directional earthquake loading. Paper presented at the 8th world conference on earthquake engineering, San Francisco, USA, 21–28 July 1984

Takiguchi K, Kokusho S, Kobayashi K, Kimura M (1980) Response of RC column to horizontal bi-directional deflection history. Paper presented at the proceeding 7th world conference on earthquake engineering, Istanbul, 9–13 September 1980

Takizawa H, Aoyama M (1976) Biaxial effects in modelling earthquake response of RC structures. Earthq Eng Struct Dyn 4:523–552

Taylor AW, El-Bahy A, Stone WC, Kunnath SK (1996) Effect of load path on damage to concrete bridge piers. Paper presented at the U.S./Japan Natural Resources Development Program (UJNR). Wind and Seismic Effects. Joint Meeting of the U.S./Japan Cooperative Program in Natural Resources Panel on Wind and Seismic Effects, Gaithersburg, 12–15 May 1998

Tsuno K, Park R (2004) Experimental study of reinforced concrete bridge piers subjected to bi-directional quasi-static loading. Struct Engrg Struct 21(1):11s–26s

Varum H (2003) Seismic assessment, strengthening and repair of existing buildings. Thesis, University of Aveiro, Aveiro

Vicente R, Rodrigues H, Costa A, Varum H, Silva JARM D (2010) Masonry enclosure walls: lessons learnt from the recent Abruzzo Earthquake. Paper presented at the Resultados da pesquisa 14th European conference on earthquake engineering, Ohrid, Macedonia, 30 August–3 September 2010

Vicente R, Rodrigues H, Varum H, Costa A, Silva JM (2012) Performance of masonry enclosure walls: lessons learned from recent earthquakes. Earthq Eng Eng Vibrat 11(1):23–24

Zahn FA, Park R, Priestly MJN (1989) Strength and ductility of square reinforced concrete column sections subjected to biaxial bending. ACI Struct J 86:123–131

3

Wave propagation in laminated composite plates

Kishori Lal Verma

Abstract

In this article, the dispersion of propagation waves in an arbitrary direction in laminated composite plates is studied in the framework of elasticity. Three-dimensional field equations of elasticity are considered, and the characteristic equation is obtained on employing the continuity of displacements and stresses at the layers' interfaces. Obtained characteristic equation is further simplified by making use of the properties of the block matrices. Some important particular cases such as of free waves on reducing plates to single layer and the surface waves when thickness tends to infinity are also discussed. Numerical results are also obtained and represented graphically.

Keywords: Dispersion, Waves, Elasticity, Laminated, Composites plates

Introduction

The growing practical applications of laminated composite materials and their uses because of advanced high strength, high modulus layered structures must go a careful inspection to sort out manufacturing defects and in-service degradation in the design of engineering structures. Composite materials are also becoming indispensible in modern industries such as aerospace, infrastructure, and energy because of their ability to customize the physical properties in addition to light weight, directional stiffness, and long fatigue life. The potential of such advanced laminated composites and fiber-reinforced materials as important structural members, wherein an internal layer provides structural strength and the outside layer provide protection against the surrounding media, has motivated an extensive amount of research in the field of mechanics. To modify composite structures or in order to meet particular demands, the design of structures has encouraged the studies of impact and wave propagation in such materials.

Also the lightweight, high strength, and corrosion resistance of fiber-reinforced polymers (FRP) make them ideally suited for quick and effective structural repairs. Investigations on the protection of historical and monumental buildings made of masonry material through the adoption of FRP composite provisions have inspired a wide scientific literature on the subject, with many theoretical as well as experimental contributions and developments of original approaches, techniques, and numerical tools in addressing the problem of masonry constructions; the planning and design of adequate protection techniques are studied in Foti (2013a), Baratta and Corbi (2005, 2010, 2011), Baratta et al. (2008). Foti (2011) studied the behavior fiber-reinforced concrete structures. They obtained results of some tests for an approach to a broader testing on the possibility of using fibers from polyethylene terephthalate (PET) bottles to increase the ductility of the concrete. The use of recycled waste PET bottle fibers for the reinforcement of concrete is studied by Foti (2013b). A series of tests have been performed with the aim to define the best solution in strengthening a deteriorated structure with a rheoplastic mortar reinforcement by Foti and Vacca (2013) and three types of possible structural reinforcing renovation on reinforced concrete pillars have been considered, with special attention to adhesion between materials with different chemical-physical and mechanical characteristics. The crack patterns obtained on the specimens have been analyzed to demonstrate the relevance of an appropriate thickness of the reinforcement to obtain an effective mechanical behavior of the reinforced concrete element over time.

Since composites consist of different materials, they are inhomogeneous and anisotropic. As a consequence, different mechanical properties among the constituents and the environmental changes can create residual stresses, which may lead to interface de-bonding in the structures, leading to failure of the system. Therefore, it is of interest to investigate the feasibility of non-destructively monitoring mechanical aging in composites. The study of wave propagation and vibration in plates, half-spaces, and laminates is an area of considerable recent research activity. The

Correspondence: klverma@netscape.net
Department of Mathematics, Government Post Graduate College, Hamirpur, Himachal Pradesh 177005, India

propagation of the plane waves in unbounded homogeneous layered media is well known, and many methods have been proposed. On the dynamic behavior of anisotropic plates, extensive review of the plate theories can be found in Achenbach (1973), Auld (1973), Fedorov (1968), Nayfeh (1995), and Wooster (1973), and wave propagation problems in periodically layered anisotropic media have been considered by the authors Norris (1992, 1993), Wang and Rokhlin (2002), and Braga and Hermann (1988). Several problems on the theories of laminated and composite plates have been considered, and their dynamic behavior was studied by authors Reddy (1987, 1990), Liu et al. (1990), Postma (1955), Rytov (1956), and Sun et al. (1968). There also have been a reasonable number of investigations of such advanced materials, and their analysis was reported by Sve (1971), Jones (1975), Graff (1991), and Reddy (1997). Nayfeh (1991) developed a transfer matrix technique to obtain the dispersion relation curves of elastic wave propagating in multilayered anisotropic media, i.e., composite laminate. Detailed review on the wave propagation in layered anisotropic media/anisotropic laminates is given by Liu and Xi (2002). Yamada and Nasser (1981) have studied harmonic wave's propagation direction in orthotropic composites, and Verma (1999) considered the similar problem in thermoelastic heat conducting material.

Similar formulations using Floquet's theorem are employed to analyze the wave motions of two-phase slender periodic structures that has been made by Tourafte (1986) and Tassilly (1987). Dispersion equation has been represented by the vanishing of a 12 × 12 determinant. Yamada and Nasser (1981) considered an orthotropic, periodically layered composite with two layers in a unit cell, and the dispersion equation corresponds to a 12 × 12 characteristic determinant. These approaches lead to important results but had limitation due to the size of the matrices that must be manipulated when the unit cell has more than two layers, or when the layers are anisotropic. In this paper, the above said limitation has been resolved by simplifying the obtained characteristic equation using the basic properties of block matrices. Thus, in order to solve the problem numerically, it is sufficient to consider the 6 × 6 determinant only instead of taking the 12 × 12 determinant, thus reducing the numerical work considerably. Further, some of the important particular cases of free waves and the surface waves are also derived and discussed from the obtained results.

Following Yamada and Nasser (1981), in this article, the propagation of waves in layered laminated composites, where the direction of the corresponding harmonic waves makes an arbitrary angle with respect to the layers, is examined. Three-dimensional field equations of elasticity are considered for this study, and the corresponding characteristic equation on employing the continuity of displacements at the layers' interface is obtained. Obtained characteristic equation is further simplified using the basic properties of block matrices, and in order to solve the problem numerically, it is sufficient to consider the determinant of the 6 × 6 matrix instead of taking the determinant 12 × 12 matrix. Some important particular cases such as of free waves on reducing plates to single layer and the surface waves when thickness tends to infinity are also discussed. Numerical results are also obtained and represented graphically.

Methods

Formulation

Consider a set of Cartesian coordinate system $x_i = (x_1, x_2, x_3)$ in such a manner that x_3- axis is normal to the layering. The coordinate axes x_1, x_2, and x_3 of the model are chosen as to be analogous with the principal axes x, y, and z, with z-axis being normal to the plate. The displacement field vector $\mathbf{u} = (u_1, u_2, u_3)$ satisfies the basic field equation of motion for an infinite generally anisotropic medium in the absence of body forces, which is

$$\sum_{j=1}^{3}\left(\frac{\partial \tau_{ij}(\mathbf{u})}{\partial x_j}\right) = \rho\frac{\partial^2 u_i}{\partial t^2} \qquad i = 1, 2, 3. \tag{2.1}$$

Constitutive relations for anisotropic materials

$$\tau_{ij} = C_{ijkl}e_{kl}, \; i, j, k, l = 1, 2, 3. \tag{2.2}$$

ρ is the density, t is time, u_i is the displacement in the x_i direction; τ_{ij}, and e_{kl} are the stress and strain tensor respectively; and the fourth order tensor of the elasticity C_{ijkl} satisfies the (Green) symmetry conditions:

$$c_{ijkl} = c_{klij} = c_{ijlk} = c_{jikl}. \tag{2.3}$$

Strain–displacement relation

$$e_{ij} = \frac{1}{2}\left(\partial u_i/\partial x_j + \partial u_j/\partial x_i\right). \tag{2.4}$$

In addition, at the interface between two layers the tractions, displacements must be continuous.

For orthotropic media, the stress equation for an orthotropic material in the symmetric coordinate system can be expressed by

$$\begin{bmatrix} \tau_{11} \\ \tau_{22} \\ \tau_{33} \\ \tau_{23} \\ \tau_{13} \\ \tau_{12} \end{bmatrix} = \begin{bmatrix} c_{11} & c_{12} & c_{13} & 0 & 0 & 0 \\ c_{12} & c_{22} & c_{23} & 0 & 0 & 0 \\ c_{13} & c_{23} & c_{33} & 0 & 0 & 0 \\ 0 & 0 & 0 & c_{44} & 0 & 0 \\ 0 & 0 & 0 & 0 & c_{55} & 0 \\ 0 & 0 & 0 & 0 & 0 & c_{55} \end{bmatrix} \begin{bmatrix} e_{11} \\ e_{22} \\ e_{33} \\ 2e_{23} \\ 2e_{13} \\ 2e_{12} \end{bmatrix}. \tag{2.5}$$

Analysis

For harmonic waves propagating in an arbitrary direction, the displacement components u_1, u_2, and u_3 are written as

$$(u_1, u_2, u_3) = \{U_1(x_3), U_2(x_3), U_3(x_3)\}e^{i\xi(l_1x_1+l_2x_2+l_3x_3-ct)}, \tag{3.1}$$

where ξ is the wave number, c is the phase velocity ($= \omega/\xi$), $i = \sqrt{-1}, \omega$ is the circular frequency, l_1, l_2, and l_3 are the direction cosine defining the propagation direction.

U_j are the constants related to the amplitudes of displacement, Floquet's theory requires functions U_j (j = 1, 2, 3) to have the same periodicity as the layering. Hence, the problem is reduced to that of one pair of layers, where

$$U_j = \bar{U}_j e^{-i\xi(l_3+\alpha)x_3}, j = 1, 2, 3, \tag{3.2}$$

where $\bar{U}_j$ are constants. On substitution of Equations (3.2) into Equations (2.1), via (2.2) to (2.4) and specializing the equations for orthotropic media, it follows that

$$M_{mn}(\alpha)\bar{U}_n = 0, m, n = 1, 2, 3, \tag{3.3}$$

where

$$\begin{aligned} &M_{11} = \left(l_1^2 + l_2^2\bar{c}_{66} + \alpha^2\bar{c}_{55} - \zeta^2\right), M_{12} = (\bar{c}_{12} + \bar{c}_{66})l_1l_2, \\ &M_{13} = -(\bar{c}_{13} + \bar{c}_{55})l_1\alpha, \\ &M_{22} = \left(l_1^2\bar{c}_{66} + l_2^2\bar{c}_{22} + \alpha^2\bar{c}_{44} - \zeta^2\right), \\ &M_{23} = -(\bar{c}_{23} + \bar{c}_{44})l_2\alpha, \\ &M_{33} = \left(l_1^2\bar{c}_{55} + l_2^2\bar{c}_{44} + \alpha^2\bar{c}_{33} - \zeta^2\right), \end{aligned} \tag{3.4}$$

where $\zeta^2 = \frac{\rho c^2}{c_{11}}$. The existence of nontrivial solutions for $\bar{U}_j$ (j = 1, 2, and 3) demands the vanishing of the determinant in Equation (3.3) and yields the sixth degree polynomial equation

$$\Delta\alpha^6 + F_1\alpha^4 + F_2\alpha^2 + F_3 = 0, \tag{3.5}$$

where

$$\Delta = c_{33}c_{44}c_{55},$$

$$\begin{aligned} F_1 = &\left[\left(c_{22}c_{33} - 2c_{23}c_{44} - c_{23}^2\right)c_{55} + c_{33}c_{44}c_{66}\right]l_2^2 \\ &+ \left[\left(c_{33} - 2c_{13}c_{55} - c_{13}^2\right)c_{44} + c_{33}c_{55}c_{66}\right]l_1^2 \\ &- \left(c_{33}c_{44} + c_{33}c_{55} + c_{44}c_{55}\right)\zeta^2 \end{aligned}$$

$$\begin{aligned} F_2 = &\left[\left(c_{33} - 2c_{13}c_{55} - c_{13}^2\right)c_{66} + c_{44}c_{55}\right]l_1^4 \\ &+ \left[\left(c_{22}c_{33} - 2c_{23}c_{44} - c_{23}^2\right)c_{66} + c_{22}c_{55}c_{44}\right]l_2^4 \\ &+ [(-c_{12}^2c_{33} - 2(c_{33}c_{44} - c_{66}c_{23}c_{55} - c_{12}c_{44}c_{55} \\ &\quad +c_{13}c_{22}c_{55} - 2c_{44}c_{55}c_{66} - c_{13}c_{44}c_{66} \\ &\quad +c_{12}c_{33}c_{66} - c_{12}c_{13}c_{44} - c_{13}c_{23}c_{66} - c_{12}c_{23}c_{55} \\ &\quad -c_{12}c_{13}c_{23}) - c_{13}^2c_{22} + c_{22}c_{33} - c_{23}^2]l_1^2l_2^2 \\ &+ \begin{bmatrix} (2c_{13}c_{55} - c_{66}c_{33} - c_{55}c_{44} - c_{44} \\ -c_{33} - c_{66}c_{55} + c_{13}^2)l_1^2 \\ +(2c_{23}c_{44} + c_{23}^2 - c_{22}c_{33} - c_{22}c_{55} \\ -c_{66}c_{44} - c_{55}c_{44} - c_{33}c_{66})l_2^2 \\ +(c_{33} + c_{44} + c_{55})\zeta^4 \end{bmatrix}\zeta^2 \end{aligned}$$

$$\begin{aligned} F_3 = &\left(c_{55}l_1^2 + c_{44}l_2^2 - \zeta^2\right)\left\{\left((1 + c_{66})l_1^2 + (c_{22} + c_{66})l_2^2\right)\zeta^2 - \zeta^4\right. \\ &\left.+\left[\left(2c_{22}c_{66} + c_{12}^2 - c_{22}\right)c_{55}\right]l_1^2l_2^2 - c_{22}c_{66}l_2^4 - c_{66}l_1^4\right\} \end{aligned}$$

Equation (3.2) can be rewritten as

$$(U_1, U_2, U_3) = \sum_{q=1}^{8}\left(\bar{U}_{1q}, \bar{U}_{2q}, \bar{U}_{3q}\right)e^{-i\xi\left(l_3+\alpha_q\right)x_3} \tag{3.6}$$

For each α_q, q = 1, 2,....6, we can use the Equations (3.3) and express the displacements ratios

$$\frac{D_1}{D} = \frac{\bar{U}_{2q}}{\bar{U}_{1q}} = \gamma_q, \frac{D_2}{D} = \frac{\bar{U}_{3q}}{\bar{U}_{1q}} = \delta_q, q = 1, 2 \ldots.6. \tag{3.7}$$

$$\begin{aligned} &D_1\left(\alpha_q\right) = M_{13}\left(\alpha_q\right)M_{23}\left(\alpha_q\right) - M_{12}\left(\alpha_q\right)M_{33}\left(\alpha_q\right), \\ &D_2\left(\alpha_q\right) = M_{12}\left(\alpha_q\right)M_{23}\left(\alpha_q\right) - M_{13}\left(\alpha_q\right)M_{22}\left(\alpha_q\right) \\ &D_3\left(\alpha_q\right) = 0, D\left(\alpha_q\right) = M_{22}\left(\alpha_q\right)M_{33}\left(\alpha_q\right) - {M^2}_{23}\left(\alpha_q\right). \end{aligned} \tag{3.8}$$

Therefore, the solution is

$$(U_1, U_2, U_3) = \sum_{q=1}^{8}\left(1, \gamma_q, \delta_q\right)\bar{U}_{1q}e^{-i\xi\left(l_3+\alpha_q\right)x_3}. \tag{3.9}$$

In view of the continuity of the displacement components, tractions across the interface of the two layers, the following conditions must be satisfied:

$$u_j^{\mathrm{I}}\Big|_{x_3=0^-} = u_j^{\mathrm{II}}\Big|_{x_3=0^+}, \tag{3.10}$$

$$\tau_{3j}^{\mathrm{I}}\Big|_{x_3=0^-} = \tau_{3j}^{\mathrm{II}}\Big|_{x_3=0^-}, \tag{3.11}$$

where superscripts I and II refer to layers one and two, respectively; 0^+ and 0^- are values of x_3 near zero.

Because of the periodicity of the deformation and stress fields, additional conditions obtained are

$$u_j^{\mathrm{I}}\Big|_{x_3=\bar{h}_1} = u_j^{\mathrm{II}}\Big|_{x_3=-\overset{+}{h}_2}, \tag{3.12}$$

$$\tau_{3j}^{\mathrm{I}}\Big|_{x_3=\bar{h}_1} = \tau_{3j}^{\mathrm{II}}\Big|_{x_3=-\overset{+}{h}_2}, j = 1, 2, 3. \tag{3.13}$$

Following Yamada and Nasser (1981) on substituting the displacements, stress components into Equations (3.10) to (3.11) and linear homogeneous equations for twelve constants $U_{11}^I, U_{12}^I, \ldots, U_{15}^{\mathrm{II}}$ and U_{16}^{II} are obtained. For nontrivial solutions, the determinant of the coefficients must vanish. This yields the following characteristic equation:

$$\det\begin{pmatrix} P_{jk} & -\bar{P}_{jk} \\ Q_{jk} & -\bar{Q}_{jk} \end{pmatrix} = 0 \;\; j, k = 1, 2 \ldots 6. \tag{3.14}$$

The entries of the 6 × 6 matrices $P_{jk}, \bar{P}_{jk}, Q_{jk}$ and $\bar{Q}_{jk}$ are

$$\begin{aligned} &P_{1j} = 1, P_{2j} = \gamma_j^I, P_{3j} = \delta_j^I, P_{4j} = b_{1j}^I c_{55}^I, P_{5j} = b_{2j}^I c_{44}^I, \\ &P_{6j} = b_{3j}^I, \bar{P}_{1j} = 1, \bar{P}_{2j} = \gamma_j^{\mathrm{II}}, \bar{P}_{3j} = \delta_j^{\mathrm{II}}, \bar{P}_{4j} = \eta b_{1j}^{\mathrm{II}} c_{55}^{\mathrm{II}}, \\ &\bar{P}_{7j} = \eta b_{2j}^{\mathrm{II}} c_{44}^{\mathrm{II}}, \bar{P}_{6j} = \eta b_{3j}^{\mathrm{II}}, Q_{jk} = P_{jk} E_k^-, \bar{Q}_{jk} = \bar{P}_{jk} E_k^+ \end{aligned} \tag{3.15a}$$

where

$$\begin{aligned} E_j^- &= e^{-iQ\left(l_3+\alpha_j^{(1)}\right)h_1/h}, Q = \xi(h_1 + h_2), \\ E_j^+ &= e^{-iQ\left(l_3+\alpha_j^{\mathrm{II}}\right)h_2/h}, \eta = c_{11}^{\mathrm{II}}/c_{11}^I b_{1j}^{(m)} \\ &= l_1\delta_j^{(m)} - \alpha_j^{(m)} b_{1j}^{(m)} = l_1\delta_j^{(m)} - \alpha_j^{(m)}, \\ b_{3j}^{(m)} &= \bar{c}_{13}^{(m)} l_1 + \bar{c}_{23}^{(m)} l_2\gamma_j^{(m)} - \bar{c}_{33}^{(m)} \alpha_j^{(m)} \delta_j^{(m)}, \\ \bar{c}_{jk}^{(m)} &= c_{jk}^{(m)}/c_{11}^{(m)} \end{aligned} \tag{3.15b}$$

From Equation (3.14), we have

$$\det\left[P_{jk}\right] \det\left(\left[-\bar{Q}_{jk}\right] - \left[Q_{jk}\right]\left[P_{jk}\right]^{-1}\left[-\bar{P}_{jk}\right]\right) = 0 \tag{3.16a}$$

which implies that either

$$\det\left[P_{jk}\right] = 0, \tag{3.16b}$$

or

$$\det\left(\left[-\bar{Q}_{jk}\right] - \left[Q_{jk}\right]\left[P_{jk}\right]^{-1}\left[-\bar{P}_{jk}\right]\right) = 0. \tag{3.16c}$$

If equation (3.16b) holds true, then the problem reduces to a free wave propagation of single plate of thickness h_1, and in this case, $\left(\left[-\bar{Q}_{jk}\right] - \left[Q_{jk}\right]\left[P_{jk}\right]^{-1}\left[-\bar{P}_{jk}\right]\right)$ will not exist as P_{jk} singular. On the other hand, if P_{jk} is nonsingular, $[P_{jk}]^{-1}$ exists and accordingly

$$\det\left(\left[-\bar{Q}_{jk}\right] - \left[Q_{jk}\right]\left[P_{jk}\right]^{-1}\left[-\bar{P}_{jk}\right]\right) = 0. \tag{3.17a}$$

Similarly, Equation (3.14) can also be written as

$$\det\left[-\bar{Q}_{jk}\right] \det\left(\left[P_{jk}\right] - \left[-\bar{P}_{jk}\right]\left[-\bar{Q}_{jk}\right]^{-1}\left[Q_{jk}\right]\right) = 0, \tag{3.17b}$$

which implies that either

$$\det\left[-\bar{Q}_{jk}\right] = 0, \tag{3.17c}$$

or

$$\det\left(\left[P_{jk}\right] - \left[-\bar{P}_{jk}\right]\left[-\bar{Q}_{jk}\right]^{-1}\left[Q_{jk}\right]\right) = 0. \tag{3.17d}$$

If Equation (3.17b) holds true, then the problem reduces to a free wave propagation of single plate of thickness h_2, and $\left(\left[-\bar{Q}_{jk}\right] - \left[Q_{jk}\right]\left[P_{jk}\right]^{-1}\left[-\bar{P}_{jk}\right]\right)$ will not exist as $\bar{Q}_{jk}$ is singular.

On the other hand, if $\bar{Q}_{jk}$ is nonsingular, therefore

$$\det\left(\left[-\bar{Q}_{jk}\right] - \left[Q_{jk}\right]\left[P_{jk}\right]^{-1}\left[-\bar{P}_{jk}\right]\right) = 0. \tag{3.18}$$

In order to solve the problem numerically, it is sufficient to consider either Equation (3.17a) or Equation (3.18) for composite plates, and for free plate, Equation (3.16b) or Equation (3.17b) can be considered.

Particular cases

Free waves

When layer I = II and $h_1 = h_2$ (say h), then the thickness of the layer is $2h$, considering the origin at the middle of the plate, then the above analysis reduces to a single plate. In this case, the eight roots of Equation (3.5) can be arranged in four pairs as $\alpha_{j+1} = -\alpha_j$, (j = 1, 3, 5).

It is observed from Equation (3.4) that M_{13} and M_{23} are odd functions of α_j, and the other M_{ij} are even functions of α_j. On employing stresses and free surfaces conditions,

$$\sigma_{3j} = 0, x_3 = \pm h, j = 1, 2, 3, \tag{4.1}$$

corresponding relations to (3.7), we have

$$\gamma_{q+1} = \gamma_q \text{ and } \delta_{q+1} = -\delta_q \tag{4.2}$$

Hence, from (3.16)

$$\begin{aligned} b_{1q+1} &= -b_{1q}, b_{2q+1} = -b_{2q}, b_{3q+1} = b_{3q}, b_{1j} \\ &= l_1\delta_j - \alpha_j, b_{2j} = l_2\delta_j - \alpha_j\gamma_j, \end{aligned} \tag{4.3}$$

$$b_{3j} = \bar{c}_{13}l_1 + \bar{c}_{23}l_2\gamma_j - \bar{c}_{33}\alpha_j\delta_j, \bar{c}_{jk} = c_{jk}/c_{11}, \quad (4.4)$$

$$\det[P_{jk}] = 0. \quad (4.5a)$$

After simple mathematical manipulations, this equation reduces to

$$\Delta_E\alpha^6 + A_{E1}\alpha^4 + A_{E2}\alpha^2 + A_{3E} = 0 \quad (4.5b)$$

$$A_{E1} = (c_1c_6F_{11} - c_6F_{13}^2 + c_5F_{13}F_{23} - c_5^2F_{33} - c_2F_{23}^2 - 2c_1c_5F_{12} + c_5F_{21}F_{23} + c_2c_6F_{33})$$

$$A_{E2} = (F_{23}^2F_{11} - F_{22}F_{13}^2 + c_6F_{11}F_{33} + c_1F_{11}F_{22} - c_1F_{12}^2 + 2F_{12}F_{13}F_{23} + 2c_5F_{12}F_{33} + c_2F_{22}F_{33})$$

$$A_{E3} = (F_{11}F_{22} - F_{12}^2)F_{33}, \Delta_E = (c_2c_6 - c_5^2)c_1,$$

where A_{Ej}, j = 1,2,3 and Δ_E are parameters defined in Verma (2008).

Equation (4.5) is the corresponding characteristic equation for free waves in elastic plate. Further, if thickness $d(=h_1 + h_2) \to \infty$, in Equation (4.5), then the problem reduces to that of surfaces waves.

On applying appropriate boundary conditions on the plate outer boundaries, a large variety of important physical problems can be solved.

Higher symmetry materials

Results for higher symmetry materials such as transversely isotropic, cubic, and isotropic can be obtained as special cases with the following restrictions for transverse isotropy symmetry:

$$c_{33} = c_{22}, c_{13} = c_{12}, c_{55} = c_{66}, c_{22} - c_{23} = 2c_{44}; \quad (4.6)$$

for cubic symmetry,

$$c_{11} = c_{22} = c_{33}, c_{12} = c_{13} = c_{23}, c_{44} = c_{55} = c_{66}; \text{ and} \quad (4.7)$$

for isotropic symmetry,

$$c_{11} = c_{22} = c_{33} = \lambda + 2\mu, \quad c_{12} = c_{13} = c_{23} = \lambda, c_{44} = c_{55} = c_{66} = \mu. \quad (4.8)$$

Results and discussion

Using Equation (3.17a), numerical results are presented to exhibit the dependence of dispersion with the angle of propagation. The materials chosen for this purpose is aluminum epoxy composite as layer I (h_1 = 0.6) and carbon steel as layer II (h_2 = 0.4).

Since the distinction among the wave mode types of thermoelastic waves in anisotropic plates is somewhat artificial, as the elastic wave modes are generally coupled, they are referred to as *quasi-longitudinal, quasi-transverse*, and *quasi-shear horizontal* modes. For a wave to propagate in the direction of higher symmetry, some wave types revert to pure modes and lead to a simple characteristic equation of lower order, and consequently, the loss of pure wave modes for general propagation direction in such cases. Here, Figure 1 depicts the dispersion curves with the direction cosines of propagation l_1 = 0.259, l_2 = 0.542, and l_3 = 0.799, whereas dispersion curves with the direction cosines of propagation l_1 = 0.195, l_2 = 0.515, and l_3 = 0.834 are shown in Figure 2. Similarly, dispersion

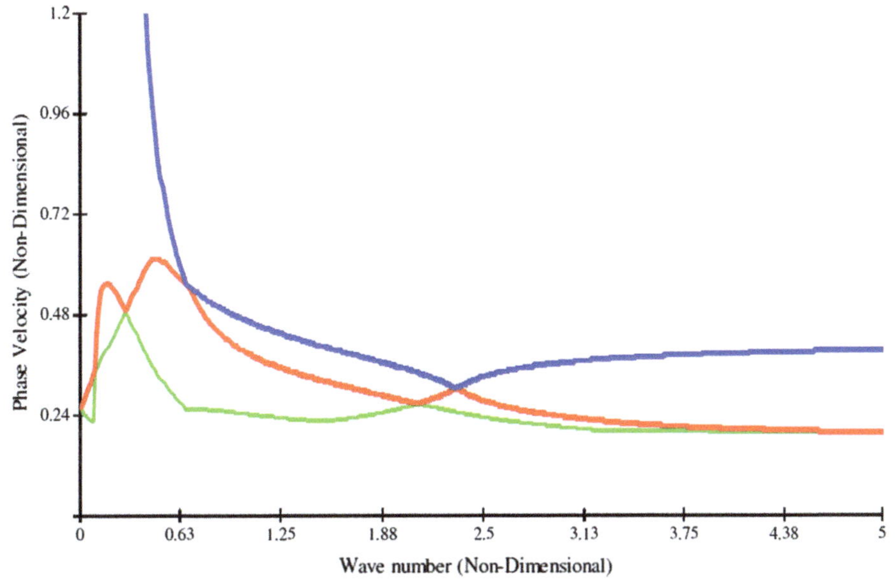

Figure 1 Phase velocity versus wave number for the direction cosine l_1 = 0.259, l_2 = 0.542, and l_3 = 0.799.

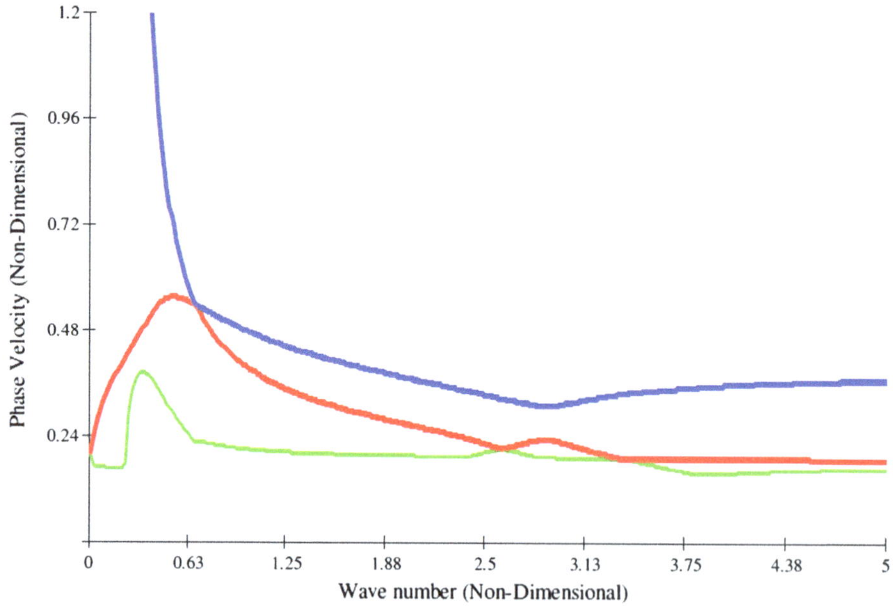

Figure 2 Phase velocity versus wave number for the direction cosine l_1 = 0.195, l_2 = 0.515, and l_3 = 0.834.

curves with the direction cosines of propagation $l_1 = 0.125$, $l_2 = 0.707$, and $l_3 = 0.696$ are shown in Figure 3.

The dispersion behavior of wave speed modes with different angles of propagation is demonstrated in Figures 1, 2 3. It is observed that at zero wave number limits, each figure (Figures 1, 2 3) displays three wave speeds corresponding to one quasi-longitudinal and two quasi-transverses. It is apparent that the largest value corresponds to the quasi-longitudinal. At low wave number limits, modes are found to be highly influenced and also vary with direction. From these figures, it is also observed that at low wave number limits, wave speed modes are dispersive. It is also observed that with the change in the propagation direction, lower modes appear to be highly influenced than the higher modes, where a small change is noticed. Thus, at low values of the wave number, only the lower modes got affected, and the little change is observed at the relatively high values of the wave number. Thus, the low value region of the wave number is found to be of more physical

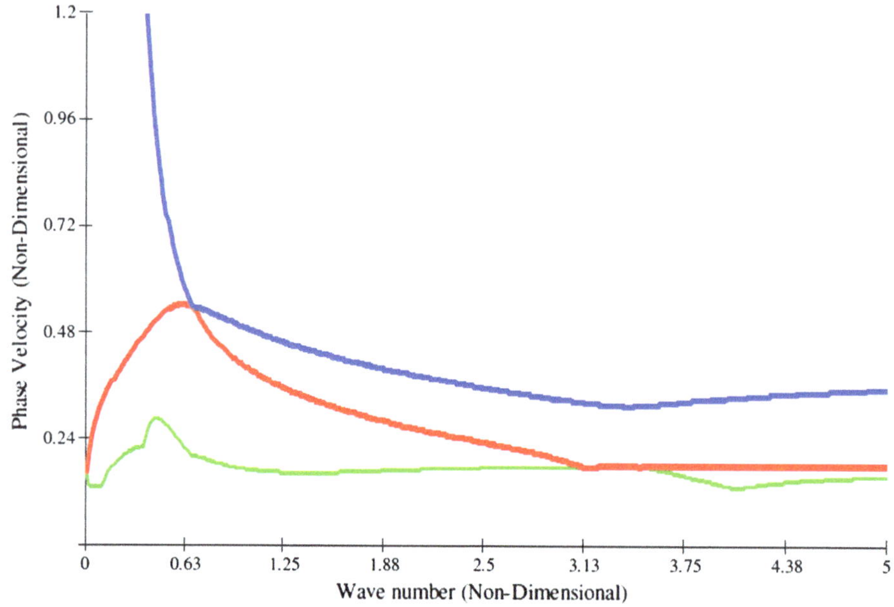

Figure 3 Phase velocity versus wave number for the direction cosine l_1 = 0.125, l_2 = 0.707, and l_3 = 0.696.

interest. Further, at high wave number limits, there is no effect in the laminated composite plates. A small change is observed in these mode values as wave number increases, and other higher modes appear; one of the modes seemed to be associated with quick change in the slope.

Conclusions

The propagation of waves in layered laminated composites, where the direction of the corresponding harmonic waves makes an arbitrary angle with respect to the layers, is studied. Three-dimensional field equations of elasticity are considered for this study and the corresponding characteristic equation is obtained on employing the continuity of displacements at the layers' interface. Obtained characteristic equation is further analytically simplified using the basic properties of block matrices, and in order to solve the problem numerically, it is sufficient to consider the determinant of the 6 × 6 matrix instead of taking the determinant 12 × 12 matrix, which reduces the numerical work considerably. Important cases such as of free waves and surface waves on reducing plates to single layer and when thickness tends to infinity are also discussed. It is found that at zero wave number limits, each figure displays three wave speeds corresponding to one quasi-longitudinal and two quasi-transverses. It is apparent that the largest value corresponds to the quasi-longitudinal. At low wave number limits, modes are found to be highly influenced and also vary with the direction. It is observed that with the change in the propagation direction, lower modes appear to have more influence than the higher modes where a small change is noticed. Therefore, at the lower value region of the wave number, it is found to be of more physical interest than at high wave number limits, on the laminated composites plates.

The study may find applications in damage identification in laminated composite, as the laminated composites possess characteristics that make them particularly useful for applications in nondestructive evaluation of such defects in plate-like structures and provide a means of inspection of an otherwise inaccessible area. In order to use waves in ultrasonic nondestructive applications, it is necessary to investigate the phenomenon of scattering of these waves in arbitrary directions.

Future work will be dedicated on enhancing the current models, especially for the heat-conducting materials in the context of generalized thermoelasticity, viscoelasticity, and orthotropic piezoelectric material. Further steps will be taken to develop a 3D model to observe the wave propagation behavior along different directions, including models of damage, to evaluate the sensitivity of auxetic laminates for damage tolerance applications. From the experimental point of view, Lamb wave propagation will be shortly evaluated for different classes of auxetic laminates with special stacking layer sequences, for comparison with the developed models. It is important to realize that, although the results presented here have been derived for a periodically layered composite with two layers in a unit cell, Floquet theory can be applied to the mathematical modeling of periodically layered structures. This can be done by finding the appropriate linear combination of Floquet waves that satisfy a set of boundary conditions prescribed at planes normal to the layering. Applying appropriate boundary conditions on the plate outer boundaries, a large variety of important physical problems can be solved, which include free waves in layered plates and in periodic media constructed from a repetition of the layered plate.

Competing interests

The author declares that he has no competing interests.

Authors' information

KL Verma gained the Ph.D. in Engineering from the Nagoya Institute of Technology, Nagoya, Japan and Masters in Philosophy and M.Sc. in Mathematics from Panjab University Chandigarh (UT), India and is presently working as an associate professor in Mathematics, Government Post Graduate College Hamirpur (HP), India. Area of specialization is in solid mechanics.

Acknowledgments

The author is thankful to the reviewers for their useful suggestions for the improvement of this work.

References

Achenbach JD (1973) Wave propagation in elastic solids. In: Lauwerier HA, Koiter WT (ed) Applied mathematics and mechanics. North-Holland, New York
Auld BA (1973) Acoustic fields and waves in solids, volume 1. Wiley, New York
Baratta A, Corbi O (2005) On variational approaches in NRT continua. Int J Solids Struct 42:5307–5321
Baratta A, Corbi O (2010) An approach to masonry structural analysis by the no-tension assumption-part I: material modelling, theoretical setup, and closed form solutions. Appl Mech Rev 63(4):1–17
Baratta A, Corbi O (2011) On the statics of no-tension masonry-like vaults and shells: solution domains, operative treatment and numerical validation. Annals of Solid and Structural Mechanics 2(2–4):107–122
Baratta A, Corbi I, Corbi O (2008) Stress analysis of masonry structures: arches, walls and vaults. Proceedings of the 6th international conference on structural analysis of historic construction: preserving safety and significance, Bath, UK :2-4. July 2008
Braga AMB, Hermann G (1988) Plane waves in anisotropic layered composite. In: Mal AK, Ting TC (ed) Wave propagation in structural composites. ASME, New York
Fedorov FI (1968) Theory of elastic waves in crystals. Plenun Press, New York
Foti D (2011) Preliminary analysis of concrete reinforced with waste bottles PET fibers. Construct Build Mater 25:1906–1915
Foti D (2013a) On the numerical and experimental strengthening assessment of tufa masonry with FRP. Mech Adv Mater Struc 20(2):163–175
Foti D (2013b) Use of recycled waste pet bottles fibers for the reinforcement of concrete. Compos Struct 96:396–404
Foti D, Vacca S (2013) Mechanical behavior of concrete columns reinforced with rheoplastic mortar Materiales de Construcción/Comportamiento mecánico de columnas de hormigón armado reforzadas con mortero reoplástico. doi:10.3989/mc.2012.03512. in press
Graff KF (1991) Wave motion in elastic solids. Dover Publications, Inc., New York
Jones RM (1975) Mechanics of composite materials. Scripta Book Co, Washington
Liu GR, Xi ZC (2002) Elastic waves in anisotropic laminates. CRC Press, Boca Raton
Liu GR, Tani J, Watanabe K, Ohyoshi T (1990) Lamb wave propagation in anisotropic laminates. J Appl Mech 57:923–929

Nayfeh AH (1991) The general problem of elastic wave propagation in multilayered anisotropic media. J Acoust Soc Am 89:1521–1531
Nayfeh AH (1995) Wave propagation in layered anisotropic media. North-Holland, Amsterdam
Norris AN (1992) Dispersive plane wave propagation in periodically layered anisotropic media. Proceeding of Royal Irish Academic Sect. A 92:49–67
Norris AN (1993) Waves in periodically layered media: a comparison of two theories. J Appl Math 53(5):1195–1209
Postma GW (1955) Wave propagation in a stratified medium. Geophysics 20:780–806
Reddy JN (1987) A generalization of two-dimensional theories of laminated plates. Commun Appl Numer Methods 3:173–180
Reddy JN (1990) On refined theories of composite laminate. Meccanica 25:230–238
Reddy JN (1997) Mechanics of laminated composite plates: theory and analysis. CRC Press, Boca Raton
Rytov SM (1956) Acoustical propagation of a thinly laminated medium. Sov Phys Acoust+ 2:68–80
Sun CT, Achenbach JD, Herrmann G (1968) Continuum theory for a laminated medium. J Appl Mech 35:467–475
Sve C (1971) Time harmonic waves traveling obliquely in a periodically laminated medium. J Appl Mech 38:447–482
Tassilly E (1987) Propagation of bending waves in a periodic beam. Int J Eng Sci 25(1):85–94
Tourafte M (1986) Floquet waves in a body with slender periodic structure. Wave Motion 8:485–495
Verma KL (1999) Propagation direction in layered orthotropic elastic composites in generalized thermoelasticity. In: Hetnarski RB (ed) Proceeding of third international congress on thermal stresses, June 13–17, 1999. Cracow University of Technology, Cracow, Poland
Verma KL (2008) On the wave propagation in layered plates of general anisotropic media. IJMPES 2(4):198–204
Wang L, Rokhlin SI (2002) Floquet wave homogenization of periodic anisotropic media. J Acoust Soc Am 112(1):38–45
Wooster WA (1973) Tensors and group theory for the physical properties of crystals. Clarendon Press, Oxford
Yamada M, Nasser SN (1981) Propagation direction in layered transversely isotropic elastic composite. J Compos Mater 15:531–542

4

An experimental study on tuned liquid damper for mitigation of structural response

Emili Bhattacharjee, Lipika Halder* and Richi Prasad Sharma

Abstract

This paper investigates the performance of unidirectional tuned liquid damper (TLD) that relies upon the motion of shallow liquid in a rigid tank for changing the dynamic characteristics of a structure and dissipating its vibration energy under harmonic excitation. A series of experimental tests are conducted on a scaled model of structure-tuned liquid damper systems to evaluate their performance under harmonic excitation. One rectangular and one square TLD with various water depth ratios are examined over different frequency ratios, and time histories of accelerations are measured by precisely controlled shaking table tests. The behaviour of TLD is also studied by changing the orientation of the rectangular TLD subjected to the given range of harmonic excitation frequencies. The effectiveness of TLD is evaluated based on the response reduction of the structure. From the study, it is found that for each TLD, there exists an optimum water depth that corresponds to the minimum response amplitude, and the maximum control of vibration is obtained under resonance condition with the attachment of TLD.

Keywords: Harmonic excitation, Tuned liquid dampers, Liquid sloshing, Energy dissipation, Response control, Selection of TLD parameters

Introduction

Due to rapid urbanisation and industrialisation, there is an increasing demand of high-rise buildings. These buildings inspired the use of high-strength, light-weight materials and the increase in the use of welded connections and light facades that serve as exterior walls without contributing to the structural strength. These developments have led to the construction of flexible buildings with reduced structural damping. As a result, the sensitivity of these buildings to dynamic excitations such as earthquake and wind has increased. Thus, it is essential to search for vibration-suppressing devices to counteract undesirable vibration in the structures. These devices may be passive, active, semi-active or hybrid types. The passive device, tuned liquid damper (TLD), is a type of tuned mass damper (TMD) where the mass is replaced by liquid (generally water). A conventional TMD needs frictionless rubber bearings, special floor for installation, springs, dashpots and other mechanical components which increase the cost of this device. However, the dead weight of the mass has no other functional use.

Although TLD is usually a rigid tank with shallow water in it, it promises to be most suitable, since existing water tanks in buildings may be used as TLD without adversely affecting its functional use and also include low cost and maintenance. The working principle of TLD is based on sloshing of the liquid to absorb a portion of the dynamic energy of the structure subjected to seismic motion and thus controlling the structural vibration.

Several research works have been carried out to find the applications of TLD in reducing the seismic vibration of the structure (Banerji et al. 2000; Banerji et al. 2010; Bauer 1984; Fujino et al. 1988; Koh et al. 1994; Modi and Welt 1988; Tamura et al. 1995). Sun et al. (1992) successfully developed an analytical model for TLD based on shallow-water wave theory. Based on this theory, Banerji et al. (2000) have conducted numerical studies on TLD and concluded that TLD can be very effective in reducing vibration of the structure if the design parameters are approximately set. Koh et al. (1995) studied the behaviour of rectangular liquid dampers under horizontal acceleration of arbitrary time history.

* Correspondence: erlhalder@yahoo.co.in
Department of Civil Engineering, National Institute of Technology, Agartala, Jirania, Barjala 799055, India

Figure 1 Experimental set-up.

Figure 2 TLD scale model performance test.

The theoretical model to predict the motion of shallow liquid in a rectangular tank was developed for sinusoidal excitation, and numerically, they concluded that the effectiveness of dampers is dependent on the frequency content of the earthquake spectrum and the positions where the dampers are placed. Reed et al. (1998) investigated the behaviour of TLDs through laboratory experiments and numerical modelling. In his study, the major focus was the large-amplitude excitations, whereas most of the experimental works conducted in the past (Fujino et al. 1992; Koh et al. 1995) were for TLDs subjected to small-amplitude excitations. It is observed that the response of a TLD to large-amplitude excitations would be significantly different from that for small-amplitude excitations due to larger probabilities of surface-wave breaking occurrence.

The objectives of this study is to reduce structural response by installing a model of TLD attached to the structure subjected to sinusoidal external excitation and to study the effects of various parameters, which affect the structural response. These parameters include the ratio of water depth to tank length, called water depth ratio, the ratio of sloshing frequency to structural natural frequency, called tuning ratio, and the ratio of excitation frequency to natural frequency of the structure, called excitation frequency ratio.

Methods

Experimental set-up and procedure

Figure 1 shows arrangement of the structure over a unidirectional shaking table. The shaking table is arranged to impose horizontal motions on the structure. The size of the table in plan is 1 × 1 m, and it weighs approximately 140 kg. The range of maximum displacement is ±100 mm. The maximum operating frequency is in between 0 and 10 Hz. The required excitation frequency is applied to the structure by means of a microprocessor-based three-phase precision AC drive. The TLD is placed centrally on the structure as shown in Figure 2. The TLD tanks are made up of acrylic sheet, having 4-mm-thick sidewalls and base plate. The structural model is made up of mild steel plate with enough thickness to ensure a rigid floor, supported on four high-tensile steel rods with a size of 6 × 6 × 500 mm. The load on the slab is transferred to the columns through two cross beams having the same dimensions as the columns. The columns are connected to the slab and base plates by welding. The natural frequency of the structural model without tank was 1.9 Hz. The motion imposed on the structure is harmonic, with control over the amplitude and the frequency of the oscillations.

Table 1 TLD parameters

Case number	Type of TLD		Container dimensions		
	Shape	Symbol	Length (cm)	Width (cm)	Depth (cm)
Case 1	Rectangular	TLD_1	30	20	30
Case 2	Rectangular	TLD_2	20	30	30
Case 3	Square	TLD_3	30	30	40

Table 2 Experimental cases

TLD type	Water depth ratio	Base displacement amplitude (mm)	External frequency (Hz)	Excitation frequency ratio (β)
TLD_1	0.05, 0.1, 0.15, 0.2,0.25, 0.3	3	1.4, 1.6, 1.8, 1.9, 2.0, 2.2, 2.3, 2.4	0.75, 0.85, 0.95, 1.0, 1.05, 1.15, 1.2, 1.3
TLD_2	0.05, 0.1, 0.15, 0.2, 0.25, 0.3	3	1.4, 1.6, 1.8, 1.9, 2.0, 2.2, 2.3, 2.4	0.75, 0.85, 0.95, 1.0, 1.05, 1.15, 1.2, 1.3
TLD_3	0.05, 0.1, 0.15, 0.2, 0.25, 0.3	3	1.4, 1.6, 1.8, 1.9, 2.0, 2.2, 2.3, 2.4	0.75, 0.85, 0.95, 1.0, 1.05, 1.15, 1.2, 1.3

In this study, the depth ratio (Δ) is defined as the ratio of water depth (h) to tank length (L) in the direction of shaking. The tuning ratio, which is defined as the ratio of fundamental sloshing frequency (f_w) to the natural frequency of the structure (f_s), is maintained by filling water up to the desired level in the TLD.

According to the linear wave theory given by Lamb (1932), the fundamental natural frequency of water sloshing motion, f_w, can be calculated as follows:

$$f_w = \frac{1}{2\pi}\sqrt{\left(\frac{\pi g}{L}\right)\tan\left(\frac{\pi h}{L}\right)}, \quad (1)$$

where, g = gravitational acceleration, h = still water depth and L = length of the tank in the direction of sloshing motion.

The excitation frequency ratio, β, which is the ratio of excitation frequency to structural natural frequency, is controlled by varying external frequencies.

The measured response parameters are displacement and acceleration of the structure along the direction of force. The displacement response is measured by attaching Brüel & Kjær Deltatron 4507–001 accelerometers (Brüel & Kjær Sound & Vibration Measurement A/S, Nærum, Denmark) at the base and top of the structure. PULSE 3560B computerised data acquisition and multi-analyser system is used to acquire and analyse the experimental data. In each set of experiments, the TLD-structure system is subjected to harmonic sinusoidal base motions with different excitation frequencies. The external sinusoidal force is applied to the structure by means of an induction motor mounted on the shaking table. The external excitation amplitude is accordingly maintained constant by keeping the displacement amplitude of the shaking table constant.

Selection of TLD parameters

The response of a structure with a TLD attached and subjected to a base excitation will depend on the characteristics of the TLD-structure system. A TLD may be considered as properly designed if it significantly reduces a structure's response for a particular ground motion for a given set of values of water depth ratio, Δ, and excitation frequency ratio, β. In this study, the vibration reduction of the structure has been observed by considering different values of the water depth ratio, Δ, and several frequency ratios, β, which is shown in Tables 1 and 2.

Results and discussion

Investigations are conducted to study the dynamic behaviour of a structure with TLD when subjected to harmonic base motion given to the shaking table. The harmonic ground motion is defined by its excitation frequency and amplitude of ground motion. As the harmonic motion

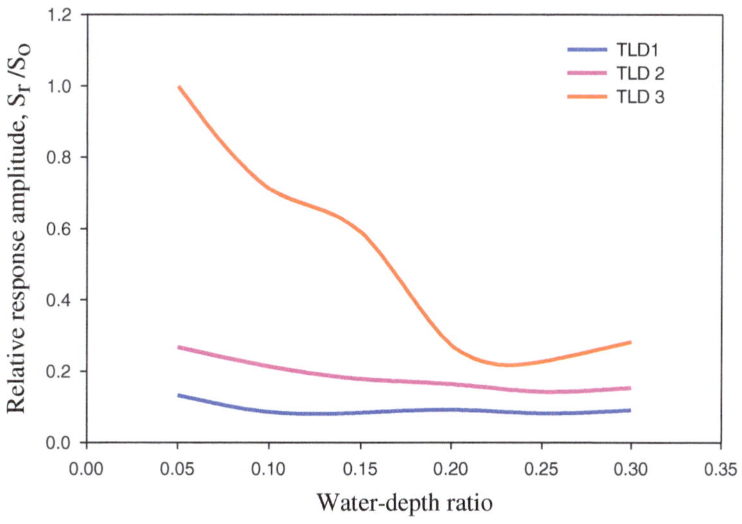

Figure 3 Structural response amplitude versus TLD water depth ratio in resonance condition.

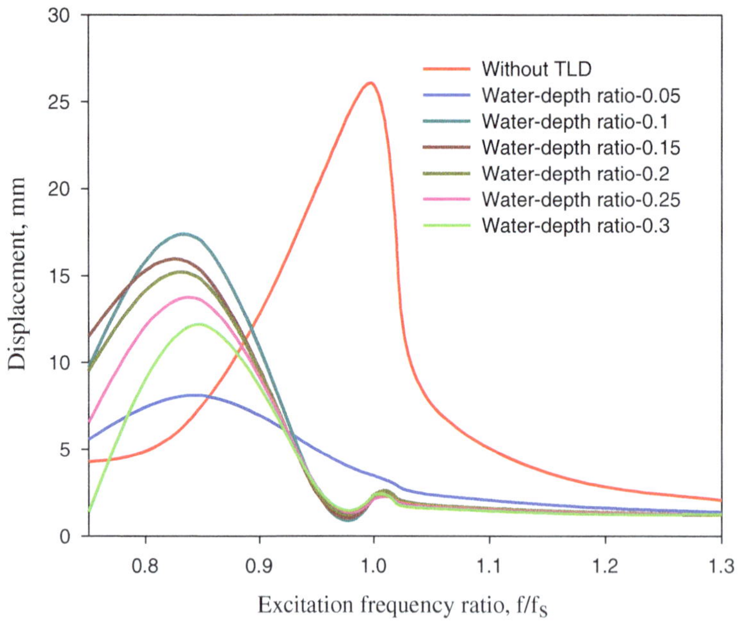

Figure 4 **Displacement response of structure for varying frequency ratios with different water depth ratios for TLD_1.**

consists of a single frequency, the study of the structure with TLD will provide an understanding of the behaviour of TLD-structure system for this kind of motion. In this study, displacement and acceleration of the structure with and without TLDs are measured by means of accelerometer and data acquisition system, considering various excitation frequencies. The results of this experimental study have been plotted as the relative response ratio as a function of water depth ratio and displacement as a function of frequency ratio. Time versus acceleration graphs are also plotted for different conditions of TLDs.

Effect of water depth ratio on structural response

Different water depth ratios, which is water depth (*h*) to tank length (*L*), varying from 0.05 to 0.3 are considered, and corresponding maximum structural response has been shown in Figure 3. The three curves in Figure 3 show the relationship between the structural response

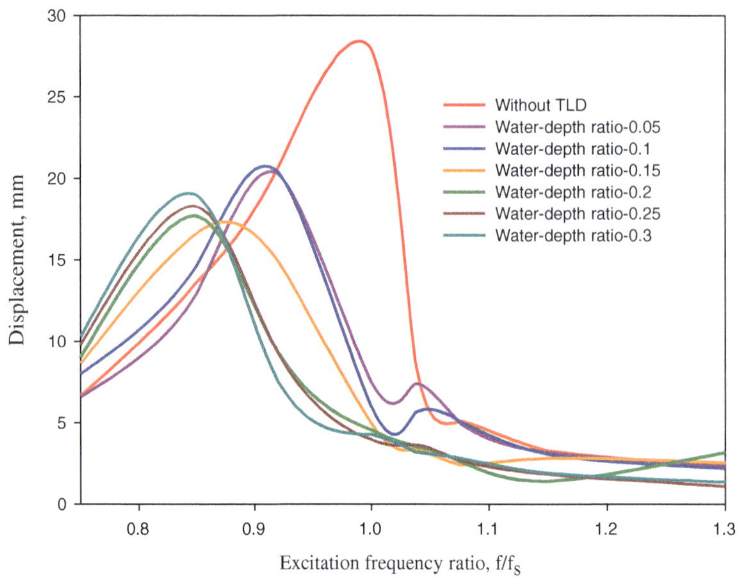

Figure 5 **Displacement response of structure for varying frequency ratios with different water depth ratios for TLD_2.**

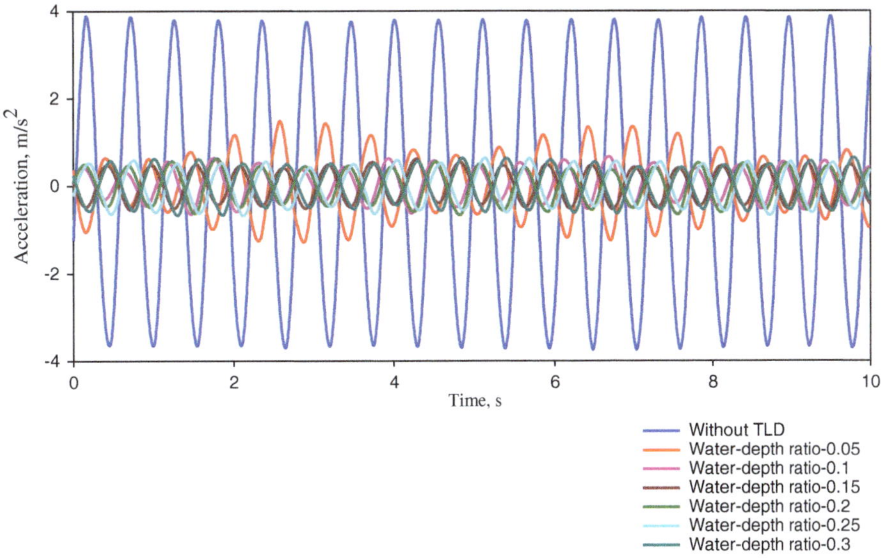

Figure 6 **Time histories of structural acceleration with and without TLD (for TLD_1, β = 0.95).**

amplitude in resonance condition and the corresponding water depth ratio in the three TLDs considered. The horizontal axis is the water depth ratio, and the vertical axis is the relative response amplitude S_r/S_0, the latter being the ratio of maximum structural response amplitude with TLD to the maximum response amplitude without TLD. From Figure 3, it can be clearly observed that for each case, there exists optimum water depth that corresponds to the minimum response amplitude. These values are 7.5, 5 and 7.5 cm, respectively, for TLD_1, TLD_2 and TLD_3. From Figure 3, it is also observed that TLD_3, that is, the square TLD, is less effective in comparison with TLD_1 and TLD_2, that is, the rectangular TLDs for controlling response of the structure. For this reason, further studies have been carried out with rectangular TLDs, that is, TLD_1 and TLD_2.

Effect of various external frequencies on structural response

The displacement response of the structure for various excitation frequency ratios with different water depth ratios has been shown in Figures 4 and 5. Several external frequency ratios ranging from 0.75 to 1.3 and water depth ratios ranging from 0.05 to 0.3 are considered in this

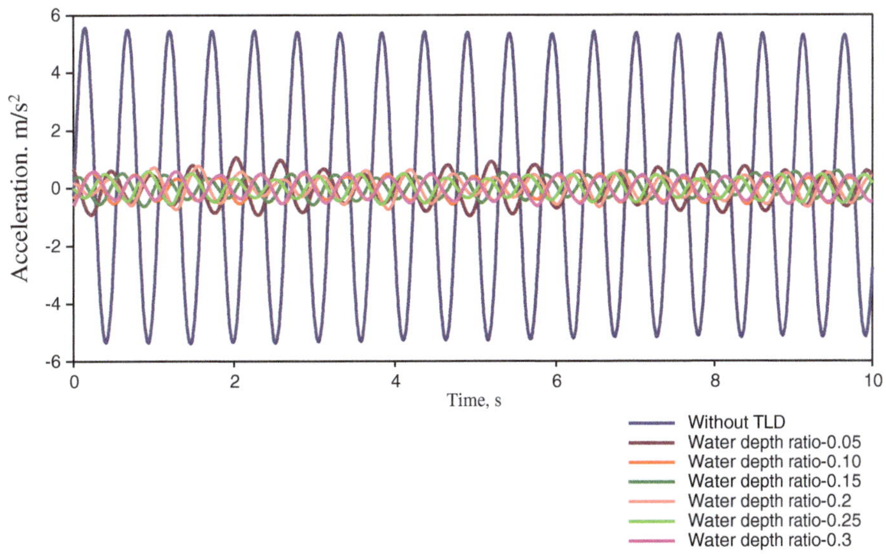

Figure 7 **Time histories of structural acceleration with and without TLD (for TLD_1, β = 1.0).**

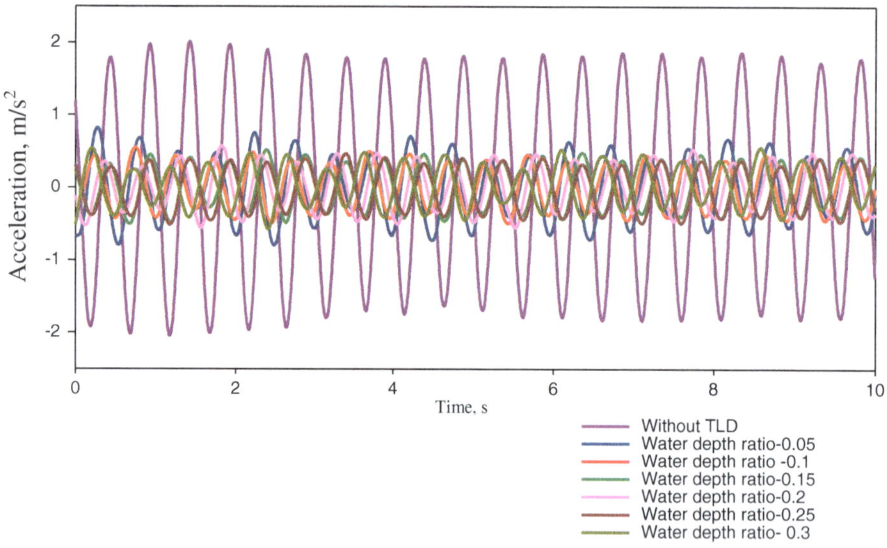

Figure 8 Time histories of structural acceleration with and without TLD (for TLD_1, β = 1.05).

study, and the corresponding structural response has been observed. From Figures 4 and 5, it is easily observed that at the initial stage, when the excitation frequency is lower than the resonance frequency, the peak structural response typically increases with increasing water depth ratio. However, at the region of resonance ($f/f_s \approx 1$), the response amplitude reduces drastically upon attachment of the TLD. Similarly, when excitation frequencies become higher than the resonance frequency, no efficient control is observed. Thus, maximum control of response is obtained when the structure is subjected to resonance frequency, and the reduction of response, considering the resonance condition, is obtained at nearly 86.6% and 73.2% for TLD_1 and TLD_2, respectively.

Typical plots of acceleration time histories of structure for excitation frequency ratios (f/f_s) 0.95, 1 and 1.05 are shown in Figures 6, 7 and 8 for TLD_1. In these cases, the acceleration response at the top of the structure has been observed, considering the varying water depth ratios. From Figure 6, it is seen that for the frequency ratio of 0.95, the maximum acceleration of the structure without TLD is as about 3.7 m/s^2, which is reduced to

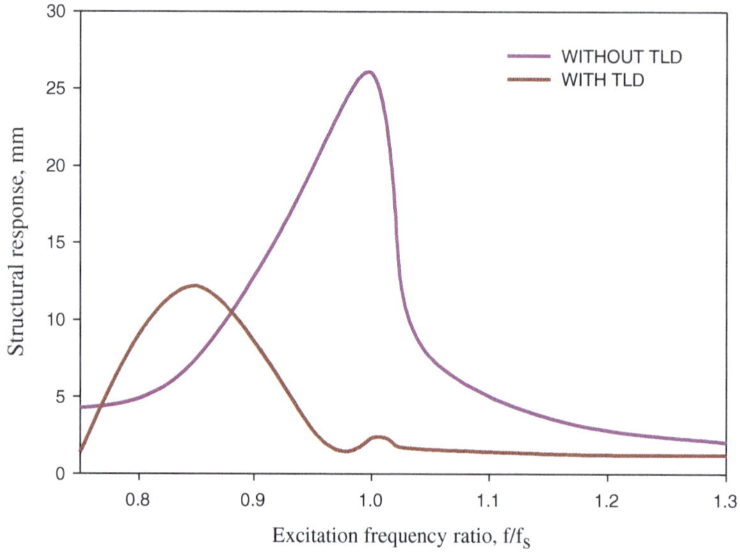

Figure 9 Displacement response for various excitation frequency ratios considering f_w/f_s = 1 for TLD_1.

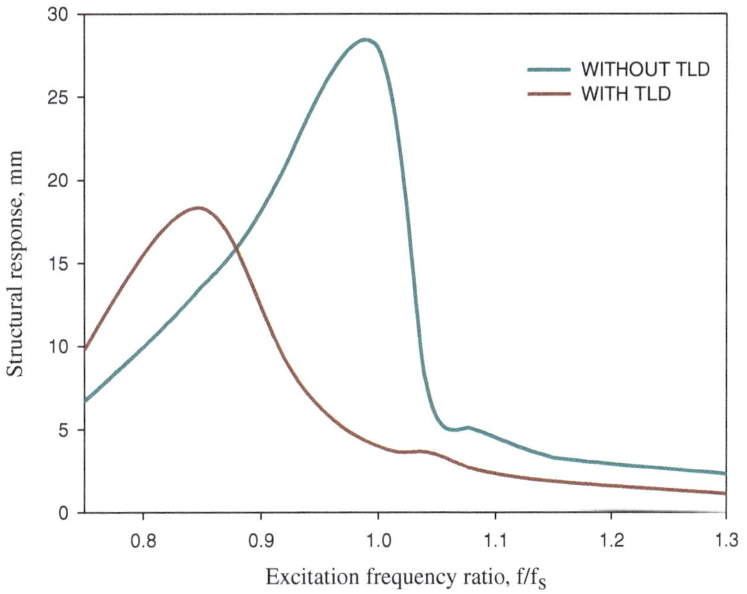

Figure 10 **Displacement response for various excitation frequency ratios considering f_w/f_s = 1 for TLD_2.**

nearly 1.4 m/s^2 for the water depth ratio of 0.05 and within the range of 0.4 and 0.6 m/s^2 for other water depth ratios. Figure 7 shows the acceleration response for the excitation frequency ratio of 1.0. In this case, the maximum acceleration of the structure without TLD has been observed as about 5.5 m/s^2, which is reduced to about 1 m/s^2 for the water depth ratio of 0.05 and within the range of 0.5 and 0.7 m/s^2 for other water depth ratios. A similar trend has been observed in Figure 8 which shows the acceleration response for the excitation frequency ratio of 1.05. Here, the maximum acceleration of the structure without TLD is found at 2 m/s^2, which is reduced to about 0.8 m/s^2 for the water depth ratio of 0.05 and in between 0.4 and 0.55 m/s^2 for other water depth ratios. Hence, from the mentioned observations, it is clear that the optimum control in peak acceleration for a particular frequency ratio is obtained with higher water depth ratios, and maximum reduction in response is obtained when the frequency ratio becomes unity.

Effect of tuning ratio on structural response

The tuning ratio of a rectangular TLD, as defined earlier, is the ratio of the fundamental linear sloshing frequency (f_w) to the natural frequency of the structure (f_s). By convention, a TLD implies that this tuning ratio is unity. However, earlier experimental studies have shown that the optimum response control is obtained when the tuning ratio is close to unity. Figures 9 and 10 show the comparison between response control of two rectangular tanks (TLD_1 and TLD_2) where the depth of liquid (90 mm for TLD_1 and 50 mm for TLD_2) in each damper is adjusted such that the fundamental sloshing frequency is almost tuned to the natural frequency of the structure. The effectiveness of TLDs, ψ, is measured in terms of the reduction of structural displacement with TLDs compared to the corresponding value without TLDs:

$$\psi = \frac{(x_0 - x_{TLD})}{x_{TLD}} \times 100\%, \qquad (2)$$

where x_{TLD} and x_0 are the peak displacement values with and without TLDs, respectively. Here, the effectiveness of TLD_1 and TLD_2 has been found as 53.1% and 34.6%, respectively. Therefore, it is observed that TLD_1 has better performance than TLD_2.

Conclusions

The present study focused on the implementation of a tuned liquid damper for mitigation of structural response. A set of experiments were carried out for studying the sloshing phenomenon in a rectangular and a square tank under harmonic loading condition. Different water depth ratios varying from 0.05 to 0.3 and several excitation frequency ratios varying from 0.75 to 1.3 were considered. The effect of tuned condition ($f_w/f_s \approx 1$) on structural response is also studied. The responses of two systems, that is, with and without TLD, are evaluated and presented in graphical and tabular forms. From this study, it has been observed that among all the water depth ratios for a given range of excitation frequency ratios, there exists optimum water depth that corresponds to the minimum response amplitude for each damper. These values are 7.5, 5 and 7.5 cm, respectively, for TLD_1, TLD_2, and TLD_3. It is seen that the square

TLD is less effective in comparison with the rectangular TLD for the controlling response of the structure.

At the initial stage, when the excitation frequency is lower than the resonance frequency, the peak structural response typically increases with increasing water depth ratio. However, at the region of resonance ($f/f_s \approx 1$), the response amplitude reduces drastically upon attachment of the TLD. Similarly, when the excitation frequencies become higher than the resonance frequency, no efficient control is observed. Thus, maximum control of response is obtained when the structure is subjected to resonance frequency, and the reduction of response, considering the resonance condition, is obtained at nearly 86.6% and 73.2% for TLD_1 and TLD_2, respectively. In case of tuned condition, the effectiveness of TLD_1 and TLD_2 has been found as 53.1% and 34.6%, respectively. Therefore, it is observed that TLD_1 has better performance in comparison withTLD_2. From this study, it has been found that TLD can be successfully used to control the response of the structure.

Competing interests

The authors declare that they have no competing interests.

Authors' contributions

EB prepared the experimental model, carried out the experiments and drafted the manuscript. LH had given the idea, participated in designing the experimental model, helped in the interpretation of experimental output and also helped in drafting the manuscript. RPS has guided during acquisition of data and also has checked the manuscript before submission. All authors read and approved the final manuscript.

Authors' information

EB has done her M.Tech. in the field of structural engineering from NIT Agartala in 2012. LH is an assistant professor in NIT Agartala and doing her Ph.D. in the field of earthquake engineering. She has published papers in the field of wind engineering and building materials. RPS is a professor in NIT Agartala and has published papers in the field of structural and geotechnical engineering.

References

Banerji P, Murudi M, Shah AH, Popplewell N (2000) Tuned liquid dampers for controlling earthquake response of structures. Earthquake Engineering and Structural Dynamics 29:587–602

Banerji P, Samanta A, Chaban SA (2010) Earthquake vibration control of structures using tuned liquid dampers: experimental studies. International Journal of Advanced Structural Engineering 2(2):133–152

Bauer HF (1984) Oscillations of immiscible liquids in a rectangular container: a new damper for excited structures. Journal of Sound and Vibration 93(1):117–133

Fujino Y, Pacheco BM, Chaiseri P, Sun LM (1988) Parametric studies of tuned liquid damper (TLD) using circular containers by free-oscillation experiments. Struct Eng/Earthquake Eng, JSCE 5:381–391

Fujino Y, Sun L, Pacheco MB, Chaiseri P (1992) Tuned liquid damper (TLD) for suppressing horizontal motion of structures. Journal of Engineering Mechanics 118(10):2017–2030

Koh CG, Mahatma S, Wang CM (1995) Reduction of structural vibrations by multiple mode liquid dampers. Engineering Structures 17(2):122–128

Koh CG, Mahatma S, Wang CM (1994) Theoretical and experimental studies on rectangular liquid dampers under arbitrary excitations. Earthquake Engineering and Structural Dynamics 23:17–31

Lamb H (1932) Problems of periodic motion. Laminar motion, diffusion of vorticity. Oscillating plane. Periodic tidal force; feeble influence of viscosity in rapid motions. In: Hydrodynamics, 6th edn. Cambridge University Press, London, England, pp 619–621

Modi VJ, Welt F (1988) Damping of wind induced oscillations through liquid sloshing. Journal of Wind Engineering and Industrial Aerodynamics 30:85–94

Reed D, Yu J, Yeh H, Gardarsson S (1998) Investigation of tuned liquid dampers under large amplitude excitation. Journal of Engineering Mechanics 124(4):405–413

Sun LM, Fujino Y, Pacheco BM, Chaiseri P (1992) Modeling of tuned liquid damper (TLD). Journal of Wind Engineering and Industrial Aerodynamics 43(1–3):1883–1894

Tamura Y, Fujii K, Ohtsuki T, Wakahara T, Kohsaka R (1995) Effectiveness of tuned liquid dampers under wind excitation. Engineering Structure 17(9):609–621

5

Confinement of concrete columns with unplasticized Poly-vinyl chloride tubes

Pramod Kumar Gupta

Abstract

In this paper an experimental study is presented to investigate the effectiveness of UPVC tube for confinement of concrete columns. UPVC tubes having 140 mm, 160 mm and 200 mm external diameters were used to confine the concrete having compressive strength 20 MPa, 25 MPa and 40 MPa. The concrete has been designed using IS code 10262-1982(Reaffirmed −2004) (BIS:10262−1982, Reaffirmed −2004). The testing of the specimens was carried out on a displacement controlled Instron make Universal Testing Machine of 2500 kN capacity. During the experiments mode of deformation and corresponding load-compression curves were recorded and obtained results are compared with the existing models for confined concrete available in the literature. It is found that the predicted capacities of columns using different models are within ±6% of the experimental capacities. It is found that UPVC tubes can be effectively used for confinement of the concrete columns and to enhance their load capacity, ductility as well as energy absorbing capacity.

Keywords: UPVC tube; Confined concrete columns; Confinement; Ductility

Introduction

Strengthening or repairing of deteriorating concrete columns has now become one of the areas of study in construction industry across the world. One of the deficiencies in concrete columns is the lack of lateral confinement and low energy absorption capacity (Schneider, 1998). Researchers are working on the techniques by which such deficiencies can be minimized during construction stage. The strength, ductility and energy absorption capacity of new concrete columns during construction can be enhanced by providing external confinement by employing UPVC tubes. These tubes can be used as formwork during construction and their after as an integral part of column. There are a number of confinement materials to be used for confining concrete columns. Structural steel tubes are extensively studied for such purpose in past two decades and as a result numerous literatures ((Oliveira et al. 2009); (Gupta et al. 2007)) are available.

Fiber reinforced polymer composite tubes are used for confinement of concrete by many researchers with performing experimental studies. Confined concrete columns having different length/diameter ratios were tested by Mirmiran & Sahaway (1997). They also compared different analytical models suggested by different researchers for confinement of concrete and proposed a modified analytical model suitable exclusively for FRP tube confinement. Tao et al. (2004), Fitzwilliam & Bisby (2010) and Bisby & Ranger (2010) tested FRP-confined RC columns with a length/diameter ratio up to 20. Tamuzs et al. (2007a) also tested FRP-confined RC columns with a length/diameter ratio of 4, but little slenderness effect existed in these columns as a result of their relatively small slenderness and end conditions. In addition, De Lorenzis et al. (2004) and Tamuzs et al. (2007b and 2008) tested slender FRP-confined concrete columns without steel reinforcement subjected to concentric compression to investigate the solidity of these columns.

A very few studies are available in which PVC tubes are used for confinement of concrete column specimens. Way back Kurt (1978) conducted experiments to see increase and decrease in strength of concrete core due to confinement and slenderness ratio respectively. In last one decade Marzouck & Sennah (2002) conducted experimental study on concrete filled PVC tubes with 100mm diameter concrete core and 3 mm thickness of tube. The lengths of to the concrete columns. They reported that as the slenderness ratio increases, the

Correspondence: spramod_3@yahoo.com
Department of Civil Engineering, I.I.T. Roorkee, Roorkee, India

Table 1 Physical properties of UPVC pipes

Parameter	Value
Density	13-14.5 KN/m^3
Elastic modulus	3380 MPa
Flexure strength	65.5 MPa
Poisson ratio	0.38
Ultimate tensile strength	27.5- 52 MPa
Breaking elongation	134%
Thermal conductivity	0.14-0.28 kcal/m.hr°c
Service life	>50 years

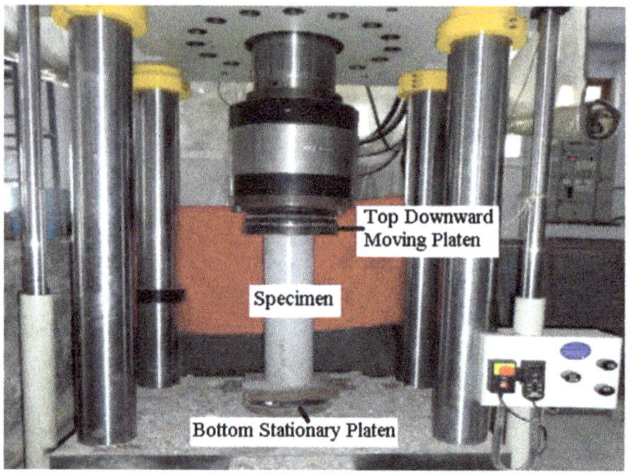

Figure 1 Test-setup for performing experimental study.

tubes were 758, 562, 416 and 270 mm. They found that the PVC tube provides significant lateral confinement compressive strength of concrete filled PVC tube decreases.

In view of these facts, in the present study the author has tried to strengthen the concrete column by using the UPVC tube as outer confining layer in order to increase the load carrying capacity, ductility and durability of the column. A comparative test research program to strengthen columns by UPVC tube has been put forward. In order to evaluate the behaviour of the proposed UPVC tube columns more objectively and thoroughly, an experimental study is performed. The test results are presented and discussed from strength, ductility and energy absorption capacity point of view. Since use of UPVC for confinement of concrete is relatively new and theoretical work available in this area is very limited. Therefore, different theoretical models available in literature that were originally proposed for confinement developed by other means are used to assess the confinement of the concrete columns with UPVC tubes.

Experimental programme

Materials

Un plasticized Poly Vinyl Chloride (UPVC) Tube

In this paper, Un-plasticized Poly-Vinyl-Chloride (UPVC) was chosen to confine concrete after considering following points

1. Polymer is one of the most stable substances when high corrosion resistance is required. The anti-corrosion characteristics of UPVC have been proven to be outstanding. Since the last century, polyethylene copolymer coatings have been used of the protection of external surface of onshore and offshore pipelines.
2. UPVC tube can also be used as a protective layer against mechanical damage caused by the severe environment. Physical properties of the UPVC are given in Table 1. The thermal conductivity of UPVC is only 0.45% that of a steel tube; it can provide a stable curing condition for the core concrete to achieve high performances and high durability. It can withstand those mechanical damages which occur in the exposure of high energy environments like marine and saline.

Finally the service life of UPVC tube is longer than 50 years, and with the mechanical supporting of core concrete UPVC can serve longer as a protective layer of the concrete structure.

Concrete

Three series of concrete M20, M25, M40 were designed as per code IS:10262–1982 (Reaffirmed –2004) (BIS:10262-1982, Reaffirmed-2004) to fill the UPVC tubes. Portland Pozzolona Cement (according to IS1489-1991 Part-1) was used to obtain concrete. Its compressive strength was 20 MPa at 3 days and 45 MPa at 28 days. Local sand with a fineness modulus of 2.7 was used as fine aggregate and saturated surface dry crushed aggregate was used as coarse aggregate. The concrete mix proportions and the compressive strength at 28 days are given in Table 2.

Table 2 Concrete mixture proportions

Grade of mix	Water/cement ratio	Water (Kg/m^3)	Cement (Kg/m^3)	Sand (Kg/m^3)	Aggregate (Kg/m^3)	Compressive strength at 28 days f_{ck}, (MPa)
M20	0.50	215	428	619	1040	23.6
M25	0.47	215	458	564	1033	28.6
M40	0.40	215	537	590.7	934.5	43.5

Table 3 Details of specimens

Specimen	D/t ratio	D x t x L mm	Confinement parameter $\alpha=\frac{A_p f_{py}}{A_c f_{ck}}$	Residual strength at 30 mm compression as % of peak load
T160M20PC-1	37.6	160 x 4.25 x 500	0.20	42.2
T160M20PC-2	37.6	160 x 4.25 x 500	0.20	49.9
T200M20PC-1	34.1	200 x 5.85 x 500	0.23	44.4
T200M20PC-2	34.1	200 x 5.85 x 500	0.23	54.1
T140M25PC-1	35.8	140 x 3.9 x 500	0.17	45.2
T140M25PC-2	35.8	140 x 3.9 x 500	0.17	45.05
T140M40PC-1	35.8	140 x 3.9 x 500	0.11	42.70
T140M40PC-2	35.8	140 x 3.9 x 500	0.11	42.70

Testing of specimens

Eight UPVC tubes with the outer diameters (D) of 140 mm, 160 mm and 200 mm were cut in 500 mm length (L). The wall thickness (t) of UPVC tubes were 3.9 mm, 4.25 mm and 5.85 mm for 140 mm, 160 mm and 200 mm diameters respectively. Typical nomenclatures for specimens are used to designate them. For example specimen T160M20PC-1: T160 – stands for tube of 160 mm external diameter, M20 – stands for grade of concrete, PC – stands for Plain concrete, 1-stands for first specimen. Therefore, T160M20PC-1 indicates specimen made with tube of diameter 160 mm, filled with plain concrete having compressive strength 20 MPa and number one. The test setup used for performing experiments is shown in Figure 1. The stress–strain curves of specimens were obtained by performing compression tests on INSTRON make UTM machine with a capacity of 2500 KN. The experiment was performed using a displacement control method with cross head speed of 1 mm/min. The upper jaw of the machine was moved in the downward direction while bottom jaw was kept stationary. The rate was applied continuously without a jerk load. Loads and displacements were measured continuously by data collection computer system. The frequency of the data collection is 0.4 Hz.

Details of column specimens are given in Table 3. To judge the confinement effect of UPVC tube constraining factor α was applied and defined as (Golla 2012, Lakumalla 2012, Wang & Yang, 2010);

$$\alpha = \frac{A_p f_{py}}{A_c f_{ck}} \tag{1}$$

where Ap and fpy are the cross-sectional area and yield stress of UPVC tube, repectively; and Ac and fck are the area and compressive strength of the core concrete respectively. UPVC tubes were tested for the tensile strength of their material with the sample as shown in Figure 2. This sample was tested in the TINIUS OLSEN machine of the material testing laboratory of Metallurgy Department of Indian Institute of Technology Roorkee with a loading rate of 50 mm/min. Figure 2(b) shows the original sample used for the testing. The stress–strain curve is shown in Figure 3(a). The ultimate strength (fpy) was recorded as 42.22 MPa. Figure 3(b) shows the arrangement for testing of pipes to determine bursting water pressure. The bursting pressure can be used as confining pressure for core concrete.

The axial displacement of all specimens was controlled in the range from 0 to 30 mm. It was found that the residual strengths of all specimens were still greater than 42% of peak load at the axial displacement of 30 mm (see Table 3).

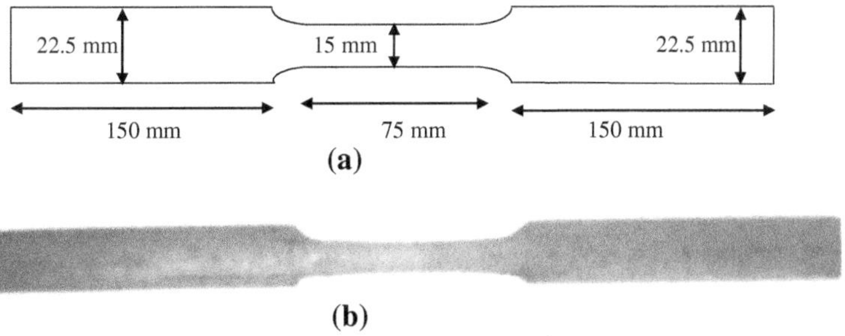

Figure 2 Typical specimen for tension test of UPVC tube (a) Dimensions, (b) Photograph.

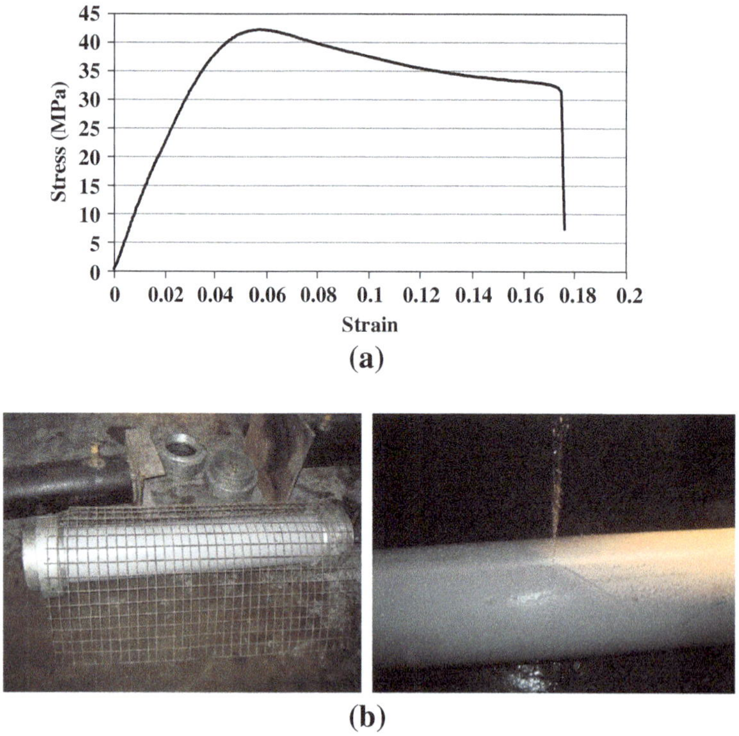

Figure 3 (a) Typical tensile stress–strain curve of UPVC tube material. (b) Typical testing arrangement to find brusting (confining) pressure for UPVC pipes.

Experimental results and discussions

Mode of failure

All specimens exhibited a good ductility and still retained the integrity when the displacement reached 30 mm. The failure pattern of all specimens was shear failure as shown in Figure 4(a-d). A typical characteristic of shear failure is that the core concrete is damaged by shear stress in one direction due to weak confinement effect of the tube. The shear crack direction can be judged by the appearance of the specimen. The same type of mode of failure was also observed by Wang & Yang (2010) in their experimental program. They also found shear type failure for specimens having $\alpha \leq 0.204$. For all specimens tested here α is between 0.11 and 0.23. (see Table 3). Shear failure is brittle type failure. To avoid the shear failure of the specimens, the diameter to wall thickness ratio (D/t ratio) of the tube should be taken sufficiently lower.

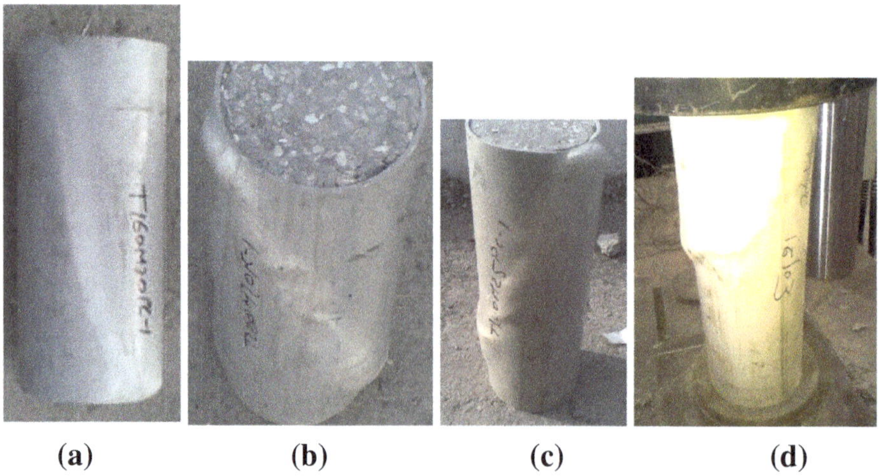

Figure 4 Typical failure modes of specimens: (a) T160M20, (b) T200M20, (c) T140M25, (d) T140M40.

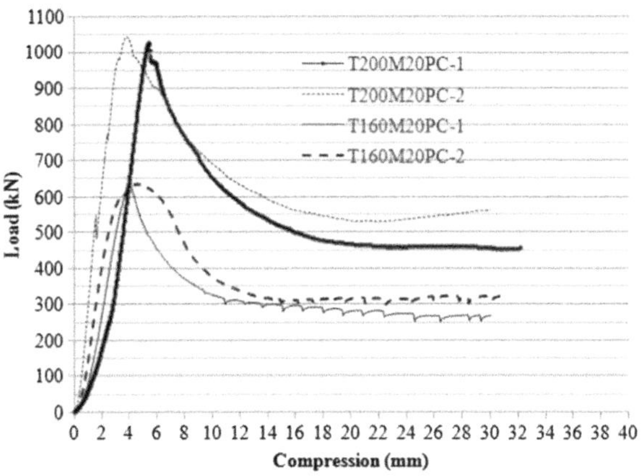

Figure 5 Comparison of load-compression variations of different diameter tubes filled with same concrete grade M20.

Load-compression curve

Figures 5 and 6 show the load-compression curves of all the specimens. Load-compression curves are classified according to the specimens with the same core concrete and different tubes, or with the same tube and different core concretes respectively.

From Figure 5 it is clear that as diameter of tube increases the peak load increase but the ductility decreases for the specimens having different diameters and filled with concrete having same compressive strength. From Figures 5 and 6 it is also clear that the tube diameter to thickness ratio (D/t) affects the post-peak behaviour of the curve. The absolute values of the slope of the curves decline by the decrease in tube diameter/thickness ratio, where as the change of the residual strength ratios of the specimens after peak stress is generally just the opposite. The reason is that the confinement effect of UPVC tube enhances as the tube thickness increases and confined concrete shows enhanced ductility.

It can be found from Figure 6 that the slope of the onset section of the curves goes down with the decrease in grade of concrete. The reason can be attributed to the difference of the elastic modulus between UPVC tube and core concrete. The elastic modulus of UPVC tube is only between 9% and 15% that of the concrete core. Therefore, when the grade of concrete decreases; the proportion of UPVC elastic modulus in a confined concrete specimen increases; and the elastic modulus of the confined concrete decreases.

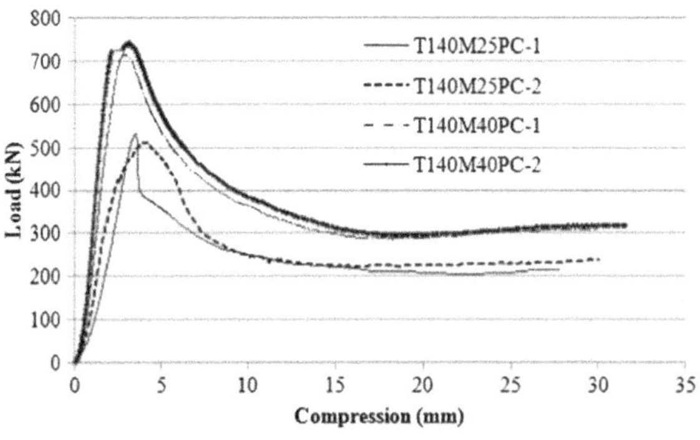

Figure 6 Comparison of load-compression variations of same diameter tubes filled with different concrete grades.

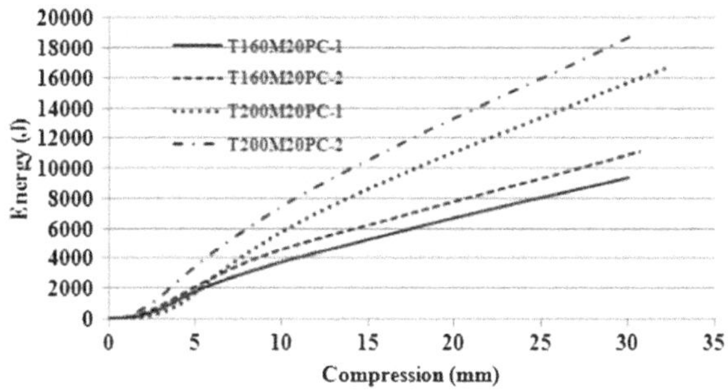

Figure 7 Comparison of energy-compression variations of different diameter tubes filled with same concrete grade M20.

From Figure 6 it is also clear that concrete strength affects the post-peak behaviour of the curve significantly. Absolute values of the slope of the curve increase with increase in concrete strength because the core concrete becomes more brittle. The residual strength of specimen decreases with increase in concrete strength.

Figures 7 and 8 shows the energy-compression curves for different specimens. It is clear from Figure 7 that as diameter of tube increases the energy absorbing capacity also increases.

Ductility

To assess the confinement effect on the ductility and energy absorbing capacity of the specimens, the ductility ratios and energy absorbing ratios of all of the column specimens were calculated and the results are tabulated in Table 4. According to (Wu, 2004), ductility can be assessed by the ductility ratio μ, which is defined as the ratio of displacement. The ductility ratio at different

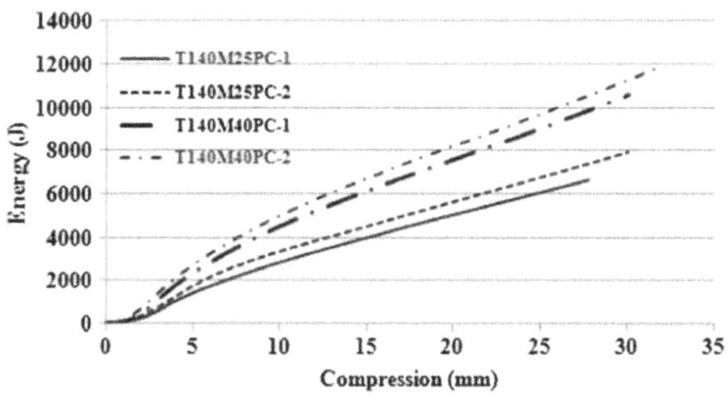

Figure 8 Comparison of energy-compression variations of same diameter tubes filled with different concrete grades.

Table 4 Details of experimental results in terms of displacement, ductility ratios μ, energy absorption and energy absorption ratio ξ

Specimen	Ultimate load (kN)	Axial displacement (mm)					Ductility ratio		Energy absorbed (Jule)					Energy ratio	
							$\mu = \Delta_u/\Delta_y$	$\mu_{70} = \Delta_{70}/\Delta_y$						$\xi = E_u/E_y$	$\xi_{70} = E_{70}/E_y$
		Δ_u	Δ_{90}	Δ_{80}	Δ_{70}	Δ_y	**1.03**	**2.65**	E_u	E_{90}	E_{80}	E_{70}	E_y		
T160M20PC-1	638	4.02	4.52	5.86	10.35	3.9	1.31	3.74	1172	1478	2163	3846	1150	1.02	3.34
T160M20PC-2	640	4.7	6.65	8.15	13.48	3.6	1.28	3.56	1840	3031	3803	5725	1154	1.59	4.96
T200M20PC-1	1026	5.37	6.17	8.45	14.96	4.2	1.27	6.74	2072	2841	4703	8628	1025	2.02	8.42
T200M20PC-2	1044	3.83	5.22	8.27	20.35/30	3.02	1.01	1.40	2221	3586	6187	13515	1403	1.58	9.63
T140M25PC-1	530	3.52	3.63	4.62	4.88	3.49	0.92	1.96	800	861	876	1238	800	1.00	1.55
T140M25PC-2	511	3.06	5.3	5.94	6.5	3.32	1.01	1.95	1231	1831	2114	2322	855	1.44	2.72
T140M40PC-1	714	2.96	3.85	4.58	5.68	2.92	1.29	2.48	1076	1679	2120	2703	1042	1.03	2.59
T140M40PC-2	742	3.06	3.99	4.76	5.88	2.37	1.03	2.65	1397	2059	2544	3165	800	1.55	3.52

stages of load-compression curve is calculated to ascertain the behaviour of the specimens during axial compression.

$$\mu_0 = \frac{\Delta U}{\Delta Y}, \quad \mu_{90} = \frac{\Delta 90}{\Delta Y}, \quad \mu_{80} = \frac{\Delta 80}{\Delta Y}, \quad \mu_{70} = \frac{\Delta 70}{\Delta Y} \tag{2}$$

Where ΔU is the Ultimate displacement at Ultimate strength and ΔY is the Yield displacement at yield point. The ultimate axial displacement ΔU is defined as the point at which the load is maximum (peak) and designated as Pu. The other load levels are defined with the loads equivalent to 90%, 80% and 70% of maximum load and designated as 0.9 Pu, 0.8 Pu and 0.7 Pu. The displacements corresponding to these load levels are designated as, Δ90, Δ80 and Δ70 respectively. The yield displacement ΔY is the yield point of an equivalent bilinear response curve that provides an equal area to that of the envelope curve, as shown in Figure 9. The ductility ratios at different load levels are calculated and summarised in Table 4.

The energy absorption capacity of the composite system indicates the effectiveness of the UPVC confinement. The energy absorption (E in Joules) was calculated by finding the area under the load-compression curve upto different load levels.

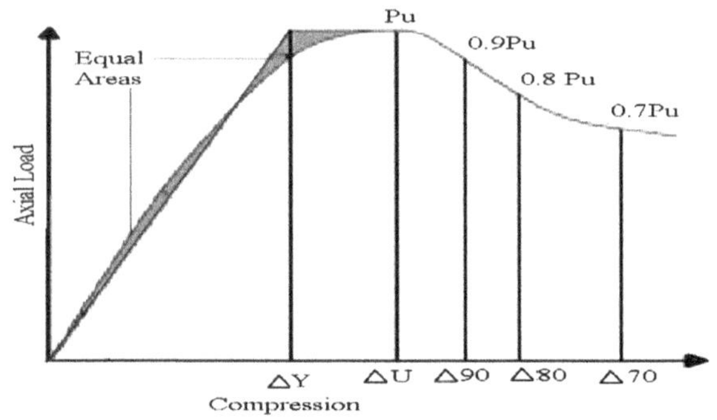

Figure 9 **Definition of Ductility and load capacity at different stages of load-Compression curve.**

The ductility ratios μ and the energy absorption E of all specimens were calculated and the results are shown in Table 4. It can be seen from Table 4 that μ_0 and μ_{70} are higher for richer mixes (Sp. T140M40PC) as compared to lean mixes (Sp. T140M25PC). Moreover, the percentage residual strength of specimen T140M40PC is lower than of specimen T140M25PC. This shows that in the initial stage of compression, ductility is higher and turns lower in the advance stage of loading for specimen T140M40PC. The ductility ratio μ_{70} for specimens varies between 1.4 and 6.74. On the other hand energy ratio ξ_{70} varies between1.55 and 9.63. The ductility ratio μ_0 for specimens varies between 1.01 and 1.31. The corresponding energy absorption ratio ξ_0 varies between 1.0 and 2.02. It is also clear from Table 4 that as D/t ratio of the UPVC tube decreases the ductility and energy absorbing capacity increases in general and consequently Δ_0, Δ_{50}, Δ_{70}, Δ_{80} and ξ_0, ξ_{50}, ξ_{70} and ξ_{80} increases.

It can be seen from Table 4 that generally the ductility ratio μ decreases with increase in grade of concrete due to the brittleness of core concrete. The energy absorption E increases with increase of concrete strength which is obvious.

Confinement

Confinement modeling

The confinement provided by transverse reinforcement, external jacket, steel tube and UPVC tube is of a passive type. The confining pressure is developed only after the surrounding member undergoes hoop elongation (Poisson's effect in concrete). The mechanics of confinement is therefore dependent mainly on two factors, one the tendency of concrete to dilate and second the radial stiffness of the confining member to restrain the dilation. Therefore, two conditions which are strain compatibility

Table 5 Comparison of experimental capacity in kN with the predicted capacity with models

Specimens	Experimental	Richart, et al.	Newmn & newman	Mander et. Al	Ahmad & shah	Saatcioglu and razvi
T160M20PC-1	638.8	598.1	640.9	663.8	604.7	669.7
T160M20PC-2	640.4	598.1	640.9	663.8	604.7	669.7
T200M20PC-1	1026.2	960.7	1029.1	1066.7	972	1076.3
T200M20PC-2	1044.3	960.7	1029.1	1066.7	972	1076.3
T140M25PC-1	530.2	520.7	557.3	575.6	525.5	574.6
T140M25PC-2	511.7	520.7	557.3	575.6	525.5	574.6
T140M40PC-1	714.0	725.1	771.7	789.9	730	779
T140M40PC-2	742.6	725.1	771.7	789.9	730	779

between the concrete core and the UPVC tube as well as equilibrium of forces in the free-body diagram for any sector of the confined section must always be satisfied.

In past, researchers have attempted to develop a generalized model to quantify the effects of confinement on strength and ductility of concrete. Among the models developed till now, the one proposed by Mander et al. (1988) is most widely used. This model was originally developed for conventional reinforced-concrete (RC) columns. A more theoretical approach was taken by Ahmad & Shah (1982) whose model is proved effective for conventional RC columns. In this section, these models are compared in terms of the basic characteristics of UPVC tube confinement.

Confinement effectiveness is defined as f'_{cc}/f'_{co} where f'_{cc} = compressive strength of confined concrete, f'_{co} = compressive strength of unconfined concrete. The confining pressure f_r can be calculated with the internal diameter (D_{int}), tensile strength (f_{py}) and wall thickness (t) of the UPVC tube;

$$f_r = \frac{2 f_{py} t}{D_{\text{int}}} \tag{3}$$

The confining pressures for different pipes were also obtained experimentally. Figure 3(b) shows the arrangement for the hydrostatic pressure test. It was observed that the values were comparable with the obtained values using Equation (3). The confining pressure obtained with experiments and Equation (3) can be used to obtain the load carrying capacity of specimens using different models because the axial compressive strain at peak load shall be very small (of the order of 0.002-0.004) and UPVC tube shall not contribute in axial stiffness at this small strain.

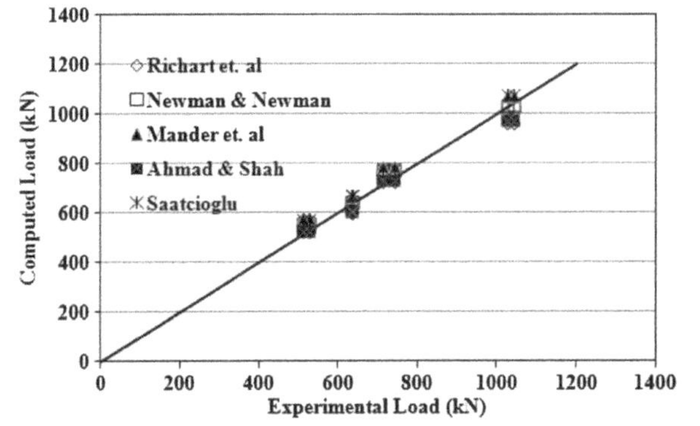

Figure 10 Graphical Representation of experimental and predicted load capacity of different specimens.

Richart et al. (1928) suggested the following relationship between confinement effectiveness and the confinement ratio f_r/f'_{co}

$$f'_{cc}/f'_{co} = 1 + k_1\left(f_r/f'_{co}\right) \tag{4}$$

where k_1 = 4.1 = effectiveness coefficient. Newman & Newman (1971) suggested the following nonlinear relationship, since confinement is not effective at high levels of lateral pressure:

$$k_1 = 3.7\left(f_r/f'_{co}\right)^{-0.14} \tag{5}$$

More recently, Saatcioglu & Razvi (1992) suggested the following coefficient based on the test data from Richart et al. (1928):

$$k_1 = 6.7(f_r)^{-0.17} \tag{6}$$

The model proposed by Mander et al. (1988) suggests a nonlinear relationship based on the ultimate strength surface developed by Elwi & Murray (1979), which reduces to the following form for circular sections:

$$f'_{cc}/f'_{co} = \left(2.254\sqrt{\left(\left(1 + 7.94 f_r/f'_{co}\right).-2 f_r/f'_{co} - 1.254\right)}\right) \tag{7}$$

Ahmad & Shah (1982) model uses a strength criterion based on the octahedral failure theory, the three stress invariants, and test data for reinforced concrete specimens, solving the octahedral equations for the confined strength of concrete, the model can be given as;

$$f'_{cc}/f'_{co} = \left(1 + 4.2556\, f_r/f'_{co}\right) \quad if\ f_r/f'_{co} < 0.68 \tag{8a}$$

$$f'_{cc}/f'_{co} = \left(1.7757 + 3.1171\, f_r/f'_{co}\right) \quad if\ f_r/f'_{co} > 0.68 \tag{8b}$$

Table 5 presents the comparison of experimental load capacity of all specimens with the predicted ones using above discussed different models. It is clear that the predicted capacities are within ±6% of the experimental capacities. The graphical representation of the predicted load capacities and experimental ones are depicted in Figure 10. It can be concluded from Figure 10 and Table 5 that the model proposed by Richart et al. (1928) predicts conservative capacities. On the other hand model proposed by Saatcioglu & Razvi (1992) calculates capacities on higher side.

Conclusions

This paper proposes a new kind of composite column, which is obtained by filling unplasticized Poly-Vinyl Chloride (UPVC) tube with concrete. Based on the test results following conclusions may be drawn;

1. Confinement of concrete columns with UPVC tubes improves their compressive strength and ductility. The improvement in strength and ductility is dependent on the concrete strength and geometrical properties of the tubes.
2. Failure pattern of all the specimens are shear type failure.
3. Confinement effect of UPVC is compared with already developed models in the literature and results are varying within ±6%.
4. Compressive strength of concrete affects the post-peak behaviour of the load-compression curve significantly. Absolute value of the slope of the curve increases with increase in concrete strength due to the increased brittleness.
5. The slope of the onset section of the curves goes down with decrease in grade of concrete.
6. The tube diameter/thickness ratio affects the post-peak behaviour of the curve; Absolute value of the slope of the curve declines by decrease in tube diameter/ thickness ratio.

Competing interests

The author declare that he has no competing interests.

Acknowledgements

Financial support provided by Building Materials & Technology Promotion Council (BMTPC), Ministry of Housing & Urban Poverty Alleviation, Government of India for this research work is acknowledged. Author is also thankful to his masters' students Golla Ramesh and Nagraju Lakumalla for conducting experiments.

References

Ahmad SH, Shah SP (1982) Complete triaxial stress-strain curves for concrete, ASCE. J Struct Engrg 108(4):728–742

BIS: 10262-1982 (Reaffirmed–2004) Recommended guidelines for concrete mix design. Bureau of Indian Standards, New Delhi, India

Bisby LA, Ranger M (2010) Axial-flexural interaction in circular FRP confined reinforced concrete columns. Construct Build Mater 24(9):1672–1681

De Lorenzis L, Tamuzs V, Tepfers R, Valdmanis V, Vilks U (2004) Stability of CFRP-confined columns. In: Proceedings, First International Conference on Innovative Materials and Technologies for Construction and Restoration, 6–9 June, 2004. Universita degli Studi di, Lecce, Italy, pp 327–342

Elwi AA, Murray DW (1979) A 3D hypoelastic concrete constitutive relationship. J Engrg Mech ASCE 105(4):623–641

Fitzwilliam J, Bisby LA (2010) Slenderness effects on circular CFRPconfined reinforced concrete columns. Journal of Composites for Construction, ASCE 14(3):280–288

Golla R (2012) *Experimental investigation into behaviour of concrete filled PVC tubes*. M. Tech. dissertation. Indian Institute of Technology, Roorkee, India

Gupta P, Sarda S, Kumar M (2007) Experimental and computational study of concrete filled steel tubular columns under axial loads. J Constructional Steel Research 63:182–193

Kurt CE (1978) Concrete filled structural plastic columns. J Str Division 104(no. ST1): pp. 55–63

Lakumalla N (2012) *Study on reinforced cement concrete filled UPVC pipes as columns*. (M. Tech. dissertation 2012). Indian Institute of Technology Roorkee, Roorkee, India

Mander J, Priestley MJ, Park RJ (1988) Theoretical stress-strain model for confined concrete. J Struct Engrg, ASCE 114(8):1804–1826

Marzouck M, Sennah K (2002) Concrete filled PVC tubes as compression members. Proc. International congress on challenges of concrete construction at Scotland, U.K

Mirmiran A, Sahaway M (1997) Behaviour of Concrete columns Confined by Fiber Composites. Journal of Structural Engineering ASCE 125(5):583–590

Newman K, Newman JB (1971) Failure theories and design criteria for plain concrete. Proc. *Int. Civ. Engrg. Mat. Conf. on Struct., Solid Mech. and Engrg. Des.* Wiley Interscience, New York, pp pp. 936–995

Oliveira W, Nardin S, Debs A, Debs M (2009) Influence of concrete strength and length/diameter on the axial capacity of CFT columns. J Constructional Steel Research 65:2103–2110

Richart FE, Brandtzaeg A, Brown RL (1928) A study of the failure of concrete under combined compressive stresses. Engg. Experimental Station Bull. No. 185. University of Illinois, Urbana III

Saatcioglu M, Razvi SR (1992) Strength and ductility of confined concrete. J Struct Engrg, ASCE 118(6):1590–1607

Schneider SP (1998) Axially loaded concrete filled steel tubes. J Str Engg 124(10):1125–1138

Tamuzs V, Valdmanis V, Gylltoft K, Tepfers R (2007a) Behavior of CFRPconfined concrete cylinders with a compressive steel reinforcement. Mech Compos Mater 43(3):191–202

Tamuzs V, Tepfers R, Zile E, Valdmanis V (2007b) Stability of round concrete columns confined by composite wrappings. Mech Compos Mater 43(5):445–452

Tamuzs V, Valdmanis V, Tepfers R, Gylltoft K (2008) Stability analysis of CFRP-wrapped concrete columns strengthened with external longitudinal CFRP sheets. Mech Compos Mater 44(3):199–208

Tao Z, Teng JG, Han LH, Lam L (2004) Experimental behaviour of FRPconfined slender RC columns under eccentric loading. In: Proceedings, Second International Conference on Advanced Polymer Composites for Structural Applications in Construction, University of Surrey, Guildford, UK., pp 203–212

Wang J, Yang Q (2010) Experimental study on Mechanical Properties of concrete confined with plastic Pipe. *Title no. 107-M17*. ACI Mater J: 107,132–137

Wu YF (2004) The Effect of Longitudinal Reinforcment on the Cyclic Shear Behaviour of Glass Fiber Reinforced Gypsum Wall Panels:Tests. Eng Struct 26(11):1633–1646

6

Toughness study on fly ash based fiber reinforced concrete

A Sofi*, K Swathy and G Srija

Abstract

The objective of the present investigation is to study the toughness of steel fiber-reinforced fly ash concrete based on the Japan Society of Civil Engineers (JSCE) approach. Fly ash is also considered as a hazardous waste due to the probable leaching of potentially toxic substances into the surface water, ground water, and soil. The ash content of the Indian coal (30% to 50%) contributes to these large volumes of fly ash. This paper highlights about the behavior of concrete when fly ash and steel fiber are added in concrete. Fiber-reinforced concrete is a concrete containing fibrous material which increases its structural integrity. The addition of random fibers to concrete considerably improves its structural characteristics such as static flexural strength, impact strength, tensile strength, ductility, and flexural toughness (Qian and Stroeven, Cem. Concr. Res. 30:63–68, 2000). Fly ash has been used by replacing cement in percentages, and steel fibers are added by volume of concrete in different percentages. Grooved type of steel fibers of aspect ratio 50 was used in this study. Flexural strength test was carried out for the specimens, and its results were highlighted. The toughness factor as measured by the JSCE approach is reported, and there is a good correlation between the steel fibers added in various percentages such as 1.5%, 2%, and 2.5% and the calculated toughness factor.

Keywords: Fly ash; Steel fiber; Flexural strength; Toughness factor

Introduction

Fly ash is a residual material of energy production using coal, which has been found to have numerous advantages for its use in the concrete industry. Some of the advantages include improved workability, reduced permeability, increased ultimate strength, reduced bleeding, and reduced heat of hydration Qian and Stroeven (2000). The use of fly ash in concrete is found to affect strength characteristics adversely. One of the ways to compensate for the early-age strength loss associated with the usage of fly ash is by incorporating fibers, which have been proved very efficient in enhancing the strength characteristics of concrete (Stroeven and Babut 1986).

For long-term strength and toughness and high stress resistance, steel fiber-reinforced concrete (SFRC) is increasingly being used in structures such as flooring, housing, precast tunneling, heavy duty pavement, and mining (Yao et al. 2003). The addition of steel fibers significantly improves many of the engineering properties of mortar and concrete, notably impact strength and toughness. The enhanced performance of fiber-reinforced concrete compared to its unreinforced counterpart comes from its improved capacity to absorb energy during fracture. While a plain unreinforced matrix fails in a brittle manner at the occurrence of cracking stresses, the ductile fibers in fiber-reinforced concrete continue to carry stresses well beyond matrix cracking, which helps maintain structural integrity and cohesiveness in the material (Bentur and Mindess 1990).

Poor toughness, a serious shortcoming of high strength concrete, can be overcome by reinforcing with short discontinuous fibers. Fibers primarily control the propagation of cracks and limit the crack width. Further, if properly designed, fibers undergo a pullout process, and the frictional work needed for pullout leads to a significantly improved energy absorption capability. This energy absorption attribute of SFRC is often termed as toughness (Pierre et al. 1999). The importance of fiber geometry and matrix strength on the toughness characteristics of SFRC has been clearly established by earlier researchers (Soroushian et al. 1992).

* Correspondence: sofime@gmail.com
VIT University, Vellore, Tamil Nadu 632014, India

Table 1 Mix proportions

Mix	Water (l)	Cement (kg)	Fly ash (kg)	Fine aggregate (kg)	Coarse aggregate (kg)	Fiber content (kg)			
						0%	1.5%	2%	2.5%
0% fly ash	180.00	405.48	-	669.59	1,088.35	-	0.216	0.288	0.361
20% fly ash	198.78	324.38	81.09	663.60	1,078.60	-	0.216	0.288	0.361
40% fly ash	202.74	243.28	162.19	656.60	1,067.23	-	0.216	0.288	0.361
60% fly ash	202.74	162.19	243.28	650.60	1,057.48	-	0.216	0.288	0.361

The most common method to measure toughness is to use the load-deflection curve obtained using a simply supported beam loaded at the third points (four-point bending). The addition of steel fibers at high dosages, however, has potential disadvantages in terms of poor workability and increased cost. In addition, due to the high stiffness of steel fibers, microdefects such as voids and honeycombs could form during placing as a result of improper consolidation at low workability levels (Bayasi and Zeng 1993). The two widely used standard test methods are the American standard testing method (ASTM) C 1018: standard test method and the Japan Society of Civil Engineers' (JSCE) standard SF-4 method (Banthia and Nandakumar 2003).

In this work, JSCE SF-4 standard technique has been followed, i.e., the area under the load-deflection plot up to a deflection of span/150 is obtained. From this measure of flexural toughness, a flexural toughness factor (FT) is calculated. It may be noted that FT has the unit of stress such that its value indicates, in a way, the post-matrix cracking residual strength of the material when loaded to a deflection of span/150. The chosen deflection of span/150 for its calculation is purely arbitrary and is not based on serviceability considerations.

Experimental program

Materials

The cement used was OPC 43 grade of specific gravity 3.15. Initial and final setting times of the cement were 140 and 205 min, respectively Indian Standard Designation IS12269-1987 (2003). Fly ash specific gravity was found to be 2.53. Aggregate used was dry and clean, natural, river aggregate. Sizes of aggregates used were 20 and 12 mm. Specific gravity of coarse aggregate was 2.62, and specific gravity of fine aggregate was 2.63.

Mix design

For each cubic meter of concrete, mix proportion given in Table 1 has been adopted and mix designation has been given in Table 2. Mixture design is made in accordance with the Indian standard code 10262–2009 for M20 grade of concrete. Fresh concretes containing 20%, 40%, and 60% fly ash as cement replacement in mass basis were prepared by modifying the reference Portland cement concrete. Fresh fiber-reinforced concretes containing 1.5%, 2.0%, and 2.5% of steel fibers in volume basis were prepared. The procedures for mixing the fiber-reinforced concrete involved the following steps: first, the gravel and sand were placed in a concrete mixer and dry mixed for 1 min. Second, the cement and fiber were spread and dry mixed for 1 min. Third, the mixing water was added and mixed for approximately 2 min. Finally, the freshly mixed fiber-reinforced concrete was cast into specimens mold and vibrated simultaneously to remove any air remain entrapped. After casting, each of the specimens was allowed to stand for 24 h in laboratory before demolding. Demolded specimens were stored in water at 23 ± 2°C until testing days.

Testing methods

For experimental work, cement was replaced by fly ash by 20% and 40% and 60% by mass basis. Steel fibers were added at three different percentages of 1.5%, 2%, and 2.5% (Table 1). Experimental investigation of fresh mix properties of fly ash fiber-reinforced concrete was conducted. Specimens of size 500 × 100 × 100 mm were used to calculate flexural strength and flexural toughness factor. All prisms were tested for flexural strength after 28 days. Flexural testing machine was

Table 2 Mix designation

Mixture number	Fly ash content (%)	Steel fiber content (%)
A1	0	0
A2	0	1.5
A3	0	2
A4	0	2.5
B1	20	0
B2	20	1.5
B3	20	2
B4	20	2.5
C1	40	0
C2	40	1.5
C3	40	2
C4	40	2.5
D1	60	0
D2	60	1.5
D3	60	2
D4	60	2.5

Table 3 Average toughness factor and flexural strength

Mix number	Deflection	Toughness factor	Average toughness factor	Equivalent flexural strength (Mpa)
A1	Center	0.272	0.170	4.12
	L/150	0.08		
	End	0.150		
A2	Center	0.215	0.219	4.20
	L/150	0.243		
	End	0.200		
A3	Center	0.455	0.298	4.68
	L/150	0.193		
	End	0.248		
A4	Center	0.831	0.408	4.82
	L/150	0.408		
	End	0.227		
B1	Center	0.27	0.230	3.84
	L/150	0.21		
	End	0.22		
B2	Center	0.33	0.340	4.10
	L/150	0.35		
	End	0.33		
B3	Center	0.543	0.541	4.56
	L/150	0.546		
	End	0.542		
B4	Center	0.680	0.649	4.84
	L/150	0.684		
	End	0.585		
C1	Center	0.18	0.195	3.17
	L/150	0.27		
	End	0.18		
C2	Center	0.20	0.212	4.15
	L/150	0.17		
	End	0.26		
C3	Center	0.098	0.260	4.48
	L/150	0.29		
	End	0.38		
C4	Center	0.68	0.315	4.75
	L/150	0.12		
	End	0.14		
D1	Center	0.047	0.094	3.05
	L/150	0.16		
	End	0.072		
D2	Center	0.047	0.16	3.89
	L/150	0.16		
	End	0.072		
D3	Center	0.047	0.168	4.11
	L/150	0.16		
	End	0.072		
D4	Center	0.11	0.17	4.19
	L/150	0.13		
	End	0.078		

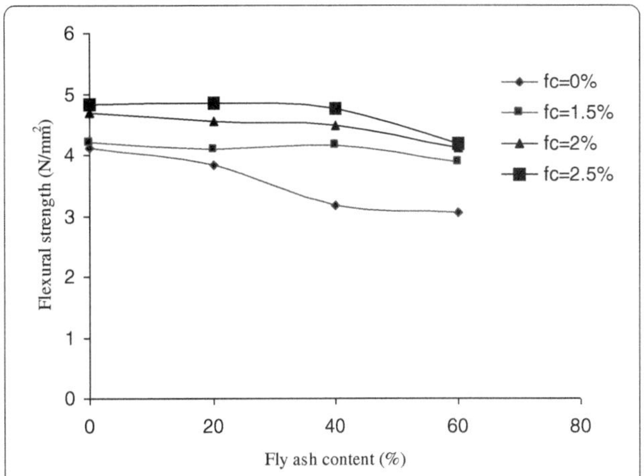

Figure 1 Flexural strength on varying fly ash content.

used to test the specimen. No special care was taken to eliminate the extraneous support settlements from the gross deflections. The applied load and deflection data were recorded manually.

Deflections were measured at the bottom of the specimen near the center and also at the end, and deflections were recorded beyond L/150. The toughness of fiber-reinforced concrete is a measure of the energy absorption capacity of the fibers and is characterized by the area under the load-deflection curve up to a specific deflection. The real effects of fiber addition can be observed as a result of the bridging stress offered by the fibers after the peak load.

Results and discussion

The equivalent flexural strength for four points bending proposed by (JCI) is given by [10] $\sigma b = [Tb \times L]/[\delta\ tb \times b \times h^2]$, where σb is the equivalent flexural strength (N/mm^2); Tb is flexural toughness (N/mm); δ tb is deflection of 1/150 of span (mm); b, d, L is width, depth, and length of section (mm). The toughness and equivalent

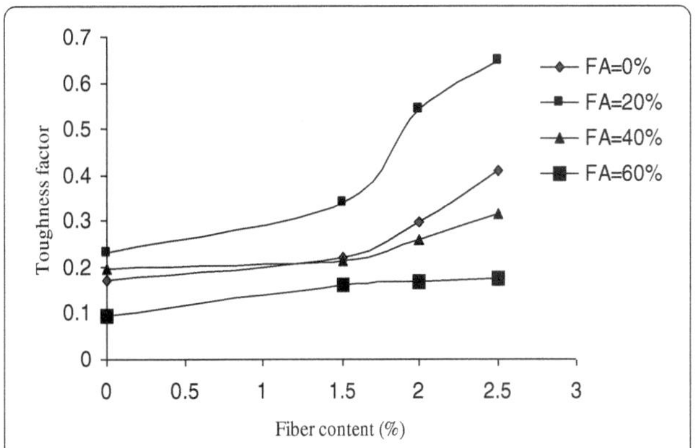

Figure 2 Toughness factor for varying fiber content.

flexural strength values calculated as per JCI specifications for various fiber concretes are given in Table 2, and their respective values are plotted in Figures 1 and 2. Equivalent flexural strength increases as fiber content increases, when 20% fly ash were added the flexural strength values are seems to be higher when compared with 40% fly ash and 60% fly ash content and toughness factor decreases when fly ash content increases.

Conclusions

The following conclusions were made from the tests results:

(1) When 2.5% fibers were added, flexural strength increases by 1.67% compared with that of the conventional concrete.
(2) When fly ash was added in addition to steel fibers, for 20% fly ash + 2.5% fiber equivalent, flexural strength increases by 0.41% compared with that of the 0% fly ash + 2.5% fiber.
(3) Equivalent flexural strength increases as fiber content increases; when 20% fly ash were added, the flexural strength values seem to be higher when compared with that of the 40% fly ash and 60% fly ash content.
(4) Toughness factor decreases by 58% when 60% fly ash + 2.5% fiber were added when compared with that of the 0% fly ash + 2.5% fiber.
(5) When 20% fly ash + 2.5% fiber were added, toughness factor increases by 59% when compared to that of the 0% fly ash + 2.5% fiber; therefore, toughness factor decreases when fly ash content increases.
(6) Therefore, it is concluded that 20% fly ash addition is the optimum dosage when steel fibers are incorporated in concrete.
(7) When fibers are added to fly ash, concrete propagation of cracks are arrested when compared to that of the conventional concrete effectively, which in turn increases flexural strength and toughness.

The characterization of flexural toughness based on the JSCE approach is very simple and is independent of the type of deflection measuring technique. No sophisticated instrumentation is required to determine the toughness factor. The determination of first crack, which is very difficult to identify, is not required in this method. From the results, it is evident that the ductility of fiber-reinforced concrete depends primarily on the fibers' ability to bridge the cracks at high levels of strain. Thus, stiffer fibers would provide better crack bridging; this explains the good performance of steel fibers.

Competing interests

The authors declare that they have no competing interests.

Authors' contributions

AS designed the research work, drafted the manuscript, analyze the results and plotted the graphs. KS and GS carried out experiments in the laboratory.

Acknowledgements

The authors would like thank to VIT University for providing support to conduct the experiments in the laboratories.

References

Banthia N, Nandakumar N (2003) Crack growth resistance of hybrid fibre reinforced cement composites. Cem Concr Compos 25:3–9

Bayasi Z, Zeng J (1993) Properties of polypropylene fiber reinforced concrete. ACI Mater J 90:605–610

Bentur A, Mindess S (1990) Fiber reinforced cementitious composites. Elsevier, London

Indian Standard Designation IS12269-1987 (2003) Specification for 53 grade ordinary Portland cement. Bureau of Indian Standards, New Delhi

IS:10262–2000, Recommended Guidelines for Concrete Mix Design. Bureau of Indian Standards (BIS), India

Japanese Concrete Institute, JCI-SF4, Methods of tests for flexural strength and flexural toughness of fiber reinforced concrete (using beam with third-point loading). pp 45–56

Pierre P, Pleau R, Pigeon M (1999) Mechanical properties of steel micro fiber reinforced cement pastes and mortars. J Mater Civil Eng 11:317–324

Qian CX, Stroeven P (2000) Development of hybrid polypropylene–steel fibre-reinforced concrete. Cem Concr Res 30:63–68

Soroushian P, Khan A, Hsu JW (1992) Mechanical properties of concrete materials reinforced with polypropylene or polyethylene fibers. ACI Mater J 89:535–540

Stroeven P, Babut R (1986) Fracture mechanics and structural aspects of concrete. Heron 31(2):15–44

Yao W, Lib J, Wu K (2003) Mechanical properties of hybrid fiberreinforced concrete at low fiber volume fraction. Cem Concr Res 33:27–30

7

Effect of silica fume on the characterization of the geopolymer materials

Hisham M Khater

Abstract

The influence of silica fume (SF) addition on properties of geopolymer materials produced from alkaline activation of alumino-silicates metakaolin and waste concrete produced from demolition works has been studied through the measurement of compressive strength, Fourier transform infrared spectroscopy, X-ray diffraction, and scanning electron microscopy (SEM) analysis. Alumino-silicate materials are coarse aggregate included waste concrete and fired kaolin (metakaolin) at 800°C for 3 h, both passing a sieve of 90 μm. Mix specimens containing silica fume were prepared at water/binder ratios in a range of 0.30 under water curing. The used activators are an equal mix of sodium hydroxide and silicate in the ratio of 3:3 wt.%. The control geopolymer mix is composed of metakaolin and waste concrete in an equal mix (50:50, wt.%). Waste concrete was partially replaced by silica fume by 1 to 10 wt.%. The results indicated that compressive strengths of geopolymer mixes incorporating SF increased up to 7% substitution and then decreased up to 10% but still higher than that of the control mix. Results indicated that compressive strengths of geopolymer mixes incorporating SF increases up to 7% substitution and then decreases up to 10% but still higher than the control mix, where 7% SF-digested calcium hydroxide (CH) crystals, decreased the orientation of CH crystals, reduced the crystal size of CH gathered at the interface, and improved the interface more effectively.

Keywords: Silica fume, Geopolymer, Waste concrete, Metakaolin

Introduction

Silica fume (SF) known as micro-silica is a by-product of the reduction of high-purity quartz with coal in electric furnaces in the production of silicon and ferrosilicon alloys. Because of its extreme fineness and high silica content, silica fume is a highly effective pozzolanic material. Silica fume is used in concrete to improve its properties like compressive strength, bond strength, and abrasion resistance; it reduces permeability and therefore helps in protecting the reinforcing steel from corrosion.

Silica fume has been used as a high pozzolanic reactive cementitious material to make high-performance concrete in the severe conditions (Kohno 1989; Gautefall 1986). This mineral admixture has highly been used in severe environmental conditions despite its several mixing and curing problems because of its acceptable early-age strength development (Cheng-Yi and Feldman 1985; Grutzeck et al. 1983). The hydration mechanism and properties of secondary C-S-H made by pozzolanic reaction have been studied by many investigators (Cheng-Yi and Feldman 1985). However, CSH formed by silica fume-calcium hydroxide reaction might be different with respectto the amount of molecular water, C/S ratio, and density (Cohen and Bentur 1988). Moreover, because of its rather different characteristics, pozzolanic gel has a high potential to contribute in reactions with other internal or external ions such as Al, Cl, and alkalies (Maage 1989; Sellevold and Nilsen 1987).

On the other hand, the search for a new environmentally friendly construction material that will match the durability of ancient concrete has provoked interest into the study of alkali-activated cementitious systems over the past decades. Alkali-activated cements refer to any system that uses an alkali activator to initiate a reaction or a series of reactions that will produce a material that possesses cementitious property (Yip et al. 2005).

Alkali-activated cement, alkali-activated slag and fly ash, and geopolymers are all considered to be alkali-activated cementitious systems; however, it is expected

Correspondence: Hkhater4@yahoo.com
Housing and Building National Research Centre (HBNRC), 87 El-Tahreer St., Dokki, Giza, P.O. Box 1770, Cairo, Egypt

that the structures of these materials are vastly different and result from different chemical mechanistic paths. It is commonly acknowledged that calcium silicate hydrate (CSH) is the major binding phase in Portland cement (Taylor 1964) and alkali-activated slags (Richardson and Cabrera 2000); however, the binding property of geopolymers is generally assumed to be the result of the formation of a three-dimensional amorphous aluminosilicate network (van Jaarsveld and van Deventer 1999a; Davidovits 1991; Phair and van Deventer 2002; Lee and van Deventer 2002).

The production of geopolymeric precursors is carried out by calcinations of aluminosilicates, natural clay materials. Their source can be also some industrial aluminosilicate waste materials. The result of the hardening mechanism is a three-dimensional zeolitic framework, unlike traditional hydraulic binders in which hardening is the result of the hydration of calcium aluminates and silicates (Davidovits 1991; Phair and van Deventer 2002; Lee and van Deventer 2002; Davidovits 1993); this circumstance is a cause of significant differences in the quality and variety of the engineering properties of the composites based on geopolymer and current cements. As a means of converting waste materials to useful products, the value of geopolymer technology lies primarily in its ability to produce a high-performance binder from materials that are rich in silica and alumina that are the bases of a geopolymerization process.

Davidovits (1991) proposed that geopolymers are formed by the polymerization of individual aluminate and silicate species, which are dissolved from their original sources at high pH in the presence of alkali metals. The resultant products are reported to have the general formula M_n-[[-[[-Si-O2]$_z$-Al-O]$_n$ wH_2O, where M is the alkali element, '–' indicates the presence of a bond, z is 1, 2, or 3, and n is the degree of polymerization. Theoretically, any alkali and alkali-earth cations can be used as the alkali element (M) in the reaction; however, the majority of the research has focused on the effect of sodium (Na^+) and potassium (K^+) ions (Phair and van Deventer 2002; van Jaarsveld and van Deventer 1999b). It has not been clearly proven whether other alkali and alkali-earth cations, including calcium, will participate in the reactions in a similar way.

The target of the present paper is to study the effect of silica fume addition in the performance of geopolymer materials and investigate its effect on the mechanical and microstructural properties of the geopolymer constituents.

Methods

Materials

Aluminosilicate materials used in this investigation are fired kaolin clay and coarse aggregate included waste concrete, both well-grinded and passing a sieve of 90 μm. An aluminosilicate kaolin material with a high kaolinite residue was utilized in the synthesis of geopolymeric binder, sourced from Al Dehesa, Sinai governorate, Egypt, and was thermally treated at 800°C for 3 h with a heating rate of 5°C·min^{-1} to produce metakaolin (Mk).

This temperature was chosen on the basis of an earlier research work (Kakali et al. 2001; Wenying et al. 2008; Cioffi et al. 2003). The most important characteristics of this calcined product (Mk) are chemical composition (SiO_2 + Al_2O_3 about 95%), fineness (surface area between 15 and 20 m^2·g^{-1}), passing a sieve of 90 μm, and its poorly crystalline nature. Raw materials used in this investigation were characterized by means of a chemical analysis as represented in Table 1.

Silica fume sourced from the Suez Cement Company (Helwan plant, Helwan, Cairo, Egypt), also known as micro-silica, is a by-product of the reduction of high-purity quartz with coal in electric furnaces in the production of silicon and ferrosilicon alloys; it was in powder form with 95% SiO_2, an average particle size of 8 μm, as represented from its laser particle size distribution (Figure 1).

Mineralogical characterization of the raw materials was done using X-ray diffraction analysis as represented in Figure 2, where coarse aggregate included waste concrete has a major content of quartz and little calcite content. Kaolin, on the other hand, is composed mainly of kaolinite and quartz while silica fume is a completely amorphous silica.

Sodium silicate solution and sodium hydroxide were used as alkali activators. Sodium hydroxide (NaOH) produced by SHIDO company (Leningradskaya Oblast, Russia) with 99% purity and liquid sodium silicate from Fisher company (Waltham, MA, USA) consists of 32% SiO_2 and 17% Na_2O with silica modulus SiO_2/Na_2O of 1.88 and density of 1.46 g·cm^{-3} ($Na_2SiO_3{\cdot}9H_2O$).

Synthesis and curing

Geopolymers were made by hand-mixing raw materials of each mixture, passing a sieve of 90 μm as represented in Table 2, with the alkaline activator solution and the mixing water for 10 min, followed by a further 5 min using a rotary mixer, and mixed at a medium speed (80 rpm) for another 30 s; the mixture was allowed to rest for 90 s and then mixed for 1 min at high speed.

All investigations involved the use of Na_2SiO_3 and NaOH in the ratio of 3:3 wt.% of dry mixes, respectively; this ratio was chosen based on previous investigations (El-Sayed et al. 2011; Abd El-Aziz 2010). The water/binder ratio was 0.30 by mass. Paste mixtures were cast into 25 × 25 × 25 mm cubic-shaped molds and sealed with a lid to minimize any loss of evaporable water.

All mixes were left to cure undisturbed under ambient temperature (23°C) for 24 h, demolded, and then some specimens were left to be water cured under room

Table 1 Chemical composition of starting materials (Mass, %)

Sample name	Si02	Al203	Fe203	CaO	MgO	Na20	K20	SO3	TiO2	P205	Loss	Total
Kaolin	56.38	27.61	1.32	0.18	0.06	0.08	0.04	0.06	3.73	0.13	10.17	99.97
Metakaolin	59.03	35.33	0.89	0.43	0.13	0.05	0.05	0.03	2.46	0.72	0.79	99.91
Waste concrete	77.63	1.57	1.19	8.61	0.24	0.24	0.12	2.31	0.01	0.03	7.61	99.56
Silica fume	94.92	0.02	1.28	0.03	0.01	0.28	0.15	0.02	-	-	3.28	99.99

temperature. At the end of the curing regime, the specimens were removed from their curing condition, dried well at 80°C for 24 h then exposed to compressive strength measurements, and then, the resulted crushed specimens were subjected to the stopping of the hydration process using the stopping solution of methanol/acetone (1:1 by volume) followed by washing with acetone as recommended by different investigators (Saikia et al. 2004; Khater 2010) to prevent further hydration and for further analysis followed by the drying of crushed specimens for 24 h at 80°C and then preserved in a well tight container until the time of testing.

Methods of investigation

Chemical analysis was carried out using the Axios PW4400 WD-XRF sequential spectrometer (Panalytical, Netherland), CuKα source with a post sample Kα filter. X-ray diffraction (XRD) patterns were collected from 0° to 50° 2θ (step size 0.02° 2θ and speed 0.4°·min^{-1}). Silica quartz was used as an internal standard. Data were identified according to the XRD software (pdf-2: database on CD-Release 2005). Particle size analysis was done using a laser scattering particle size distribution analyzer (Horiba LA-950, Kyoto, Japan). Stopping of the hydration was performed on crushed specimens by subjecting to stopping the solution of alcohol/acetone (1:1) followed by washing with acetone as recommended by different investigators (Saikia et al. 2004; Khater 2010) to prevent further hydration. The microstructure of the hardened activated specimens were studied using the Inspect S scanning electron microscope (SEM) (FEI Company, Eindhoven, Netherlands) equipped with an energy dispersive X-ray analyzer. Bonding characteristics of alkali-activated specimens were analyzed using the Jasco-6100 Fourier transformed infrared spectrometer (FTIR; Tokyo, Japan).The wave number was ranging from 400 to 4,000 cm^{-1} (Panias et al. 2007).

Results and discussion

XRD patterns of alkali-activated geopolymer mix that are made without silica fume and water cured from 1 up to 180 days is shown in Figure 3. The patterns illustrate a broad band in the region of 6° to 10° 2θ for aluminosilicate gel and broad bands in the region of 17° to 35° 2θ which characterize the glassy phase of the geopolymer constituents. These two regions are considered as a vital role in geopolymer characterization, where any increase in these regions will be reflected on the performance and efficiency of the geopolymer composition. An increase in the CSH content is also noticed with the increase of curing time as indicated from the increased broadness at 29.4° that results from the interaction of freely dissolved silica with Ca species in the matrix forming CSH, which accumulate in the open pores and transformed into crystalline form at the later curing ages.

XRD patterns of alkali-activated geopolymer mix containing various ratios of silica fume from 0% up to 10% as a partial replacement of the used waste concrete and water cured up to 180 days is shown in Figure 4.

The patterns illustrate the decreased calcite content up to 7% SF then increased again upon using 10% SF, which is due to the ability of the silica fume material with its higher specific surface area in interacting with dissolved calcium from waste concrete materials and forming CSH that positively affect the geopolymerization process by forming nucleation centers for the formation and accumulation of the geopolymer. Increasing the silica fume content beyond 7% results in a negative effect by forming agglomerates which are concentrated in a small area and so hinder the formation of both CSH and geopolymer phases. It can also be noticed that an increase of the broad band in the region of 6° to 10° 2θ for aluminosilicate gel and broad bands in the region of 17° to 35° 2θ which characterize the glassy phase of the

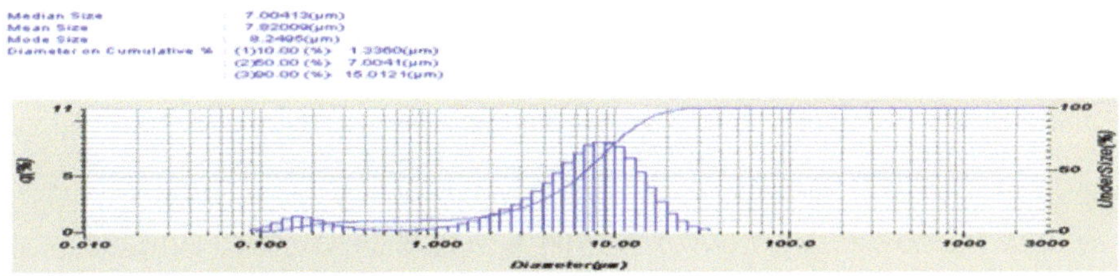

Figure 1 Laser particle size distribution of silica fume.

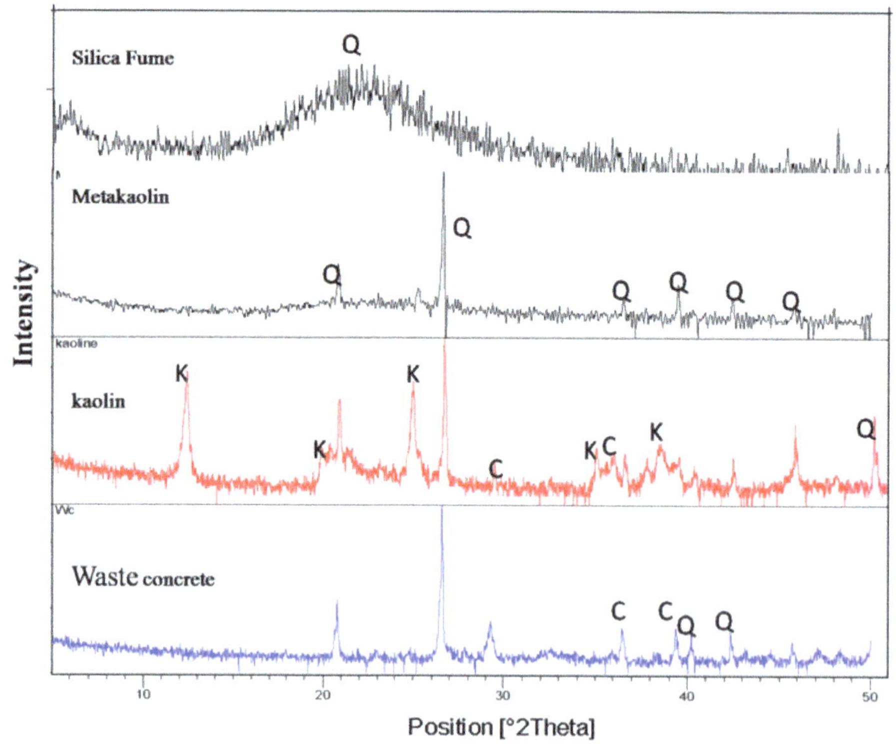

Figure 2 XRD analysis of the used raw materials. Q, quartz; FS, feldspar; C, calcite; CSH, calcium silicate hydrate.

geopolymer constituents by increasing silica fume content up to 7% while there is a decrease with the addition of 10% SF.

Morphology and microstructure of 1 and 180 days water-cured reference geopolymer specimens made without silica fume content are shown in Figure 5. Evidently, Figure 5a shows the low contribution of geopolymer constituents, where most of the geopolymer are in the oligomer state after the day of hydration; the CSH phases are also seen in the microstructure and bound most of the formed oligomer by forming nucleation sites for the accumulation of the geopolymer (Temuujin et al. 2009; Khater 2012). After 180 days of hydration, the oligomer condensation and precipitation of the formed geopolymer are clearly distinguished in Figure 5b, where a crystalline plate of geopolymer constituents fill the pore, forming a well compact and dense structure.

Table 2 Composition of the geoplymer mixes (Mass, %)

Mix no.	Waste concrete (WC), %	Metakolin (Mk), %	Silica fume (SF), %
S0	50	50	0
S1	49	50	1
S3	47	50	3
S5	45	50	5
S7	43	50	7
S10	40	50	10

Figure 6 shows representative SEM micrographs of the geopolymer binder containing 0, 5, 7, and 10 wt.% of silica fume after water was cured up to 180 days. The micrograph of the geopolymer mix made without silica fume displayed oligomer condensation and precipitation of the formed geopolymer (Figure 6a), where crystalline plates of geopolymer constituents fill the pores, forming a well compact and dense structure. Increasing the silica fume to 5% results in an increase in the content of the formed CSH as most of the liberated free lime are consumed by the added silica fume; this will be positively reflected on the morphology of the formed geopolymer leading to a well compact structure (Figure 6b).

Further addition of silica fume (7%) leads to a complete consumption of free hydrated lime and so leads to the coexistence of CSH (I) that has a CaO/SiO_2 ratio <1.5 with a fibrous structure and CSH (III) with almost-denser spheres form with CaO/SiO_2 ratio <1.5 (Taylor 1992; Diamond 1986; Malhotra and Mehta 1996; Gleize et al. 2003) as represented in Figure 6c; these compact CSH phases result in an enhancement in the microstructure performance providing an additional strength to the sample specimens in addition to its vital role in the formation of additional nucleation or polymerization-condensation centers for the precipitation of geopolymer as was indicated from the

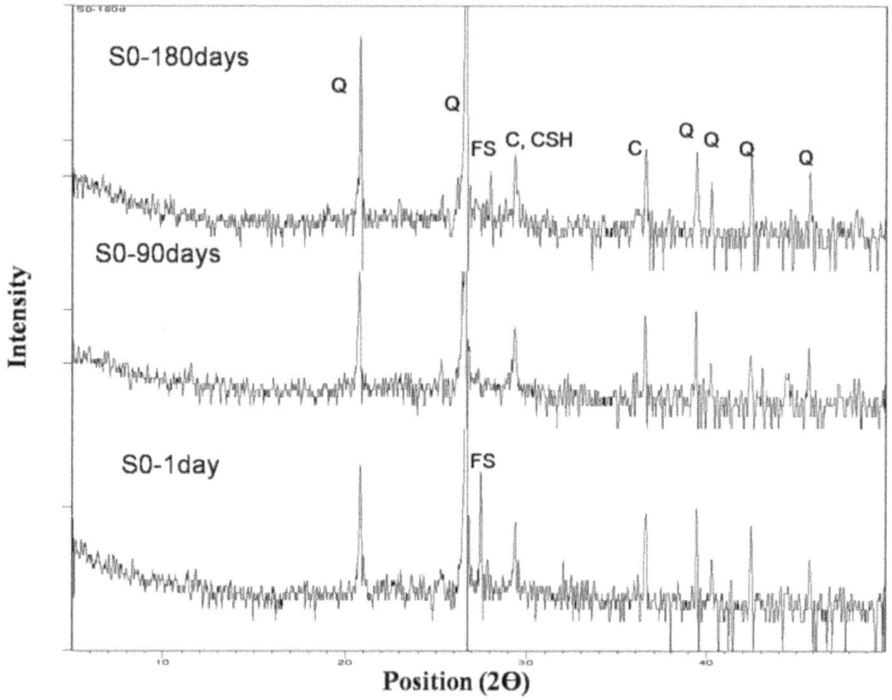

Figure 3 XRD pattern of geopolymer mix without silica fume at different curing ages. Q, quartz; FS, feldspar; C, calcite; CSH, calcium silicate hydrate.

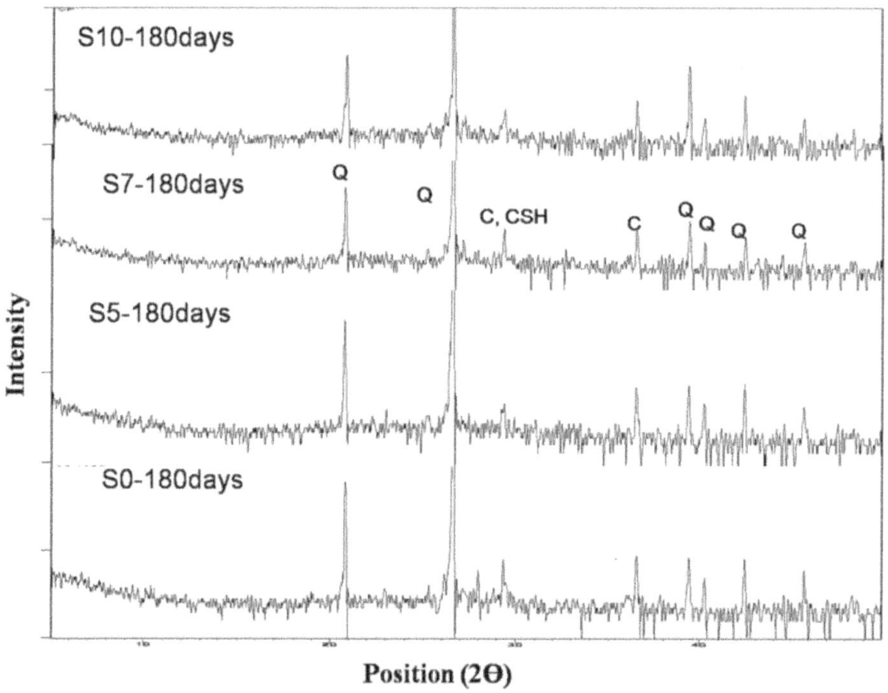

Figure 4 XRD pattern of geopolymer mixes having various ratios of silica fume, water cured at 180 days. Q, quartz; FS, feldspar; C, calcite; CSH, calcium silicate hydrate.

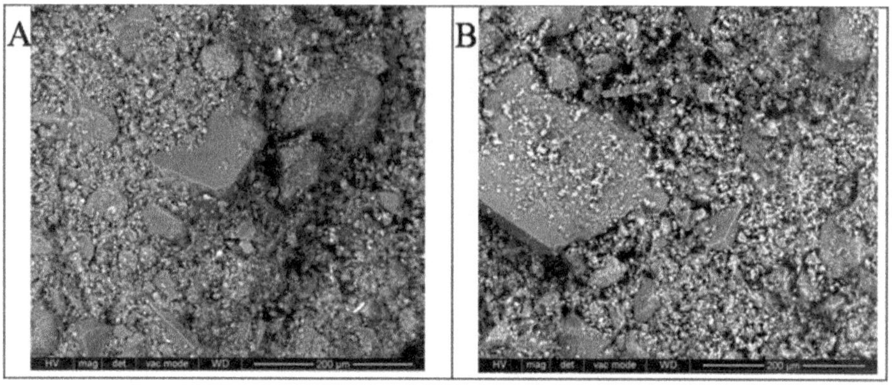

Figure 5 SEM micrographs of hydrated geopolymer mix made without silica fume. Water cured at (**A**) 1 day and (**B**) 180 days.

specimen morphology where a massive platy geopolymer layer fill mostly the matrix composition.

It should be noted that using a higher content of silica fume must be accompanied by an adjustment of the water and superplasticizer dosage in the mix in order to ensure that specimens do not suffer excessive self desiccation and cracking. However, the addition of 10% silica fume in this work did not include the addition of superplasticizer that leads to sample cracking and hinders the propagation of three-dimensional geopolymer networks as indicated from the morphology of the formed geopolymer composition (Figure 6d); also, the increase in silica content leads to the formation of a two-dimensional cross-linked poly-sialate with rode-like structure that is produced as a result of the increase in Si/Al ratio of more than three times and known by lower mechanical and hardening properties as compared with a three-dimensional network that formed at Si/Al =1.5:2 (Davidovits 1999). It can also be noticed that the presence of small geopolymeric contents which fill the pores with low conjunction with each other leading to low compaction of the formed geopolymer composite.

After 180 days of hydration, the FTIR spectra of geopolymer mix specimens containing various silica

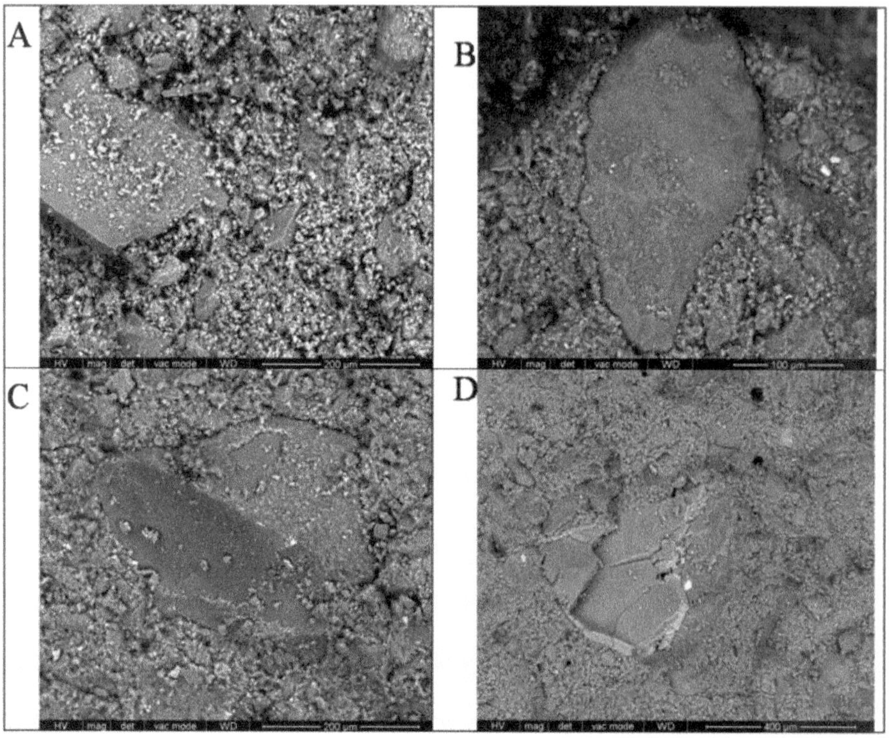

Figure 6 SEM mirographs of hydrated geopolymer mixes containing different silica fume ratios. Water cured at 180 days: (**A**) 0% SF, (**B**) 5% SF, (**C**) 7% SF, and (**D**) 10% SF.

fume contents are shown in Figure 7. Band descriptions are as follows: stretching vibration of O-H bond in the region 3,429 to 2,300 cm^{-1}, bending vibration of HOH in the region of 1,595 to 1,628 cm^{-1}, stretching vibration of CO_2 located at about 1,430 cm^{-1}, asymmetric stretching vibration of T-O-Si at about 980 to 1,100 cm^{-1}, where T=Si or Al, Si-O stretching in the region of 880 cm^{-1}, symmetric stretching vibration of Si-O-Si between 770 to 780 cm^{-1}, symmetric stretching vibration of Si-O-Si and Al-O-Si in the region 675 to 685cm^{-1}, and bending vibration of Si-O-Si and O-Si-O in the region 450 to 460 cm^{-1}.

The most characteristic difference observed between the FTIR specta of geopolymer specimens made with various silica fume concerning bands attributed to the asymmetric stretching vibrations (T-O-Si) that increase in its intensity and broadness with a silica fume addition up to 7% while the pattern of 10% is subjected to the decrease in this band broadness. The bending vibration of Si-O-Si appeared to increase in sharpness with increasing silica fume up to 7% then slightly decreased with 10% silica fume content. The bending vibration band of HOH increases in broadness with silica fume addition up to 7% and decreases sharply with further increase of SF. The broadness of HOH band implies the overlap of more bands with higher intensity which is mainly related to the combined water in CSH, and so, the increase of this peak reflects the increase of binding materials of this specimen. This band decreases sharply with an addition of 10% silica fume as a result of inclusion of added SF in the formation of two-dimensional polysialatedisiloxo cross-link (which has a two-dimensional structure with low hardness than three-dimensional one) than forming CSH which in turn acts as a seeding agent for geopolymer accumulation. The symmetric band of CO_2 located in the FTIR spectra looks nearly similar for all specimens.

Compressive strength values of hydrated geopolymer mixes containing different SF contents from 0 up to 10 wt. % as partial replacement of waste concrete at different curing times of 1, 7, 28, and 90 days at room temperature are shown in Figure 8. Evidently, the compressive strength of mixes increases with the increase of curing time as a result of progressive hydration, forming CSH as well as geopolymer formation leading to a fine and homogeneous structure. The results also indicated the increase of strength with increasing SF content up to 7%, then a decrease up to 10%. The strength increase with silica fume up to 7% is due to the highly effective pozzolanic activity of silica fume which improves its properties like compressive strength, bond strength, and abrasion resistance. Also, silica fume is assumed to be the reactant to produce secondary C-S-H by consuming calcium hydroxide. The decreased calcium hydroxide content of the cement matrix and increased amount of C-S-H gel together with filler effect of SF contribute to the safeguarding of the matrix against external ingressive ions.

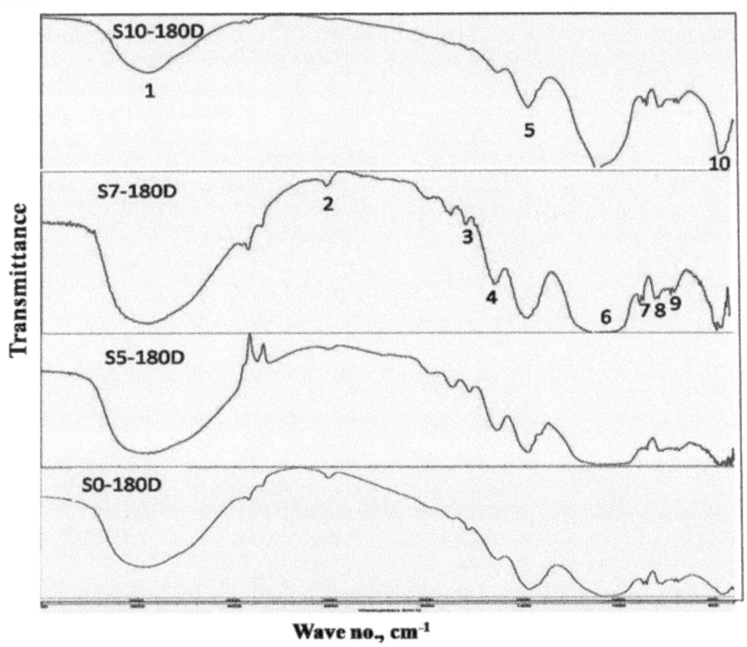

Figure 7 FTIR spectra of 180 days of water-cured geopolymer specimens having various ratios of silica fume. 1,2, stretching vibration of O-H bond; 3,4, bending vibrations of HOH; 5, stetching vibration of CO2; 6,7, asymmetric stretching vibration (T-O-Si); 8, symmetric stretching vibration of Si-O; 9, symmetric stretching vibration (Si-O-Si and Al-O-Si); 10, bending vibrations (Si-O-and O-Si-O).

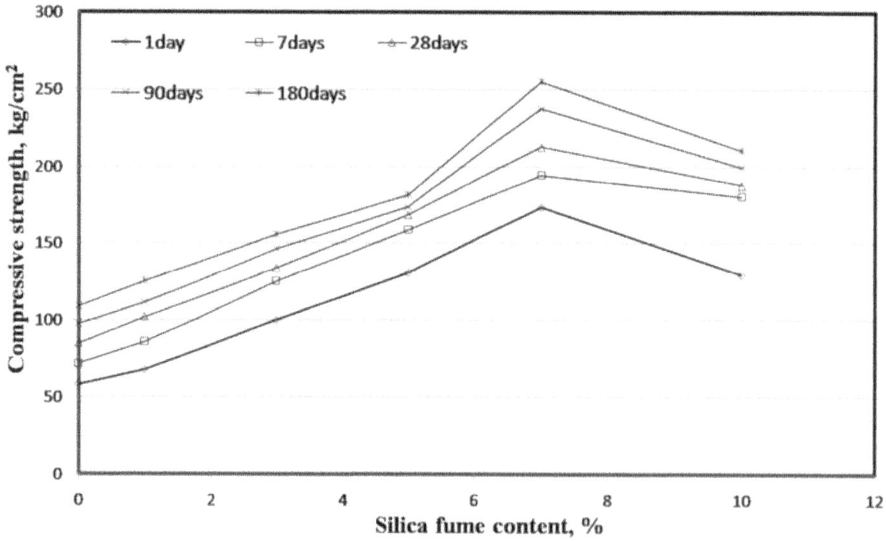

Figure 8 Compressive strength of geopolymer mixes having various ratios of silica fume.

It is known that SF has many saturated bonds, =Si–O–Si=, and little unsaturated bonds at the surface, so the reaction mechanism of silica fume with hydrated lime is as follows:

$$= \text{Si–O–Si} = +\text{H–OH} \rightarrow\ = \text{Si–OH reactly slowly}$$

$$= \text{Si–OH} + \text{Ca(OH)2} \rightarrow \text{C–S–H}$$

However, it should be noted that using a higher content of silica fume of more than 7% must be accompanied by an adjustment of the mixing water and superplasticizer dosage of the mix in order to ensure that specimens do not suffer excessive self-desiccation and cracking. However, in this work, there is no superplasticizer used, and the mixing water is in a constant ratio of 0.3% for better comparison of the reaction products. Otherwise, using this much quantities of silica fume could actually lower the strength of composites instead of improving it as indicated from the strength results.

On the other hand, there is a huge gap between the strength of mix that was made without silica fume and

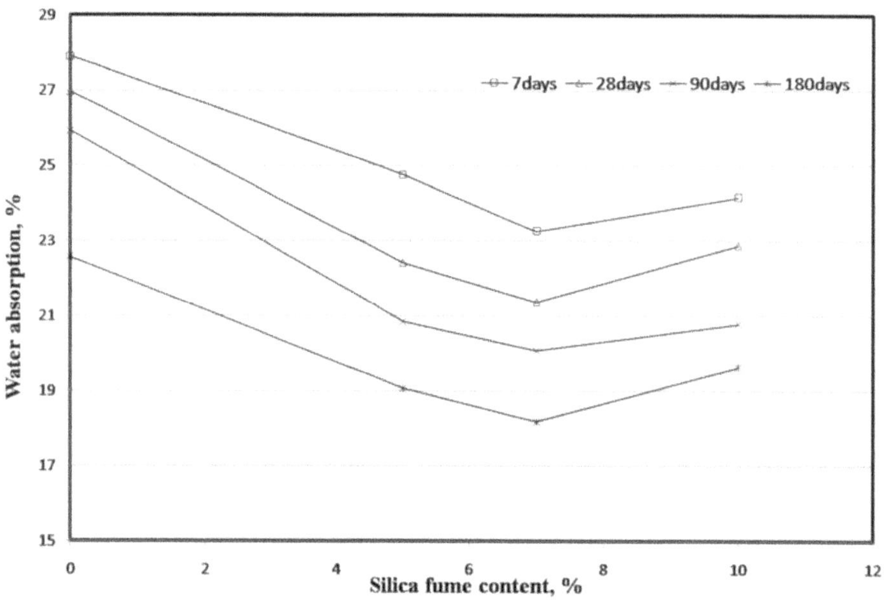

Figure 9 Water absorption of geopolymer mixes having various ratios of silica fume.

that which contains 7% SF, where silica fume represents an enrichment source of amorphous silica and enhancing geopolymerization. It is known that mechanical properties and microstructure of the geopolymer strongly depends on the initial Si/Al ratio. Better strength properties have been reported for mixtures with SiO_2/Al_2O_3 ratios in the range of 3.0 to 3.8. Changes in the SiO_2/Al_2O_3 ratio beyond this range usually result in low strength systems (De Silva et al. 2007). In spite of that, the geopolymer may be formed in the absence of alumina, forming a homogeneous polysiloxy structure.

The trend of variation of compressive strength coincides with the XRD data, FTIR spectrum, and SEM observations, where the amorphous content of geopolymer materials increase with increasing SF content as it forms precipitation sites for geopolymer accumulation and precipitation when it interacts with free lime in the medium, forming CSH. Further increase in silica fume leads to insufficient wetting of the medium that hinder the propagation of geopolymer chains and so weaken its mechanical properties; still, it exceeds that which has no silica fume content as its higher pozzolanic activity leads to the increase of the formed CSH and hence possess higher mechanical and microstructural characteristics.

Water absorption of geopolymer mixes containing SF in the ratios from 0 up to 10 wt.% at different curing ages are shown in Figure 9. Obviously, the results in Figure 9 indicate the decrease of water absorption with the increase of curing time as a result of progressive hydration with the formation of CSH and geopolymer, leading to a denser structure. Evidently, it decreases with increasing SF content up to 7% and then increases slightly up to 10%. The decease of water absorption with increasing SF is due to the fact that when a material with high specific surface area is added to a mix specimen, it acts as a micro-filler of the matrix particles, which can reduce the amount of water that filled in the voids of the blending materials. However, replacing cement with a high specific surface material would increase the wettable surface area and the amount of water adsorbed. Thus, the final water requirement will depend on which of the two above-mentioned factors will be superior. With replacement of less than 7% in this paper, the former factor may be superior for SF content (Gleize et al. 2003).

Conclusion

Alkaline activation of aluminosilicate wastes in the presence of silica fume and metakaolin using sodium hydroxide and sodium silicate in the ratio of 3:3 wt.% of the dry weight leads to the formation of an aluminosilicate geopolymer that possess an enhancement in both mechanical and microstructural properties. Silica fume addition is likely to result in the formation of nucleation CSH sites for the precipitation of the geopolymer product. Metakaolin inclusion greatly improves the geopolymeric structure. Silica fume addition up to 7% greatly enhances the geopolymerization process with the formation of a well-refined and compact matrix, while further increase of SF content leads to the decrease in the mechanical characteristics of the reaction product. Alkaline activation of these aluminosilicate wastes can help in an ideal utilization of demolition wastes in disposal landfills all over the world and producing valuable sustainable materials that can be better applied in building industry.

Competing interests
The author has no competing interests.

Acknowledgements
I wish to thank all my friends in my institute who helped in doing my investigation for my research paper such as Mr. Sherief, a researcher, who worked on the investigation of the hydration materials by scanning microscopy; also, I wish to give all my thanks to Mr. Mohamed Abd El Nabi for his kind cooperation in doing the XRD analysis. I wish to give many thanks to the editor who really helped me in editing and reviewing the paper.

References

Abd El-Aziz H (2010) Physico-chemical studies of alkali acivated slag pastes. Master Theses, Zagazig University, Zagazig, Egypt

Cheng-Yi H, Feldman RF (1985) Hydration reactions in Portland cement- silica fume blend. Cem Concr Res 15:585–592

Cioffi R, Maffucci L, Santoro L (2003) Optimization of geopolymer synthesis by calcinations and polycondensation of a kaolinitic residue. Resources, Conservation and Recycling 40:27–38

Cohen MD, Bentur A (1988) Durability of Portland cement-silica fume pastes in magnesium sulfate and sodium sulfate solutions. ACI Mater J 85(3):148–157

Davidovits J (1991) Geopolymers: inorganic polymeric new materials. J Therm Anal 37:1633–1656

Davidovits J (1993) Geopolymer cements to minimize carbon-dioxide green house warming. Ceram Trans 37:165–82

Davidovits J (1999) Chemistry of geopolymeric systems terminology. In: Davidovits J, Davidovits R, James C (ed) Proceedings of the second international conference geopolymer, 1999. Institut Géopolymère, Saint-Quentin, France, pp 9–39

De Silva P, Sagoe-Crenstil K, Sirivivatnanon V (2007) Kinetics of geopolymerization: role of Al_2O_3 and SiO_2. Cem Concr Res 37(4):512–8

Diamond S (1986) The microstructures of cement paste in concrete. Proceedings of the VIII congress on cement chemistry. Rio deJaneiro, Brazil, In, pp 122–47

El-Sayed HA, Abo El-Enein SA, Khater HM, Hasanein SA (2011) Resistance of alkali activated water cooled slag geopolymer to sulfate attack. Ceramics – Silikáty 55(2):153–160

Gautefall O (1986) Effect of CSF on the diffusion of chloride through hardened cement paste. In: Proceedings of the second international conference on the use of fly ash, silica fume, slag and natural pozzolans in concrete, 1986, Madrid, vol 2. American Concrete Institute, Publication SP-91, Detroit, pp 991–998

Gleize PJP, Müller A, Roman HR (2003) Microstructural investigation of a silica fume–cement–lime mortar. Cem Concr Compos 25:171–175

Grutzeck MW, Atkinson S, Roy Della M (1983) Mechanism of Hydration of condensed silica fume in calcium hydroxide solution, fly ash, silica fume, slag and other mineral by-products in concrete. American Concrete Institute Publication SP-79, Detroit, pp 643–665

Kohno K (1989) Relative durability properties and strengths of materials containing finely grounded silica and silica fume. In: Proceedings of the third international conference on the use of fly ash, silica fume, slag and natural pozzolans in concrete, Trondheim, Norway, 1989, American Concrete Institute Publication SP-114, Detroit 2. American Concrete Institute Publication SP-114, Detroit, pp 815–826

Kakali G, Perraki T, Tsivilis S, Badogiannis E (2001) Thermal treatment of kaolin: the effect of mineralogy on the pozzolanic activity. Applied Clay Science 20:73–80
Khater HM (2012) Calcium effect on geopolymerization of alumino silicate wastes. J MaterCiv Eng 24(1):92–102
Khater HM (2010) Influence of metakaolin on resistivity of cement mortar to magnesium chloride solution. Ceramics – Silikáty J 54(4):325–333
Lee WKW, van Deventer JSJ (2002) Structural reorganisation of class F fly ash in alkaline silicate solutions. Coll Surf A 211:49–66
Maage M (1989) Efficiency factors for condensed silica fume in concrete. In: Third international conference on fly ash, silica fume, slag and natural pozzolans in concrete, Trondheim, Norway, 1989. American Concrete Institute Publication SP 114-70, Detroit
Malhotra VM, Mehta PK (1996) Pozzolanic and cementitious materials, Advances in concrete technology,vol 1. Overseas Publishers Association, Ottawa
Panias D, Giannopoulou IP, Perraki T (2007) Effect of synthesis parameters on the mechanical properties of fly ash-based geopolymers. Colloids and surfaces A: physicochem Eng Aspects 301:246–254
Phair JW, van Deventer JSJ (2002) Characterisation of fly-ash based geopolymeric binders activated with sodium aluminate. Ind Eng Chem Res 41:4242–4251
Richardson IG, Cabrera JG (2000) The nature of C–S–H in model slag cements. Cem Concr Compos 22:259–266
Saikia N, Usami A, Kato S, Kojima T (2004) Hydration behavior of ecocement in presence of metakolin. Resource Progressing Journal 51(1):35–41
Sellevold EJ, Nilsen T (1987) Condensed silica fume in concrete: a world review. In: Malhotra VM (ed) Supplementary Cementing Materials for Concrete. CANMET, Ottawa, pp 165–243
Taylor HFW (1964) The chemistry of cements, 1st edition. Academic Press, London
Taylor HFW (1992) Cement chemistry, 2nd edition. Academic Press, London
Temuujin J, Van Riessen A, Williams R (2009) Influence of calcium compounds on the mechanical properties of fly ash geopolymer pastes. Journal of Hazardous Materials Journal of Hazardous Materials 167(1–3):82–88
van Jaarsveld JGS, van Deventer JGJ (1999a) The effect of metal contaminants on the formation and properties of waste-based geopolymers. Cem Concr Res 29(8):1189–1200
van Jaarsveld JGS, van Deventer JSJ (1999b) Effect of the alkali metal activator on the properties of fly ash-based geopolymers. Ind Eng Chem Res 38:3932–3941
Wenying G, Guolin W, Jianda W, Ziyun W, Suhong Y (2008) Preparation and Performance of Geopolymers. Journal of Wuhan University of Technology-Mater Sci Ed 23(3):285–430
Yip CK, Lukey GC, van Deventer JSJ (2005) The coexistence of geopolymeric gel and calcium silicate hydrate at the early stage of alkaline activation. Cem Concr Res 35:1688–1697

The discussion on optimization models of pure bending beam

Heting Qiao[1*] and Haiyan Li[2]

Abstract

This article is devoted to shape optimization design of pure bending beams under single loading condition. Compliance minimization with material volume constraint, the maximum stress minimization problem, and the maximum displacement are considered. In the case of trusses, it has been shown that the former two problems have the same optimal topology. The possibility of extending this result for pure bending beam problems is examined in the present work. First, the comparison of the optimum design results between the maximum displacement, the conventional mean compliance, and the maximum stress is carried out by an example of optimal cross-sectional design of a continuous beam. Then, geometric average displacement (GAD) is introduced in optimization models of linearly elastic structures. The elevated accuracy in results achieved with GAD is shown in this article.

Keywords: Compliance, Maximum stress, Maximum displacement, Geometric average displacement (GAD)

Introduction

Structural optimization is one of the most challenging research topics in the field of computational mechanics. It has received more and more attention recently because of its great potential application in many industrial areas. Its importance lies in the fact that the appropriate result of structural design is generally the most decisive factor that influences the product efficiency.

Structural optimization, especially topology optimization, is being used increasingly in aerospace vehicles, maritime carriers, wind turbine blades, and various mechanical equipment where high strength, high stiffness, and low weight are important. In such applications, the problem of selecting a suitable optimization model has been investigated for a long time. In the vast literature on structural optimization model, arguably the two most studied problems are stress-constrained weight minimization and material volume-constrained compliance minimization. Indeed, Cox (1965) succeeded to prove that results attained via compliance minimization model would be equivalent to Michell's truss (stress-constrained weight minimization result) in 1965.It has been shown a long time ago that these seemingly different problems possess equal optimal topologies (result), when the truss is subject to a single loading condition and the allowable stresses in tension and compression are equal (Dorn et al. 1964; Hemp 1973). This result has been extended by Achtziger (1996) for cases where the allowable stresses in tension and compression are not equal.

Despite these successes, most of the topology optimization problems were modeled to minimize the compliance of the structures, following the methods adopted by Bendsøe and Kikuchi. The optimization problems of minimum compliance have been widely studied in the relevant literature (Xie et al. 2012; Bendsøe 1989; Eschenauer and Olhoff 2001; Sethian and Wiegmann 2000; Xie and Steven 1993; Gerzen and Barthold 2012; Gain and Paulino 2012; Lee et al. 2012; Bruggi and Duysinx 2012).

Given all that, in most static structure design examples, the ultimate goal is to find the structures with maximum stiffness, or with minimum weight under stress constraint. Most of the current designs were modeled by minimum compliance and achieved the desired results by solving the minimum compliance problems. Although many good results have been

* Correspondence: qiaoheting311@163.com
[1]School of Mechanical Engineering, Shenyang University of Technology, Shenyang 110004, People's Republic of China
Full list of author information is available at the end of the article

attained in this way, stiffness is more accurately characterized by the maximum displacement of a structure under load. Moreover, Mela and Koski (2012) have suggested that the stress-constrained minimum weight problem and the compliance minimization problem do not have equal optimal topologies of truss under multiple loading conditions.

The main purpose of the present paper is to show that the maximum stress minimization problem and the compliance minimization problem have equal optimal results of pure bending beam under a single loading condition. Compared to them, the maximum displacement minimization problem does not have equal optimal results, and it is necessary to find the appropriate index as approximation of the maximum displacement.

The paper is organized as follows: the optimal design results attained via the different models is presented and discussed in detail in the 'Methods' section. The main result of the paper is presented in the 'Results and discussion' section, where a cantilever beam is minimized separately for the maximum displacement, maximum stress, compliance, and so on. To illustrate the application of the proposed index as appropriate approximation of the maximum displacement in the static structure optimization, a power index parameter is solved in the 'Results and discussion' section. Finally, the results are summarized with discussion in the 'Conclusions' section.

Methods

Cross-sectional design of the cantilever beam

Problem description

In order to present the differences in optimal design results attained via the maximum displacement, maximum stress, and conventional mean compliance methods, a cantilever beam was studied with varying square cross sections, with ends $x = 0$ and $x = L$, where x is the abscissa measured along the beam axis and subjected to a distributed constant line load q, as shown in Figure 1a. The equation of bending moment and bending moment diagram are given in Figure 1b. The objective of this example is to obtain the excellent mechanical performance of the cantilever beam by changing the cross-sectional areas along the x-axis.

The exact differential equation of the deflection (displacement of the y-axis) curve can be described as

$$EI(x)\frac{\partial^2 v}{\partial x^2} = M(x) = -\frac{1}{2}qx^2. \tag{1}$$

Here, v is the displacement of any point along the x-axis, q is the load line density, and E is Young's modulus of the material. Here, moments of inertia $I(x)$ can also be stated as

$$I(x) = \frac{A^2(x)}{12}. \tag{2}$$

Here, $A(x)$ is the continuously differentiable function of the area of the square cross section along the x-axis.

Then, the mechanical optimization problem can be formulated as

$$\begin{array}{ll} \text{find}: & A(x) \\ \min: & f(A) \\ \text{s.t.}: & \int_0^L A(x)dx = W_0. \end{array} \tag{3}$$

Here, $f(A)$ denotes a mechanical behavior index, and the material volume is limited by W_0. With the Lagrange multiplier method, the solution of the cross-sectional areas of cantilever beam is attained by

$$\delta_A(f(A)) + \lambda \int_0^L \delta A dx = 0, \quad \delta\lambda\left(\int_0^L A(x)dx - W_0\right) = 0. \tag{4}$$

Here, δA is the variation function of $A(x)$, and $\delta_A(f(A))$denotes the Fréchet derivative of $f(A)$ with respect to $A(x)$.The constrained optimization problem (3) is transformed into an equivalent equation (4) by making use of the Karush-Kuhn-Tucker conditions of the constrained optimization. The optimal cross-sectional area fields $A(x)$ in problem (3) can be obtained by solving Equation 4.

Minimization of the maximum displacement

In many practical cases, a commonly used design criterion is that the maximum displacement of the structure should not exceed a specified value. Thus, maximum displacement is naturally the ideal objective function of optimization models. According to the elementary theory of the beam, the maximum displacement is located in the boundary of $x = 0$ and can be written as

$$v_{\max}(A) = \int_0^L \frac{\bar{M}(x)M(x)}{EI}dx = \int_0^L \frac{6qx^3}{EA^2}dx, \tag{5}$$

where $\bar{M}(x)$ is the bending moment produced by a unit load applied on the free end of the beam ($x = 0$) and is of the form

$$\bar{M}(x) = -x. \tag{6}$$

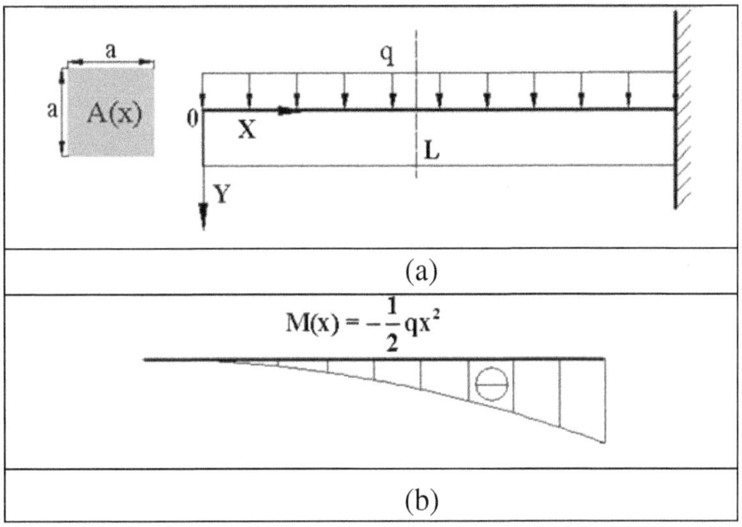

Figure 1 A Cantilever beam with varying sections. (**a**) A cantilever beam under uniformly distributed load. (**b**) Equation of bending moment.

Substituting Equations 5 and 6 into Equation 4 gives

$$\int_0^L \left(-\frac{12qx^3}{EA^3}+\lambda\right)\delta A dx = 0 \tag{7}$$

and

$$\int_0^L A dx - W_0 = 0. \tag{8}$$

Since δA is an arbitrary function, $A(x)$ can be written as

$$A = \left(\frac{12q}{\lambda E}\right)^{1/3} x. \tag{9}$$

Here, λ is an unknown variable. Substituting Equation 9 into Equation 8, the optimal cross-sectional area fields $A_{v_{\max}}(x)$. and λ can be expressed as

$$\begin{aligned} \lambda &= \frac{8qL^6}{2EM^3} \\ A_{v_{\max}}(x) &= \frac{2W_0}{L^2}x \end{aligned} \tag{10}$$

and the corresponding displacement function is

$$v_{v_{\max}} = \frac{3qL^4}{4EW_0^2}(x-L)^2. \tag{11}$$

The sectional maximum stress is found at the location on the cross section where y is the largest and can be written as

$$\sigma_{v_{\max}} = \frac{My_{\max}}{I} = \frac{6MA^{\frac{1}{2}}}{A^2} = \sqrt{\frac{9}{8}\frac{qL^{\frac{7}{2}}}{W_0^{\frac{3}{2}}}}\left(\frac{x}{L}\right)^{\frac{1}{2}} \tag{12}$$

Introducing a dimensionless parameter

$$\widetilde{x} = \frac{x}{L}, \tag{13}$$

the cross-sectional area fields and the corresponding displacement function can be expressed in the dimensionless space as

$$\widetilde{A}_{v_{\max}}(\widetilde{x}) = \frac{A_{v_{\max}}(\widetilde{x})L}{W_0} = 2\widetilde{x}, \tag{14}$$

$$\widetilde{v}_{v_{\max}}(\widetilde{x}) = \frac{v_{v_{\max}}(\widetilde{x})EW_0^2}{qL^6} = \frac{3}{4}(\widetilde{x}-1)^2, \tag{15}$$

and

$$\widetilde{\sigma}_{v_{\max}} = \frac{\sigma_{v_{\max}}(\widetilde{x})W_0^{\frac{3}{2}}}{qL^{\frac{7}{2}}} = \sqrt{\frac{9}{8}}\widetilde{x}^{-\frac{1}{2}}. \tag{16}$$

Here, the subscript $v_{\max}$ shows that the optimization objective is to minimize the maximum displacement.

Minimization of the maximum stress

In many practical cases, a commonly used design criterion is that the maximum stress of the structure should not exceed a specified value (strength criterion). Thus, maximum stress is naturally the ideal objective function of optimization models. However, the location of the maximum stress usually will not be fixed with the change of material distribution in the optimization process. Therefore, the maximum stress is an implicit function with respect to material distribution. Hence, to resolve this problem, a performance index called geometric average sectional maximum stress instead of a direct optimization of the maximum stress is proposed, which is expressed as

$$f_{\sigma_{\max}} = \lim_{n\to\infty}\left(\frac{1}{L}\int_0^L (\sigma_{\max})^n dx\right)^{1/n} = \lim_{n\to\infty}\left(\frac{1}{L}\int_0^L \left(\frac{6M}{A^{\frac{3}{2}}}\right)^n dx\right)^{1/n}. \tag{17}$$

Theoretically, geometric average displacement (GAD) tends to geometric average stress when n tends to infinity (Gentile 2003; Li and Fang 1997).

Substituting Equation 17 into Equation 4 gives

$$\lim_{n\to\infty}\int_0^L \left(-Z\frac{x^{2n}}{A^{\frac{3}{2}n+1}}+\lambda\right)\delta A dx = 0, \tag{18}$$

where

$$Z = \frac{3(3q)^n}{2L}\left(\frac{1}{L}\int_0^L \left(\frac{3qx^2}{A^{\frac{3}{2}}}\right) n dx\right)^{1/n-1}. \tag{19}$$

Since δA is an arbitrary function, $A(x)$ can be written as

$$A = \lim_{n\to\infty}\left(\frac{Z}{\lambda}\right)^{\frac{2}{3n+2}} x^{\frac{4n}{3n+2}} = Hx^{\frac{4}{3}}, \tag{20}$$

where H is an unknown variable:

$$H = \lim_{n\to\infty}\left(\frac{Z}{\lambda}\right)^{\frac{2}{3n+2}}. \tag{21}$$

Substituting Equation 20 into Equation 8 gives

$$H = \frac{7W_0}{3L^{7/3}} \qquad A_{\sigma_{\max}}(x) = \frac{7W_0}{3L^{7/3}}x^{4/3}. \tag{22}$$

The corresponding displacement function and the sectional maximum stress function are

$$v_{\sigma_{\max}} = \frac{81qL^{14/3}}{98EW_0^2}\left(3x^{4/3} - 4L^{1/3}x + L^{1/3}\right) \qquad \sigma_{\sigma_{\max}} = \sqrt{\frac{243}{343}}\frac{qL^{\frac{7}{2}}}{W_0^{\frac{3}{2}}}. \tag{23}$$

As shown in Equation 22, we can obtain an iso-stress (the sectional maximum stress function is constant) design via the maximum stress-based model. The dimensionless cross-sectional area fields, displacement function, and the sectional maximum stress function are

$$\widetilde{A}_{\sigma_{\max}}(\widetilde{x}) = \frac{A_{\sigma_{\max}}(\widetilde{x})L}{W_0} = \frac{7}{3}\widetilde{x}^{4/3} \qquad \widetilde{v}_{\sigma_{\max}}(\widetilde{x}) = \frac{81}{98}\left(3\widetilde{x}^{4/3} - 4\widetilde{x} + 1\right) \qquad \widetilde{\sigma}_{\sigma_{\max}} = \sqrt{\frac{243}{343}}. \tag{24}$$

Here, the subscript $\sigma_{\max}$ shows that the optimization objective is to minimize the maximum stress.

Minimization of the compliance

For the cantilever beam in Figure 1, the compliance can also be expressed as

$$f(A) = C = \int_0^L \frac{M(x)M(x)}{2EI}dx = \int_0^L \frac{3q^2x^4}{2EA^2}dx. \tag{25}$$

Here, the compliance is used as an optimization objective function. The optimal cross-sectional area fields $A(x)$ should obey the following necessary conditions:

$$\int_0^L\left(-\frac{3q^2x^4}{EA^3}+\lambda\right)\delta A dx = 0, \qquad \int_0^L A dx - W_0 = 0. \tag{26}$$

The cross-sectional area fields can be obtained by solving Equation 26, which is

$$A_C(x) = \frac{7W_0}{3L^{7/3}}x^{4/3} \tag{27}$$

Compared with Equation 22, the compliance minimization problem and the maximum stress minimization problem have equal optimal results in this example. To amplify this highly simplified conclusion, the contrast between the compliance minimization problem and the maximum stress minimizaion problem can be formulated as

$$\begin{cases} \lim\limits_{n\to\infty} \int_0^L \left(-Z \dfrac{M^n}{(A_{\sigma_{\max}})^{\frac{3n+2}{2}}} + \lambda_{\sigma_{\max}} \right) \delta A dx \quad Z = \dfrac{3(6)^n}{2L} \left(\dfrac{1}{L} \int_0^L \left(\dfrac{6M}{A_{\sigma_{\max}}^{\frac{3}{2}}} \right)^n dx \right)^{1/n - 1} & \int_0^L A_{\sigma_{\max}} dx - W_0 = 0 \\ \int_0^L \left(-\dfrac{12M^2}{EA_C^3} + \lambda_C \right) \delta A dx = 0, & \int_0^L A_C dx - W_0 = 0 \end{cases} \tag{28}$$

then

$$\begin{cases} A_{\sigma_{\max}} = \lim\limits_{n\to\infty} \left(\dfrac{ZM^n}{\lambda_{\sigma_{\max}}} \right)^{\frac{2}{3n+2}} = H_{\sigma_{\max}} M^{\frac{2}{3}} & H_{\sigma_{\max}} = \lim\limits_{n\to\infty} \left(\dfrac{Z}{\lambda_{\sigma_{\max}}} \right)^{\frac{2}{3n+2}} \\ A_C = \left(\dfrac{12M^2}{E\lambda_C} \right)^{\frac{1}{3}} = H_C M^{\frac{2}{3}} & H_C = \left(\dfrac{12}{E\lambda_C} \right)^{\frac{1}{3}} \end{cases} . \tag{29}$$

Substituting Equation 29 into Equation 4 gives

$$\begin{cases} H_{\sigma_{\max}} = H_C \\ A_{\sigma_{\max}} = A_C \end{cases} . \tag{30}$$

Therefore, the first example helps us draw the conclusion that in this kind of problem (pure bending beam), the optimal design generated from the compliance formulation and the maximum are identical, same to the truss problems (Cox 1965; Dorn et al. 1964; Hemp 1973).

Comparisons of the results

Comparing these results, it shows that the optimization models with compliance (maximum stress) and the maximum displacement as objective function sometimes does not give the same optimal results in this kind of problem, and we find significant differences in the cross-sectional area fields, displacement, and the sectional maximum stress function. To be more specific, from the results, the maximum displacement of the optimal design generated from maximum displacement decreases by extra 10% in comparison with the design obtained by conventional compliance (maximum stress). Furthermore, the maximum stress of the optimal design generated from maximum displacement with the extra increase of 26% is presented to demonstrate the validity of this example, as shown in Figure 2.

Results and discussion

The optimization model of the geometric average displacement

In many practical cases, a commonly used design criterion is the maximum displacement of the structure which does not exceed a specified value (stiffness criterion). Thus, maximum displacement is naturally the ideal objective function of optimization models. However, the location of the maximum displacement usually changes with the change of material distribution in the optimization process, resulting in a discontinuous maximum displacement function, especially for topology optimization. Hence, to achieve a good balance between the optimization performance and numerical cost, a performance index called GADU_{GAD}, instead of a direct optimization of the maximum displacement, is proposed, which is expressed as

$$U_{GAD} = \left(\frac{1}{|\Omega|} \int_\Omega (u(x))^n d\Omega \right)^{1/n}, \quad x \in \Omega. \tag{31}$$

Here, $|\Omega|$ denotes the area (or volume) of the design region, and the displacement of a general point can be described in terms of $u(x)$. Theoretically, GAD tends to the maximum displacement when n tends to infinity (Li and Fang 1997), i.e., $U_{GAD} \xrightarrow{n\to\infty} \max(u(x)),\ x \in \Omega$. When n is big enough, GAD is an appropriate approximation of the maximum displacement.

In order to present the validity of GAD, a cantilever beam was studied again with varying square cross sections and subjected to a distributed linear load, as shown in Figure 3a. The equation of bending moment and bending moment diagram are given in Figure 3b.

First, the conventional compliance formulation is applied under a given weight constraint. Then, the same problem in Figure 3 is solved using the minimization of the maximum displacement. Finally, to discuss the

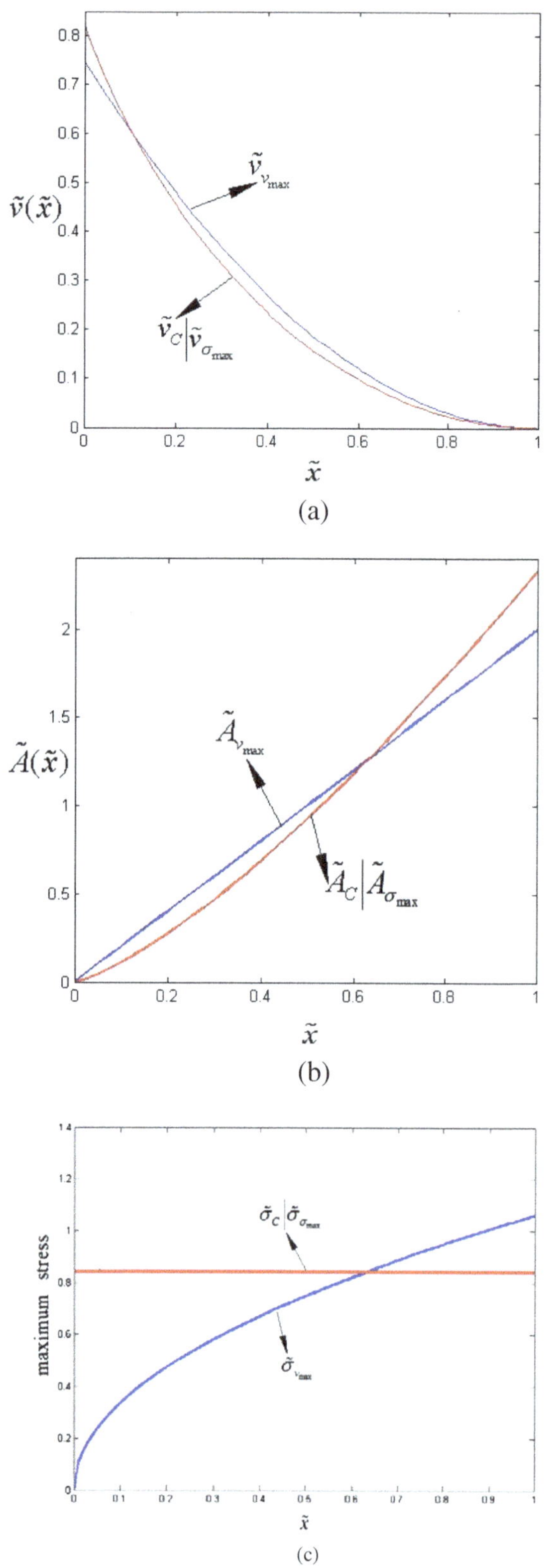

Figure 2 Comparisons of design results of objectives. (**a**) Comparisons of displacement function. (**b**) Comparisons of cross-sectional area fields. (**c**) Comparisons of the sectional maximum stress function.

influence of the objective functions, a GAD-based optimization model is carried out to find the maximum stiffness design with the different power indices n (1,2,3,4).

For the cantilever beam in Figure 3, the compliance can also be expressed as

$$f_c(A) = C = \int_0^L \frac{M(x)M(x)}{2EI} dx = \int_0^L \frac{Q^2x^6}{6EL^2A^2} dx. \tag{32}$$

The cross-sectional area fields can be obtained by substituting Equation 32 into Equation 4, which is

$$A_C(x) = \frac{3W_0}{L^3} x^2. \tag{33}$$

The corresponding displacement function and the sectional maximum stress function are

$$\begin{aligned} v_C &= \frac{2QL^5}{9EW_0^2}\left(x \ln^x + L - x - x \ln^L\right) \\ \sigma_C &= \sqrt{\frac{1}{27}} \frac{QL^{\frac{7}{2}}}{W_0^{\frac{3}{2}}}. \end{aligned} \tag{34}$$

We also obtain an iso-stress design via compliance-based model as the former example. The dimensionless cross-sectional area fields, displacement function, and the sectional maximum stress function of compliance are

$$\begin{aligned} \tilde{A}_C(\tilde{x}) &= 3\tilde{x}^2 \\ \tilde{v}_{\sigma_{max}}(\tilde{x}) &= \begin{cases} \frac{2}{9}\left(\tilde{x} \ln^{\tilde{x}} - \tilde{x} + 1\right) & \tilde{x} \neq 1 \\ \frac{2}{9} & \tilde{x} = 1 \end{cases} \\ \tilde{\sigma}_{\sigma_{max}} &= \sqrt{\frac{1}{27}}. \end{aligned} \tag{35}$$

According to Equations 28 to 30, the compliance minimization problem and the maximum stress minimization problem will have equal optimal results in this example. Wherefore, it is useless to solve this problem again via the maximum stress-based model.

For the cantilever beam in Figure 3, the maximum displacement can also be expressed as

$$f_{v_{max}}(A) = v(x)|_{x=L} = \int_0^L \frac{\bar{M}(x)M(x)}{EI} dx = \int_0^L \frac{2Qx^4}{ELA^2} dx. \tag{36}$$

The cross-sectional area fields can be obtained by substituting Equation 32 into Equation 4, which is

$$A_{v_{max}}(x) = \frac{7W_0}{3L^{\frac{7}{3}}} x^{\frac{4}{3}}. \tag{37}$$

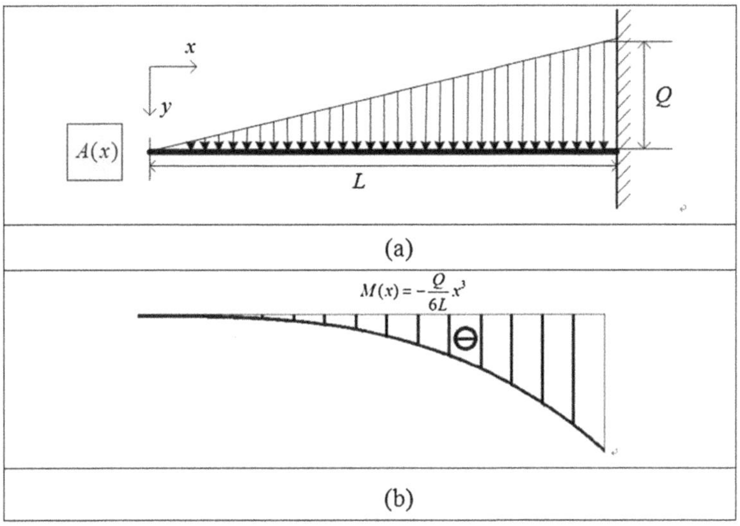

Figure 3 Cantilever beam with varying sections subjected to a distributed linear load. (**a**) A cantilever beam under uniformly distributed load. (**b**) Equation of bending moment.

The corresponding displacement function and the sectional maximum stress function are

$$v_{v_{\max}} = \frac{27QL^{\frac{11}{3}}}{686EW_0^2}\left(3x^{\frac{7}{3}} - 7L^{\frac{4}{3}}x + 4L^{\frac{7}{3}}\right)$$
$$\sigma_{v_{\max}} = \sqrt{\frac{27}{243}\frac{QL^{\frac{7}{2}}}{W_0^{\frac{3}{2}}}}x. \tag{38}$$

The dimensionless cross-sectional area fields, displacement function, and the sectional maximum stress function of compliance are

$$\widetilde{A}_{v_{\max}}(\widetilde{x}) = \frac{7}{3}\widetilde{x}^{\frac{4}{3}}$$
$$\widetilde{v}_{v_{\max}}(\widetilde{x}) = \frac{27}{686}\left(3\widetilde{x}^{\frac{7}{3}} - 7\widetilde{x} + 4\right) \tag{39}$$
$$\widetilde{\sigma}_{v_{\max}} = \sqrt{\frac{27}{243}}\widetilde{x}.$$

In order to illustrate that GAD is an appropriate approximation of the maximum displacement, the problem is modeled again to minimize GAD as follows:

$$f_{\mathrm{GAD}}^{n} = \left(\frac{1}{L}\int_0^L v^n(x)dx\right)^{\frac{1}{n}}. \tag{40}$$

Based on the optimal result generated by the maximum displacement, we can assume the optimal cross-sectional area fields obtained by GAD as follows:

$$A_{\mathrm{GAD}}^{n} = H(m)x^{\frac{m}{2}}, \tag{41}$$

where the superscript n denotes a power index parameter in Equation 40, and m is the evaluated variable. Substituting Equation 41 into Equation 4 gives

$$H(m) = \frac{(m+2)W_0}{2L^{\frac{m+2}{2}}}, \tag{42}$$

Table 1 Results obtained by geometric average displacement for varying *n*

n	m	$\widetilde{A}_{\mathrm{GAD}}$ (cross-sectional area)	$\widetilde{v}_{\mathrm{GAD}}$ (displacement function)	$\widetilde{\sigma}_{\mathrm{GAD}}$ (sectional maximum stress function)
1	3.33	$\widetilde{A}_{\mathrm{GAD}} = 4.33\widetilde{x}^{3.33}$	$\widetilde{v}_{\mathrm{GAD}} = 0.25\left(\widetilde{x}^{1.67} - 1.67\widetilde{x} + 0.67\right)$	$\widetilde{\sigma}_{\mathrm{GAD}} = 0.23\widetilde{x}^{0.5}$
2	3.09	$\widetilde{A}_{\mathrm{GAD}} = 4.09\widetilde{x}^{3.09}$	$\widetilde{v}_{\mathrm{GAD}} = 0.18\left(\widetilde{x}^{1.91} - 1.91\widetilde{x} + 0.91\right)$	$\widetilde{\sigma}_{\mathrm{GAD}} = 0.25\widetilde{x}^{0.68}$
3	2.97	$\widetilde{A}_{\mathrm{GAD}} = 3.97\widetilde{x}^{2.97}$	$\widetilde{v}_{\mathrm{GAD}} = 0.15\left(\widetilde{x}^{2.03} - 2.03\widetilde{x} + 1.03\right)$	$\widetilde{\sigma}_{\mathrm{GAD}} = 0.26\widetilde{x}^{0.77}$
4	2.74	$\widetilde{A}_{\mathrm{GAD}} = 3.74\widetilde{x}^{2.74}$	$\widetilde{v}_{\mathrm{GAD}} = 0.13\left(\widetilde{x}^{2.26} - 2.26\widetilde{x} + 1.26\right)$	$\widetilde{\sigma}_{\mathrm{GAD}} = 0.27\widetilde{x}^{0.95}$
$\infty(v_{\max})$	2.67	$\widetilde{A}_{v_{\max}} = 3.67\widetilde{x}^{2.67}$	$\widetilde{v}_{\mathrm{GAD}} = 0.12\left(\widetilde{x}^{2.33} - 2.33\widetilde{x} + 1.33\right)$	$\widetilde{\sigma}_{v_{\max}} = 0.33\widetilde{x}$

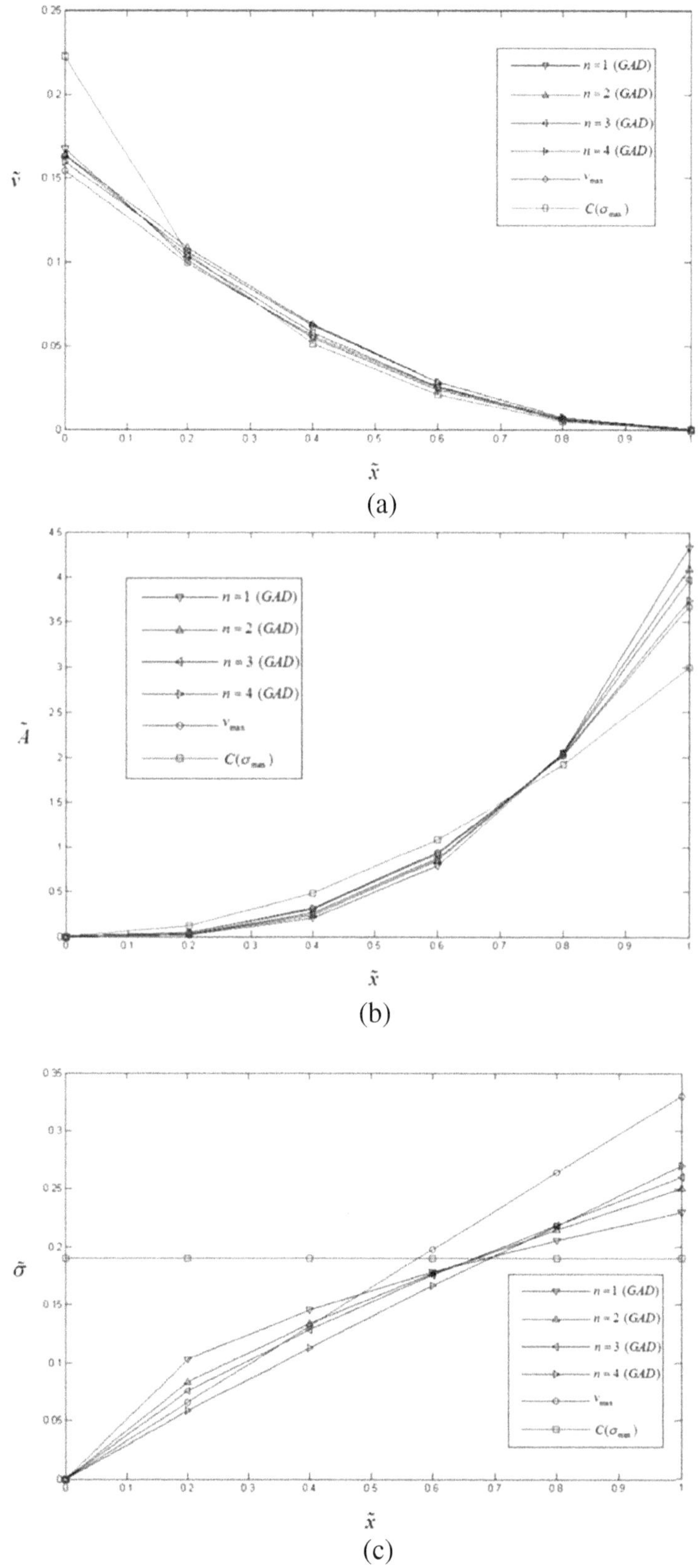

Figure 4 Comparisons of design results of objectives. (**a**) Comparisons of displacement function. (**b**) Comparisons of cross-sectional area fields. (**c**) Comparisons of the sectional maximum stress function.

and the corresponding displacement function is

$$v_{\mathrm{GAD}}=\begin{cases}\dfrac{8QL^{m+1}}{EW_0^2(4-m)(5-m)(m+2)^2}\left(x^{5-m}-(5-m)L^{4-m}x+(4-m)L^{5-m}\right) & m<4\\ \left.\begin{array}{ll}\dfrac{2QL^5}{9EW_0^2}\left(x\ln^x+L-x-x\ln^L\right) & x\neq 0\\ \dfrac{2QL^6}{9EW_0^2} & x\neq 0\end{array}\right\} & m=4\end{cases}. \tag{43}$$

The geometric average displacement can be rewritten as

$$f_{\mathrm{GAD}}^n=f_{\mathrm{GAD}}^n(n,m)=\begin{cases}\dfrac{8QL^6}{EW_0^2(4-m)(5-m)(m+2)^2}\left(\dfrac{1}{L}\displaystyle\int_0^1\left(\tilde{x}^{5-m}-(5-m)\tilde{x}+(4-m)\right)^n d\tilde{x}\right)^{\frac{1}{n}} & m<4\\ \dfrac{2QL^6}{9EW_0^2}\left(\dfrac{1}{L}\displaystyle\int_0^1\left(\tilde{x}\ln^{\tilde{x}}-\tilde{x}+1\right)^n d\tilde{x}\right)^{\frac{1}{n}} & m=4\end{cases}, \tag{44}$$

and the evaluated variable m in the optimal cross-sectional area fields can be solved by

$$\frac{\partial f_{\mathrm{GAD}}^n(n,m)}{\partial m}=0. \tag{45}$$

To facilitate the comparisons, the solutions of the optimization models with the compliance (the maximum stress), the maximum displacement, and GAD as objective functions are shown in Table 1 and Figure 4. It can be seen that the compliance design experiences large displacement under the applied force, whereas the compliance and GAD design have only very slight displacement that implies a much stiffer design. To be more specific, the maximum displacement of the optimal design generated from GAD decreases by an extra 32% in comparison with the design obtained by compliance. Furthermore, the maximum stress of the optimal design generated from GAD, and the compliance with the extra increase of 21% is presented to demonstrate the validity of this example. With the increase in power index n, the material distribution and the displacement field obtained by the GAD-based model rapidly move close to the convergence of results obtained by the maximum displacement. Since the approximate level tends to stability with the increasing power index n, an appropriate value n is required to be selected in the practical optimization process.

Conclusions

The classic test problems indicate that for the pure bending beam under single loading condition, the maximum stress minimization problem and the compliance minimization problem have equal optimal results, and the maximum displacement minimization problem and the compliance minimization problem do not have equal optimal results. This anticipated result has so far been without the proof that the test problems provide. The implication of the conclusion is that the designer can rely on finding the stress-constrained minimum weight solution by performing optimization for the compliance minimization problem, and it is necessary to propose an appropriate index as approximation of the maximum displacement for the complex problems. Through a classic example, it was shown that the solutions achieved via the model utilizing GAD rapidly move close to the convergence of results obtained by the maximum displacement.

Competing interests

The authors declare that they have no competing interests.

Authors' contributions

HQ carried out the shape optimization of pure bending beams, participated in the sequence optimization models discussion. HL conceived of the study and participated in its design examples. Both authors read and approved the final manuscript.

Acknowledgments

This research is supported by the National Basic Research Program (973 Program) of China (no. 2011CB610304), National Natural Science Foundation of China (nos 11172052 and 90816025), and the Research Fund for the Doctoral Program of Higher Education of China (grant no. 20090041110023). The financial support is gratefully acknowledged.

Author details

[1]School of Mechanical Engineering, Shenyang University of Technology, Shenyang 110004, People's Republic of China. [2]Computing Center, Northeastern University, Shenyang 110004, People's Republic of China.

References

Achtziger W (1996) Truss topology optimization including bar properties different for tension and compression. Struct Optim 12:63–74

Bendsøe MP (1989) Optimal shape design as a material distribution problem. Struct Multidiscip Optim 1(4):193–202

Bruggi M, Duysinx P (2012) Topology optimization for minimum weight with compliance and stress constraints. Struct Multidiscip Optim 46(3):369–384

Cox HL (1965) The design of structures of least weight. Pergamon, Oxford

Dorn W, Gomory R, Greenberg M (1964) Automatic design of optimal structures. Journal de mecanique 3:25–52

Eschenauer HA, Olhoff N (2001) Topology optimization of continuum structures: a review. Appl Mech Rev 54(4):331–390

Gain AL, Paulino GH (2012) Phase-field based topology optimization with polygonal elements: a finite volume approach for the evolution equation. Struct Multidiscip Optim 46:327–342

Gentile C (2003) The robustness of the p-norm algorithms. Mach Learn 53(3):265–299

Gerzen N, Barthold FJ (2012) Enhanced analysis of design sensitivities in topology optimization. Struct Multidiscip Optim 46(4):585–595

Hemp W (1973) Optimum structures. Clarendon Press, Oxford

Lee E, James KA, Martins JR (2012) Stress-constrained topology optimization with design-dependent loading. Struct Multidiscip Optim 46(5):647–661

Li XS, Fang SC (1997) On the entropic regularization method for solving min-max problems with applications. Math Meth Oper Res 46(1):119–130

Mela K, Koski J (2012) On the equivalence of minimum compliance and stress-constrained minimum weight design of trusses under multiple loading conditions. Struct Multidiscip Optim 46(5):679–691

Sethian JA, Wiegmann A (2000) Structural boundary design via level set and immersed interface methods. J Comput Phys1 63(2):489–528

Xie YM, Steven GP (1993) A simple evolutionary procedure for structural optimization. Comput Struct 49(5):885–896

Xie YM, Zuo ZH, Huang XD, Rong JH (2012) Convergence of topological patterns of optimal periodic structures under multiple scales. Struct Multidiscip Optim 46(1):41–54

9

Evaluation of Egyptian code provisions for seismic design of moment-resisting-frame multi-story buildings

Shehata E Abdel Raheem[1,2]

Abstract

Since the occurrence of the Cairo earthquake on October 1992, the design of structures for earthquakes became a major demand enforced in the Egyptian design codes. The seismic response of building structures can be estimated through utilization of a variety of analysis methods that range from simple equivalent static analysis to complex nonlinear dynamic analysis. The traditional approach is to employ equivalent static analysis methods, while current design practice is moving toward an increased emphasis on the nonlinear analysis method. The Egyptian code provisions for building seismic design adopt the traditional approach of equivalent static load method as the main method for evaluating seismic actions and recommend the response spectrum method for nonsymmetrical buildings. This study aims to evaluate the Egyptian code provisions for the seismic design of moment-resistant frame multi-story building through using nonlinear time history analysis. The analysis procedures are evaluated for their ability to predict deformation demands in terms of inter-story drifts, potential failure mechanisms and story shear force demands. The results of the analysis of the different approaches are used to evaluate the advantages, limitations, and ease of application of each approach for seismic analysis.

Keywords: Seismic design; Egyptian building code; Time history analysis; Building lateral stiffness; Diaphragm flexibility; Dynamic analysis

Introduction

Occurrences of recent earthquakes in different parts of the world and the resulting losses, especially human lives, have highlighted the structural inadequacy of buildings to carry seismic loads. The great losses due to the Cairo earthquake on October 1992 (Ms 5.4) were mainly related to the fact that at the time of construction, the buildings were designed to resist only vertical loads and had insufficient lateral resistance. Thus, the columns and beam column connections were found to have inadequate shear capacity, ductility, and confinement in plastic hinges. This earthquake illustrated the vulnerability of the building stock, especially older structures, due to design, detailing, construction, and maintenance issues (Khater 1992; Badawi and Mourad 1994; Mourad et al. 2000). So there is an urgent need for assessment of existing buildings in terms of seismic performance and to continuously upgrade the seismic codes for the design of new buildings. The design of structures for earthquakes became a major demand enforced in the Egyptian design codes that motivated the Ministry of Housing and Buildings to update the Egyptian codes regularly, taking into account the seismic loads. Since October 1992, a set of Egyptian codes have been released to prevent building collapse and/or control major damages of structural elements. Many advances in earthquake engineering have been made from the observation of the performance of real structures that have been subject to a severe earthquake. Analytical modeling, including finite element analysis, has an important role, but its limitations must be recognized. For many engineered structures, satisfactory seismic performance requires careful attention to analysis, design, and detailing and good construction practice. Safety is thus achieved by the successful integration of analysis, design, and construction.

Correspondence: shehataraheem@yahoo.com
[1]Taibah University, Madinah Munawarh, P.O. 344, Kingdom of Saudi Arabia
[2]Civil Engineering Department, Faculty of Engineering, Assiut University, Assiut 71516, Egypt

Building code restrictive seismic design provisions and building system type and configuration have remarkable implications on the seismic performance of reinforced concrete moment-framed structures. The specifications permit the designer to utilize a variety of methods for seismic analysis that range from simple equivalent static analysis to complex nonlinear dynamic analysis (UBC 1997; AIJ 1999; SEAOC 1999; ICC 2003; ECS 2004; ASCE 2005; ECP 2007, 2008). For building structures, it is common practice to utilize a simplified approach, such as equivalent static load. This approach has several shortcomings, which have been accepted due to its simplicity and a lack of alternative practical approaches. Such approach may be regarded as force based since the method's primary emphasis is on the forces within the structure. In recent years, there has been a shift of attention away from linear methods of seismic analyses to nonlinear methods which put emphasis on the displacements within the structure. Thus, nonlinear analysis methods that are capable of realistically predicting the deformations imposed by earthquakes on structures are needed. In response to this need, nonlinear static analysis procedures have appeared in national resource documents such as the ATC-40 report on seismic evaluation and retrofit of concrete buildings (ATC 1996) and the FEMA-356 pre-standard on seismic rehabilitation of buildings (FEMA 2000). Such analysis methods are useful for predicting inelastic displacement capacities while simultaneously offering a compromise between the oversimplification of linear static analysis and the inherent complexity of nonlinear dynamic analysis. The latest Egyptian code for load and forces (ECP 2008) and most of the international participating building codes depend on the traditional approach of equivalent static load method as the main method for evaluating seismic actions on symmetrical buildings (UBC 1997; AIJ 1999; ECS 2004; ECP 2008). For nonsymmetrical buildings, ECP (2008) recommended the response spectrum method to be used for building seismic analysis and design, which was considered to be a more accurate method of analysis than the equivalent static load method (Chopra 1995; Paz and Leigh 2003).

The structural response is a function of building mass, stiffness, and material damping (Ghosh and Fanella 2003; Abdel Raheem et al. 2010); however, ECP(2008) gives an empirical expression to calculate the equivalent static load of seismic action depending on the total building weight only, neglecting the effect of building stiffness and material damping on the structural response. This study aims to evaluate the way Egyptian code seismic provisions treat the consideration of seismic loads and analysis methods during the seismic design of buildings, to discuss the alternative solutions for cases wherein existing provisions do not lead to satisfactory results, and to quantify the effect of building lateral stiffness and diaphragm flexibility on the structural response so that designers can be aware of the likely impact of their decisions. The evaluation of Egyptian code seismic provisions and simplified methods is performed through comparison with a more refined approach whereas an effort is made to quantitatively assess the relative importance of various design and analysis assumptions. Nonlinear time history (TH) analysis has been performed to evaluate equivalent static load (ESL) and response spectrum (RS) analysis methods; a set of time history records have been used. A parametric study is carried out to evaluate the design parameter effects on the building seismic demands in different approaches of analysis and to assess the fundamental period, total base shear, displacements, and story drifts for the three methods of analysis; the design parameters include the building lateral stiffness and floor diaphragm in-plane stiffness. The results show that the building lateral stiffness and floor diaphragm in-plane stiffness have a significant influence on the shear and displacement demands calculated from response spectrum and nonlinear time history methods. The ESL method is overestimated and not accurate for calculating seismic action.

Methods

Egyptian code seismic provisions

Until the beginning of the last decade, consideration of seismic loading was absent from national codes in Egypt. Buildings were typically designed to resist gravity loads, and the only means of lateral load resistance was provided through the consideration of wind loads in some cases. The first official code of practice to consider seismic loading was published by the Ministry of Housing, Utilities and New Communities in 1989-the Reinforced Concrete Code (ECP 1989). However, the code overlooked a number of basic seismic considerations, including the influence of soil conditions and the dynamic characteristics of buildings. More importantly, for the Greater Cairo area, a crude approach was adopted whereby an arbitrary percentage of building weight was proposed as lateral seismic loading. The loading code issued on December 1993 (ECP 1993), following the 1992 earthquake, provided an approach for determining seismic loads for different types of structures. Although this code represented an improvement in comparison with previous regulations (ECP 1989), it still adopted significantly simplified assumptions in terms of loading considerations and design procedures. More recently, in 2004 a new loading code (ECP 2004a) was issued and dealt with most of the shortcomings present in preceding standards, particularly on the loading side. The new code largely follows the same framework adopted in EC8

(ECS 2004; UBC 1997). It introduces the concept of response spectra and codified force reduction factors for the design of structures and includes safety verifications relevant to ultimate and serviceability limit states. The Egyptian code gradually introduced ductility concepts and detailing procedures through its versions (ECP 2001, 2004b, 2007), although these aspects of the code still need considerable improvement and development.

Table 1 illustrates the base shear formulas with the related parameters to present the major changes to the seismic provisions in different Egyptian code of practice (ECP) editions released from 1993 to 2008 and the Regulations for Earthquake Resistant Design of Buildings in Egypt issued by the Egyptian Society for Earthquake Engineering (ESEE) (Sobaih et al. 1988). It has been found that there is high variation between the base shear and base moment obtained using the ECP versions specified in the different analysis methods and in comparison with the ESEE. The major difference between ECP (1993) and ECS (2004) was remarked by the new adoption of the response spectrum method that presented pseudo-acceleration anchored to peak ground acceleration (PGA). Also, the soil parameter has a big effect on the response spectrum curve. On the other hand, the two newest provisions ECP (2004a,b) and ECP (2008) have a similar base shear formulation except for the existence of the importance factor γ_I either in the response spectrum or in the base shear equation which yields final identical base shear. It is clear from the initial comparison how the base shear value obtained from ECP (2004a, 2008) is greater than the value obtained from ECP(1993).

The ESEE presented the response spectrum method in addition to the equivalent static load with many factors not taken into consideration in ECP (2004a, 2008) like M(material), R(risk), and Q(construction quality factors). All of these factors are still neglected in all ECP editions till 2008. Most recent seismic codes, including the 2008 draft of the Egyptian code for design loads on structures (ECP 2008), are developed with two performance levels: one with the intent of limiting damage during frequent moderate earthquakes, namely the serviceability limit state, and the other ensuring collapse prevention during a major earthquake, namely the ultimate limit state. Earthquake engineering research has proven that deformability is a governing factor in satisfying seismic codes' life safety requirement as well as damage limitation (FEMA 2007).

The period of vibration is a fundamental parameter in the force-based design of structures as this parameter defines the spectral acceleration and thus the base shear force to which the building should be designed. This paper takes a critical look at the way in which seismic design codes around the world have allowed the designer to estimate the period of vibration for use in both linear static and dynamic analysis (Crowley and Pinho 2010). The fundamental period of a building is a key parameter for the seismic design of a building structure using the equivalent lateral force procedure. As the building period cannot be analytically calculated before the building is designed, periods from the empirical period formulas recommended in seismic design codes or from finite element analysis with assumed mass and stiffness are used during the preliminary design stage. The form of the formulas for the RC and steel MRFs in ATC3-06 (ATC 1978; Goel and Chopra 1997) [6] was developed based on the assumption that lateral forces are distributed linearly over the height of a building and that the deflections of the building are controlled by drift limitation.

Table 1 Egyptian code provisions for seismic design development from 1992 up to 2008 (El-Arab 2011)

Parameter	Sobaih et al. (1988)	Egyptian code practice (ECP-201)	
		ECP (1993)	ECP (2008)
ESL	$V = C_s W_t$	V = Z.I.C.K.S.W	$F_b = S_d(T)\ \lambda\ W/g$
	C_s = Z.I.S.M.R.Q		S_d (T) is the response spectrum related to (a_g, S, R, T, γ_I, η)
Seismic hazard parameter	Z = (0.0, 0.02, 0.04, 0.08)g	Z = (0.1, 0.2, 0.3) g	a_g = (0.1, 0.125, 0.15, 0.2, 0.25, and 0.3)g
Importance categories and importance factor	I = 1.0, 1.3, and 1.5	I = 1 or 1.25	γ_I = 0.8, 1.0, 1.2, and 1.4 (increase of earthquake safety)
Structural resistance system	$0.67 \leq S \leq 3.20$	$0.67 \leq K \leq 1.33$	$2 \leq R \leq 7$
Site response factor	F = 1.0, 1.3, and 1.5 (related to soil class)	S =1, 1.15, or 1.3	S is related to soil class and spectrum type
Period effect	$T = (0.09H) / \sqrt{d}$ or 0.1N	$T = 0.1\ N$, $C = 1/15\ \sqrt{T} \leq 0.12$	S_d (T_1) is related to period T_1
Correction factor	N.A.	N.A.	λ = 0.85 or 1.0
Damping correction	N.A.	N.A.	$0.95 \leq \eta \leq 1.2$

N.A., Not Applicable.

Seismic analysis procedures

In the preliminary design process, equivalent static seismic forces are used to determine the design internal forces of structural members using linear elastic analyses of structure and, in turn, determine the design member strength demands. Such static seismic forces are simply determined corresponding to the elastic design acceleration spectrum divided by a structural strength reduction factor particularly called the response modification factor R (UBC 1997), the structural behavior factor q (ECS 2004), or the structural factor D_s (AIJ 1999). Usually, the elastic design spectrum, which is often related to 5% or 10% probability of exceedance in 50 years, is defined smoothly as a reasonable representation of the seismic action demand on the structure at the site of interest. The adopted strength reduction factor is thus intended to represent an expected inelastic response demand or expected damage level demand of the whole structure, which may be induced during earthquake excitation (Thuat 2012).

ESL method

According to ECP (2008), the seismic base shear force, F_b, for each horizontal direction in which the building is analyzed, shall be determined using the following expression:

$$F_b = \gamma \times S_d(T_1) \times \lambda \times W/g, \tag{1}$$

where S_d (T_1) is the ordinate of the design spectrum at period T_1; T_1 is the fundamental period of vibration of the building for lateral motion in the direction considered; W is the total weight of the building, above the foundation level; g is the gravity acceleration; γ is the importance factor of the building; and λ is the effective modal mass correction factor, the value of which is equal to λ = 0.85 for $T \leq 2\ T_c$ and n> 2 stories, where T_c is the upper limit of the period of the constant spectral acceleration branch as shown in Figure 1 and n is the number of stories.

The value of the fundamental period of vibration, T, is determined using the following expression:

$$T = C_t \times H^{3/4}, \tag{2}$$

where C_t is a factor determined according to the structural system and building material and is equal to 0.075 in the case of a moment-resistant space concrete frame, and H is the height of the building (m), from the foundation or from the top of a rigid basement. The ordinate of the design spectrum, S_d (T_1), can be determined from

$$S_d(T) = \frac{2.5}{R} \times a_g \times \gamma \times S \times \eta \left[\frac{T_c}{T}\right] \geq [0.2] \quad \times a_g \times \gamma, \tag{3}$$

where a_g is the design ground acceleration for the reference return period, T_c is the upper limit of the period of the constant spectral acceleration branch as shown in Figure 1,$\eta$is the design damping correction factor for the horizontal elastic response spectrum where a reference value of η= 1 corresponds to a normal 5% viscous damping ratio (in the case of RC buildings),S is the soil factor, γ is the importance factor and R is the reduction factor according the statical system of the structure. The total base shear, F_b, shall be determined

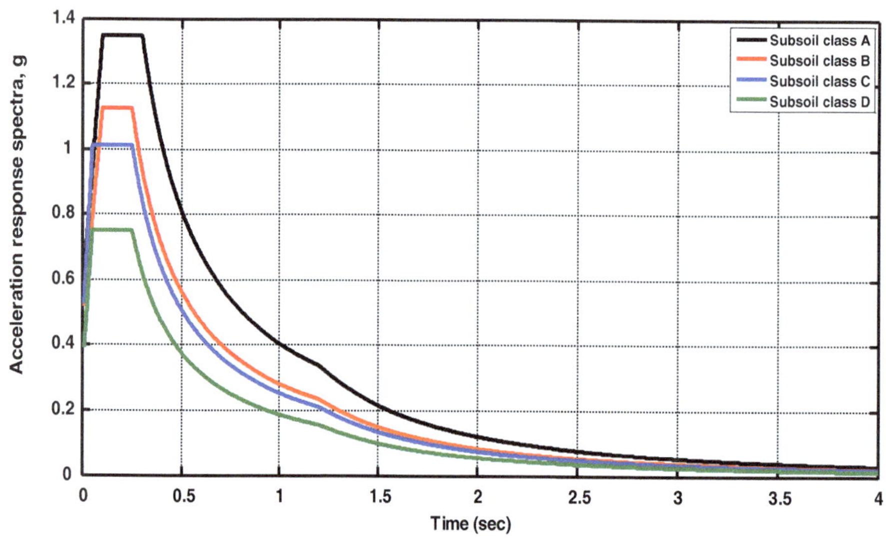

Figure 1 ECP (2008) design response spectrum for different subsoil classes.

by applying horizontal forces F_i to each story mass m_i and shall be distributed as follows:

$$F_i = \left[\frac{z_i \times W_i}{\sum_{j=1}^{n} z_j \times W_j}\right] \times F_b, \quad (4)$$

where F_i is the horizontal force acting on story i; F_b is the seismic base shear force (Equation 1); z_i and z_j are the heights of masses m_i and m_j above the foundation level, respectively; W_i and W_j are the weights of masses m_i and m_j, respectively; and n is the number of stories above the foundation level. Equation 4 gives a linear shear distribution depending only on the height of the story.

Modal RS method

The modal response spectrum analysis is applicable for all types of buildings, while the lateral force method of analysis has many restrictions on its use due to the 'fear' that it would provide un-conservative results in certain conditions; however, in spite of this disadvantage, the method is still widely used due to its ease of application (Crowley and Pinho 2010). Response spectrum analysis includes sufficient modes of vibration to capture participation of at least 90% of the structure's mass in each of two orthogonal directions (Kunnath and Kalkan 2004). Figure 1 shows the design response spectrum curve for the current case of study; it shall be noted that ECP (2008) includes a damping coefficient in the elastic response spectra equations. Hence, no damping ratio has been used in the analysis of this method. As specified in most design codes, the shape of the target elastic acceleration spectrum is characterized by the seismic intensity expressed in terms of the expected design PGA, a_g, and the effect of soil ground types expressed in terms of the response spectral periods T_b and T_c. Various values of T_b and T_c were considered corresponding to different soil ground types (from A-type of hard rock to D-type of soft soil).

Many codes recognize that the period of vibration from the simplified period-height equation is more realistic, having been directly obtained from the measured periods of vibration of buildings subject to earthquake ground motions, but that when higher modes are important (in tall and/or irregular structures), the modal response spectrum method gives a more realistic profile of the lateral forces. Hence, these codes (NBCC 2005; ASCE 2005) require the designer to check whether the modal base shear force is less than 85% of the base shear force from the equivalent lateral force method. If this is the case, then the modal forces, but not the drifts, should be multiplied by 0.85 V/V_t, where V is the base shear from the lateral force method and V_t is the base shear from the required modal combination. Even when higher modes are not important and the designers are allowed to use the linear static method, but they decide to calculate the period of vibration from the Rayleigh method, many codes apply an upper bound to the period of vibration from the Rayleigh method. This is another procedure which is used to safeguard against unrealistically high periods of vibration used in the design to lower the base shear forces (Crowley and Pinho 2010).

Nonlinear TH method

Nonlinear time history analysis is by far the most comprehensive method for seismic analysis. The earthquake record in the form of acceleration time history is input at the base of the structure. The response of the structure is computed at each second for the entire duration of an earthquake. This method differs from response spectrum analysis because the effect of 'time' is considered. That is, stresses and deformations in the structure at an instant are considered as an initial boundary condition for computation of stresses in the next step. Furthermore, nonlinearities that commonly occur during an earthquake can be included in the time history analysis. Such nonlinearities cannot be easily incorporated in response spectrum analysis. Unlike the response spectrum method, nonlinear time history analysis does not assume a specific method for mode combination. Hence, results are realistic and not conservative. Furthermore, this method is equivalent to getting 100% mass participation using response spectrum analysis. Full mass participation is necessary to generate correct earthquake forces. Usually, only 90% to 95% participation is obtained in response spectrum analysis. All types of nonlinearities can be accounted for in this analysis. This could be very important when seismic retrofit involves energy dissipation by yielding of members or plastic hinge rotation. However, this method is very expensive and time-consuming to perform. Large amounts of information are generated. Furthermore, input earthquake is never known with certainty. Hence, three to five different histories should be used, further increasing the cost.

Since the results of the time history depend mainly on the characteristic of the used acceleration time history records and the shapes of their corresponding elastic response spectra (Kunnath and Kalkan 2004), the reason of using the inelastic TH method is to verify the results obtained by other code-specific analysis procedures (ESL and RS methods) against a time history record. Nonlinear time history analysis was performed taking into consideration the P-Δ and large displacements effect. A constant damping ratio of 0.05 has been taken for RC buildings. The inelastic time history analysis is performed using the direct integration technique considering a time step of 0.005 s. Nonlinear analysis could be used to justify a design that would not satisfy the prescriptive building code requirements. Story drifts and floor accelerations are important indicators of damage to nonstructural components and overall building performance. For nonlinear seismic analyses, a total seismic mass including self-weight and floor cover 'dead load' (DL) plus 50% of 'live load' (LL) (1.0DL + 0.5LL) is considered (ASCE 2005).

Description of building structure

During the past two decades, the building environment in Egypt had extensively utilized medium-rise RC buildings having 12 stories, which is the maximum height allowed by the local authorities in most districts. These buildings are built with different configurations and structural systems having varying stiffness parameters that may have great influence on their seismic behavior. Two samples of typical buildings with 6 and 12 stories are chosen for this study as shown in Figure 2; the building's layout is essentially bisymmetric in plan, has regular plans of four equal bays with a typical bay width of 5 m in both directions, and is representative of benchmark typical buildings in current practice in Egypt. The height of every story (column height) is taken equal to 3 m, as a normal height for residential buildings. Beams are assumed on all grid lines, and the base columns are assumed to be fixed to the foundation.

The building structural elements have been first designed according to the Egyptian code of practice (ECP 2007,2008) under static loads assuming un-cracked sections for beam and slabs in the analysis. Square columns were used with different cross sections to represent the change in lateral stiffness. Also, different slab thicknesses were used to represent the change of diaphragm flexibility. These sections have been checked under seismic actions using the Egyptian code for load and forces (ECP 2008) to satisfy the Egyptian code requirements taking into consideration the effect of earthquake loads. The minimum safe column cross section under static and dynamic loads, to satisfy the Egyptian code requirements (ECP 2007, 2008), is 0.4 × 0.4 m for the 6-story building and 0.6 × 0.6 m for the 12-story building.

The materials used in the design are C300 for concrete and St52 for steel, and the material for the building structure is taken as a reinforced concrete with the following constant properties: modulus of elasticity E = 25 GPa, Poisson ratio μ = 0.2, density of 24 kN/m^3, f_c = 28 MPa, f_y = 355 MPa, and minimum steel reinforcement ratio ρ = 0.0035.

For gravity load design, dead loads include the self-weight of the structure, a typical floor cover of 1.5 kN/m^2, and partition (wall) load intensity of 1.5 kN/m^3 including plastering and assuming typical wall thickness of 0.25 m. A live load of 2.5kN/m^2 is considered. On the other hand, for seismic design purposes, a total seismic mass including self-weight and floor cover plus 50% of live load is considered. The seismic design has been carried out assuming a soil type 'C' as per referring to dense/stiff soil, an importance factor of 1.0, and a seismic zone 5B, and the shape of the spectrum is type 1 as per Egyptian zoning system with a design ground acceleration, a_g, of 0.30 g associated with the code reference probability of exceedance of 10% in 50 years.

Finite element model

A three-dimensional mathematical model of the physical structure will be used that represents the spatial distribution of the mass and stiffness of the structure to an extent that is adequate for the calculation of the significant features of the building's dynamic response. All structures are modeled and analyzed in this paper using ETABS and Sap2000 (Computers and Structures, Inc 2003, 2007, 2011a,b). The building is modeled as a three-dimensional (3D) frame structure using frame

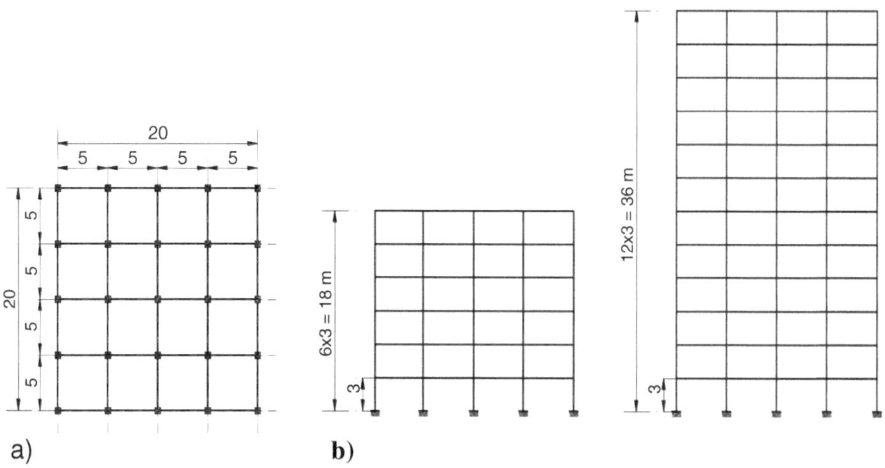

Figure 2 Configuration of a sample building model. (a) Plan configuration. **(b)** Elevation configuration for 6-story and 12-story buildings.

elements for columns, longitudinal beams, and transverse beams and shell element for slabs. The structures are assumed to have a rigid foundation; therefore, the soil foundation interaction and foundation flexibility effects are ignored.

The seismic zone considered in this study is zone 5B, and the shape of the spectrum is type 1, as shown in Figure 2. The two models are considered as residential buildings with an importance factor γ = 1. The soil is considered to be stiff soil, which presents soil class C and a soil factor S = 1.5. The reduction factor, R, is taken considering that the vertical loads and the total base shear are totally resisted by the frame structure without using shear walls or bracings (R = 6). It should be noted that ECP(2008) recommends that in the application of the ESL method, the building should meet the criteria for regularity in both plan and elevation, and with calculated structural period T not greater than 2 s or $4T_c$ (1 s for the selected soil class C). In the ESL method, according to ECP (2008), the base shear is determined as a percentage of the total building weight that gives a value of 7.2% of the total weight of the building in the 6-story building and 4.3% in the 12-story building.

Input seismic excitation

It is impossible to predict ground motion characteristics that may occur in the future at a construction site because the property of the ground motion is interrelated with many factors such as fault mechanism, seismic wave propagation from source to site, and the amplification characteristics of the ground. The important factors of ground motions affecting the structure's response results are peak ground acceleration, frequency contents, duration of ground motion, and shapes of the waveform. Egypt is a region of moderate seismicity, where infrequent moderate to large earthquakes have occurred in the past. However, there is a serious lack of strong motion records of engineering interest in the region. Therefore, the use of a large number of artificial or natural earthquake records from the literature is indispensable for the nonlinear time history analysis. The seismic design guidelines provide an acceleration response spectrum for estimating the design seismic force of a structure. Accordingly, the input ground motion applied to the dynamic response analysis of structures would be appropriate for the ground motion history which is highly related with design seismic force. A suite of nine ground motion records from seven different earthquakes (PEER 2012) are selected for the purpose of understanding the input ground motion effect, as listed in Table 2.

A suite of nine ground motion acceleration time history records, representing a wide range of intensity and frequency contents, are selected and used in the study. Those records are downloaded from the website of the Pacific Earthquake Engineering Research Center (PEER 2012). The earthquake records are listed in Table 2 by their magnitude, PGA, peak ground velocity (PGV), and peak ground displacement (PGD). Figure 3 shows the elastic pseudo-acceleration response spectra of the earthquake motions for 5% damping; the figure shows significant variation in the frequency contents and the response characteristics. The ground motion records are grouped into three levels depending on the peak ground acceleration as low (0.1 up to 0.3 g), moderate (0.4 up to 0.6 g), and high (0.7 up to 0.9 g). The records are chosen such that the period ratio (T_1/T_g and T_2/T_g; adjacent building period over the ground motion characteristic period) has a wide range.

Results and discussion

This study aims to evaluate the way Egyptian code seismic provisions treat the consideration of seismic loads and analysis methods during the seismic design of buildings, to discuss the alternative solutions for cases wherein existing provisions do not lead to satisfactory results, and to quantify the effect of building lateral stiffness and diaphragm flexibility on the structural response so that designers can be aware of the likely impact of their decisions. The evaluation of Egyptian code seismic provisions and simplified methods is performed through

Table 2 Characteristics of earthquake ground motion records used in the analysis

Level	PGA (g)	Input wave	M_w	Earthquake/station	Scale factor	EPD(km)	PGV(cm/s)	PGD(cm)	T_g(s)
Low	0.21	1MVH	6.0	North Palm Springs, 1986/Morongo Valley	1.7	10.1	40.9	15.0	1.90
	0.30	2A-GRN	6.0	Whittier Narrows, 1987/E-Grand Avenue	1.5	9.0	23.0	3.3	0.70
	0.29	3G06	6.2	Morgan Hill, 1994/Gilroy Array #6	1.0	11.8	36.7	6.1	1.20
Moderate	0.48	4CYC	6.9	Loma Prieta, 1989/Coyote Lake Dam	1.0	21.8	39.7	15.2	0.65
	0.51	5STG	6.9	Loma Prieta, 1989/Saratoga-Aloha Avenue	1.0	11.7	41.2	16.2	1.80
	0.59	6NPS	6.0	North Palm Springs, 1986/5070	0.6	8.2	73.3	11.5	1.10
High	0.60	7D-PVY	5.8	Coalinga, 1983/Pleasant Valley P.P.	1.2	17.4	34.8	8.1	0.65
	0.84	8RRS	6.7	Northridge, 1994/Rinaldi	0.6	7.1	166.1	28.8	1.05
	1.04	9CPM	7.1	Cape Mendocino, 1992/Cape Mendono	0.6	8.5	42.0	12.4	2.00

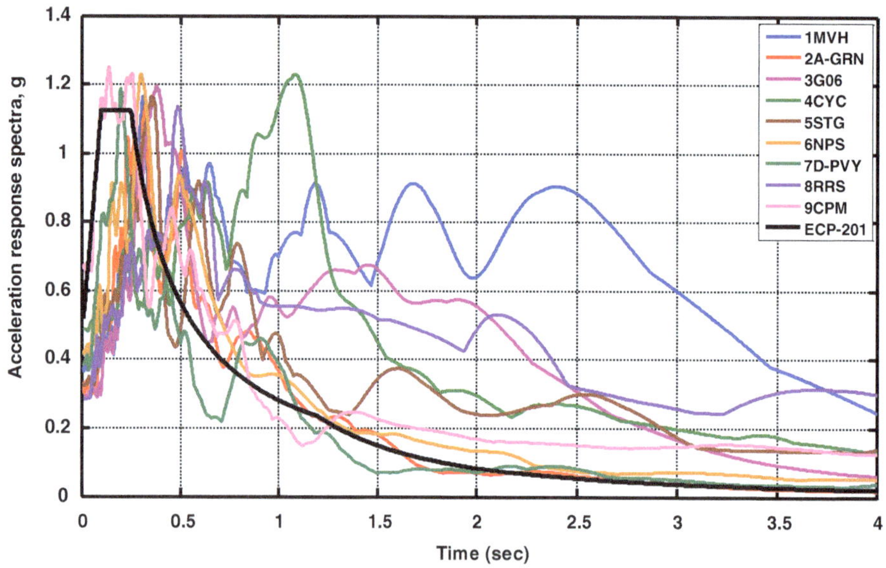

Figure 3 Associated 5%damped response spectrum for the nine records.

comparison with a more refined approach whereas an effort is made to quantitatively assess the relative importance of various design and analysis assumptions. TH analysis has been performed to evaluate ESL and RS analysis methods; a set of time history records have been used. A parametric study is carried out to evaluate the design parameter effects on the building seismic demands in different approaches of analysis and to assess the fundamental period, total base shear, displacements, and story drifts for the three methods of analysis; the design parameters include the building lateral stiffness and floor diaphragm in-plane stiffness. This paper compares seismic provisions in the Egyptian code for loads (ECP 2008) to highlight the differences in dealing with the seismic design of moment-resisting-frame buildings.

Effect of building model lateral stiffness

The 6-story and 12-story building models have been studied to evaluate the effect of the lateral stiffness on determination and evaluation of fundamental period, base shear, displacement, and story drift. A constant beam and slab sections with rigid diaphragm assumption are considered in the analysis with different column cross sections to present the change in lateral stiffness as shown in Table 3.

Natural vibration analysis

The period of vibration is a fundamental parameter in the force-based design of structures as this parameter defines the spectral acceleration and thus the base shear force to which the building should be designed. This study takes a critical look at the way in which seismic design codes around the world have allowed the designer to estimate the period of vibration for use in both linear static and dynamic analyses. In most building design projects, empirical building period formulas are used to initiate the design process (Kwon and Kim 2010).

The fundamental period of vibration, T, is a function of the stiffness of the lateral load-resisting system and the building mass. The fundamental period in ECP (2008), T, is not influenced by the change of the column cross section but depends only on the building height. Table 4 presents different fundamental periods for the buildings studied, as obtained from structural analysis using finite element models and empirical expression in ECP (2008) and other international building codes. In both 6-story and 12-story buildings, the periods computed from empirical expressions are significantly shorter than those computed from structural models especially for flexible building structures. As the building's

Table 3 Building structural element dimensions for different lateral stiffness models

Number of stories	Model	Beam size (cm)	Slab thickness (cm)	Column (cm)
6-story	LS1	25 ×60	15	40× 40
	LS2	25 ×60	15	60 ×60
	LS3	25 ×60	15	80 ×80
	LS4	25 ×60	15	100 ×100
12-story	LS5	25 ×60	15	60 ×60
	LS6	25 ×60	15	80 ×80
	LS7	25 ×60	15	100 ×100
	LS8	25 ×60	15	120 ×120
	LS9	25 ×60	15	140 ×140

Table 4 Fundamental period of the RC moment-resisting-frame building

Code	Period, *T*	Fundamental period (s)								
		LS1	LS2	LS3	LS4	LS5	LS6	LS7	LS8	LS9
3D model natural vibration analysis	1st sway mode	1.171	0.834	0.750	0.698	1.755	1.633	1.598	1.577	1.555
	2nd sway mode	0.387	0.264	0.219	0.188	0.570	0.513	0.480	0.448	0.414
	1st torsion mode	1.046	0.763	0.678	0.620	1.570	1.463	1.402	1.329	1.240
ECP (2008)	$T = 0.075H^{3/4}$		0.655					1.102		
ECP (1993)	$T = 0.1\ N$		0.600					1.200		
ICC (2003)	$T = 0.03\ H^{3/4}$		0.262					0.441		
UBC (1997)	$T = 0.02\ H^{3/4}$		0.175					0.294		
ECS (2004)	$T = 0.075\ H^{3/4}$		0.655					1.102		
NBCC (2005)	$T = 0.05\ H^{3/4}$		0.437					0.735		

H is the building's height above the foundation level, and *N* is the number of the stories.

lateral stiffness increases, the fundamental period decreases. Table 4 shows the disparity between the fundamental period of vibration from the empirical period-height equation from different codes and the period of vibration from the eigenvalue or Rayleigh analysis of a bare frame model. The fundamental period estimated by the ECP-201 empirical equation is underestimated especially for flexible models; the fundamental period reaches 179% and 159% in models LS1 and LS5, respectively. Many codes recognize that the period of vibration from the simplified period-height equation is more realistic, having been directly obtained from the measured periods of vibration of buildings subject to earthquake ground motions, but that when higher modes are important (in tall and/or irregular structures), the modal response spectrum method gives a more realistic profile of the lateral forces. However, the empirical equation should be calibrated to obtain a conservative estimate for the base shear.

Base shear demands for different lateral stiffness models

The ESL, dynamic modal RS, and TH analyses are employed, and a comparison of base shear and maximum displacement at the building top is provided as shown in Figures 4 and 5. The response is quantified by the formulation of maximum displacements and shear forces for which each response quantity result is calculated by ETABS to represent a measure of the probable maximum magnitude for that quantity. The resulting maximum and minimum responses provide a range for which the actual structural response of the building is expected to fall within. The results of the analyses show that the time history analysis (Av, average; MinEnv, minimum envelope; MaxEnv, maximum envelope of response demands) is the most appropriate method in capturing the behavior of this particular building under dynamic loading. The ESL method depends on the building weight and the fundamental period calculated from the empirical equation, and the ESL overestimates the base shear and displacement response demands. The seismic demands using ESL has more reliability for the 6-story building than that for the 12-story building (fundamental mode with longer period). The RS underestimates the seismic response demands and should be calibrated for the bare frame eigenvalue to calculate the fundamental vibration period.

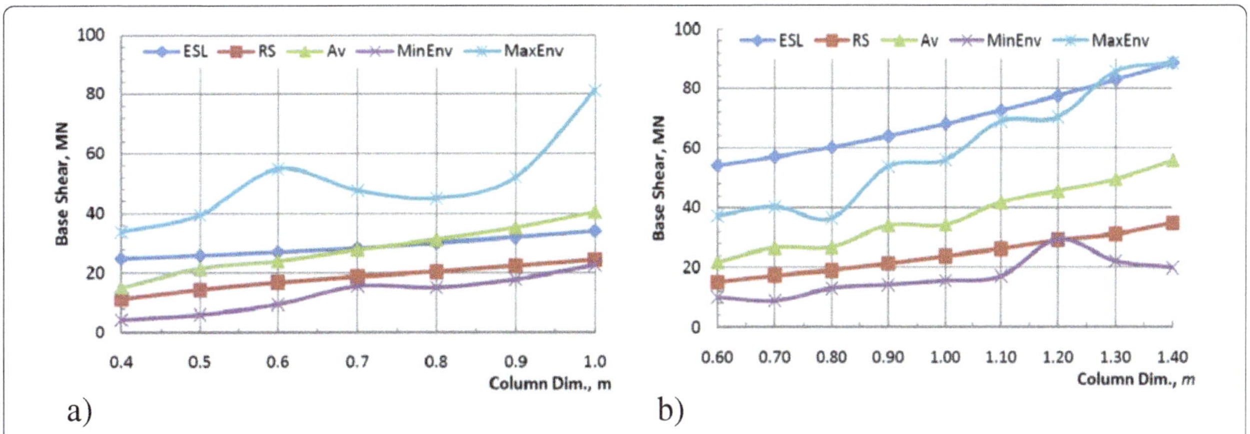

Figure 4 Story base shear calculated for different lateral stiffness models. (a) 6-story building. **(b)** 12-story building.

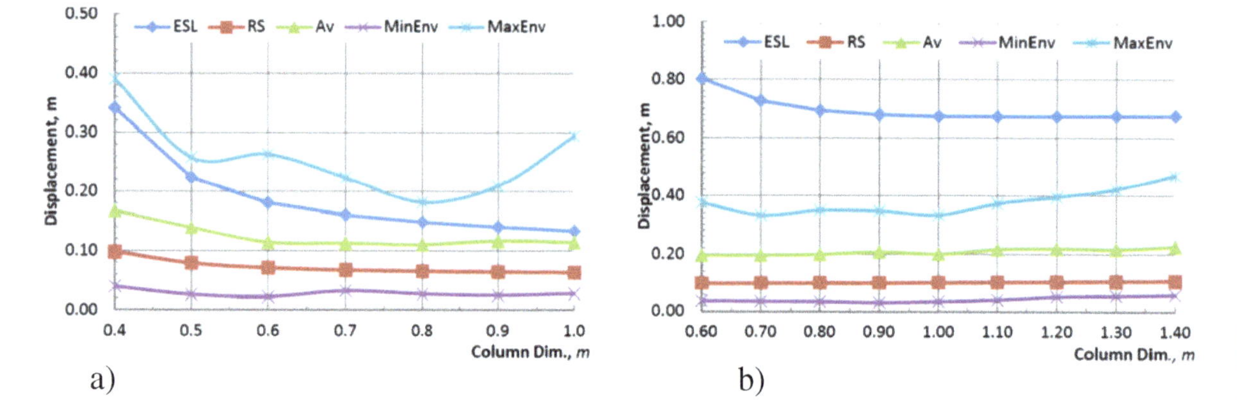

Figure 5 Displacement calculated for different lateral stiffness models. (a) 6-story building. **(b)** 12-story building.

Story shear and lateral drift demands

To show the effect of the column rigidity on the shear distribution, the displacements and the story drifts in the case of 6-story and 12-story buildings have been calculated using the three different methods of analysis (ESL, RS, and time history for the nine records 'Av, MinEnv, and MaxEnv') and shown in Figures 6 and 7. As the lateral stiffness increases, the story shear demands increase while the story displacement and inter-story drift decrease, with the assumption that the diaphragm is rigid enough to distribute uniformly the lateral loads on the vertical elements. Therefore, the displacements and the story drifts for the external and the internal columns are the same.

In the 6-story building model LS4 with stiff lateral stiffness, the story shear, displacement, and inter-story drift calculated from the ESL method closely match the average response calculated from the time history method. The story shear, displacement, and inter-story drift are significantly overestimated using the ESL method and underestimated using the RS method for flexible lateral stiffness design of buildings (LS1).

In the 12-story building, the story shear, displacement, and inter-story drift are significantly overestimated using the ESL method and underestimated using the RS method for flexible lateral stiffness design of buildings (LS5 to LS9). The tangent inclination angle of RS and Av inter-story drift show a slight change over the building height. The shear distribution along the stories displays a nonlinear pattern and depends on the column cross section.

Effect of floor diaphragm in-plane stiffness on seismic demands

The vertical distribution of forces from the simple equivalent lateral force method, which is commonly used for designing the lateral force-resisting system, is based on assumptions for elastic mode shapes of structures with rigid diaphragms (Fleischman and Farrow 2001; SEAOC 1999). Therefore, for structures with flexible diaphragms, the diaphragm design forces calculated according to the code methods may not be appropriate. When a structure with a flexible diaphragm is considered to have a rigid diaphragm, the period will be underestimated. For structures with periods on the descending branch of the spectral acceleration period plot, this may result in some structural base shears being slightly overestimated. However, the actual displacement will be greater than predicted. That is, the structure with a flexible diaphragm will have larger displacements along the diaphragm length compared with the displacements because of a rigid diaphragm assumption. This unaccounted increased displacement, because of a rigid diaphragm assumption, may increase the possibility of damage because of structural pounding with neighboring structures. It will also affect the distribution of forces between vertical lateral force-resisting 'VLFR' elements, which may be critical if they, or the elements they are connected to, do not possess sufficient strength or ductility (Sadashiva et al. 2012).

The code empirical methods under-predict the fundamental natural period of structures with flexible diaphragms. This effect on period calculation means that the design forces are likely to be overestimated, which is conservative. Fundamental natural period increased with increasing diaphragm flexibility. A critical level of diaphragm flexibility exists when the mass tributary to the lateral system and the remainder of the diaphragm mass act independently. Diaphragm flexibility effects on seismic demands are higher for short-period structures than for medium- or long-period structures For the diaphragm flexibility effect study, simple models of 6-story and 12-story building structures are developed with different floor in-plane stiffness 'diaphragm flexibility' using slab thicknesses that range from 0.15 to 0.30 along with a grid beam of 0.25× 0.60 cross section as shown in Table 5. Flexible diaphragm effects were evaluated by conducting time history analyses and comparing the responses of structures with rigid and flexible diaphragms.

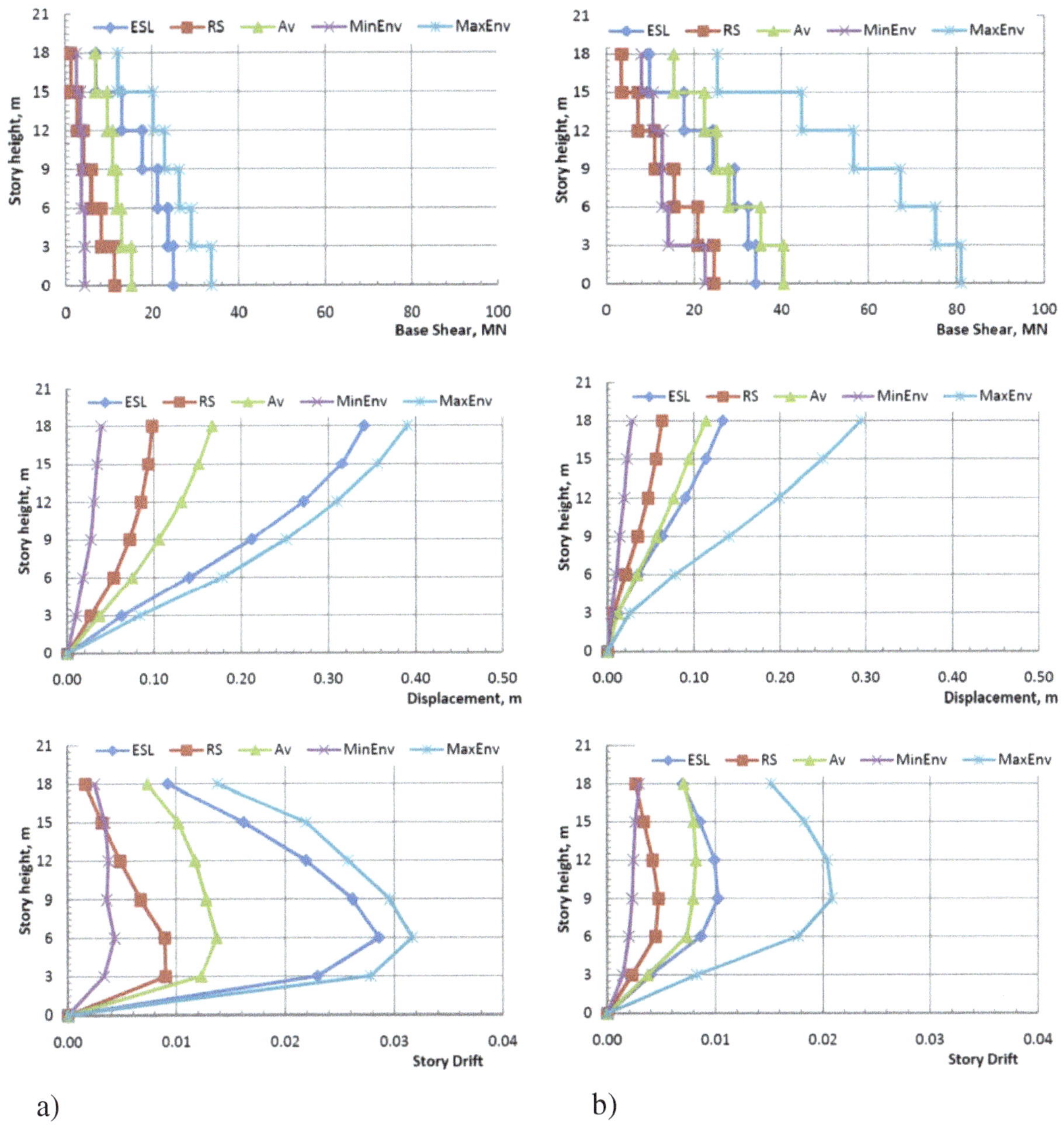

Figure 6 Effect of lateral stiffness on seismic responses for 6-story models. (a) LS1. **(b)** LS4.

Table 6a,b presents different fundamental periods, for 6-story and 12-story buildings, as obtained from structural analysis using finite element models and empirical expression in ECP (2008) and other international building codes. In both 6-story and 12-story buildings, the periods computed from empirical expressions are significantly shorter than those computed from structural models especially for flexible diaphragm. As the floor in-plan stiffness increases, the fundamental period decreases. The fundamental period estimated by the ECP-201 empirical equation is underestimated especially for flexible diaphragm models; the fundamental period reaches 179% and 159% in models DS1 and DS5, respectively. The 6-story building models DS1-1 to DS4-1 are laterally flexible; as the diaphragm stiffness increases, the fundamental period slightly decreases in contact to behavior trend with models DS1-2 to DS4-2 that are relatively laterally rigid(the fundamental period decreases as the diaphragm stiffness increases), while for the 12-story models, the fundamental period decreases as the diaphragm stiffness increases.

The base shear for the eight different models calculated from the three methods of analysis for both 6-story and 12-story buildings, is shown in Figure 8. As the floor diaphragm in-plane stiffness increases, the base shear

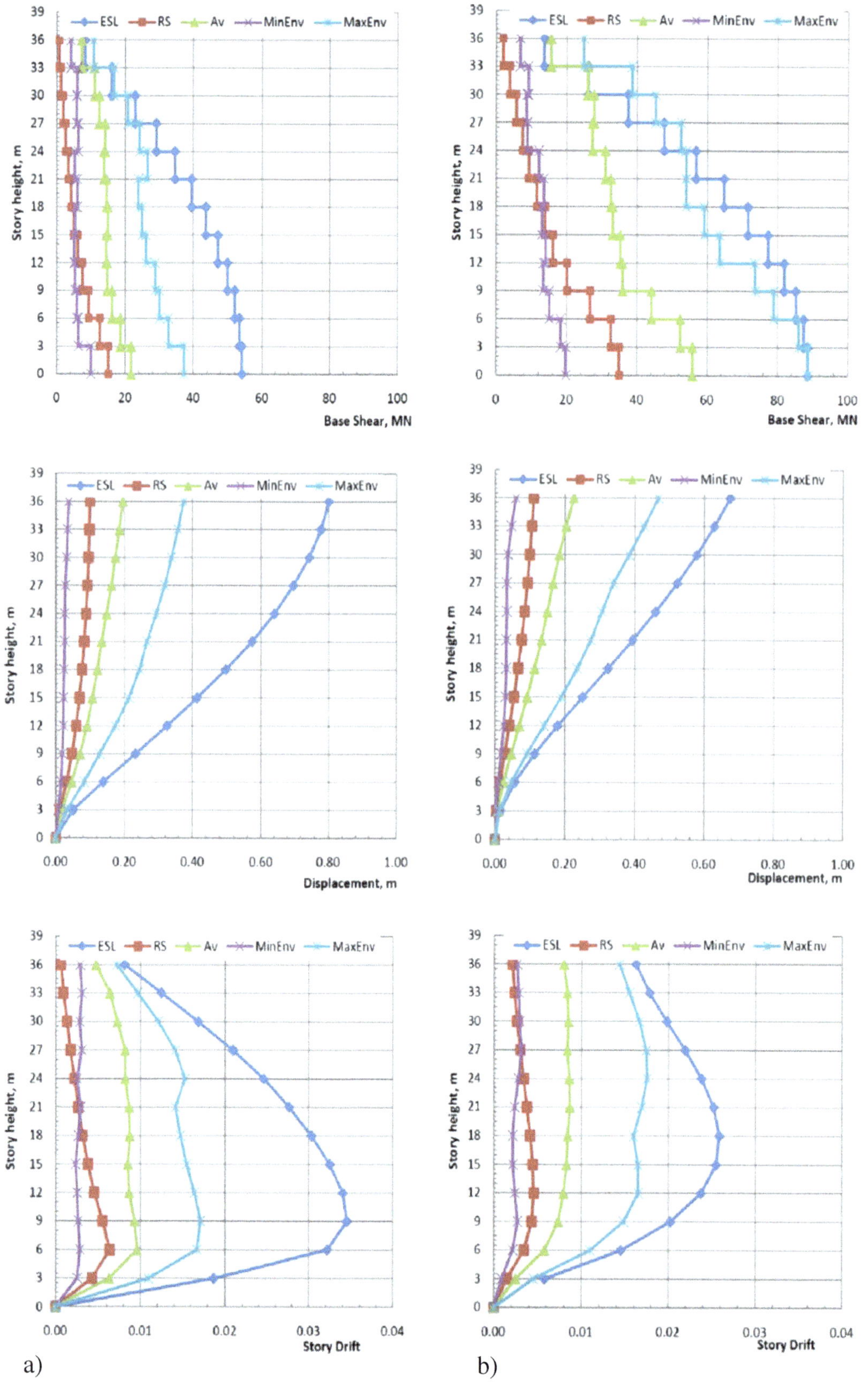

Figure 7 Effect of lateral stiffness on seismic responses for 12-story models. (a) LS5. **(b)** LS9.

Table 5 Building structural element dimensions for different floor diaphragm in-plane stiffness models

Number of stories	Model	Column (cm)	Slab thickness (cm)	Beam size (cm)
6-story	DS1-1	40 × 40	15	25× 60
	DS2-1		20	
	DS3-1		25	
	DS4-1		30	
	DS1-2	80 × 80	15	
	DS2-2		20	
	DS3-2		25	
	DS4-2		30	
12-story	DS5-1	60 × 60	15	
	DS6-1		20	
	DS7-1		25	
	DS8-1		30	
	DS5-2	120 × 120	15	
	DS6-2		20	
	DS7-2		25	
	DS8-2		30	

increases, while story displacement response displays almost no change in the case of 6-story and 12-story models except for the ESL that displays a significant decrease of story displacement demand. The base shear and story displacement responses calculated from the ESL method are significantly larger than those calculated from RS and TH methods. The base shear and story displacement calculated from RS for the different models are slightly smaller than the average response 'Av' calculated from TH methods, which means that the actual base shear on the buildings is smaller than the base shear calculated from ECP (2008).

Effect of floor diaphragm in-plane stiffness on story shear and drift demands

To show the effect of the floor diaphragm in-plane stiffness on the shear distribution, the displacements and the story drifts in the case of 6-story and 12-story buildings have been calculated using the three different methods of analysis as shown in Figures 9 and 10. In all models of 6-story and 12-story buildings, the displacement calculated from the ESL method is significantly larger than the displacement calculated from the RS and nonlinear TH methods. In the case of the RS and TH methods, the tangent

Table 6 Estimation of the fundamental period for different building models

a) 6-story models

Code	Period, T	Fundamental period (s)							
		DS1-1	DS2-1	DS3-1	DS4-1	DS1-2	DS2-2	DS3-2	DS4-2
3D model natural vibration analysis	1st sway mode	1.171	1.202	1.223	1.243	0.750	0.727	0.687	0.644
	2nd sway mode	0.387	0.399	0.408	0.417	0.219	0.216	0.209	0.199
	1st torsion mode	1.046	1.081	1.104	1.122	0.678	0.672	0.650	0.619
ECP (2008)	$T = 0.075\ H^{3/4}$	0.655							
ECP (1993)	$T = 0.1\ N$	0.600							
ICC (2003)	$T = 0.03\ H^{3/4}$	0.262							
UBC (1997)	$T = 0.02\ H^{3/4}$	0.175							
ECS (2004)	$T = 0.075\ H^{3/4}$	0.655							
NBCC (2005)	$T = 0.05\ H^{3/4}$	0.437							

b) 12-story models

Code	Period, T	Fundamental period (s)							
		DS5-1	DS6-1	DS7-1	DS8-1	DS5-2	DS6-2	DS7-2	DS8-2
3D model natural vibration analysis	1st sway mode	1.755	1.719	1.654	1.592	1.577	1.498	1.385	1.265
	2nd sway mode	0.570	0.560	0.541	0.522	0.448	0.435	0.411	0.383
	1st torsion mode	1.570	1.563	1.523	1.471	1.329	1.307	1.256	1.185
ECP (2008)	$T = 0.075\ H^{3/4}$	1.102							
ECP (1993)	$T = 0.1\ N$	1.200							
ICC (2003)	$T = 0.03\ H^{3/4}$	0.441							
UBC (1997)	$T = 0.02\ H^{3/4}$	0.294							
ECS (2004)	$T = 0.075\ H^{3/4}$	1.102							
NBCC (2005)	$T = 0.05\ H^{3/4}$	0.735							

H is the building height above the foundation level, and N is the number of the stories.

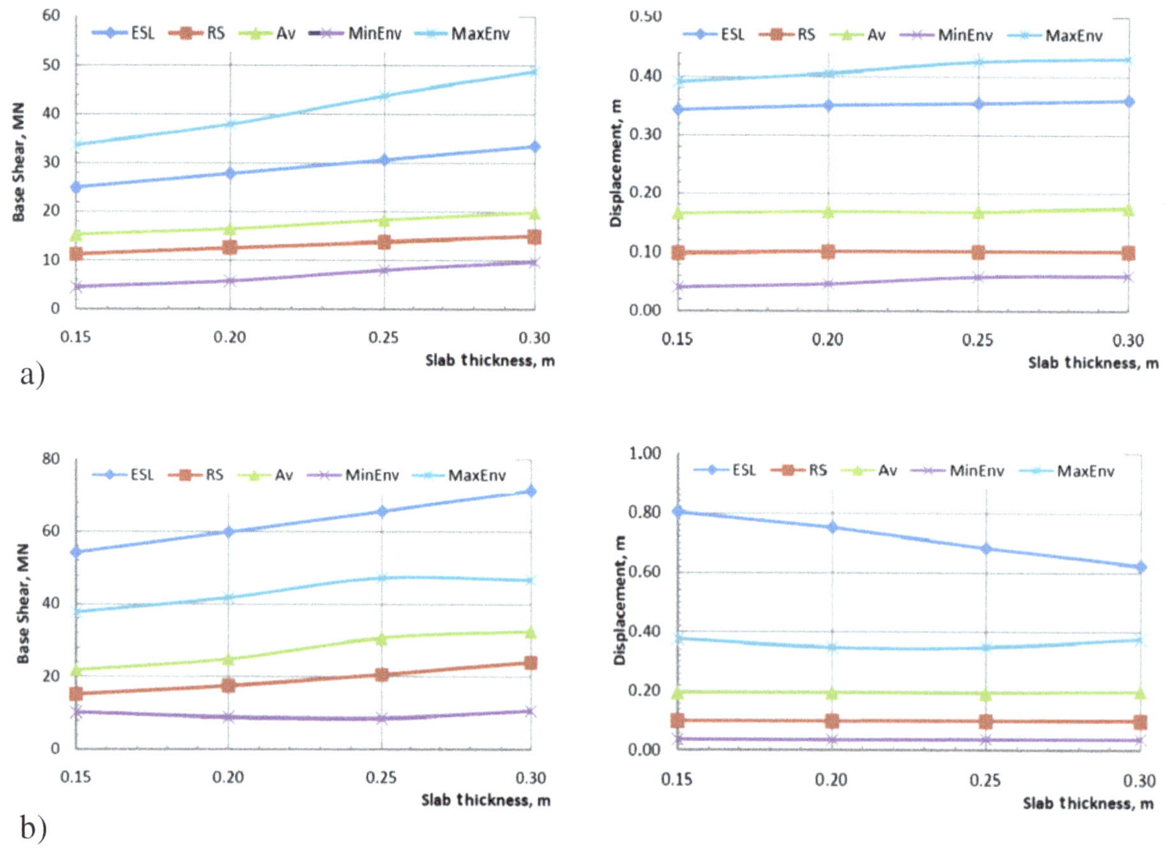

Figure 8 Floor diaphragm in-plane stiffness effect on seismic response demands. (a) 6-story building. **(b)** 12-story building.

inclination of the displacement curve and the curvature of the story drift in the case of the relatively flexible diaphragm, model DS1-1 in the 6-story building and model DS5-1 in the 12-story building, are significantly larger than those of the relatively rigid diaphragm, model DS4-1 in the 6-story building and model DS5-1 in the 12-story building. This means that the change in the floor diaphragm in-plane stiffness has a significant effect on the shear distribution and the floor displacement.

Conclusions

In this study, nonlinear TH analysis, for nine different time history records, has been used to evaluate the ESL and RS methods, adopted in the Egyptian code for load and forces (ECP 2008). These expressions depend on building weight only and do not take into consideration the effect of the building dynamic characteristics. The evaluation of Egyptian code seismic provisions and simplified methods is performed through comparison with a more refined approach whereas an effort is made to quantitatively assess the relative importance of various design and analysis assumptions that have to be made when building seismic design is taken into consideration, based on the study of the dynamic response of different buildings. The main findings of the study are summarized as follows:

The empirical expression for calculating the fundamental period of vibration by ECP (2008) underestimates the fundamental period compared to the structural model. As the lateral stiffness increases, the fundamental period for the structural model decreases; this means that the fundamental period is not only a function of building height but also a function of lateral stiffness.

For flexible diaphragm, the fundamental period calculated from the structural model is larger than the fundamental period calculated from the ECP (2008) empirical expression, which means that the change in the floor diaphragm in-plane stiffness has a significant effect on the fundamental period of vibration. It can be seen from the study that the main important role of the slab is actually to act as a deep beam in transferring the horizontal loads from the slabs to the vertical lateral force-resisting system. Increased diaphragm flexibility changes the demands on the whole structure. That is, it increases the structural period, and this affects the forces entering the structures thereby changing the force demands and displacements of the elements. It changes the distribution of forces between the vertical lateral force-resisting elements. Diaphragm flexibility resulted

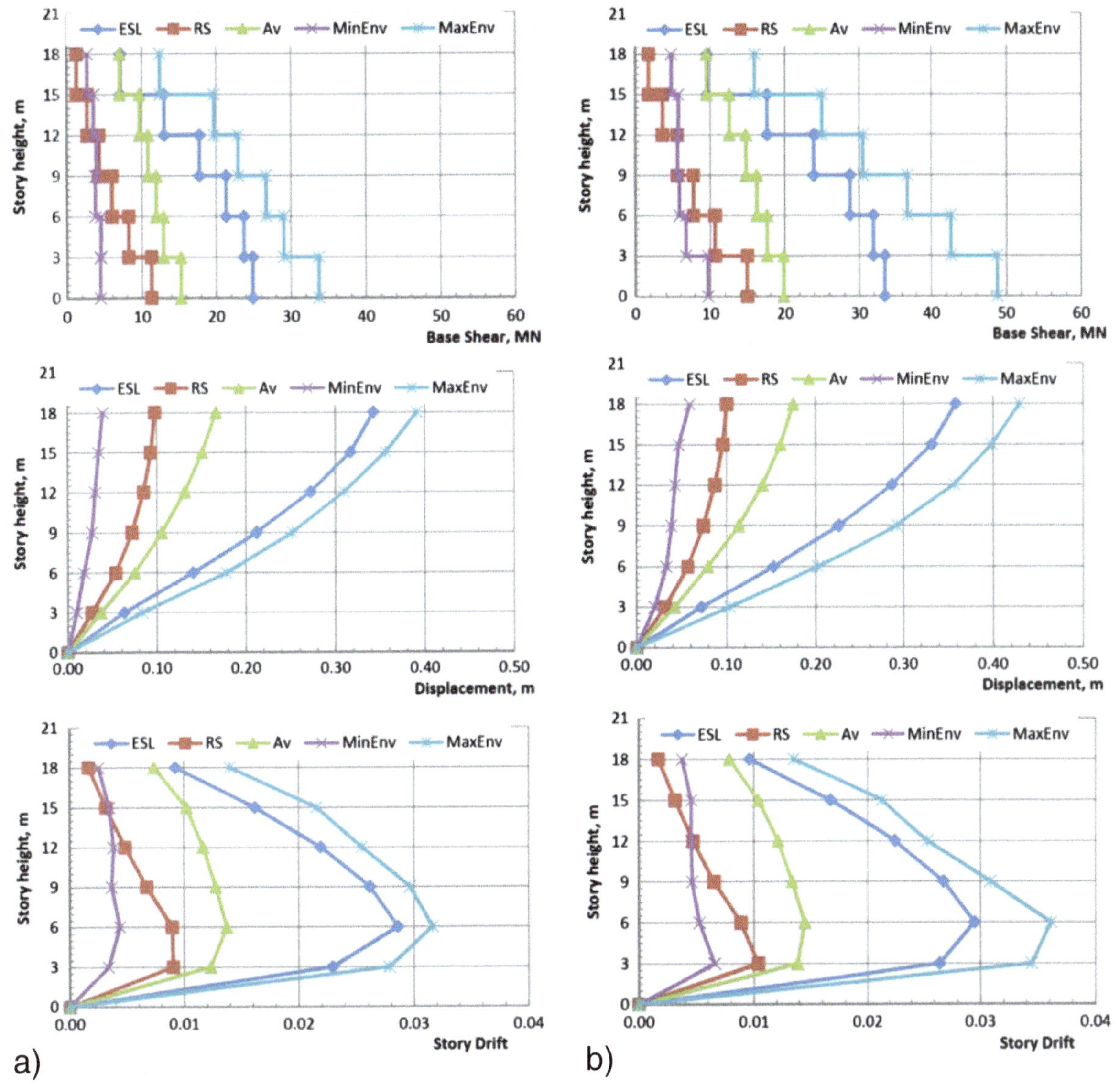

Figure 9 Effect of floor diaphragm in-plane stiffness on seismic response demands for 6-story building. (a) DS1-1. (b) DS4-1.

in the interior fames to resist a larger proportion of the lateral load than the exterior frames. Floor deflections increased because of diaphragm flexibility, with higher increases obtained at the first floor than at other floors.

The fundamental natural period of structures increased because of diaphragm flexibility. The flexible diaphragms result in greater floor displacements compared with rigid diaphragms of equivalent buildings. Also, the shears on the interior frames with larger tributary areas increased because of diaphragm flexibility. The above diaphragm flexibility effects amplified when the number of spans was increased or when the number of stories was decreased.

The code empirical methods under-predict the fundamental natural period of structures with flexible diaphragms. This effect on period calculation means that the design forces are likely to be overestimated, which is conservative. Fundamental natural period increased with increasing diaphragm flexibility. A critical level of diaphragm flexibility exists when the mass tributary to the lateral system and the remainder of the diaphragm mass act independently. Diaphragm flexibility effects on seismic demands are higher for short-period structures than for medium- or long-period structures.

The increase of the lateral stiffness and/or floor diaphragm in-plane stiffness increases the total building base shear. However, the increase of RS and/or nonlinear TH base shear is different from the increase of ESL base shear, which depends only on the building weight. This means that the base shear is not only a function of the building weight but also a function of the building rigidity.

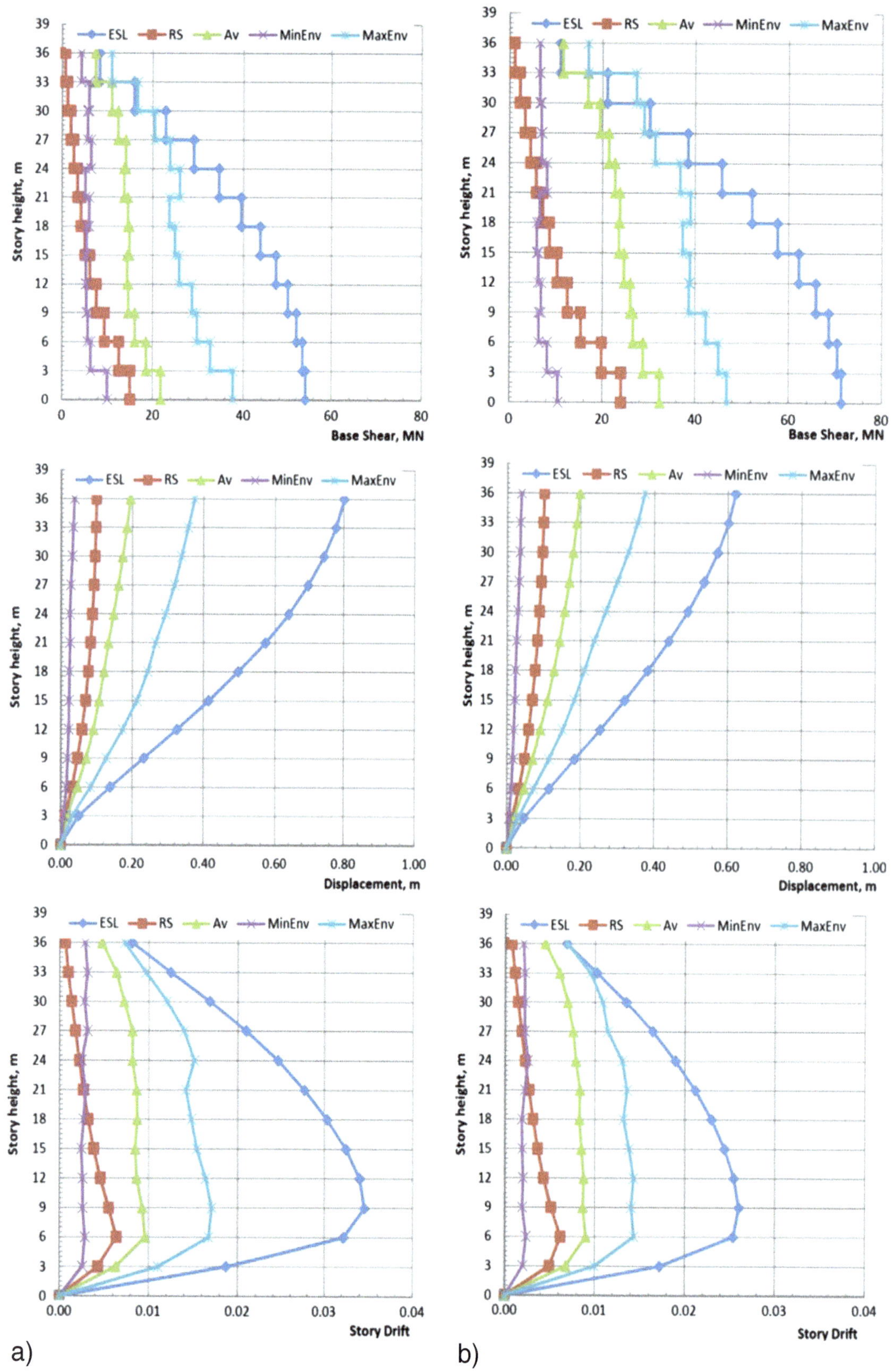

Figure 10 Effect of floor diaphragm in-plane stiffness on seismic response demands for 12-story building. (a) DS5-1. **(b)** DS8-1.

The nonlinear TH base shear, in all studied models, is smaller than the ESL base shear. This means the ECP-201 empirical expressions for calculating the base shear are overestimated. The results show that the ESL method is overestimated and not accurate for calculating seismic action.

The displacements and story drifts calculated from the ESL method don't significantly change with the change of the column and/or floor diaphragm in-plane stiffness. On the contrary, the displacements and story drifts calculated from RS and TH methods change more significantly with the change of the lateral and floor diaphragm in-plane stiffness. This means that the linear shear distribution assumed in ECP-201 is not adequately accurate and depends on the building rigidity and the linear shear distribution is inconvenient.

The building lateral stiffness and floor diaphragm in-plane stiffness have a significant effect on the story shear, overturning moment, story displacement, and inter-story drift demands calculated from the response spectrum and nonlinear time history methods.

It should also be highlighted that the lessons learnt from the catastrophic consequences of recent earthquakes, revisions in the seismic design code, and the developments in the material and workmanship characteristics are mandatory for significant improved quality of newer constructions in Egypt.

A future investigation as an extension of this research based on the present study is recommended for the evaluation of the Egyptian code under the effect of low, medium, and high ductility demands; the pushover analysis for calculating the seismic demands and its compatibility with the Egyptian code methods; and the effect of masonry infill on the seismic behavior of buildings designed according to the Egyptian code.

Competing interests

The author declares that he has no competing interests.

Author's information

Shehata E. ABDEL RAHEEM, Assoc. Prof. of Earthquake, Bridge, Structural and Geotechnical Engineering; 2009, Civil Engineering Department, Faculty of Engineering, Assiut University, Egypt. Doctor of Engineering; 2003 and Japan Society for the Promotion of Science fellow; 2006, Hokkaido University, Japan. Alexander von Humboldt fellow; 2005, Kassel University, Germany. He has published more than 60 papers in the field of structural engineering, structural analysis, earthquake engineering, soil structure interaction, seismic pounding, structural control, and nonlinear dynamic of offshore structures. He is a previous member of the Japan Society of Civil Engineers. He reviewed many papers for international journals. He is currently working as an engineering counselor at the general project management of Taibah University. He is Assoc. Prof. of Earthquake, Structural and Geotechnical Engineering, Faculty of Engineering, Taibah University, Madinah Munawarh, Kingdom of Saudi Arabia.

Acknowledgements

The author is grateful to the editor and reviewers for his interest and comments to improve the quality of the manuscript.

References

Abdel Raheem KA, Abdel Raheem SE, Soghair HM, Ahmed MH (2010) Evaluation of seismic performance of multistory buildings designed according to Egyptian code. Journal of Engineering Sciences, Assiut University 38(2):381–402

AIJ (1999) Design guidelines for earthquake resistant reinforced concrete structures. Architectural Institute of Japan, Tokyo

ASCE (2005) Building code requirements for structural concrete (ASCE318-05) and commentary (ASCE318R-05). American Concrete Institute, Farmington Hills

ATC (1978) Tentative provisions for the development of seismic regulations for buildings. Report No. ATC3-06. Applied Technology Council, Palo Alto

ATC (1996) Seismic evaluation and retrofit of concrete buildings. Report No. ATC-40. Applied Technology Council, Redwood City

Badawi HS, Mourad SA (1994) Observations from the 12 October 1992 Dahshour earthquake in Egypt. Natural Hazards Journal 10(3):261–274

Chopra AK (1995) Dynamics of structures theory and application to earthquake engineering. University of California at Berkeley, Berkeley

Computers and Structures, Inc. (2003) SAP2000: integrated software for structural analysis and design. CSI, Berkeley

Computers and Structures, Inc (2007) SAP2000 advanced 11.0.8. Static and dynamic finite element analysis of structures. CSI, Berkeley

Computers and Structures, Inc. (2011a) ETABS, integrated finite element analysis and design of structures. User's manual, CSI, Berkeley

Computers and Structures, Inc (2011b) Extended three dimensional analysis of building systems (ETABS). Computer software package, version 9.7.4. CSI, Berkeley

Crowley H, Pinho R (2010) Revisiting Eurocode 8 formulae for periods of vibration and their employment in linear seismic analysis. Earthquake Engineering and Structural Dynamics 39:223–235

ECP (1989) ECP-203: Egyptian code for design and construction of reinforced concrete structures, ECPCS-203. Housing and Building National Research Center. Ministry of Housing, Utilities and Urban Planning, Cairo

ECP (1993) ECP-201: Egyptian code for calculating loads and forces in structural work and masonry. Housing and Building National Research Center. Ministry of Housing, Utilities and Urban Planning, Cairo

ECP (2001) ECP-203: Egyptian code for design and construction of reinforced concrete structures, ECPCS-203. Housing and Building National Research Center. Ministry of Housing, Utilities and Urban Planning, Cairo

ECP (2004a) ECP-201: Egyptian code for calculating loads and forces in structural work and masonry. Housing and Building National Research Center. Ministry of Housing, Utilities and Urban Planning, Cairo

ECP (2004b) ECP-203: Egyptian code for design and construction of reinforced concrete structures, ECPCS-203. Housing and Building National Research Center. Ministry of Housing, Utilities and Urban Planning, Cairo

ECP (2007) ECP-203: Egyptian code for design and construction of reinforced concrete structures, ECPCS-203. Housing and Building National Research Center. Ministry of Housing, Utilities and Urban Planning, Cairo

ECP (2008) ECP-201: Egyptian code for calculating loads and forces in structural work and masonry. Housing and Building National Research Center. Ministry of Housing, Utilities and Urban Planning, Cairo

ECS (2004) Eurocode 8 - design of structures for earthquake resistance. Part 1: general rules, seismic actions and rules for buildings. EN1998-1. European Committee for Standardization, Brussels

El-Arab IE (2011) Seismic analysis of existing school buildings using different Egyptian seismic provisions. The twelfth East Asia-Pacific conference on structural engineering and construction. Procedia Engineering 14:1906–1912

FEMA (2000) Prestandard and commentary for the seismic rehabilitation of buildings.FEMA-356. American Society of Civil Engineers (ASCE)/Federal Emergency Management Agency, Washington, DC

FEMA (2007) NEHRP recommended provisions for new buildings and other structures: training and instructional materials. FEMA 451B. Federal Emergency Management Agency, Washington, DC

Fleischman RB, Farrow KT (2001) Dynamic behavior of perimeter lateral-system structures with flexible diaphragms. Journal of Earthquake Engineering and Structural Dynamics 30:745–763

Ghosh SK, Fanella DA (2003) Seismic and wind design of concrete buildings: 2000 IBC, ASCE 7–98, ACI 318–99, 3rd edn. Kaplan Publishing

Goel RK, Chopra AK (1997) Period formulas for moment-resisting frame buildings. Journal of Structural Engineering-ASCE 123(11):1454–1461

ICC (2003) International building code- IBC. International Code Council, Country Club Hills

Khater M (1992) Reconnaissance report on the Cairo, Egypt earthquake of October 12, 1992. Technical report NCEER-92-0033. National Center for Earthquake Engineering Research, Buffalo, p 52

Kunnath SK, Kalkan E (2004) Evaluation of seismic deformation demands using non-linear procedures in multistory steel and concrete moment frames. ISET Journal of Earthquake Technology 41(1):159–181

Kwon O, Kim ES (2010) Evaluation of building period formulas for seismic design. Earthquake Engineering and Structural Dynamics 39:1569–1583

Mourad SA, Elattar AG, Megahid HA, Hosny AH, Elthahaby KM, Riad S (2000) Assessment of seismic effects on structures in Egypt and measures for mitigation. An Arabic study for the National Academy for Scientific Research and Technology. National Academy for Scientific Research and Technology, Cairo

NBCC (2005) National building code of Canada, 12th edn. Canadian Commission on Building and Fire Codes, National Research Council of Canada (NRCC), Ottawa

Paz M, Leigh WE (2003) Structural dynamics: theory and computation, 5th edn. Springer

PEER (2012) PEER Strong Motion Database. http://peer.berkeley.edu/smcat/search.html (accessed July 10, 2010)

Sadashiva VK, MacRae GA, Deam BL, Spooner MS (2012) Quantifying the seismic response of structures with flexible diaphragms. Earthquake Engineering and Structural Dynamics 41(10):1365–138

SEAOC (1999) Recommended lateral force requirements and commentary, 7th edn. Seismology Committee, Structural Engineers Association of California, Sacramento, California

Sobaih et al (1988) Regulations for earthquake-resistant design of buildings in Egypt. Egyptian Society for Earthquake Engineering (ESEE), Cairo

Thuat DV (2012) Strength reduction factor demands for building structures under different seismic levels. The Structural Design of Tall and Special Buildings. doi:10.1002/tal.1018

UBC (1997) Uniform building code, vol. 2: structural engineering design provisions. International Conference of Building Officials (ICBO), Whittier

10

On a sequence of approximate solutions: bending of a simply supported square plate

Kaza Vijayakumar

Abstract

The solution from Kirchhoff's theory for a simple textbook problem of bending of a simply supported square plate under doubly sinusoidal vertical load is re-examined concerning the exact solution of the three-dimensional problem. Well-known deficiencies in widely used classical theories such as Reissner's theory and first-order shear deformation theory (FSDT) along with Kirchhoff's theory are highlighted. These theories are based on the stationary property of relevant total potential. They do not provide proper initial solutions for the generation of a sequence of approximate solutions converging to the exact solution. The aim of the present work is to eliminate deficiencies in these classical theories. A sixth-order system consisting of three second-order equations is initially formulated with the aid of a normal strain ignored in FSDT. These second-order equations, uncoupled from transverse deflection in Kirchhoff's theory, govern in-plane displacements and transverse stresses. A supplementary problem consisting of a fourth-order system of equations is included to rectify inadequacies of polynomial expansions in the thickness-wise distribution of approximate solutions. More realistic and practical edge conditions are considered. An auxiliary problem is formulated, and its solution is used in the generation of a proper sequence of two-dimensional problems. Higher-order corrections to primary solutions are proposed through an iterative procedure.

Keywords: Plates; Bending; Isotropy; Elasticity

Introduction

One finds several theories reported in the literature for the analysis of bending of plates within the classical small deformation theory of elasticity. These theories are generally based on making suitable assumptions about thickness-wise distribution of displacements and/or stresses (or strains) to derive two-dimensional plate equations from three-dimensional equations. They are primarily intended to overcome inherent limitations in the theory of Kirchhoff (1850) which is simple and continuously used to obtain design information. For a brief survey of most of these theories, one may refer to the following reviews: Reissner (1985), Lo et al. (1977, 1978), Lewinski (1986, 1987), Blocki (1992), Kienzler (2002), and Batista (2010).

Kirchhoff's theory consists of a single-variable model in which in-plane displacements are expressed in terms of gradients of vertical deflection $w_0(x, y)$ from the assumption of zero transverse shear strains. Solution for w_0 is such that zero face shear conditions and edge support conditions are satisfied. It is governed by a fourth-order equation associated with two edge conditions instead of three edge conditions required in a 3-D problem. A consequence of this lacuna is the well-known Poisson-Kirchhoff boundary conditions paradox unresolved satisfactorily over the past 16 decades.

Reissner (1985) in his article felt that inclusion of transverse shear deformation effects was directly (or by implication) presumed to be the key in resolving Poisson-Kirchhoff's boundary conditions paradox. In his pioneering work, Reissner (1944) brought out the use of transverse shear stresses in deriving a sixth-order plate theory. Equations governing (average) displacements in his sixth-order theory (Reissner 1945) consist of Kirchhoff's fourth-order equation and a second-order equation coupled through edge conditions. They correspond to plate element equations governing stress resultants consistent in the reduction of 3-D equations through calculus of variations. After Reissner (1944, 1945), several sixth- and higher-order shear deformation theories were reported. Governing equations derived either from using calculus of variations or from

Correspondence: kazavijayakumar@gmail.com
Department of Aerospace Engineering, Indian Institute of Science, Bangalore 560012, India

statics correspond to errors in satisfying in-plane equilibrium equations and/or constitutive relations.

The assumption of zero transverse shear strains is discarded, resulting in a three-variable model of the widely used first-order shear deformation theory (FSDT), based on Hencky (1947). In this theory, w_0 and in-plane displacements $[u, v] = z[u_1(x, y), v_1(x, y)]$ are coupled in governing differential equations and boundary conditions. In Reissner's sixth-order theory (Reissner 1945) and FSDT, applied in-plane shear is combined with transverse shear resulting in a 'torsion-type' problem in flexure instead of in-plane shear combining with reactive vertical shear implied in Kelvin and Tait's (Love 1934) physical interpretation of contracted boundary condition in Kirchhoff's theory. In fact, torsion problem is associated with flexure problem (Vijayakumar 2011b), whereas flexure problem (unlike that directly or indirectly implied in the energy methods) is independent of the torsion problem. This torsion problem is, however, an approximation to the associated torsion problem in the bending of plates and results in unrealistic vertical deflection of the plate. In fact, a solution of the associated torsion problem is to nullify the effect due to applied or reactive edge stress τ_{xy} in the flexure problem (In this connection, a pertinent observation is that Kirchhoff's theory and FSDT are valid for the analysis of hard and soft, simply supported plates, respectively). This can be inferred from numerical results reported by Lewinski (1990) in which *higher-order theories give decreasing values of vertical deflection*. Recently, Kirchhoff's theory is modified such that the resulting sixth-order theory of bending and St. Venant's theory of torsion are mutually exclusive to each other (Vijayakumar 2009).

In the present author's recently developed theory designated as 'Poisson's theory of plates in bending', reactive transverse shear stresses are initially determined independent of material constants retaining their thickness-wise parabolic distributions. Dependence on material constants is through higher-order corrections to these stresses. These higher-order corrections are determined through an iterative procedure based on the equilibrium of a 3-D infinitesimal element instead of on the use of the stationary property of total potential in the energy methods. Dependence of analysis on vertical deflection $w_0(x, y)$ is eliminated (one should note that the coupling of analysis with w_0 in energy methods is due to the work done by the applied transverse stresses during deformation and also the root cause for Poisson-Kirchhoff's boundary conditions paradox). Also, the coupling between flexure problem and associated torsion problem is eliminated through an iterative method (Vijayakumar 2011a). The problem at each stage of iteration is defined by a sixth-order system of equations using a solution at the preceding stage of iteration. In the earlier investigations (Vijayakumar 2009, 2011a,b), in-plane displacements are expressed in terms of gradients of two functions, ψ and φ, in which ψ is related to w and φ is required to decouple bending and torsion problems. The function φ is an auxiliary plane harmonic function from zero rotation $(v_{,x} - u_{,y})$ denoted by ω_z about z-axis. The function φ was introduced earlier by Reissner (1947) as a 'stress function' governed by a 'wave equation' with imaginary velocity of propagation. Kirchhoff's theory, FSDT, Reissner's theory, and other shear deformation theories and reported higher-order theories based on energy principles are definitely not useful (explained later) in the generation of a proper sequence of 2-D problems converging to a 3-D problem.

In Kirchhoff's theory and shear deformation theories including the widely used FSDT, vertical deflection w_0 (average displacement in Reissner's sixth-order theory) is same in the face parallel planes. However, it is known physically that neutral plane deflection has to be higher than face deflection. This is due to the fact that an elastic medium is on either side of the neutral plane, whereas it is on one side of each of the top and bottom faces. Proper estimation of neutral plane deflection is dependent on normal strain ε_z ignored in Kirchhoff's theory and shear deformation theories. In an earlier investigation (Vijayakumar 2011a), it is shown that the expansion of displacements in polynomials of thickness coordinate z is not adequate to properly estimate face and neutral plane deflections. This fact is overlooked in the analysis of even isotropic homogeneous plates through the widely used FSDT and other shear deformation theories. A solution to a supplementary problem is required to obtain neutral plane deflection. It is, however, observed that the error in the estimation of face deflection is much higher than that of neutral plane deflection. However, it is desirable to provide uniform approximation to deformations through thickness of the plate. It appears to the author's knowledge that no suitable 2-D modeling which gives more or less the same percentage of approximation to thickness-wise distribution of a displacement variable is reported until now. Proper higher-order theories are to reduce only maximum error in the estimation of a physical variable to a desirable level.

Variational methods based on the principle of virtual work consist of virtual increments in displacement variables. Due to arbitrary virtual increments, one derives static equations governing real displacements independent of these virtual displacements which do not participate in these equations. If a 3-D variable, for example, $u(x, y, z)$ is expressed in infinite series with sum on $(i, j, k = 0, 1, 2, \ldots\ldots)$

$$u(x, y, z) = \left(\Sigma \alpha_i x^i\right)\left(\Sigma b_j y^j\right)\left(\Sigma c_k z^k\right)$$

then determination of real constants (a_i, b_j, c_k) from relevant algebraic equations generated in the numerical methods such as finite difference methods, finite element methods, etc. is presumably valid for solutions of such 3-D variables. However, expansion of displacements based on polynomials of thickness coordinate z is not adequate as mentioned earlier in obtaining proper solutions of displacements. Moreover, sequence of 2-D problems derived from variational principles do not converge to 3-D problems since

$$\delta\left[\left(\Sigma a_i x^i\right)\left(\Sigma b_j y^j\right)\left(\Sigma c_k z^k\right)\right] \neq \Sigma z^k \delta\left[\left(\Sigma a_{ik} x^i\right)\left(\Sigma b_{jk} y^j\right)\right],$$
$$(i, j, k = 0, 1, 2, \cdots\cdots).$$

Hence, it is necessary to consider equilibrium equations of a 3-D infinitesimal element (instead of plate element) to generate proper sequence of 2-D problems governing 2-D displacement variables.

The condition zero ω_z decoupling flexure from the torsion problem is satisfied in Kirchhoff's theory. By imposing this condition in FSDT, the sum of in-plane normal strains ($\varepsilon_x + \varepsilon_y$) in isotropic plates is governed by a second-order equation due to applied transverse loads. Using this sum of in-plane strains together with condition $\omega_z = 0$ gives uncoupled second-order equations governing in-plane displacements. Reactive transverse shear stresses are gradients of ($\varepsilon_x + \varepsilon_y$) uncoupled from w_0.

In Poisson's theory mentioned earlier, in-plane distribution of ($\varepsilon_x + \varepsilon_y$) is denoted by $\psi(x, y)$ so that the gradients $\psi_{,x}$ and $\psi_{,y}$ give in-plane distributions of transverse shear stresses independent of material constants. The function ψ is related to normal strain ε_z, and $\psi = 0$ implies $\varepsilon_z = 0$ along the supported edge. The condition $\varepsilon_z = 0$ in place of $w_0 = 0$ is mathematically valid and more so at point supports along the wall of the plate.

In the bending problem of a simply supported square plate under doubly sinusoidal vertical load, accuracy achieved in the estimation of face deflection (Vijayakumar 2011a) is much less than that of neutral plane deflection at the end of the first stage of iteration. This is likely to persist even in higher-order approximations. This deficiency is rectified earlier (briefly described and used later in the present analysis) through solution of an auxiliary bending problem (Vijayakumar 2013). This auxiliary problem is based on consideration of the influence of $\sigma_z = \pm zq/2$ satisfying face load conditions and associated constant transverse shear stresses on the preliminary solutions of primary bending problems. These transverse stresses are treated as virtual quantities since they do not participate in the static in-plane equilibrium equations as the above-mentioned virtual incremental displacements in the application of principle of virtual work. Constant transverse shear stresses, however, contribute in the integrated equilibrium equations, and linear σ_z contributes in the in-plane normal stress–strain relations. These transverse stresses are absent in Kirchhoff's theory.

Bending of a simply supported square plate

A square plate bounded within $0 \leq X, Y \leq a$, $Z = \pm h$ planes with reference to the Cartesian coordinate system (X, Y, Z) is considered. The material of the plate is homogeneous and isotropic with elastic constants E (Young's modulus), ν (Poisson's ratio), and G (Shear modulus) that are related to each other by $E = 2(1 + \nu)\, G$. For convenience, coordinates X, Y, Z and displacements (U, V, W) in non-dimensional form $x = X/a$, $y = Y/a$, $z = Z/h$, $(u, v, w) = (U, V, W)/h$ and half-thickness ratio $\alpha = (h/a)$ are used. With the above notation and $\rightleftarrows$ indicating interchange, equilibrium equations in stress components are (with 3-D stress components as functions of coordinates x, y, and z)

$$\alpha\left(\sigma_{x,x} + \tau_{xy,y}\right) + \tau_{xz,z} = 0 \qquad \rightleftarrows(x, y) \tag{1a}$$

$$\alpha\left(\tau_{xz,x} + \tau_{yz,y}\right) + \sigma_{z,z} = 0 \tag{1b}$$

in which the suffix after ',' denotes the partial derivative operator.

The plate is subjected to asymmetric load $\sigma_z = \pm (q_0/2) \sin(\pi x/a) \sin(\pi y/a)$ and zero shear stresses along $z = \pm 1$ faces. In a simply supported plate, conditions along x (and y) constant edges are

$$\sigma_x = 0 \quad \rightleftarrows(x, y), v = 0 \quad \rightleftarrows(v, u), w = 0 \tag{2}$$

In displacement-based models, stress components are expressed in displacements via six stress–strain constitutive relations and six strain–displacement relations. These relations within the classical small deformation theory of elasticity are as follows:

Constitutive relations:

$$E\varepsilon_x = \sigma_x - \nu\left(\sigma_y + \sigma_z\right) \quad \rightleftarrows(x, y) \tag{3a}$$

$$E\varepsilon_z = \sigma_z - \nu\left(\sigma_x + \sigma_y\right) \tag{3b}$$

$$\left[\tau_{xy}, \tau_{xz}, \tau_{yz}\right] = G\left[\gamma_{xy}, \gamma_{xz}, \gamma_{yz}\right] \tag{4}$$

Strain–displacement relations:

$$\left[\varepsilon_x, \varepsilon_y, \varepsilon_z\right] = \left[u_{,x}, v_{,y}, w_{,z}\right] \tag{5}$$

$$\left[\gamma_{xy}, \gamma_{xz}, \gamma_{yz}\right] = \left[u_{,y} + v_{,x}, u_{,z} + w_{,x}, v_{,z} + w_{,y}\right] \tag{6}$$

Methods

Analysis consists of exact solutions of 3-D equations of bending problem with and without torsion, 2-D equations from Poisson's theory and extended Poisson's theory.

Exact solutions

With reference to the proper sequence of 2-D problems converging to a 3-D problem, an exact solution of the above-mentioned problem of a simply supported square plate presented briefly earlier (Vijayakumar 2011a) is described here to assess solutions of the present 2-D problems. Equilibrium equations in displacements in the 3-D problem are

$$\alpha^2 \Delta u + u_{,zz} + \frac{1}{1-2\nu}\alpha\left(\alpha\, u_{,xx} + \alpha\, v_{,yx} + w_{,zx}\right) = 0 \quad \rightleftarrows (x, y), (u, v) \tag{7}$$

$$\alpha^2 \Delta w + w_{,zz} + \frac{1}{1-2\nu}\left(\alpha\, u_{,xz} + \alpha\, v_{,yz} + w_{,zz}\right) = 0 \tag{8}$$

Displacements are expressed in the form with $\beta = \sqrt{2}\ \alpha\ \pi$

$$u = (A_1 \sinh\beta z + A_2 z \cosh\beta z) \cos\pi x \sin\pi y \quad \rightleftarrows (x, y), (u, v) \tag{9}$$

$$w = (C_1\ \cosh\ \beta z + C_2 z \sinh\ \beta z) \sin\ \pi x \sin\ \pi y \tag{10}$$

From satisfying the above equilibrium equations, one gets $C_2 = \sqrt{2}A_2 = (2\alpha\ \pi\ A_1 - \beta\ C_1)\ /\ (3\text{–}4\ \nu)$.

Zero shear stresses and vertical load condition along faces give

$$C_1 = \frac{\beta \tanh \beta + 2(1-\nu)}{\beta \tanh \beta - (1-2\nu)}\sqrt{2}A_1 = (1+\nu)\frac{\beta \sinh \beta + 2(1-\nu)\cosh \beta}{\beta(\sinh \beta \cosh \beta - \beta)}(q_0/2E) \tag{11}$$

From the above solutions, one obtains with $\nu = 0.3$ and $\alpha = 1/6$ (i.e., plate thickness ratio $2h/a = 1/3$)

$$(E/2q_0)\ w(a/2, a/2, 0) = 3.49$$
$$(E/2q_0)\ w(a/2, a/2, 1) = 2.37$$

The above estimates correspond to those from the associate torsion problem since $\omega_z \neq 0$ in Equations 7 and 8. In the case $\omega_z = 0$ decoupling bending and torsion problems, equilibrium equations are in terms of (u, v, σ_z) given by

$$E'\alpha^2 \Delta u + \mu\alpha\sigma_{z,x} + \tau_{xz,z} = 0 \quad \rightleftarrows (x, y)$$
$$\alpha\left(\tau_{xz,x} + \tau_{yz,y}\right) + \sigma_{z,z} = 0$$

Estimated vertical deflection parameters from solutions of the above equations are

$$(E/2q_0)\ w(a/2, a/2, 0) = 4.49$$
$$(E/2q_0)\ w(a/2, a/2, 1) = 4.17$$

Estimates from Kirchhoff's theory and FSDT

The above face deflection parameter from Kirchhoff's theory is 2.27. It is the same in all face parallel planes. Correction due to coupling with torsion is 1.45 (Vijayakumar 2011b), whereas it is 1.42 from FSDT and other sixth-order theories (Lewinski 1990). With reference to numerical values reported in Lewinski's article, the above correction is less than 1.23 in Reissner's 12th-order and other higher-order theories. It clearly shows that shear deformation theories and other higher-order theories do not lead to the solutions of bending problems. It is to be noted that vertical deflection w_0 from FSDT is usually considered to be w_{0N}, but its estimated value (3.69) of the deflection parameter is much nearer the w_{0F}.

Poisson's theory

In Kirchhoff's theory, the basic variable is $w_0(x, y)$, and $[u, v]$ are from $[\gamma_{xz}, \gamma_{yz}] \equiv 0$ in the plate. In FSDT, w_0 is associated with $z[u_1, v_1]$ through $[\gamma_{xz}, \gamma_{yz}]$. In these theories and other shear deformation theories, σ_z is neglected in constitutive relations. In the present analysis, as in Kirchhoff's theory and FSDT, σ_z is neglected initially in the semi-inverted constitutive relations

$$\sigma_x = E'\left(\varepsilon_x + \nu\, \varepsilon_y\right) + \mu\sigma_z \quad \rightleftarrows (x, y) \tag{12a}$$

$$\varepsilon_z = -\mu\, e + (1-2\,\nu\,\mu)\sigma_z/E \tag{12b}$$

in which $e = (\varepsilon_x + \varepsilon_y)$, $E' = E\ /\ (1 - \nu^2)$ and $\mu = \nu\ /\ (1 - \nu)$.

In-plane variables (u_1, v_1) uncoupled from w_0 are basic variables as in earlier investigations (Vijayakumar 2009, 2011a, b). Due to the condition $\omega_z = 0$ required to decouple bending and torsion problems, one obtains reactive transverse stresses from thickness-wise integration of equilibrium equations

$$\tau_{xz} = E' f_2(z) \alpha\, e_{1,x} \quad \rightleftarrows (x, y) \tag{13}$$

$$\sigma_{z3} = -E' f_3(z) \alpha^2 \Delta\, e_1 \tag{14}$$

in which $f_2 = (1 - z^2)\ /\ 2$ and $f_3 = (z - z^3/3)\ /\ 2$.

Constitutive relation gives $\varepsilon_z = -\mu\, f_1\, e_1$ with $f_1 = z$ and $e_1 = \varepsilon_{z1}$. From satisfying face load condition, one gets the equation governing e_1 as

$$(2/3) E' \alpha^2 \Delta\, e_1 + q = 0 \tag{15}$$

In Poisson's theory, e_1 in the above equation is replaced by $\psi_2(x, y)$, and it is solved with edge condition $\psi_2 = 0$ in the simply supported plate problem.

As mentioned earlier, the function ψ_2 is related to normal strain ε_z and $\psi_2 = 0$ implies $\varepsilon_z = 0$.

Equations governing u_1 and v_1 are given by

$$\alpha(u_{1,x} + v_{1,y}) = \psi_2, \alpha(u_{1,y} - v_{1,x}) = 0 \quad (16)$$

so that

$$\alpha^2 \Delta u_1 = \alpha \psi_{2,x}, \alpha^2 \Delta v_1 = \alpha \psi_{2,y} \quad (17)$$

Homogeneous equations show that u_1 and v_1 are conjugate harmonic functions coupled through conditions (2) on u_1 and v_1 along x (and y) constant edge

$$u_1 = 0 \quad \text{or} \quad E' u_{1,x} = T_x(y) \quad \rightleftarrows (x, y), (u, v) \quad (18a)$$

$$v_1 = 0 \quad \text{or} \quad 2G\, v_{1,x} = T_{xy}(y) \rightleftarrows (x, y), (u, v) \quad (18b)$$

In the above equations, T_x and T_{xy} are prescribed stress distributions.

From zero face shear conditions, one obtains $w_0(x, y)$ in terms of known $[u_1, v_1]$ in the form

$$\alpha w_0(x, y) = -\int [u_1 dx + v_1 dy] \quad (19)$$

The vertical deflection thus obtained corresponds to face deflection (note that zero face shear conditions do not participate in the 3-D domain equations unlike in Kirchhoff's theory). Its zero value along the edge requires simply a support to prevent vertical deflection of intersection of the face with the wall of the plate. Such a support also ensures zero deflection of the neutral plane along its edge. It is because the face and neutral plane deflections (w_{0F}, w_{0N}) are one and the same since $w(x, y, z)$ with $(w_2 + \varepsilon_{z1}) = 0$ can be expressed as

$$\begin{aligned} w(x, y, z) &= w_{0F}(x, y) + f_2(z)\, w_2 \\ &= w_{0N}(x, y) - z^2 \varepsilon_{z1}/2 \end{aligned} \quad (20)$$

However, $f_2(z)$ of the second-order correction w_2 to the vertical deflection w is parabolic; thereby, the applied or reactive transverse shear stress along an edge is parabolic.

Equations 15 and 17 form a sixth-order system to determine ψ_2, u_1, and v_1. Relevant three edge conditions correspond to those required by Poisson. Hence, the theory is designated as 'Poisson's theory of plates in bending,' parallel to Kirchhoff's theory. It is significant to note that the condition $w_0 = 0$ prescribed along segments of the edge has no effect on the displacements $[u_1, v_1, w_0]$ obtained from the present Poisson's theory. It is to be noted that Poisson's theory gives an additional value of 0.27 to $w(x, y, 1)$ due to ε_{z1} to the earlier w_{0F} estimate from Kirchhoff's theory.

Numerical results and discussion

After the first stage of iteration involving $f_3(z)[u_3, v_3]$ in the earlier work (Vijayakumar 2011a), higher-order correction to w_0 uncoupled from torsion is 1.26 so that total correction to the value from Kirchhoff's theory is about 1.54. Neutral plane deflection w_{0N} is corrected from the solution of a supplementary problem. Here the total correction over face deflection is about 0.66 giving a value of 4.46 which is very close to the exact value 4.49. Hence, it is safe to conclude that second-order corrections in the displacements and transverse stresses from the iterative method serve the purpose of assessing data from Kirchhoff's theory and FSDT.

Solution for w_0 consists of satisfying (i) zero face shear conditions and (ii) edge support conditions in w_0. In Kirchhoff's theory, both of these requirements are met through the single variable w_0 governed by a fourth-order equation. In Reissner (1945), the average vertical displacement satisfies condition (ii), and in FSDT, condition (i) is not satisfied but its effect is included through shear energy correction factor. In any case, both of them are approximations to associated torsion problem (Vijayakumar 2011b). The present analysis provides some clarity with regard to the preliminary solution of primary flexure problem. It shows that determining in-plane displacements, bending stresses, and reactive transverse stresses from integration of equilibrium equations is uncoupled from w_0. However, edge condition $w_0 = 0$ or contracted stress resultant $V_x = V_{xo}$ in Kirchhoff's theory is different from the condition along x (and y) = constant edge

$$\varepsilon_z = 0 \quad \text{or} \quad \tau_{xz2} = T_{xz} \qquad \rightleftarrows (x, y)$$

Error in the estimated value (=3.81) of w_{0F} is relatively high compared to the accuracy achieved in neutral plane deflection. In the earlier work (Vijayakumar 2011a), estimation of w_{0F} is further improved by modifying in-plane displacements (u^*_3, v^*_3) such that (τ_{xz2}, τ_{yz2}) are independent of ε_{z1}. In the present example, correction to the face deflection changes to 1.43 so that face deflection value is 3.97 (=2.27 + 0.27 + 1.43) which is under 4.7% from the exact value. The correction 1.43 to the face deflection is due to ε_{z3} from the constitutive relation. Determination of ε_{z3} in terms of (σ_{z3}, u_3, v_3) involves lengthy algebra and arithmetical work.

The above analysis consists of a basic sixth-order system and a supplementary fourth-order system. Since the second-order corrections are mainly due to inclusion of σ_{z1} in the in-plane constitutive relations, it is much simpler to find its effect from the following extended Poisson's theory without consideration of higher-order displacement components (u_3, v_3).

Improved estimates

The analysis presented here eliminates the assumptions in Kirchhoff's theory. Preliminary solution consists of a combination of solutions of four problems presented elsewhere (Vijayakumar 2013) in the analysis of bending of symmetric laminates with anisotropic plies in which the analysis of the present title problem is included as illustrative example. The first one is a solution of an auxiliary problem to account for thickness-wise linear distribution of $\sigma_z = z\ q(x, y)\ /\ 2$ neglected in the constitutive relations in Kirchhoff's theory and shear deformation theories. The second one is concerned with determining $[u_{1c}, v_{1c}]$ due to $\sigma_z = z\ q(x, y)\ /\ 2$ in the in-plane constitutive relations. In the case of a simply supported plate, $[u_{1c}, v_{1c}]$ are governed by a fourth-order system with homogeneous in-plane edge conditions. The third one consists of solution of a fourth-order system of finding $[u_1, v_1]_b$ due to transverse stresses obtained in the second problem. Finally, the fourth one consists of solution of a fourth-order system of a supplementary problem to obtain a more or less uniform approximation of thickness-wise distributions of deformations in the plate.

Auxiliary problem

One finds that σ_{z1} is neglected in shear deformation theories. Moreover, transverse shear stress conditions along the edges are specified with reference to reactive (statically equivalent) stresses in the form of parabolic distributions, though mathematically valid but not practical. Kirchhoff's theory is based on assuming identically zero transverse shear strains and neglecting σ_{z1} in the constitutive relations. The former assumption results in highly constrained deformations of the plate.

In view of the above observations, transverse shears along faces of the plate are initially assumed in the form $[\tau_{xz0}, \tau_{yz0}] = [T_{xz}(x, y), T_{yz}\ (x, y)]$. It is more reasonable and practical to assume that they are the same in face parallel planes instead of $w_0(x, y)$. Correspondingly, $[\tau_{xz0}, \tau_{yz0}]$ along constant x (and y) edges are

$$\tau_{xz0} = T_{xz}(y) \qquad \rightleftarrows(x, y) \tag{21}$$

They are more realistic edge conditions and more practical if they are independent of in-plane coordinates. In view of Equation 1b, $[\tau_{xz0}, \tau_{yz0}]$ are expressed as

$$\left[\tau_{xz0}, \tau_{yz0}\right] = \alpha\left[\psi_{0,x}, \psi_{0,y}\right] \tag{22}$$

so that $\psi_0(x, y)$ is governed by

$$\alpha^2\Delta\psi_0 + \sigma_{z1} = 0 \tag{23}$$

in which Δ is the plane Laplace operator $(\partial^2/\partial x^2 + \partial^2/\partial y^2)$ and $\sigma_{z1} = q/2$ with $\sigma_z = z\ \sigma_{z1}$.

The above equation is to be solved with edge condition either $\psi_0 = 0$ or its normal gradient equal to the applied transverse shear stress along the edge. It is to be noted that the assumed face stresses $[T_{xz}, T_{yz}]$ correspond to $[\tau_{xz0}, \tau_{yz0}]$ in Equation 21 consistent with applied or reactive vertical shears along the edges.

The condition $\psi_0 = 0$ is different from the usual condition $w_0 = 0$ in Kirchhoff's theory. The function ψ_0 is related to normal strain ε_z which is proportional to Δw_0. If the plate is free of applied transverse stresses, that is, the plate is subjected to bending and twisting moments only, $\psi_0(x, y)$, thereby, $\Delta w_0 \equiv 0$ from Equation 2 (In Kirchhoff's theory, the tangential gradient of Δw_0 is proportional to the corresponding gradient of applied τ_{xy} along the edge of the plate implied from Kelvin and Tait's physical interpretation of the contracted transverse shear condition). This Laplace equation is not adequate to satisfy two in-plane edge conditions. One needs its conjugate harmonic functions implying $v_{1,x} = u_{1,y}$ to express in-plane displacements in the form

$$[u_1, v_1] = -\alpha z\left[w_{0,x} + \varphi_{0,y}, w_{0,y} - \varphi_{0,x}\right] \tag{24}$$

The function φ_0 was introduced earlier by Reissner (1945) as a stress function in satisfying Equation 2.

After finding w_0 and φ_0 from solving in-plane equilibrium equations (Equation 1a,b), correction w_{0c} to w_0 from zero face shear conditions is given by

$$w_{0c} = \int\left[\varphi_{0,y}dx - \varphi_{0,x}dy\right] \tag{25}$$

One should note here that $w(x, y) = w_0 + w_{0c}$ is the face deflection. It does not satisfy the prescribed $w = 0$ edge condition. As mentioned in Poisson's theory, its zero value along the edge requires only a support to prevent the vertical deflection of intersection of the face with the wall of the plate. Correction w_{0c} is obviously due to the tangential gradient of τ_{xy} in vertical shear $V_{x(y)}$. In Kirchhoff's theory, $w(x, y)$ is a bi-harmonic function. It is obtained in the simply supported (hard type) plate satisfying $w = 0$ along with in-plane condition along the x (and y) constant edge

$$w_{,x} = 0 \quad \text{or} \quad E'w_{,xx} = T_x(y) \qquad \rightleftarrows(x, y) \tag{26}$$

In Kirchhoff's theory, $w(x, y)$ obtained above corresponds to neutral plane deflection w_{0N} with edge condition $w_{0N} = 0$. Obviously, this edge condition is not required to obtain $[u_1, v_1]$ and face deflection. With reference to the present analysis, it is relevant to note the following observations: In the preliminary solution, σ_z, $\alpha[\tau_{xz}, \tau_{yz}]$, $\alpha^2[\sigma_x, \sigma_y, \tau_{xy}]$ are of O(1). As such, estimation of in-plane stresses, thereby, in-plane displacements, are not dependent on w_0. The support condition $w_0 = 0$ along the plate edge does not play any role in the estimation of in-plane displacements. In fact, Kirchhoff's theory gives only a lower bound for α for the validity of small deformation theory. Validity of its application is a questionable proposition for small values of α even slightly higher than this lower bound since the theory gives highly constrained solutions for displacements. Shear deformations are included in FSDT in which face shear conditions are not satisfied and Reissner's theory in which edge condition on w_0 is not satisfied. Both of these theories give more or less the same maximum vertical deflection. It shows that satisfying either edge condition on w_0 or face shear conditions has equal effect on deformations. In each of these theories, both conditions are satisfied only in the limit. Limiting process, however, nullifies the effect due to the applied or reactive edge stress τ_{xy} in the bending problem. Moreover, linear variation of σ_z in z is neglected in constitutive relations. The present analysis shows that this linear σ_z results in more flexible plate deformations in conjunction with zero face shear conditions.

In 3-D problems of homogeneous plates, $w_0(x, y, 0)$ and $w_0(x, y, 1)$ are the neutral plane deflection w_{0N} and face deflection w_{0F}, respectively. Both of them are 2-D functions and they are from the thickness-wise integration of ε_z in a 2-D problem. Determination of w_{0N} used as an unknown variable is dependent on edge support condition and that of w_{0F} on zero face shear conditions. The present analysis is with reference to finding w_{0F} and the need to satisfy that edge condition on w_{0N} does not exist.

$[u_{1c}, v_{1c}]$ due to transverse stresses $[\tau_{xz0}, \tau_{yz0}, z\sigma_{z1}]$

With reference to the second problem mentioned earlier, the in-plane displacements to be determined are $z(u_1, v_1)_c$. Here transverse stresses are expressed in the form

$$\tau_{xz} = \tau_{xz0} + f_2 \tau_{xz2c} \tag{27}$$

$$\tau_{yz} = \tau_{yz0} + f_2 \ \tau_{yz2c} \tag{28}$$

$$\sigma_z = z\,\sigma_{z1} + f_3 \sigma_{z3c} \tag{29}$$

in which $(\tau_{xz2}, \tau_{yz2}, \sigma_{z3})_c$ yet to be determined are corrections due to $(u_1, v_1)_c$. *Since the added terms are dependent on material constants, the above transverse shear stresses are no longer applied stresses along the edges.*

In order to keep σ_{z3c} as a free variable, the function $f_3(z)$ in Equation 29 is replaced by $f_3^*(z) = (f_3 - z/3)$ so that $f_3^*(\pm 1) = 0$. Then the face load condition is satisfied with

$$\sigma_z = z\,(q/2) + [f_3^* \sigma_{z3c}] \tag{30}$$

In further analysis, it is convenient to express σ_z with $\beta_1 = 1/3$ in the form

$$\sigma_z = [q/2 - \beta_1 \sigma_{z3c}]z + f_3(z)\,\sigma_{z3c} \tag{31}$$

In-plane distributions $[u_1, v_1]_c$ are modified in the form with

$$u_{1c}* = u_{1c} + \tau_{xz0}/G - \alpha\, w_{0,x} \qquad \rightleftarrows (x, y), (u, v) \tag{32}$$

The introduction of $w_0(x, y)$ term in the above equation is to avoid its participation in the static in-plane equilibrium equations. Modified transverse shear stresses from stress–strain and strain–displacement relations are

$$\tau_{xz2c} = G u_{1c} + \tau_{xz0} \qquad \rightleftarrows (x, y), (u, v) \tag{33}$$

Substitution of the above shear stresses in Equation 1b with $\sigma_z = z[q/2 - \beta_1 \sigma_{z3c}]$ and $e_{1c} = (\varepsilon_{x1} + \varepsilon_{y1})_c$ gives

$$G e_{1c} = \beta_1 \sigma_{z3c} \tag{34}$$

For the use of $[u_1^*, v_1^*]_c$ in the integration of equilibrium equations (Equation 1a,b), displacements $[u_1, v_1]_c$ due to the requirement of $v_{1c,x} = u_{1c,y}$ to decouple bending and torsion problems are expressed in the form

$$[u_1, v_1]_c = -\alpha\, [\psi_{1,x}, \psi_{1,y}]_c \tag{35}$$

in the case of hard and/or soft simply supported edge conditions.

Contributions of ψ_{1c} and w_0 in $[u_1, v_1]_c^*$ are one and the same in giving corrections to $w(x, y, z)$ and transverse stresses (in fact, the contribution of w_0 is through strain–displacement relations in static equilibrium equations and through constitutive relations in through-thickness integration of equilibrium equations). Since w_0 does not participate in the in-plane static equations, its contribution is through $[u_1, v_1]_c$ in the integrated in-plane equilibrium equations. Hence, w_0 in $[u_1, v_1]_c^*$ is replaced by ψ_{1c} (so as to be independent of w_0 used in strain–displacement relations) so that $[u_1, v_1, \varepsilon_{x1}, \varepsilon_{y1}, \gamma_{xy1}]_c^*$ are

$$u_{1c}* = (2\,u_{1c} + \gamma_{xz0}) \qquad \rightleftarrows (x, y), (u, v) \tag{36}$$

$$\varepsilon_{x1c}* = \left(2\,\varepsilon_{x1c} + \alpha\,\gamma_{xz0,x}\right) \qquad \rightleftarrows (x, y) \tag{37a}$$

$$\gamma_{xy1c}* = \left[2\,\gamma_{xy1c} + \alpha\left(\gamma_{xz0,y} + \gamma_{yz0,x}\right)\right] \tag{37b}$$

From the integration of Equation 1a using the above strains in Equation 12a along with $v_{1,x} = u_{1,y}$, reactive transverse stresses are

$$\tau_{xz2c} = \alpha\,(E' e_{1c}* + \mu\sigma_{z1})_{,x} \qquad \rightleftarrows (x, y) \tag{38}$$

$$\alpha\left(\tau_{xz2,x} + \tau_{yz2,y}\right)_c* + \sigma_{z3c} = 0 \tag{39}$$

One equation governing in-plane displacements (u_1, v_1), noting that σ_{z3} from Equation 29 is negative of the one from Equation 34 due to $(f_{3,zz} + f_1) = 0$, is given by

$$\alpha\,\beta_1\left(\tau_{xz2,x} + \tau_{yz2,y}\right)_c = G\,e_{1c}$$

From Equations 29, 34, and 39, one gets

$$E'\beta_1\alpha^4\Delta\Delta\,(2\,\psi_{1c} + \psi_0/G) + Ge_{1c} - \mu\,\beta_1\alpha^2\Delta\sigma_{z1} = 0 \tag{40}$$

The above equation becomes a fourth-order equation in ψ_{1c} to be solved with two in-plane conditions along constant x (and y) edges

$$\sigma_{x1c}{}^* = 0 \quad \rightleftarrows(x, y), v_{1c}{}^* = 0 \quad \rightleftarrows(u, v) \tag{41}$$

After some algebra with $G = E/2(1 + \nu)$, $\mu = \nu/(1 - \nu)$ and $\psi_{1c} = c_{1c}$ sin (πx) sin (πy), the equation governing c_{1c} becomes

$$\left[\beta^2 + \frac{3(1-\nu)}{4}\right] c_{1c} - \frac{(1+\nu)}{8}(2 - \nu)(2q_0/E) = 0 \tag{42}$$

From the above equation, one gets $c_{1c} = 0.26(2q_0/E)$ with $\nu = 0.3$ and $\alpha = 1/6$.

Face deflection w_{0c} from the integration of transverse shear strain–displacement relations is

$$w_0 = \left\{\left[(1+\nu)/2\beta^2\right](2q_0/E) + c_{1c}\right\} \tag{43}$$

from which $w_{0c} = 1.39(2q_0/E)$ in the present example.

Primary displacements $z(u_1, v_1)_b$

In the absence of higher-order in-plane displacement terms, transverse stresses in the bending problem are

$$\tau_{xz2b} = \tau_{xz0} + \tau_{xz2c} \quad \rightleftarrows(x, y) \tag{44}$$

$$\sigma_{z3b} = \sigma_{z1} + \sigma_{z3c} \tag{45}$$

Since the added terms are dependent on material constants, above transverse shear stresses are no longer applied stresses along the edges but they are completely known. By expressing $[\tau_{xz2}, \tau_{yz2}]_b$ as gradients of a function ψ_{2b}, one obtains

$$\alpha\psi_{2b} = \int (\tau_{xz2b}dx + \tau_{yz2b}dy) \tag{46}$$

Equations governing u_{1b} and v_{1b} as in the above Poisson's theory are given by

$$\alpha(u_{1b,x} + v_{1b,y}) = (G/E')\psi_{2b}, \quad \alpha\,(u_{1b,y} - v_{1b,x}) = 0 \tag{47}$$

so that

$$\alpha^2\Delta\,u_{1b} = (G/E')\alpha\,\psi_{2b,x}, \tag{48a}$$

$$\alpha^2\Delta\,v_{1b} = (G/E')\alpha\,\psi_{2b,y} \tag{48b}$$

In the case of a simply supported plate, solutions for $[u_1, v_1]_b$ from the above pair of second-order equations are coupled through conditions along x (and y) constant edges

$$u_{1b,x} = 0, v_{1b} = 0 \quad \rightleftarrows(u, v) \tag{49}$$

w_{0b} due to (u_{1b}, v_{1b})

Here it is convenient to take $(u_{1b}, v_{1b}) = -\alpha[\psi_{1,x}, \psi_{1,y}]_b$ as in Kirchhoff's theory so that one obtains the equation governing ψ_1 as

$$E'\alpha^4\Delta\Delta\,\psi_{1b} = 3\left[q/2 + G\,e_{1c}/\beta_1\right] \tag{50}$$

The above equation is expressed in the form

$$\alpha^4\Delta\Delta\,\psi_{1b} = \frac{3(1-\nu)(1+\nu)}{4}(2q/E) - \frac{3(1-\nu)}{2}\alpha^2\Delta\,\psi_{1c} \tag{51}$$

from which

$$c_{1b} = \left[\frac{3(1-\nu)(1+\nu)}{4}(2q/E) + \frac{3(1-\nu)}{2}\beta^2 c_{1c}\right]/\beta^4 \tag{52}$$

In the present example, the estimated c_{1b} is $2.42(2q_0/E)$, (=2.27 + 0.15). It is also equal to w_{0b} from the integration of strain–displacement relations.

The estimated face deflection $(E/2q_0)\,w_F = 4.08$ (=2.27 + 0.27 + 1.39 + 0.15) which is fairly close to the exact value (4.17). *It shows that the solution to w from the present analysis provides proper correction to the estimation of face deflection.* It is underestimated by 2.16%.

Supplementary problem

Corrective in-plane displacements in the supplementary problem are assumed in the form

$$u_s = u_{1s}\sin(\pi\,z/2) \quad \rightleftarrows(u, v) \tag{53}$$

In-plane stresses are

$$\sigma_{1si} = \left(E'\varepsilon_{1sj} + \mu\sigma_{z1}\right)\sin(\pi\,z/2) \tag{54}$$

With the above in-plane stresses along with $[\tau_{xz}, \tau_{yz}] = [\tau_{xz2}, \tau_{yz2}]_s$ cos (π $z/2$) and $\sigma_{z3} = \sigma_{z3s}$ sin (π $z/2$), the integration of equilibrium equations gives

$$\tau_{xz2s} = -(2/\pi)\,\alpha(E'e_{1s} + \mu\sigma_{z1})_{,x} \quad \rightleftarrows(x, y) \tag{55}$$

$$\sigma_{z3s} = (2/\pi)^2\alpha^2\Delta\,(E'e_{1s} + \mu\sigma_{z1}) \tag{56}$$

In-plane distributions u_{1s} and v_{1s} are added as corrections to the known in-plane displacements $(u_1{}^*, v_1{}^*)$ so that (u, v) in the supplementary problem are

$$u = (u_1{}^* + u_{1s})\sin(\pi z/2) \quad \rightleftarrows(u, v) \tag{57}$$

in which $u_1{}^* = (u_{1c}{}^* + u_{1b})$.

By equating σ_{z3s} (from the integration of Equation 1a,b with s variables) with $\beta_1\sigma_{z3}$ (from static equations with * variables), one equation governing $[u_{1s}, v_{1s}]$ is

$$(2/\pi)^2\alpha^2\Delta(E'e_{1sj} + \mu\sigma_{z1}) = \beta_1\sigma_{z3} \tag{58}$$

By expressing $[u_{1s}, v_{1s}] = -\alpha\ [\psi_{1s,x}, \psi_{1s,y}]$ due to the second equation $v_{1s,x} = u_{1s,y}$, Equation 58 becomes a fourth-order equation in ψ_{1s} to be solved with two in-plane conditions along constant x (and y) edges in the present example

$$\sigma_{x1s} = 0 \qquad \rightleftarrows(x, y), v_{1s} = 0 \qquad \rightleftarrows(v, u) \tag{59}$$

ψ_{1s} is related to e_{1c} by the equation

$$E'(4/\pi^2)\ \alpha^4\Delta\Delta\psi_{1s} = Ge_{1c} \tag{60}$$

c_{1s} is related to c_{1c} by the equation

$$(E/2q_0)\ c_{1s} = \left[\frac{1-v}{2}\pi^2 - \beta_1{}^2\right](E/2q_0)\ c_{1c} \tag{61}$$

so that $c_{1s} = 0.55(2q_0/E)$ with $v = 0.3$ and $\alpha = 1/6$. Neutral plane deflection $w_{0N} = 4.36(2q_0/E)$, (=2.27 + 0.15 + 1.39 + 0.55) is underestimated by 2.90% from the exact value (4.49).

Evaluation of neutral plane deflection involves u_{1c} and u_{1s} associated with $f_2(z)$ and cos ($\pi z/2$) distributions of transverse shear stresses, respectively. As such, the solution of the present auxiliary problem provides second-order corrections to the solutions of the primary problem. It is to be noted that the wide variation of percentage of errors from 0.67 to 4.80 in w along the vertical line from the neutral plane to the face plane at the first stage of iteration using (u_3, v_3) in Poisson's theory is reduced from 2.16 to 2.90 from the face plane to the neutral plane without using (u_3, v_3) in extended Poisson's theory. One iteration using (u_3, v_3) is expected to reduce this gap even further along with decreased percentage of errors. *A significant implication of this observation is that solution of the present auxiliary problem is necessary so as to obtain a more or less uniform approximation to thickness-wise distributions of deformations.*

The fourteenth-order system of equations in the analysis of a simply supported plate presented here forms the initial set in the iterative procedure. Moreover, *the analysis clearly establishes that the second-order corrections in Kirchhoff's theory are mainly due to inclusion of $\sigma_z = z\ \sigma_{z1}$ in the constitutive relations and not due to consideration of shear deformations presumed in FSDT and Reissner's theory.* It is possible to decrease the percentage of errors in the estimation of $w(x, y\ z)$ by considering the influence of in-plane displacements thus obtained in estimation of transverse stresses. Successive iterative steps consist of a pair of fourth-order system of equations forming a proper sequence of sets of 2-D problems converging to a 3-D problem.

Conclusions

Estimation of transverse stresses in the preliminary solution is independent of material constants through Poisson's theory in the analysis of primary bending problem defined from Kirchhoff's theory. Hence, it is unaltered in the analysis of bending of homogeneous and laminated plates with orthotropic and anisotropic materials. Application of transverse shear stress independent of thickness coordinate z along the edges of the plate is more practical instead of stress resultants which have no unique z-distributions. Coupling of $w_0(x, y)$ in obtaining in-plane displacements which are the root cause for the Poisson-Kirchhoff boundary condition paradox is eliminated. In addition, inclusion of the tangential gradient of applied in-plane shear stress in transverse shear in Kirchhoff's theory is removed. Vertical stress σ_z linear in z neglected in Kirchhoff's theory and shear deformation theories gives primary second-order corrections instead of shear deformations. Thickness-wise distributions of displacements in terms of polynomials in z are not adequate to reduce 3-D problems into a sequence of 2-D problems.

Competing interests

The author declare that he has no competing interests.

Acknowledgement

The author expresses his sincere thanks to Dr. N. Bhushan Mandava, President, Mandava Associates, LLC, Washington, D.C., USA, for his help to initiate the author to academic work.

References

Batista M (2010) The derivation of the equations of moderately thick plates by the method of successive approximations. Acta Mech 210:159–168

Blocki J (1992) A higher-order linear theory for isotropic plates-I: theoretical considerations. I J Sol Struc 29(7):825–836

Hencky H (1947) Über die Berücksichtigung der Schubverzerrung in ebenen Platten. Ingenieur-Archiv 16:72–76

Kienzler R (2002) On consistent plate theories. Arch Appl Mech 72:229–247. doi:10-1007/s00419-002-0220-2.

Kirchhoff G (1850) Über das Gleichgewicht und die Bewegung einer elastischen Scheibe. Journal für reins und angewandte Mathematik 40:51–88

Lewinski T (1986) A note on recent developments in the theory of elastic plates with moderate thickness. Eng Trans 34(4):531–542

Lewinski T (1987) On refined plate models based on kinematical assumptions. Ingenieur- Archiv 57(2):133–146

Lewinski T (1990) On the twelth-order theory of elastic plates. Mech Res Comm 17(6):375–382

Lo KH, Christensen RM, Wu EM (1977) A higher-order theory of plate deformation. J Appl Mech 44:663–676

Lo KH, Christensen RM, Wu EM (1978) Stress determination for higher-order plate theory. I J Sol Struc 14:655–662

Love AEH (1934) A treatise on mathematical theory of elasticity, 4th edn. Cambridge University Press, Cambridge, pp 458–463

Reissner E (1944) On the theory of bending of elastic plates. J Math Phys 23:184–191

Reissner E (1945) The effect of transverse shear deformations on the bending of elastic plates. J Appl Mech 12:A69–A77

Reissner E (1947) On bending of elastic plates. Quarterly of Appl Math 5(1):55–68

Reissner E (1985) Reflections on the theory of elastic plates. Appl Mech Rev 38:1453–1464

Vijayakumar K (2009) A new look at Kirchhoff's theory of plates. AIIA J 47:1045–1046. doi:10.2514/1.38471.

Vijayakumar K (2011a) A relook at Reissner's theory of plates in bending. Arch App Mech 81:1717–1724. doi:10.1007/s00419-011-0513-4.

Vijayakumar K (2011b) Modified Kirchhoff's theory of plates including transverse shear deformations. Mech Res Comm 38(3):211–213. doi:10.1016/j.mechrescom.2011.02.007.

Vijayakumar K (2013) On uniform approximate solutions in bending of symmetric laminates. Computers, Materials, & Continua, CMC, in press

11

Soil-structure interaction modeling effects on seismic response of cable-stayed bridge tower

Shehata E Abdel Raheem[1,2*] and Toshiro Hayashikawa[3]

Abstract

A nonlinear dynamic analysis, including soil-structure interaction, is developed to estimate the seismic response characteristics and to predict the earthquake response of cable-stayed bridge towers with spread foundation. An incremental iterative finite element technique is adopted for a more realistic dynamic analysis of nonlinear soil-foundation-superstructure interaction system under great-earthquake ground motion. Two different approaches to model soil foundation interaction are considered: nonlinear Winkler soil foundation model and linear lumped-parameter soil model. The numerical results show that the simplified lumped-parameter-model analysis provides a good prediction for the peak response, but it overestimates the acceleration response and underestimates the uplift force at the anchor between superstructure and pier. The soil bearing stress beneath the footing base is dramatically increased due to footing base uplift. The predominant contribution to the vertical response at footing base resulted from the massive foundation rocking rather than from the vertical excitation.

Keywords: Soil-structure interaction, Cable-stayed bridge towers, Linear lumped-parameter model, Uplift, Seismic design, Nonlinear Winkler soil model, Time history analysis

Introduction

Long-span bridges, such as cable-stayed bridge structural systems, provide a valuable environment for the nonlinear behavior due to material and geometrical nonlinearities of the structures of relatively large deflection and forces (Ali and Abdel-Ghaffar 1995; Abdel Raheem and Hayashikawa 2003; Committee of Earthquake Engineering 1996). The Hyogoken-Nanbu (Kobe, Japan) earthquake on 17 January 1995 led to an increased awareness concerning the response of highway bridges subjected to earthquake ground motions. The dynamic analyses and ductility design have been reconsidered by Japan Road Association (Committee of Earthquake Engineering 1996; Japan Road Association 1996a, 1996b). Necessity has arisen to develop more efficient analysis procedures, which can lead to a thorough understanding and a realistic prediction of the nonlinear dynamic response of the bridge structural systems, to improve the bridge seismic performances.

Current design practice does not account for the nonlinear behavior of soil-foundation interface primarily due to the absence of reliable nonlinear soil-structure interaction modeling techniques that can predict the permanent and cyclic deformations of the soil as well as the effect of nonlinear soil-foundation interaction on the response of structural members. Safe and economic seismic designs of bridge structures directly depend on the understanding level of seismic excitation and the influence of supporting soil on the structural dynamic response. Long-span bridges are more susceptible to relatively severe soil-structure interaction effect during earthquakes when compared to buildings due to their spatial extent, varying soil condition at different supports, and possible incoherence in the seismic input. The necessity of incorporating soil-structure interaction in the design of a wide class of bridge structures has been pointed out by several post-earthquake investigations, experimental, and analytical studies (Committee of Earthquake Engineering 1996; Vlassis and Spyrakos 2001; Trifunac and Todorovska 1996; Megawati et al. 2001). Recently, several cable-stayed bridges have been constructed on relatively soft ground, which results in a great demand to evaluate the effect of soil-structure interaction on the seismic behavior of bridges and properly reflect it on seismic design to accurately

* Correspondence: shehataraheem@yahoo.com
[1]Taibah University, Medina, Saudi Arabia
[2]Civil Engineering Department, Faculty of Engineering, Assiut University, Assiut 71516, Egypt
Full list of author information is available at the end of the article

capture the response, enhance the safety level, and reduce design and construction costs.

Due to material and geometrical nonlinearities, the dynamic characteristics of soil-structure system are changed during a severe earthquake. These nonlinearities are sometimes treated using an equivalent linear model (Abdel Raheem et al. 2002; Chaojin and Spyrakos 1996; Spyrakos 1997). However, the dynamic characteristics of soil-structure system are dependent on the soil stress level during a severe earthquake (Ahn and Gould 1992; Gazetas and Dobry 1984; Kobayashi et al. 2002). Foundation rocking and uplifting are important for short-period structures on a relatively soft soil site (Yim and Chopra 1984; Harden et al. 2006). For nonlinear structures, effects of soil-structure interaction due to rocking can result in significantly larger ductility demands under certain conditions (Tang and Zhang 2011; Zhang and Tang 2009). These findings necessitate the evaluation of structural responses considering soil-structure interaction using foundation models that can realistically capture the nonlinear force, displacement behaviors, and energy dissipation mechanisms. Nonlinear foundation movements and associated energy dissipation may be utilized to reduce force and ductility demands of a structure, particularly in high-intensity earthquake events (Raychowdhury 2011).

The purpose of this study is to assess the effects of soil-structure interaction on the seismic response and dynamic performance of the cable-stayed bridge towers with spread foundation. An incremental iterative finite element technique for a more realistic dynamic analysis of nonlinear soil-foundation-superstructure system subjected to earthquake ground motion is developed. Two different modeling approaches, nonlinear Winkler soil model and linear lumped-parameter soil model, for the soil foundation interaction are investigated. The seismic responses of the nonlinear Winkler soil model are compared to those of the simplified linear lumped-parameter model. In the lumped-parameter soil model, the soil-structure interaction is simulated with translational, rotational, and their coupling linear springs acting at the centroid of the spread foundation at footing base level. While in the nonlinear Winkler soil model, the soil structure interaction is simulated with continuous spring system (Winkler) along the embedded depth of tower pier and underneath the spread foundation. Soil strain-dependent material nonlinearity is considered through hysteretic element, while geometrical nonlinearity by basemat uplift is considered through a gap element. The massive pier of the tower model activates rocking response of the spread foundation under strong earthquake ground motion, hence, results in the foundation uplift and yield of the underlying soil (Kawashima and Hosoiri 2002; Gelagoti et al. 2012). The results of the nonlinear Winkler-soil-model approach are compared with those from the linear lumped-soil-model approach.

Methods

Finite element analysis procedure

Three-dimensional beam element tangent stiffness

A nonlinear dynamic finite element technique is developed to analyze the elastoplastic dynamic response of frame structures under strong earthquake excitation, in which the nonlinear three-dimensional beam elements are employed. A three-dimensional beam element has six degrees of freedom at each node, in which all couplings among bending, twisting, and stretching deformations for the beam element are incorporated according to the geometrical nonlinear beam theory. The element nodal displacement vector in local coordinate system is given as follows:

$$\mathbf{d} = \left\{\bar{u}_i \;\; \bar{v}_i \;\; \bar{w}_i \;\; \bar{\theta}_{xi} \;\; \bar{\theta}_{yi} \;\; \bar{\theta}_{zi} \;\; \bar{u}_j \;\; \bar{v}_j \;\; \bar{w}_j \;\; \bar{\theta}_{xj} \;\; \bar{\theta}_{yj} \;\; \bar{\theta}_{zj}\right\}^T. \tag{1}$$

The displacements u, v, and w of a general point (x, y, and z) in the beam can be written as follows:

$$u = \bar{u} - y{\cdot}d\bar{v}/dx - z{\cdot}d\bar{w}/dx, v = \bar{v}, w = \bar{w}, \tag{2}$$

where $\bar{u}$, $\bar{v}$, and $\bar{w}$ are the displacements of a point in the centroid axis of a beam corresponding to x-, y-, and z-axes, respectively. $\bar{\theta}$ is the cross-sectional rotation about the x-axis. The displacement transformation matrix of the beam element relates the element internal displacement to the nodal point displacement:

$$\bar{\mathbf{u}} = \left\{\bar{u} \;\; \bar{v} \;\; \bar{w} \;\; \bar{\theta}\right\}^T = \left[\bar{\mathbf{N}}_x \, \bar{\mathbf{N}}_y \, \bar{\mathbf{N}}_z \, \bar{\mathbf{N}}_\theta\right]^T \mathbf{d}, \tag{3}$$

$$\begin{aligned}
\bar{\mathbf{N}}_x &= [N_1\; 0\; 0\; 0 \quad 0\; 0\; N_2\; 0\; 0\; 0 \quad 0\;\; 0],\\
\bar{\mathbf{N}}_y &= [0\;\; N_3\; 0\;\; 0 \quad 0\;\; N_4\; 0\;\; N_5\; 0\;\; 0 \quad 0\;\; N_6],\\
\bar{\mathbf{N}}_z &= [0\;\; 0\;\; N_3\; 0 - N_4\; 0\;\; 0\;\; 0\;\; N_5\; 0 - N_6\; 0],\\
\bar{\mathbf{N}}_\theta &= [0\; 0\;\; 0\;\; N_1 \quad 0\;\; 0\;\; 0\; 0\;\; 0\;\; N_2\;\; 0\;\; 0].
\end{aligned} \tag{4}$$

The shape functions N_i (i = 1 to 6) are assumed for the local element displacements and given as follows:

$$\begin{aligned}
&N_1 = 1 - \zeta, N_2 = \zeta, N_3 = 1 - 3\zeta^2 + 2\zeta^3, \zeta = x/l,\\
&N_3 = (\zeta - 2\zeta^2 + \zeta^3)l, N_5 = 3\zeta^2 - 2\zeta^3, N_6 = (-\zeta^2 + \zeta^3)l,
\end{aligned} \tag{5}$$

in which l is the original length of the beam element.

$$\begin{aligned}
\frac{d\bar{u}}{dx} &= \mathbf{B}_x\mathbf{d}, \frac{d^2\bar{v}}{dx} = \mathbf{B}_2\mathbf{d}, \frac{d^2\bar{w}}{dx} = \mathbf{B}_3\mathbf{d}, \frac{d\bar{v}}{dx} = \mathbf{B}_4\mathbf{d}, \frac{d\bar{w}}{dx}\\
&= \mathbf{B}_5\mathbf{d}\;, \frac{d\theta}{dx} = \mathbf{B}_6\mathbf{d}.
\end{aligned} \tag{6}$$

The strain in the deformed configuration $\boldsymbol{\varepsilon}_t$ can be expressed in terms of the displacement at the equilibrium state at any specified point in the beam cross section (x, y, and z) that defines the state of strain at any arbitrary point.

The strain displacement equation (Green-Lagrange representation of axial strain $\boldsymbol{\varepsilon}_t$) is given in the following:

$$\varepsilon_t = \frac{du_t}{dx} + \frac{1}{2}\left(\frac{du_t}{dx}\right)^2 + \frac{1}{2}\left(\frac{dv_t}{dx}\right)^2 + \frac{1}{2}\left(\frac{dw_t}{dx}\right)^2, \qquad (7)$$

where u_t, v_t, and w_t are the element displacement vectors in x-, y-, and z-axis directions, respectively; it is assumed that $du_t/dx << 1$. Thus, $(du_t/dx)^2$ is ignored compared to its linear term. The normal strain $\boldsymbol{\varepsilon}_t$ is then calculated. The unknown strain in $t + 1$ configuration can be written incrementally from configuration t as follows:

$$\varepsilon_{t+1} = \frac{du_{t+1}}{dx} + \frac{1}{2}\left(\frac{dv_{t+1}}{dx}\right)^2 + \frac{1}{2}\left(\frac{dw_{t+1}}{dx}\right)^2, \qquad (8)$$

$$\begin{aligned}\delta\varepsilon_{t+1} &= \frac{d\delta u}{dx} + \frac{dv_t}{dx}\cdot\frac{d\delta v}{dx} + \frac{dw_t}{dx}\cdot\frac{d\delta w}{dx} \\ &\quad + \frac{1}{2}\delta\left(\frac{dv}{dx}\right)^2 + \frac{1}{2}\delta\left(\frac{dw}{dx}\right)^2 \\ &= (\mathbf{B}_L + \mathbf{B}_{NL})\delta\mathbf{d},\end{aligned} \qquad (9)$$

$$\begin{aligned}\mathbf{B}_L &= \mathbf{B}_1 - y\mathbf{B}_2 - z\mathbf{B}_3, \\ \mathbf{B}_{NL} &= \left(\mathbf{B}_4^T\mathbf{B}_4 + \mathbf{B}_5^T\mathbf{B}_5\right)/2, \\ \mathbf{B}_1 &= \mathbf{B}_x + \frac{dv_t}{dx}\mathbf{B}_4 + \frac{dw_t}{dx}\mathbf{B}_5.\end{aligned} \qquad (10)$$

From the principle of energy, the external work is equivalent to the internal work. The equilibrium equation in the deformed configuration $t + 1$ can be expressed in terms of the principle of virtual displacements:

$$\delta U - \delta W = 0., \qquad (11)$$

$$\int_V \sigma_{t+1}\,\delta\,\varepsilon_{t+1}\,dV + \int_l T_{sv}\cdot(d\theta/dx)dx + \frac{1}{2}\int_l GJ(d\theta/dx)^2dx - \int_S f_{t+1}^T\,\delta\,\mathbf{d}_{t+1}\,dS = 0, \qquad (12)$$

where U is the internal work, W is the external virtual work, σ_{t+1} is the axial Couchy stress, V is the volume, δ is the virtual quantity, f is the external applied force, and θ is the cross-sectional rotation about the x-axis. The unknown configuration is $t + 1$ which can be written incrementally from configuration t consistently with a nonlinear Lagrangian scheme as follows:

$$\int_V (\sigma_t + \sigma)\,\delta\,\varepsilon_{t+1}\,dV + \int_l T_{sv}\frac{d\theta}{dx}dx + \frac{1}{2}\int_l GJ\left(\frac{d\theta}{dx}\right)^2 dx - \int_S \left(f_t^T + f^T\right)\,\delta(\mathbf{d}_t + \mathbf{d})\,dS = 0, \qquad (13)$$

where σ is the second Piola-Kirchhoff axial stress:

$$\begin{aligned}&\int_V \{\sigma\,\mathbf{B}_L + \sigma_t\mathbf{B}_{NL}\}dV + \frac{1}{2}\int_l GJ\left(\frac{d\theta}{dx}\right)^2 dx - \int_S f^T\delta\mathbf{d}\,dS \\ &= \int_S f_t^T\delta\mathbf{d}\,dS - \int_V \sigma_t\mathbf{B}_L dV - \int_l T_{sv}\frac{d\theta}{dx}dx,\end{aligned} \qquad (14)$$

$$\begin{aligned}&\int_V \{\sigma\,\mathbf{B}_L + \sigma_t\mathbf{B}_{NL}\}dV + 0.5\int_l GJ(d\theta/dx)^2dx \\ &= \delta\mathbf{d}^T\int \{E_1\mathbf{B}_1^T\mathbf{B}_1 - E_2\left(\mathbf{B}_2^T\mathbf{B}_1 + \mathbf{B}_1^T\mathbf{B}_2\right) \\ &\quad + E_3\mathbf{B}_2^T\mathbf{B}_2 - E_4\left(\mathbf{B}_3^T\mathbf{B}_1 + \mathbf{B}_1^T\mathbf{B}_3\right) + E_5\mathbf{B}_3^T\mathbf{B}_3 \\ &\quad + E_6\left(\mathbf{B}_2^T\mathbf{B}_3 + \mathbf{B}_3^T\mathbf{B}_2\right)\}\mathbf{d}dx \\ &\quad + \delta\mathbf{d}^T\int_l F_x\left(\mathbf{B}_4^T\mathbf{B}_4 + \mathbf{B}_5^T\mathbf{B}_5\right)dx\mathbf{d} \\ &\quad + \delta\mathbf{d}^T\int_l GJ\left(\mathbf{B}_6^T\mathbf{B}_6\right)\mathbf{d}dx\end{aligned} \qquad (15)$$

$$\begin{aligned}&\int_V \sigma_t\,\mathbf{B}_L dV + \int_l T_{sv}(d\theta/dx)dx \\ &= \delta\mathbf{d}^T\int_V \sigma_t\left(\mathbf{B}_1^T - y\mathbf{B}_2^T - z\mathbf{B}_3^T\right)dV + \delta\mathbf{d}^T\int_l T_{sv}\mathbf{B}_6^T\,dx \\ &= \delta\mathbf{d}^T\int_l \left(F_x\mathbf{B}_1^T - M_z\mathbf{B}_2^T - M_y\mathbf{B}_3^T + T_{sv}\mathbf{B}_6^T\right)dx,\end{aligned} \qquad (16)$$

$$\begin{aligned}&E_1 = E_T\int dydz, \quad E_2 = E_T\int ydydz, \quad E_3 = E_T\int y^2dydz, \\ &E_4 = E_T\int zdydz, \quad E_5 = E_T\int z^2ydydz, \quad E_6 = E_T\int zydydz, \\ &F_x = \int \sigma_t dydz, \quad M_z = \int \sigma_t ydydz, \quad M_y = \int \sigma_t zdydz\end{aligned} \qquad (17)$$

which can be determined through numerical integration over the cross section of fiber segments. The tangent stiffness can be written as follows:

$$\begin{aligned}k_t = \int \{&E_1\mathbf{B}_1^T\mathbf{B}_1 - E_2\left(\mathbf{B}_2^T\mathbf{B}_1 + \mathbf{B}_1^T\mathbf{B}_2\right) \\ &+ E_3\mathbf{B}_2^T\mathbf{B}_2 - E_4\left(\mathbf{B}_3^T\mathbf{B}_1 + \mathbf{B}_1^T\mathbf{B}_3\right) + E_5\mathbf{B}_3^T\mathbf{B}_3 \\ &+ E_6\left(\mathbf{B}_2^T\mathbf{B}_3 + \mathbf{B}_3^T\mathbf{B}_2\right) + F_x\left(\mathbf{B}_4^T\mathbf{B}_4 + \mathbf{B}_5^T\mathbf{B}_5\right) \\ &+ GJ\left(\mathbf{B}_6^T\mathbf{B}_6\right)\}dx.\end{aligned} \qquad (18)$$

In order to capture the spread of plastic zone in individual element, the beam element is divided along its length and over its cross section directions. The stiffness quantities of the section are calculated based on the stress states of integration points over the cross section. The element stiffness quantities are then obtained by integrating along the length of the beam, where the plastic development of the material is automatically considered. The stiffness matrix calculations of the elements are completed by a numerical integration procedure. After obtaining the stiffness matrix of the elements with respect to the local coordinate system, the element coordinates are transformed from the local coordinate system to the global coordinate system. The stiffness matrix of the elements is then assembled using standard procedures and by getting the global stiffness matrix. In the nonlinear incremental analysis, the structure tangent stiffness matrix, which is assembled from the element tangent stiffness matrices, is used to predict the next incremental displacements under a loading increment.

Soil-structure interaction formulation

The interaction between the soil and the structure is simulated with translational, rotational, and their coupling spring system and equivalent viscous damping. The spring constants of the lumped-parameter model in both bridge axis and right-angle directions are calculated based on foundation geometry and soil profile underneath and along the embedded depth of the foundation, as specified in Japanese Highway Specification (Committee of Earthquake Engineering 1996; Japan Road Association 1996a). Soil properties from the standard penetration test (SPT) data and logs of boreholes at the tower site are used to determine the coefficients of vertical and horizontal subgrade reactions that are used in a spread foundation design. A set of spring elements along all the six degrees of freedom and the coupling of lateral and rotational directions constitutes the linear lumped soil model, and it is connected to the foundation centroid node at the footing base level. The influence of the soil flexibility in the global dynamic response is considered through the assembling of the stiffness matrix of the soil spring system using standard procedures getting the globe stiffness matrix (Abdel Raheem 2009; Abdel Raheem et al. 2002, 2003).

In the nonlinear Winkler soil model, three types of soil resistance displacement models could describe the soil characteristics: the first type represents the lateral soil pressure against the pier and the corresponding lateral pier displacement relationship, the second type represents the skin friction and the relative vertical displacement between the soil and the pier relationship, and the third type describes the bearing stress beneath the spread foundation and settlement relationship. All three types assume the soil behavior to be nonlinear and can be developed from the basic soil parameters. The soil-structure interaction is simulated with nonlinear spring-dashpot system along the embedded depth of the pier. Both strain-dependent material nonlinearity and geometrical nonlinearity by basemat uplift are considered through a beam on nonlinear Winkler soil element connected in series with gap element spring system (Abdel Raheem et al. 2002, 2003; Kobayashi et al. 2002). The spring constants in both bridge axis directions are calculated based on foundation geometry and soil profile underneath and along the embedded depth of the foundation as specified in Japanese Highway Specification (Committee of Earthquake Engineering 1996; Japan Road Association 1996a). The nonlinear soil-foundation model is described as the Winkler model that has nonlinear springs distributed along the embedded depth in the three directions of the local coordinate system. The reaction displacement model of the tensionless foundation at base level is considered.

The contribution of soil reactions to the foundation element equivalent nodal forces vector and the element tangent stiffness of the foundation, $\{f_{\text{soil}}\}$ and $[k_{\text{soil}}]$, respectively, can be given as follows:

$$\{f_{\text{soil}}\} = -\int_{l} [N]^T \{r\} dx \tag{19}$$

and

$$[k_{\text{soil}}] = -\frac{\delta\{f_{\text{soil}}\}}{\delta\{d\}} = \int_{l} [N]^T \frac{\delta\{r\}}{\delta\{d\}} dx = \int_{l} [N]^T [k][N] dx, \tag{20}$$

where

$$\delta\{r\} = \begin{bmatrix} k_x & & \\ & k_y & \\ & & k_z \end{bmatrix} \begin{Bmatrix} \delta\bar{u} \\ \delta\bar{v} \\ \delta\bar{w} \end{Bmatrix} = [k]\delta\{\bar{u}\} = [k][N]\{d\}. \tag{21}$$

Equation of motion

The governing nonlinear dynamic equation of the tower structure response can be derived using the principle of energy, i.e., the external work is absorbed by the internal, inertial, and damping energy for any small admissible motion that satisfies compatibility and boundary condition. By assembling the element dynamic equilibrium equation at time $t + \Delta t$ over all the elements, the incremental finite element method dynamic equilibrium equation (Abdel Raheem 2009; Chen 2000) can be obtained as follows:

$$[M]\{\ddot{u}\}^{t+\Delta t} + [C]\{\dot{u}\}^{t+\Delta t} + [K]^{t+\Delta t}\{\Delta u\}^{t+\Delta t} = \{F\}^{t+\Delta t} - \{F\}^{t}, \tag{22}$$

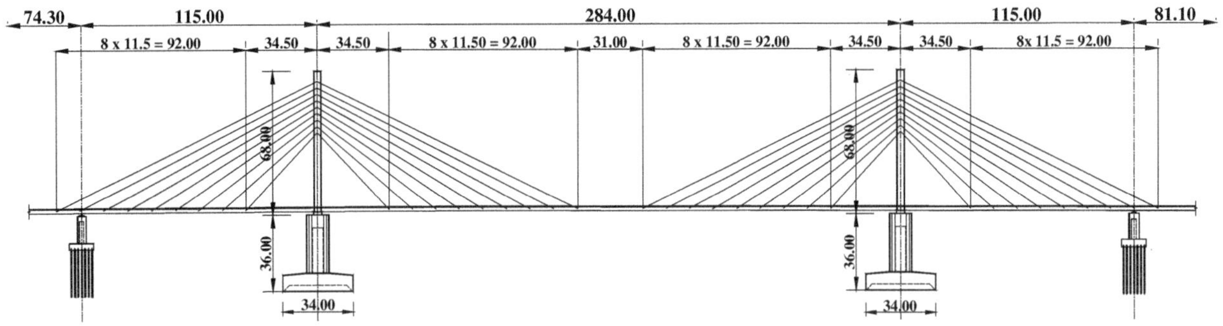

Figure 1 Schematic representation of bridge elevation (*m*).

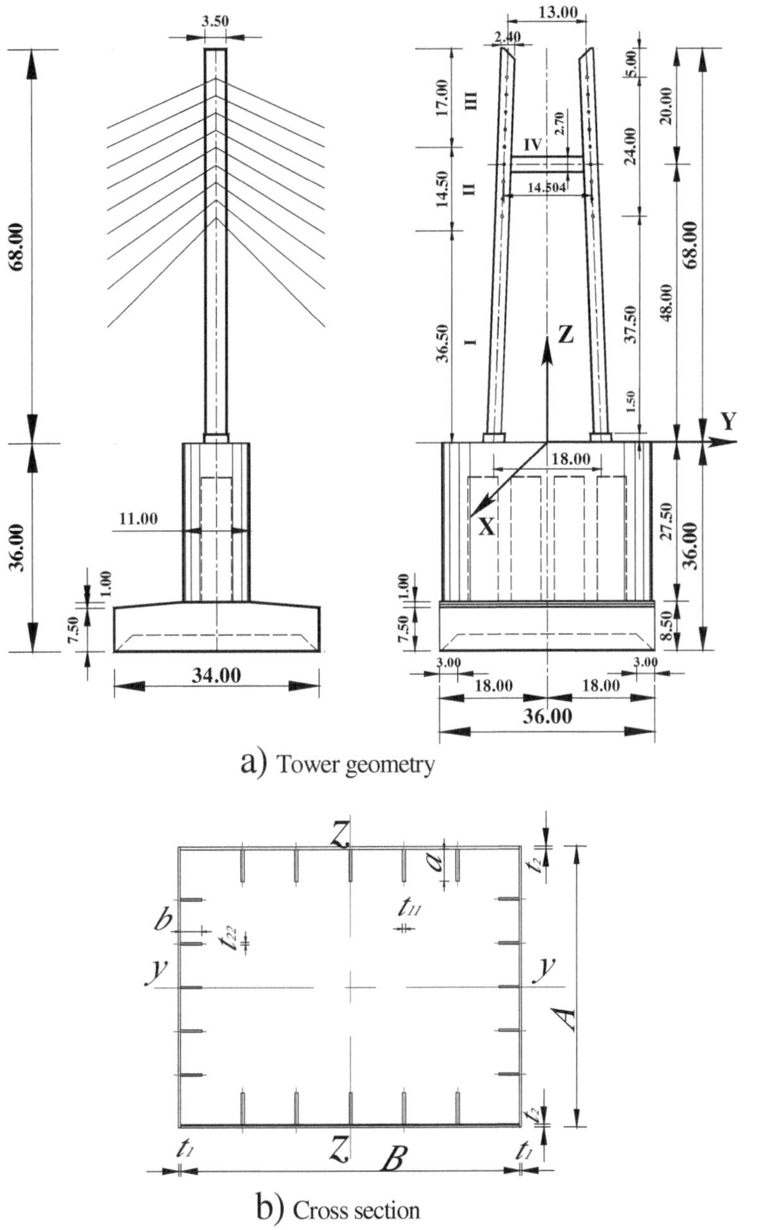

Figure 2 Tower geometry (a) and cross section (b) of steel tower of cable-stayed bridge.

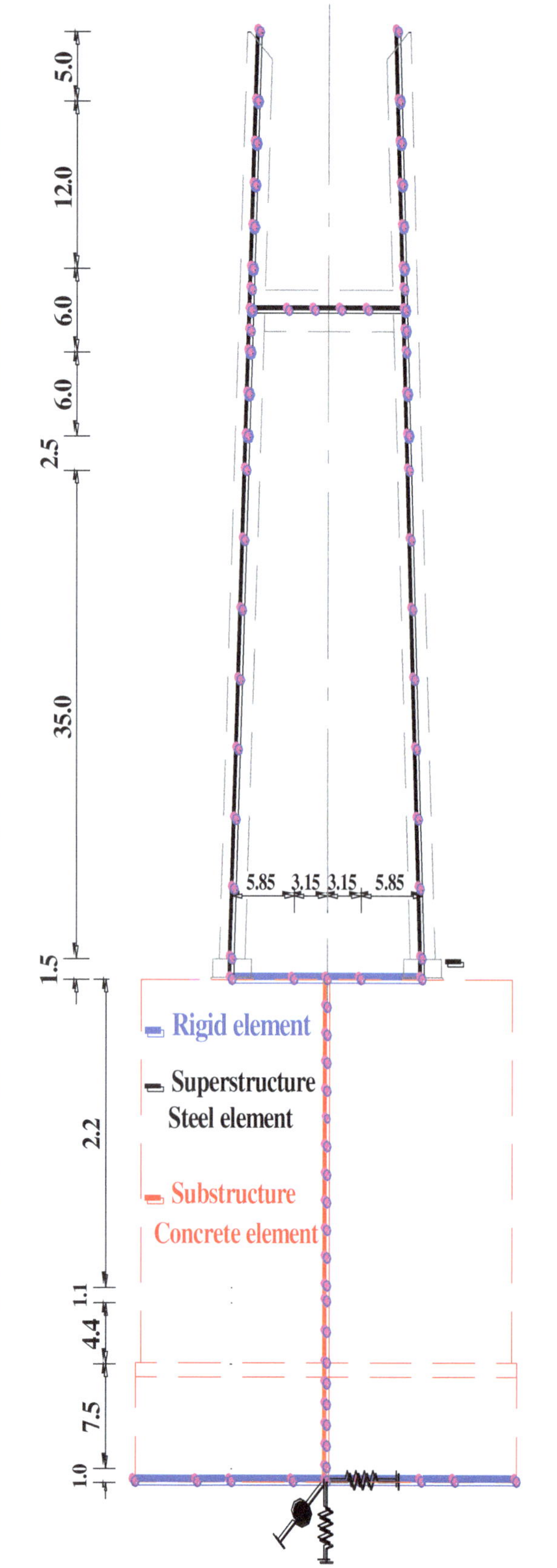

Figure 3 Finite element model of bridge tower.

Table 1 Cross section dimension of different tower regions

Tower parts	Cross section dimensions (cm)							
	Outer dimension				Stiffener dimension			
	A	*B*	t_1	t_2	*a*	*b*	t_{11}	t_{22}
I	240	350	2.2	3.2	25	22	3.6	3.0
II	240	350	2.2	3.2	22	20	3.2	2.8
III	240	350	2.2	2.8	20	20	2.8	2.2
IV	270	350	2.2	2.6	31	22	3.5	2.4

where [**M**], [**C**], and $[\mathbf{K}]^{t+\Delta t}$ are the system mass, damping, and tangent stiffness matrices at time $t + \Delta t$, respectively; $\ddot{u}$, $\dot{u}$ and Δu are the accelerations, velocities, and incremental displacement vectors at time $t + \Delta t$, respectively. $\{\mathbf{F}\}^{t+\Delta t} - \{\mathbf{F}\}^{t}$ is the unbalanced force vector. It can be noticed that the dynamic equilibrium equation of motion takes into consideration the different sources of nonlinearities (geometrical and material nonlinearities), which affect the calculation of the tangent stiffness and internal forces. The implicit Newmark step-by-step integration method is used to directly integrate the equation of motion, and then it is solved for the incremental displacement using the Newton–Raphson iteration method, in which the stiffness matrix is updated at each increment to consider the geometrical and material nonlinearities and to speed the convergence rate.

The tower structure damping mechanism is adapted to the Rayleigh damping; the damping coefficient is taken to be 2% for steel materials of tower superstructure and 10% for concrete materials of embedded pier substructure (Committee of Earthquake Engineering 1996; Japan Road Association 1996a, 1996b). Vibration periods of 2.5 and 0.50 s are considered to represent a broad range of high participation modes and the softening that takes place as the columns and soil yield. A common design approach of maintaining elastic behavior in the substructure is considered to avoid inelastic behavior below the ground surface, where the damage would be difficult to detect or to repair. A nonlinear dynamic analysis computer program is developed based on the above-mentioned formulation to predict the vibration behavior of framed structures as well as the nonlinear response under earthquake loadings. The program has been validated through a comparison with different commercial software EDYNA, DYNA2E, and DYNAS (Japanese software).

Input excitation

In the dynamic response analysis, the seismic motion by an inland direct-strike type earthquake that was recorded during the 1995 Hyogoken-Nanbu earthquake of high intensity but short duration is used as an input ground motion to assure the seismic safety of the bridges.

Table 2 Stiffness values of linear lumped-parameter soil model

Stiffness values	Lateral stiffness K_L (kN/m)	Rotational stiffness K_R (kN m/rad)	Coupling stiffness K_C (kN/rad)	Vertical stiffness K_V (kN/m)	Torsional stiffness K_T (kN·m/rad)
Bridge axis	1.798×10^7	2.778×10^9	2.747×10^7	3.084×10^7	3.075×10^9
Right-angle axis	1.739×10^7	3.454×10^9	2.552×10^7	3.071×10^7	3.075×10^9

The horizontal and the vertical accelerations recorded at the station of JR-Takatori observatory (Committee of Earthquake Engineering 1996; Japan Road Association 1996a, 1996b) are used for the dynamic response analysis of the cable-stayed bridge tower at a soil of type II, where the fundamental period of the site ranges from 0.2 to 0.6 s, and the mean value of shear wave velocity of 30 m of the soil deposit ranges from 200 to 600 m/s (Japan Road Association 1996a). It is considered to be capable of securing the required seismic performance during the bridge service life. The selected ground motion has maximum acceleration intensities of its components of 642 (N-S), 666 (E-W), and 290 gal (U-D). The kinematic interaction can be modeled by a transfer function, which can be used to modify a free-field ground motion so that an effective input motion for a soil-structure system can be obtained. A simple model based on analytical modeling of rigid foundation is adopted, in which the effective seismic motion is obtained starting from the free-field motion by an approximate analytical solution (Harada et al. 1981; Ganev et al. 1995) for embedded foundation with its base resting on a stiffer layer.

Finite element modeling of tower structure

The steel tower of a three-span continuous-cable-stayed bridge located in Hokkaido, Japan is considered, in which the main span length is equal to 284 m. Figure 1 shows a schematic representation of the bridge elevation. The towers are H-shaped steel structures as shown in Figure 2. Each tower is founded on a spread foundation on good soil layer, and it consists of two steel legs and horizontally connected beam. The cross section of each leg has a hollow rectangular shape with an interior stiffener; the cross-sectional size varies over the height of the tower. Cable-stayed bridges represent highly redundant and mechanically nonhomogeneous structural systems; the tower, deck, and cable stays affect the structural response in a wide range of vibration modes (Ali and Abdel-Ghaffar 1995; Abdel Raheem and Hayashikawa 2003). The tower is taken out of the cable-stayed bridge, and a finite element model is constructed based on the design drawings, as shown in Figure 3. A flexural fiber element is developed for the tower characterization; the element incorporates both geometric and material nonlinearities. A bilinear model is adopted to describe the stress–strain relationship of the element. The yield stress and the modulus of elasticity are equal to 355 MPa (SM490) and 200 GPa, respectively; the plastic region strain hardening is 0.01. The nonlinearity of inclined cables is idealized using the equivalent modulus approach (Karoumi 1999). The nonlinearity of the cables originates with an increase in the loading followed by a decrease in the cable sag, resulting in an increase of the cable apparent axial stiffness. In this approach, each cable is simulated by a truss element with equivalent tangential modulus of elasticity E_{eq} that is given by (Ernst 1965) as follows:

$$E_{eq} = E/\{1 + EA(wL)^2/12T^3\}, \tag{23}$$

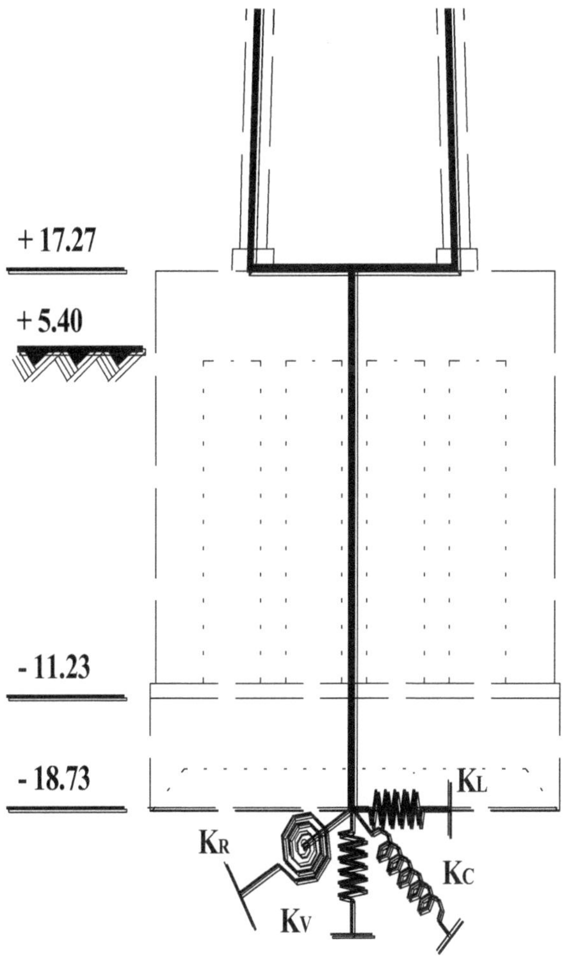

Figure 4 Linear lumped-parameter soil model.

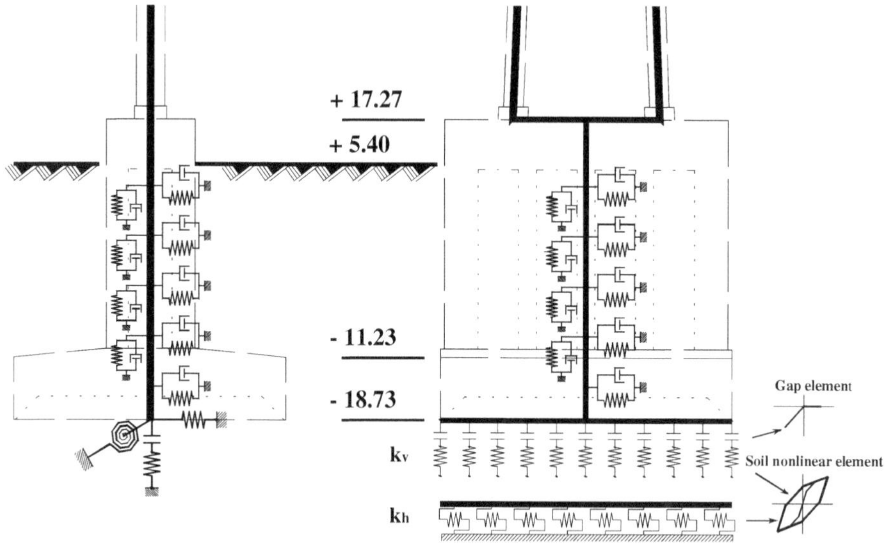

Figure 5 Winkler model for soil-structure interaction.

where E is the material modulus of elasticity, L is the cable horizontal projected length, w is the weight per unit length of the cable, A is the cross section of the area of the cable, and T is the tension force in the cable. This tower has nine cables in each side of the tower. The stiffening girder dead load is considered to be equivalent to the vertical component of the pretension force of the cables acting at their joints and the two vertical components at stiffening girder-tower connection at the substructure top level. The tower steel superstructure has a rectangular hollow steel section of different dimensions along the different parts of the tower, as indicated in Table 1. In the numerical analysis, steel superstructure is modeled as a three-dimensional elastoplastic beam element, while concrete substructure is modeled as an elastic element.

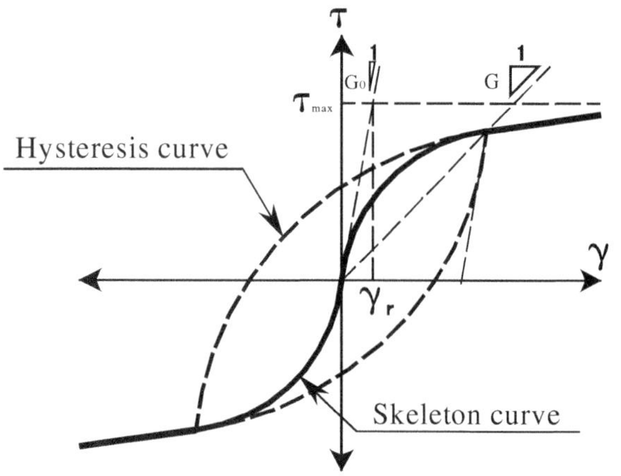

Figure 6 Hardin-Drnevich model and Masing rule hysteresis loop.

Modeling of soil-structure interaction

A crucial goal of current seismic foundation design, particularly as entrenched in the respective codes (European Committee for Standardization 1998), is to avoid full mobilization of strength in the foundation by guiding failure to the aboveground structure. The designer must ensure that the (difficult to inspect) belowground support system will not even reach a number of thresholds that would statically imply failure: mobilization of the soil bearing capacity, significant foundation uplifting, and sliding or any of their combination are prohibited or severely limited. To this end, following the norms of capacity design, overstrength factors and safety factors (explicit or implicit) are introduced against those failure modes.

Linear lumped-parameter soil model

In the linear lumped-parameter soil model, the interaction between the soil and the structure is simulated with translational, rotational, and their coupling spring system. The spring stiffness values in both bridge axis and right-angle directions are given in Table 2. The stiffness is calculated based on foundation geometry and soil profile underneath and along the embedded depth of the foundation, as specified in Japanese Highway Specification (Japan Road Association 1996a, 1996b) and as shown in Figure 4.

Nonlinear Winkler model for soil-structure interaction

The soil-structure interaction is simulated with nonlinear spring-dashpot system along the pier embedded depth. Strain-dependent material nonlinearity is implemented using the nonlinear Hardin-Drnevich soil model (Hardin and Drnevich 1972a, 1972b). Geometrical nonlinearity

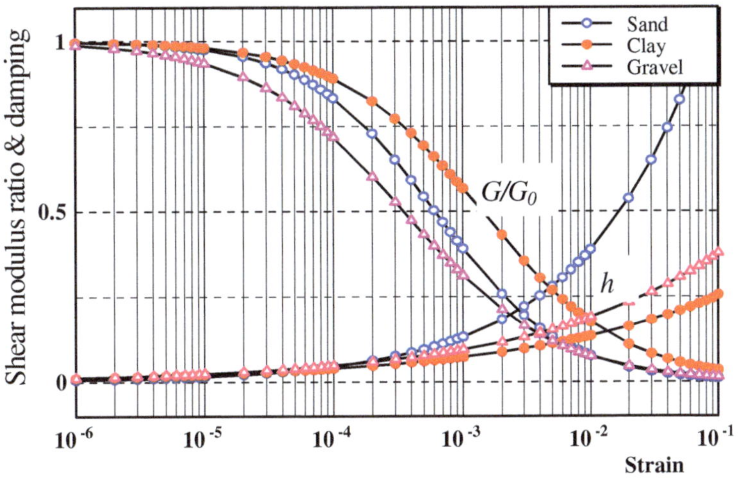

Figure 7 Strain-dependent soil material nonlinearity (rigidity and damping).

by basemat uplift is considered through nonlinear soil element connected in series with gap element spring system, as shown in Figure 5. The spring constants in both bridge axis directions are calculated based on foundation geometry and soil profile underneath and along embedded depth of foundation, as specified in Japanese Highway Specification (Japan Road Association 1996a, 1996b). Soil properties from the SPT data and logs of boreholes at the tower site are used to determine the coefficients of vertical and horizontal subgrade reactions. The subgrade reaction coefficients are obtained from the ground stiffness corresponding to the deformation caused in the ground during an earthquake.

Soil nonlinearity idealization One of the most important factors in the analysis of soil-foundation interactive behavior is the nonlinear constitutive laws of the soil (material nonlinearity). In this study, the Hardin-Drnevich model is proposed to represent the soil material nonlinearity that is often used for its capacity to trace the degradation of stiffness. The parameters used to define the skeleton curve and family of hysteresis stress–strain curves are indicated in Figure 6. The skeleton curve is expressed as follows:

$$\tau = G_0\gamma/\left(1+|\gamma/\gamma_r|\right), \gamma_r = \tau_{\max}/G_0, \tag{24}$$

where G_0 is the initial shear modulus, τ is the generalized soil shear stress, $\tau_{\max}$ is the shear stress at failure, γ_r is the reference strain, and γ is the generalized strain. The hysteretic curve can be constructed using the Masing rule (Masing 1926) and is given as follows:

$$\tau \pm \tau_m = G_0(\gamma \pm \gamma_m)/\left\{1+\left|(\gamma \pm \gamma_m)/2\gamma_r\right|\right\}, \tag{25}$$

where τ_m and γ_m indicate the coordinates of the origin of the curve, that is, the point of the most recent load reversal. The hysteresis curve is the same in shape as the skeleton curve but is enlarged twice. The nonlinear dynamic soil parameters including the dynamic shear moduli and the damping ratios for the employed soil models in this study are modulated based on the shear strain-dependent relationships for gravel, sand, and clay shown in Figure 7. Soil exhibits nonlinear nature even

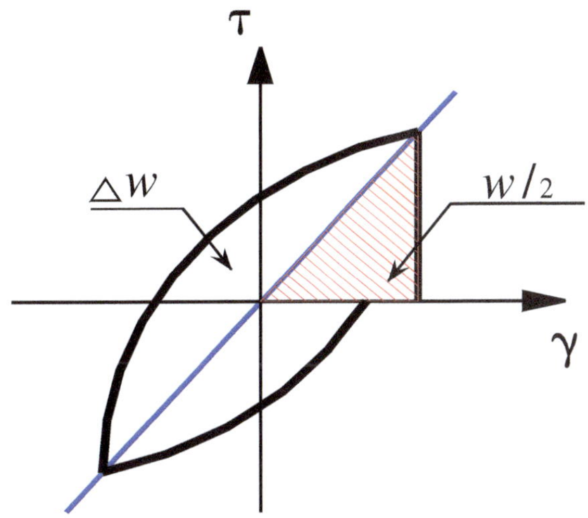

Figure 8 Hysteretic damping of soil.

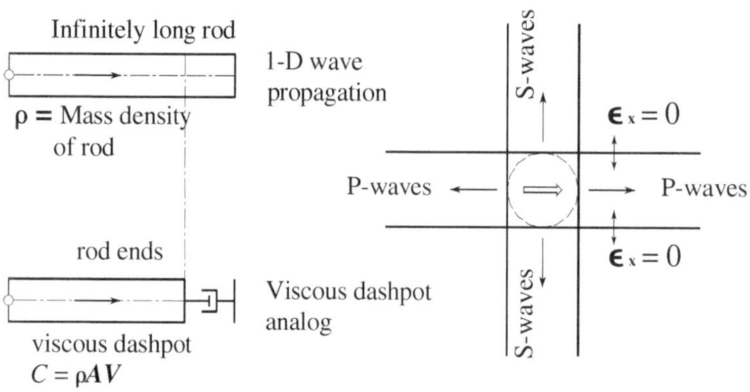

Figure 9 One-dimensional radiation-damping model.

at small strains. The shear modulus (G) can be described as follows:

$$G/G_0 = 1/(1 + \gamma_m/\gamma_r). \tag{26}$$

The soil element stiffness is idealized by the Winkler model. For practical use, frequency-independent spring coefficients are computed based on Japanese Specification for Highway Bridges (Japan Road Association 1996a, 1996b). Each spring consists of a gap element and a soil element. The gap element transmits no tensile stress, which can express the geometrical nonlinearity of basemat uplift.

Soil damping idealization The hysteretic damping characteristic of the soil, which resulted from the deformations produced by interaction with the pier, is represented by nonlinear viscous dashpots. The damping ratio of the soil dashpot strain-dependent material nonlinearity is described by a simple relationship between the shear modulus and damping, as shown in Figure 8.

$$\begin{aligned} h &= (\Delta w/w)/2\pi \\ &= (2/\pi)\left[(2G_0/G)\left\{(\gamma_r/\gamma_m) - (\gamma_r/\gamma_m)^2\log(1 + g_r/g_m)\right\} - 1\right]. \end{aligned} \tag{27}$$

The material-damping ratio h is defined as follows:

$$h = C_m/C_r, C_r = 2(k/m)^{0.5}, \tag{28}$$

in which C_m is the coefficient of material damping, C_r is the coefficient of material critical damping, and k and m

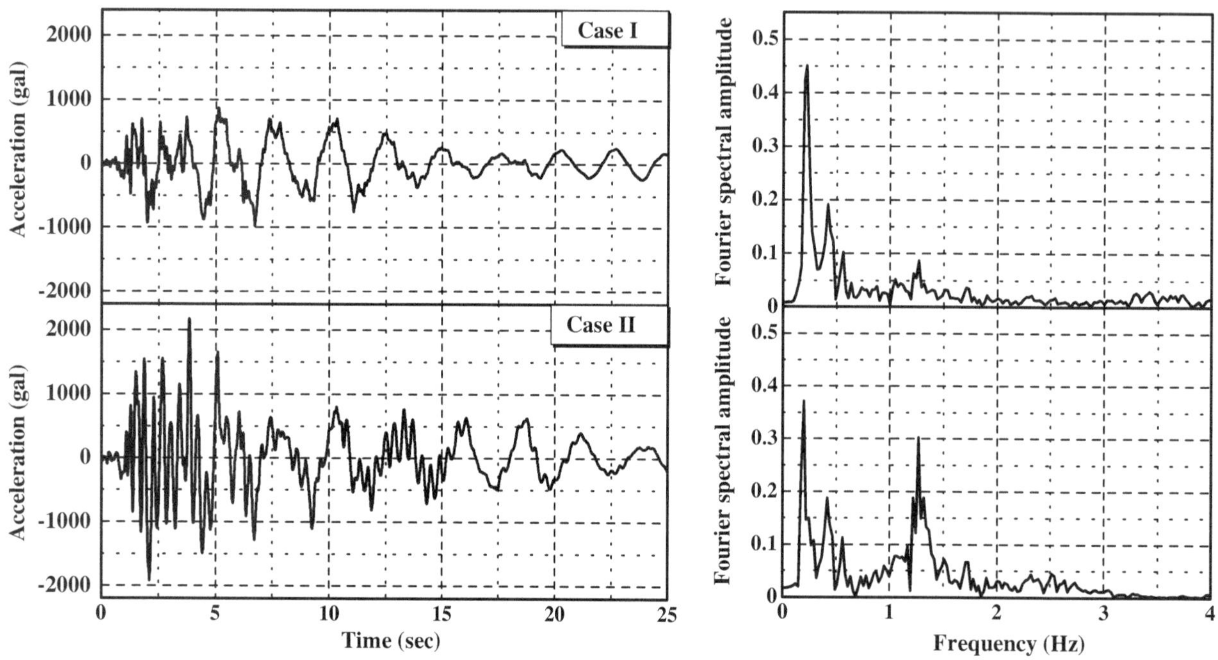

Figure 10 Acceleration time history and response spectra at tower top.

are the soil spring stiffness and pier mass per unit length, respectively. The coefficient of material damping of the soil C_{soil} is obtained as follows:

$$C_{soil} = 2h_{max}(1 - k/k_0)(k/m)^{0.5}. \quad (29)$$

The radiation-damping characteristic of soil is represented through an approximation of a para-axial boundary, where viscous dampers can be used to represent a suitable transmitting boundary for many applications involving both dilatational waves and shear waves. A one-dimensional viscous boundary model is selected for this study. It is assumed that a horizontally moving pier cross section would solely generate one-dimensional P waves traveling in the direction of shaking and one-dimensional S waves in the direction perpendicular to shaking, as shown in Figure 9. Based on the previous assumption, the coefficient of viscous dashpot that will absorb the energy of the waves originating at soil-pier interface is evaluated.

Results and discussion

To study the effects of soil-foundation-structure interaction on the seismic behavior of the cable-stayed bridge towers, two different approaches for soil-foundation-interaction modeling (nonlinear Winkler model and linear lumped-parameter model) are considered. The soil-foundation-superstructure model with Winkler hypothesis for soil idealization is compared to a simplified lumped-parameter model, and dynamic response simulation is conducted by applying acceleration at the base level of the foundation. The following two different cases of soil idealization are analyzed:

Case I Seismic response with soil material and geometrical nonlinearities: nonlinear Winkler model.

Case II Seismic response of tower model with linear lumped-parameter soil model.

Soil-structure interaction modeling effects on global responses

The acceleration time history at the tower top, where the tower acceleration response is significantly decreased due to degradation of soil stiffness and energy dissipation through soil hysteresis, clarifies the effects of soil nonlinearity. The reduction of acceleration response of Winkler model (case I) related to soil nonlinearities (material and geometrical) approaches about 60% of that of the simplified model (case II), as illustrated in Figure 10. Moreover, the acceleration time history of case II is characterized by large peak acceleration that is associated with a short duration impulse of high frequency (acceleration spike). The Fourier acceleration response spectrum clarifies the significant contribution of high-frequency modes to the tower seismic response.

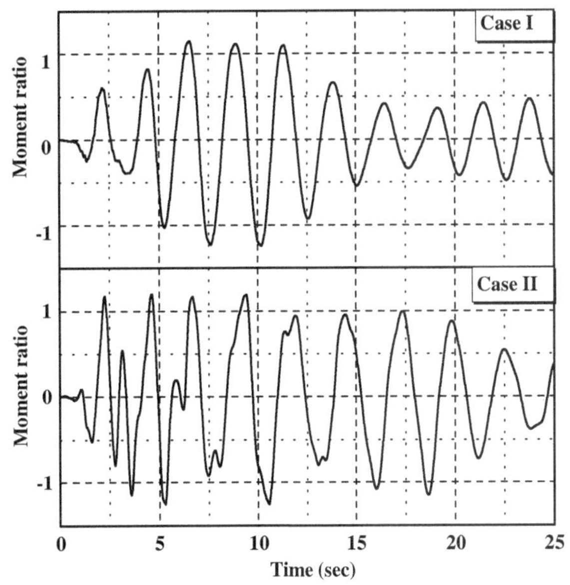

Figure 11 Moment ratio time history at tower superstructure base.

The displacement, moment ratio, and vertical force time histories of Winkler model (case I) at the tower superstructure base show the response nature to strong excitation. The moment ratio is defined as the ratio of moment demand to the moment capacity at the first yield of the cross section. The large pulse in the ground motion produces two or three cycles of large force response with rapidly decaying amplitudes of displacement, moment, and force after the peak excursions. The predominant contribution to the vertical force response

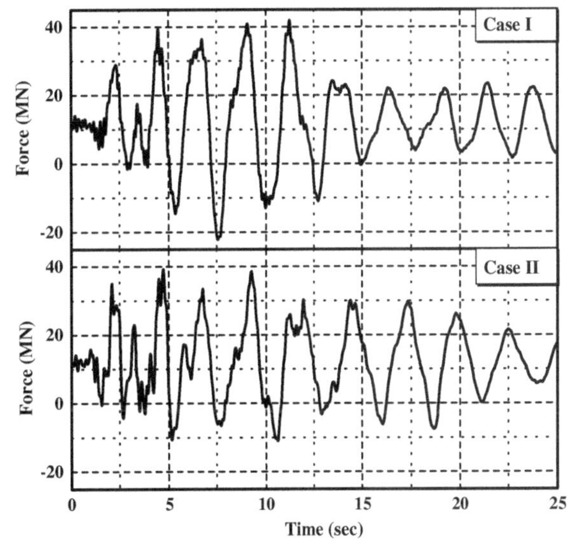

Figure 12 Vertical force time history at tower superstructure base.

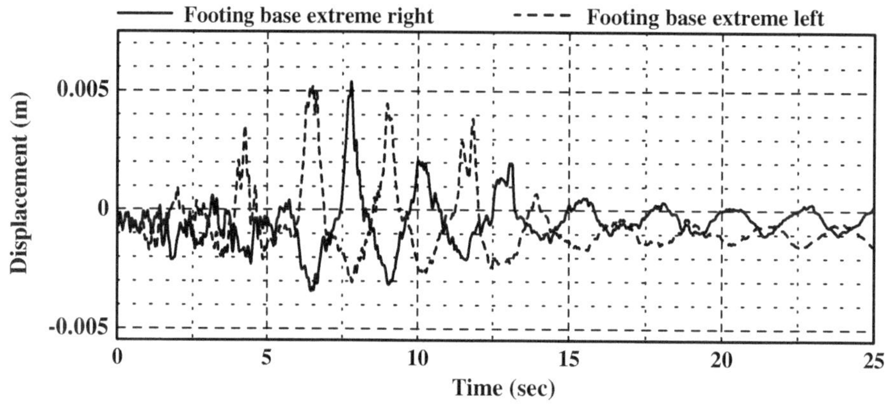

Figure 13 Vertical displacement time history of footing extreme right and left at footing base level.

at the tower superstructure base results from the in-plane rocking vibration rather than from the vertical excitation, and it is dominated by long-period in-plane vibration and slightly affected by high-frequency vertical excitation and vibration, as shown in Figures 11 and 12. However, the lumped-parameter-model analysis (case II) for tower superstructure base moment and axial force displays slight attenuation after peak response excursions, and the time history response is characterized by high-frequency spike. The simplified lumped-parameter-model analysis provides a good prediction for peak response but overestimates the acceleration response and underestimates the uplift force at the anchor between superstructure and pier. It can be obviously seen that foundation uplifting and plasticity have a significant effect on the reduction of seismic force of bridge tower. The foundation allowed uplifting, and yielding in design will protect tower pier and superstructures. The value of reduction of seismic force increases with the increase of seismic intensity.

Soil nonlinearity effects on tower response

The uplift of the tower footing is investigated through vertical deformation and force resistance of the basement soil at the extreme right and left sides of the footing base (case I). Two aspects of foundation response to seismic excitation, rocking deformation and the accumulation of the permanent settlement can be distinguished. The permanent settlement was found to be the less significant in the total vertical displacement at the footing base level, as shown in Figures 13 and 14. It can be concluded that separation of the soil from the structure occurs under large dynamic loads leading to changes in the predominant vibration of the system. As a result of the decrease in soil support at the sidewalls of the foundation by stiffness degradation due to hysteresis, the stress caused by the structural weight on the bottom soil is increased during earthquakes. The reduction of the foundation width in contact during uplift induces an increase of the stresses under the foundation. This leads to a larger soil yielding, which itself modifies the uplift behavior of the foundation.

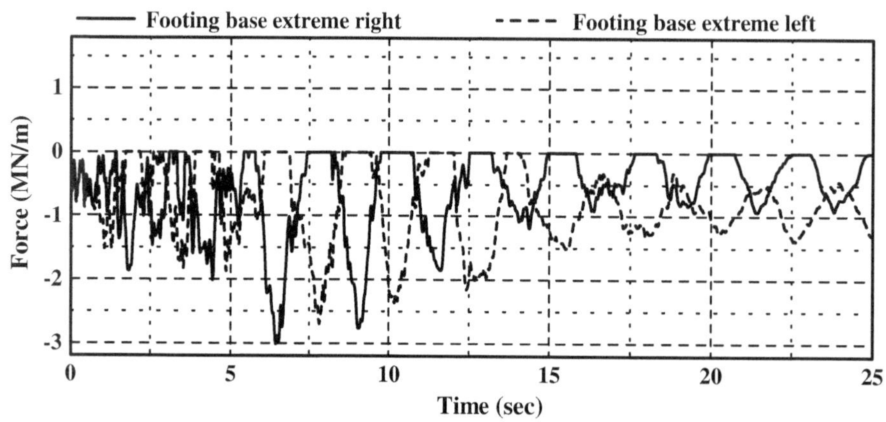

Figure 14 Vertical force per unit length time history of footing extreme right and left at footing base level.

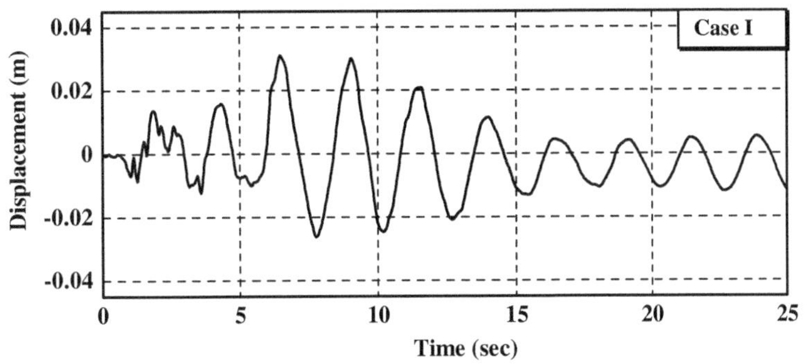

Figure 15 In-plane displacement time history at superstructure base.

The maximum uplift of not more than 5 mm clarifies that the tower foundation almost has sufficient structural loading to minimize the uplift. The soil layer at the footing base level has sufficient bearing capacity to prevent foundation excessive settlement.

Since the foundation structure is very rigid compared to the underlying foundation soil, the rocking motion of the foundation is significantly observed, which affects the characteristics of the foundation motion. The limited shear strength of soil also causes sliding of the foundation block, which could alter the characteristics of the foundation motion. In addition to the inertial soil-foundation interaction due to the existence of foundation mass, the size and stiffness of the foundation could also affect the foundation motion characteristics. From superstructure base (steel superstructure and concrete substructure connection joint) displacement time history (case I) as shown in Figure 15, it could be seen that the displacement could reach around 3.0 cm for soil-structure interaction model of the tower with both material and geometrical nonlinearities; this effect is totally ignored for fixed base model assumption. The limited shear strength of basement soil causes sliding of the foundation, which could be underestimated by elastic and linear soil assumption, as shown in Figure 16.

From the Fourier spectra study of tower acceleration response at different levels of tower for nonlinear soil-foundation-superstructure interaction model (case I), it is shown that there is amplification of different modes over a wide frequency range, as seen in Figures 17 and 18. The in-plane superstructure base response spectrum is larger than that at the footing base at a spectral frequency less than 2.0 Hz because of amplification induced by flexible superstructure and massive rigid substructure interaction, while at high frequency above 2.0 Hz, the response spectra are slightly attenuated due to inertial interaction. The tower top response spectra are significantly amplified at low-frequency range and are almost totally attenuated at high-frequency range due to tower superstructure flexibility, as seen in Figure 18. The massive foundation has the effect of amplifying the response over a wide frequency band. The vertical acceleration response at the footing base level shows relatively high-frequency amplification as the response spectra within the frequency range of 2 to 3 Hz are slightly amplified at the superstructure base level, and they are dramatically amplified at the tower top. On the other hand, the response spectrum at high-frequency range is attenuated by superstructure flexibility filter of the tower response. The nonlinear seismic response of bridge

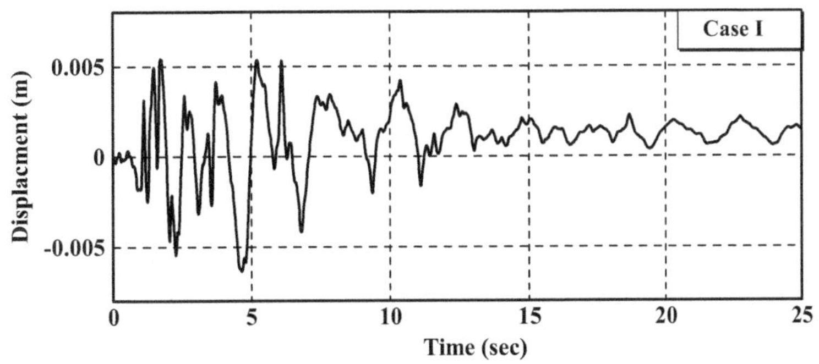

Figure 16 Foundation sliding time history at footing base level.

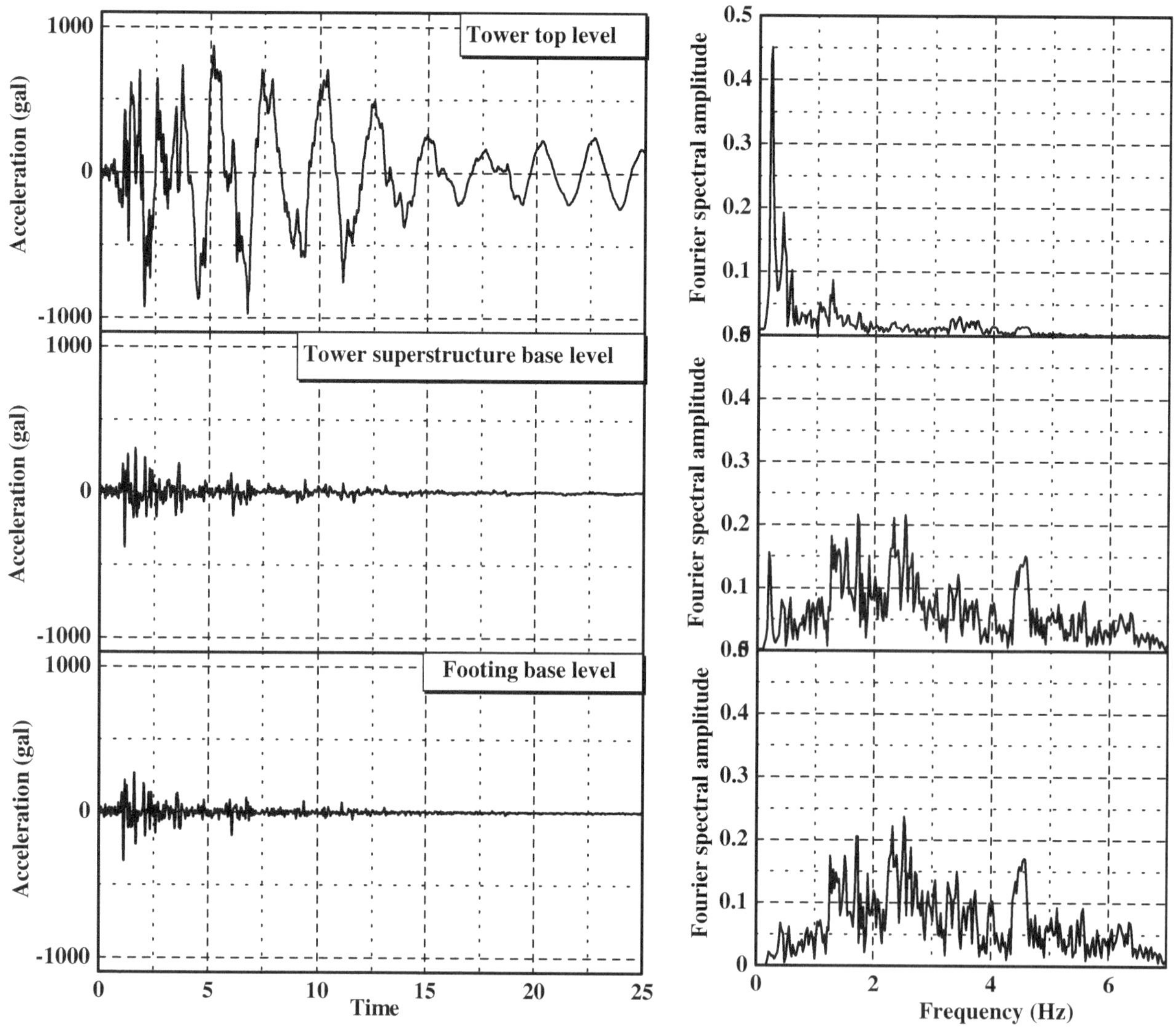

Figure 17 In-plane acceleration time history and response spectra (case I).

piers is distinctly different from that of the linear response. There is a great difference whether it is in vibration amplitude or in frequency property. The nonlinear properties of foundations make the stiffness of the structure low, the response of rotational angle increase, and the response of bending moment decrease.

Conclusions

A finite element model to study the effects of soil-foundation-superstructure interaction on the seismic response of a cable-stayed bridge tower supported on spread foundation was presented, in which an incremental iterative technique was adopted for a more realistic analysis of the nonlinear soil-foundation-superstructure interaction. The problem was analyzed, employing the finite element method and taking account of material (soil, steel superstructure) and geometric (uplifting and P-Δ effects) nonlinearities. Two different modeling approaches for soil foundation interaction, nonlinear Winkler model and linear lumped-parameter model, were investigated. The seismic responses of Winkler soil model were compared to those of the simplified linear lumped-parameter model. In the lumped-parameter soil model, the soil-structure interaction is simulated with translational, rotational, and their coupling spring system acting at centroid of spread foundation at the footing base level while in the Winkler soil model, the soil-structure interaction is simulated with continuous spring system (Winkler) along embedded depth of tower pier and underneath the spread foundation. Soil yielding under the foundation and along the embedded depth was modeled using the Hardin-Drnevich model to express nonlinear soil

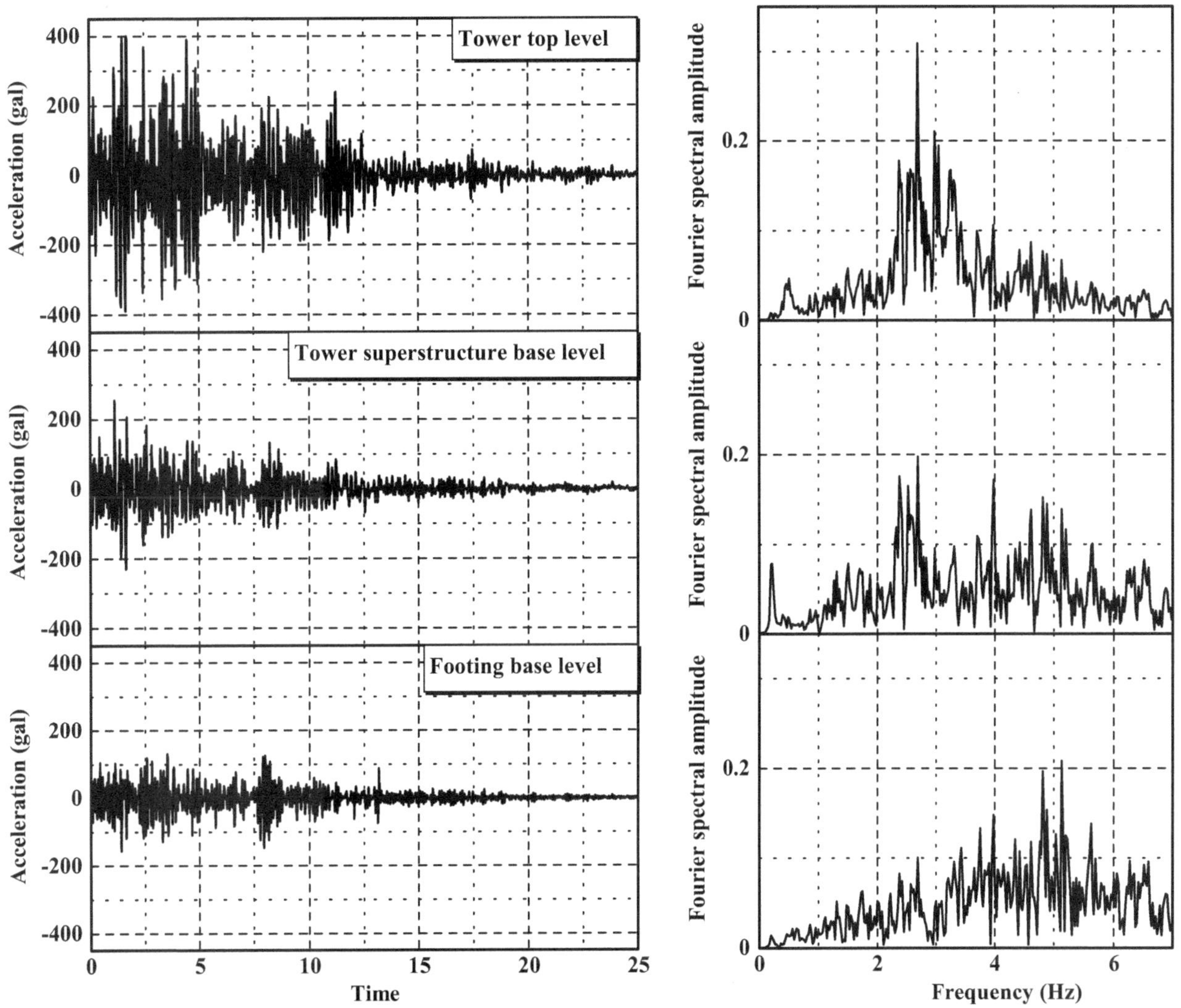

Figure 18 Vertical acceleration time history and response spectra (case I).

characteristics. The contact nonlinearity induced by the uplift of the foundation was integrated using gap element. Radiation damping associated with wave propagation was accounted for implicitly through viscous damping while the energy dissipation through soil material nonlinearity was explicitly modeled. The interaction effects generated by the normal and tangential resistance of the soil against all active sides of the footing were taken into account. The following conclusions can be drawn as follows:

The analysis with simplified linear lumped-parameter model provides a good prediction for the peak response but overestimates the acceleration response and underestimates the uplift force at the anchor between superstructure and pier. The nonlinear soil-foundation interaction effect is very remarkable when uplifting and yielding of supporting soil are considered. The analysis using the nonlinear interaction Winkler model shows that the foundation uplift alters tower displacement seismic response and reduces structural member forces. Compared with the linear analysis, the stiffness of the tower-soil system degrades in each cycle after considering uplifting and yielding. The nonlinear analysis displays larger rotational angles and smaller bending moments compared with the linear lumped-parameter-model analysis. The inclusion of massive foundation and nonlinearity of soil effects leads to the amplification of higher modes of vibration and activates the high-frequency translational motion of the input ground motion and generates foundation-rocking responses. The rocking vibration dominates the lateral bearing stress for the soil along the embedded depth of the tower pier. The predominant contribution to the vertical response at the footing base resulted from the massive foundation rocking rather than from the vertical excitation; the permanent settlement is found to be less significant. The soil bearing stress demand beneath the spread foundation is significantly increased due to footing base uplift. The bearing stress under the foundation should be checked considering both soil material and geometrical nonlinearities with an appropriate factor of safety to avoid catastrophic foundation failure. The soil at the foundation level should have sufficient bearing capacity to prevent foundation excessive settlement.

From the nonlinear Winkler soil model, it is concluded that the soil interaction effect is very remarkable when uplifting and yielding of supporting soil are considered. Compared with the linear lumped soil model, the stiffness of tower foundation-soil model degrades in each cycle after considering uplifting and yielding. It is shown that the nonlinear analysis can get larger rotational angles and smaller bending moments compared with the linear analysis. The nonlinear modeling of soil-foundation interface shows that the soil structure interaction effect may play an important role in altering the force and displacement demand, indicating the necessity for consideration of foundation behavior in the modern design codes to accomplish a more economic yet safe structural design. For more general and versatile conclusions, different input excitations and different ground conditions at bridge construction site should be considered.

Competing interests

The authors declare that they have no competing interests.

Authors' contributions

SEAR carried out the finite element modeling and numerical simulations for the extensive parametric study, which is presented in the research work, and drafted the manuscript. SEAR and TH discussed the idea and the important system parameters, which are included for the numerical study and also have checked the manuscript before submission. Both authors read and approved the final manuscript.

Authors' information

SEAR is an associate professor in Earthquake, Bridge, Structural and Geotechnical Engineering since 2009 at the Civil Engineering Department in Assiut University, Egypt. He acquired his Doctorate in Engineering in 2003 and was a Japan Society for the Promotion of Science fellow in 2006 in Hokkaido University, Japan. He was an Alexander von Humboldt fellow in 2005 in Kassel University, Germany. He has published more than 60 papers in the field of structural engineering, structural analysis, earthquake engineering, soil structure interaction, seismic pounding, structural control, and nonlinear dynamic of offshore structures. He is a previous member of the Japan Society of Civil Engineers. He reviewed many papers for international journals. He is currently working as an engineering counselor at the general project management of Taibah University, Medina, Saudi Arabia. TH is a professor in Earthquake, Bridge, Structural and Geotechnical Engineering since 2004. He acquired his Doctorate in Engineering in 1984 in Hokkaido University, Japan. He was a visiting research fellow at Princeton University, USA in 1985. His research works/interests include Response of Cable-Stayed Bridges under Great Earthquake Ground Motion; Seismic Isolation Design; Energy Dissipation System and Performance-Based Design of Steel Structures; Damage Identification of Highway Bridge; and Bridge Management System Vibration Control and Intelligential Structure of Steel Bridges. He is a fellow member of the Japan Society of Civil Engineers. He reviewed many papers for international journals.

Author details

[1]Taibah University, Medina, Saudi Arabia. [2]Civil Engineering Department, Faculty of Engineering, Assiut University, Assiut 71516, Egypt. [3]Graduate School of Engineering, Hokkaido University, Sapporo, Japan.

References

Abdel Raheem SE (2009) Pounding mitigation and unseating prevention at expansion joint of isolated multi-span bridges. Engineering Structures 31 (10):2345–2356

Abdel Raheem SE, Hayashikawa T (2003) Parametric study on steel tower seismic response of cable-stayed bridges under great earthquake ground motion. Structural Engineering and Earthquake Engineering - JSCE 20(1):25–41

Abdel Raheem SE, Hayashikawa T, Hashimoto I (2002) Study on foundation flexibility effects on steel tower seismic response of cable-stayed bridges under great earthquake ground motion. Journal of Construction Steel, JSSC 10:349–354

Abdel Raheem SE, Hayashikawa T, Hashimoto I (2003) Effects of soil-foundation-superstructure interaction on seismic response of cable-stayed bridges tower with spread footing foundation. Journal of Structural Engineering, JSCE 49A:475–486

Ahn K, Gould PL (1992) Interactive base-isolation foundation system: I. Finite element formulation. Journal of Engineering Mechanics, ASCE 118 (10):2048–2058

Ali HM, Abdel-Ghaffar AM (1995) Modeling the nonlinear seismic behavior of cable-stayed bridges with passive control bearings. Computers & Structures 54(3):461–492

Chaojin X, Spyrakos CC (1996) Seismic analysis of towers including foundation uplift. Engineering Structures 18(4):271–278

Chen CN (2000) Efficient and reliable solutions of static and dynamic nonlinear structural mechanics problems by an integrated numerical approach using DQFEM and direct time integration with accelerated equilibrium iteration schemes. Applied Mathematical Modelling 24(8–9):637–655

Committee of Earthquake Engineering (1996) The 1995 Hyogoken-Nanbu earthquake, investigation into damage to civil engineering structures. Japan Society of Civil Engineers, Tokyo

Ernst JH (1965) Der E-modul von seilen unter berucksichtigung des durchhanges [in German]. Der Bauingenieur, Berlin 40(2):52–55

European Committee for Standardization (1998) EC8: design provisions for earthquake resistance of structures, part 5: foundations, retaining structures and geotechnical aspects. European Committee for Standardization, Brussels

Ganev T, Yamazaki F, Katayama T (1995) Observation and numerical analysis of soil-structure interaction of reinforced concrete tower. Earthquake Engineering and Structural Dynamics 24(4):491–503

Gazetas G, Dobry R (1984) Horizontal response of piles in layered soils. Journal of the Geotechnical Engineering Division - ASCE 110(1):20–40

Gelagoti F, Kourkoulis R, Anastasopoulos I, Gazetas G (2012) Rocking isolation of low-rise frame structures founded on isolated footings. Earthquake Engineering and Structural Dynamics 41:1177–1197

Harada T, Kubo K, Katayama T (1981) Dynamic soil-structure interaction analysis by continuum formulation method. Report of the Institute of Industrial Science, University of Tokyo 29(5):139–194

Harden C, Hutchinson T, Moore M (2006) Investigation into the effects of foundation uplift on simplified seismic design procedures. Earthquake Spectra 22(3):663–692

Hardin BO, Drnevich VP (1972a) Shear modulus and damping in soil: measurement and parameter effects. Journal of the Soil mechanics and Foundation Engineering Division ASCE 98(6):603–624

Hardin BO, Drnevich VP (1972b) Shear modulus and damping in soils: design equations and curves. Journal of the Soil Mechanics and Foundation Division ASCE 98(7):667–692

Japan Road Association (1996a) Specification for highway bridges - part V seismic design. Japan Road Association, Tokyo

Japan Road Association (1996b) Reference for highway bridge design, specification for highway bridges - part IV substructures. Japan Road Association, Tokyo

Karoumi R (1999) Some modeling aspects in the nonlinear finite element analysis of cable supported bridges. Computers & Structures 71(4):397–412

Kawashima K, Hosoiri K (2002) Effect of nonlinear rocking response of direct foundations on the hysteretic behavior of bridges. Journal of Structural Mechanics and Earthquake Engineering - JSCE 703/I-59:97–111
Kobayashi T, Yoshikawa K, Takaoka E, Nakazawa M, Shikama Y (2002) Time history nonlinear earthquake response analysis considering materials and geometrical nonlinearity. Nuclear Engineering and Design 212(1–3):145–154
Masing G (1926) Eigenspannungeu und verfertigung beim Messing. In: Proceedings of the 2nd international congress on applied mechanics, Zurich
Megawati K, Higashihara H, Koketsu K (2001) Derivation of near-source ground motions of the 1995 Kobe (Hyogo-ken Nanbu) earthquake from vibration records of the Akashi Kaikyo bridge and its implications. Engineering Structures 23(10):1256–1268
Raychowdhury P (2011) Seismic response of low-rise steel moment-resisting frame (SMRF) buildings incorporating nonlinear soil–structure interaction (SSI). Engineering Structures 33:958–967
Spyrakos CC (1997) Soil-structure-water interaction of intake-outlet towers allowed to uplift. Soil Dynamics and Earthquake Engineering 16(2):151–159
Tang Y, Zhang J (2011) Probabilistic seismic demand analysis of a slender RC shear wall considering soil–structure interaction effects. Engineering Structures 33:218–229
Trifunac MD, Todorovska MI (1996) Nonlinear soil response - 1994 Northridge, California, earthquake. Journal of Geotechnical Engineering 122(9):725–735
Vlassis AG, Spyrakos CC (2001) Seismically isolated bridge piers on shallow soil stratum with soil structure interaction. Computers & Structures 79 (32):2847–2861
Yim SCS, Chopra AK (1984) Dynamics of structures on 2-spring foundation allowed to uplift. Journal of Engineering Mechanics 110(7):1124–1146
Zhang J, Tang Y (2009) Dimensional analysis of structures with translating and rocking foundations. Soil Dynamics and Earthquake Engineering 29 (10):330–346

Span-to-depth ratio effect on shear strength of steel fiber-reinforced high-strength concrete deep beams using ANN model

Uday Naik* and Sunil Kute

Abstract

The paper predicts the shear strength of high-strength steel fiber-reinforced concrete deep beams. It studies the effect of clear span-to-overall depth ratio on shear capacity of steel fiber high-strength deep beams using artificial neural network (ANN8). The three-layered model has eight input nodes which represent width, effective depth, volume fraction, fiber aspect ratio and shear span-to-depth ratio, longitudinal steel, compressive strength of concrete, and clear span-to-overall depth ratio. The model predicts the shear strength of high-strength steel fiber deep beams to be reasonably good when compared with the results of proposed equations by researchers as well as the results obtained by neural network (ANN7) which is developed for seven inputs excluding span-to-depth ratio. The developed neural network ANN8 proves the versatility of artificial neural networks to establish the relations between various parameters affecting complex behavior of steel fiber-reinforced concrete deep beams and costly experimental processes.

Keywords: Artificial neural network; Deep beam; High strength; Steel fiber; Shear span ratio; Span to depth ratio; Volume fraction; Fiber aspect ratio

Introduction

The deep beams differ from ordinary beams in their geometrical proportions and nowadays are widely used as transfer girders in multistory buildings to avoid columns for providing free space at different floor levels for parking. The ACI Committee 318 (2002), section 11.7.1 defines the deep beam based on two criteria: as clear span-to-depth ratio ($L/D < 4$) and shear span-to-depth ratio ($a/d < 2$). The IS: 456 (2000) describes deep beam having clear span-to-depth ratio less than 2. ACI-ASCE Committee 426 classifies a beam with shear span-to-depth ratio (a/d) less than 1.0 as deep beam and a beam with a/d exceeding 2.5 as an ordinary shallow beam. Any beam in between these two limits is categorized as a moderate deep beam. The concrete has inherent microcracks present in it which turns into macrocracks as the members are subjected to loading. The cracks are formed within the shear span when the principal tensile stress exceeds the tensile strength of concrete. The addition of randomly dispersed discrete steel fibers into concrete mix improves the post-cracking tensile strength of RC beams. The addition of fibers in concrete results in the reduction of crack formation and propagation (Anant and Modhera 2008). The high-strength concrete has relatively more brittleness and lack of ductility than normal strength concrete. This can be overcome by additions of fibers in high-strength concrete (HSC). It is important to understand better and predict the shear behavior of steel fiber-reinforced concrete (SFRC) deep beam for its wider applications. The lack of rational design equations in various building codes, shear failure modes, and various parameters affecting the load-carrying mechanism makes the shear behavior of steel fiber-reinforced concrete beams more complex, particularly in deep beams, and a topic for research (Roberts and Ho 1982). Many researchers in the past have studied the addition of steel fibers in concrete to improve the flexural as well as shear capacity of normal and deep beams. The studies support the use of steel fibers as shear reinforcement of reinforced concrete (RC) beam. The previous research is mainly carried on SFRC shallow

* Correspondence: prec_upn@yahoo.com
Department of Civil Engineering, K.K. Wagh Institute of Engineering and Research Centre, Nasik Pin 422003, Maharashtra, India

beams ($a/d > 2.5$). The proposed equations are also generally based on tests carried on a certain type of fibers. The behavior of SFRC deep beam in shear needs to be investigated for different parameters for generalized shear strength equation if steel fibers are to be used in actual structural members (Al Ta'an and Al-Feel 1990; Mau and Hsu 1987; Kotsovos 1987; Lim 1987; Narayanan and Darwish 1987, 1988; Manicka Selvem and Kuruvilla 1988; Kwak 2002; Misrayah and Banthia 2002; Shah and Mishra 2004; Gohnert 2006; Gustavo 2006). The present study aims at proving the effectiveness of artificial neural network to study the effect of clear span-to-overall depth ratio on shear capacity of steel fiber in high-strength deep beams for L/D limit as per IS: 456 (2000) and compare the results obtained with the equations suggested by previous researchers. It also highlights the importance of span-to-depth ratio (L/D) in predicting the shear strength of deep beams.

Methods

Artificial neural network

The neural networks comprise a large number of highly interconnected elements called 'neurons'. The neural network is configured for a specific application. The main advantage is that they can solve problems that do not have an algorithmic solution or for which the solution is too complex. In recent past, artificial neural network (ANN) has been successfully applied in various disciplines of engineering analysis including the field of civil engineering. Sanand and Saka (2001) predicted the ultimate shear strength of RC deep beams using the artificial network. The results developed from ANN are more accurate than by ACI code and strut and tie method. Adhikari and Mutsuyoshi (2006) developed ANN for shear strength of SFRC beams and found that the accuracy of the network result increases with increase in number of inputs. It was found that five-input network predicts shear strength more closely than four-input networks. The accuracy of the results is found not biased with concrete strength, shear span-to-depth ratio, and beam depth. Arafa et al. (2011) showed that the developed ANN model has higher potential in predicting the ultimate shear strength of both normal and high-strength deep beams when compared with calculated values using ACI code.

The major concerns of structural analysis problems like predicting different properties of concrete and load-carrying capacity have been solved mostly by using back-propagation neural network. The processing units are grouped into layers of input, hidden, and output processing units. The networks are trained so that a particular input leads a specific target output. The log-sigmoid transfer function is used in multilayer networks that are trained using the backpropagation algorithm because this function is differentiable. This transfer function takes the input and squashes the output into the range 0 to 1, according to the expression:

$$a = 1/(1 + exp^{-n}) \quad \text{where } a = \text{logsig}\,(n), n = \text{net input.} \tag{1}$$

The networks are adjusted based on the comparison of output and target. The batch training of network proceeds by making weight and bias changes based on the entire set of input vectors. A stable set of weights adaptively evolves will produce good results.

Experimental data

The development of neural network requires reliable data which include input parameters affecting the system and corresponding output. These data can be experimental test results, reliable empirical data, or theoretical results.

The experimental test results of 118 tests on SFRC normal and high-strength concrete deep beams, carried by various researchers (Narayanan and Darwish 1987, 1988; Kwak et al. 2002; Shah and Mishra 2004; Gustavo 2006) have been used to develop the ANN8 model in the present study where the number stands for number of input parameters affecting the shear strength of SFRC deep beams. The data consists of 90 test results for normal strength concrete (NSC) and 28 test results for HSC steel fiber-reinforced beams for different values of span-to-overall depth ratio. The basic parameters that control the shear strength of SFRC beams, based on previous research works, are beam width (b), effective depth (d), fiber volume (V_f), fiber aspect ratio (l_f/d_f), shear span ratio (a/d), compressive strength of concrete (f_c) and longitudinal reinforcement area (A_{st}), and clear span-to-depth ratio (L/D). The shear span-to-depth ratio (a/d) of 45 beams, out of total data, has shear span-to-depth ratio equal to or less than 2. The total data base also includes 19 beams having clear span-to-depth ratio less than or equal to 2 to satisfy the condition for deep beam as per

Table 1 Statistic for NSC and HSC SFRC beams

Parameter	Min	Max	Mean
Width (b), mm	50	300	175
Depth (d), mm	126	610	286
Fiber volume (V_f),%	0.25	2	1.125
Fiber aspect ratio (l_f/d_f)	40	133	84.27
Shear span ratio (a/d)	0.5	5.72	2.47
Main steel (A_{st}),%	0	4.5	2.47
Compressive strength (f_c), MPa	19	99.6	62.34
Clear span-to-depth ratio (L/D)	1.5	9.88	5.43

IS: 456 (2000). Table 1 shows the statistics for input parameters for SFRC beams as per Additional file 1.

Preprocessing of data

The learning rate of networks is significantly affected by the data presented to artificial neural networks. To ensure that all variables receive equal attention during training, the data scaling is important. Hence, the data needs preprocessing before presented for training. The scaling of the inputs is done in the range of [−1, +1] to improve the speed of learning process as these values fall in the region of sigmoid. The normalization of input and output data before presenting them to network is done by linear normalization function within the values of 0 and 1 and given by

$$S = [2(P - P_{\min})/(P_{\max} - P_{\min})] - 1. \qquad (2)$$

Training of ANN

The MATLAB (R2009a) package is used for the development of present ANN8 model. The backpropagation is most successful and widely used in civil engineering. The collected data is divided into three units, i.e., training, validation, and testing. In the present case, training set comprises 80% of data base and remaining data base is equally divided into validation (10%) and testing sets (10%).

The values of network parameters considered in this approach are as follows: number of hidden layers = 1, number of hidden units = 20, learning rate = 0.5, momentum factor = 0.5, and learning cycles = 1,000. The root mean square error (R^2) was 0.924. The coefficient indicates a significant correlation.

Topology of network

The topology or architecture is obtained using trial and error method. The architecture of the network ANN8 is as shown in Figure 1.

Performance of the proposed model

The network is trained for 80% of data collected by gradually increasing the number of neurons in the hidden layer. The performance was monitored during training process as sum squared error over all the training data. The training process stops when the maximum number of epochs is reached.

A comparison between experimental shear strength and corresponding values obtained by ANN8 model for 118 normal and high-strength SFRC ordinary and deep beams is shown in Figure 2. It shows that predicted shear strength is in good agreement with the experimental results.

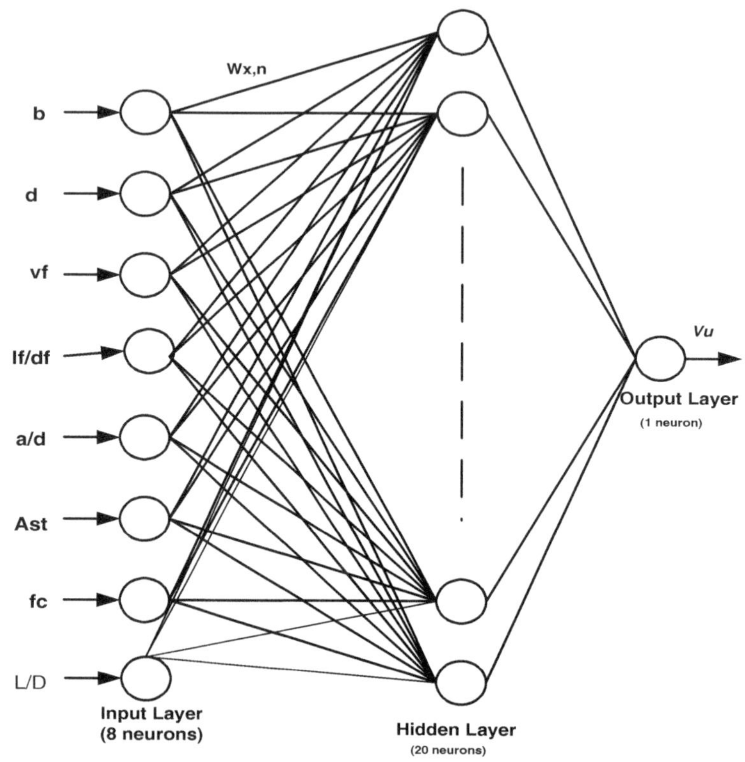

Figure 1 Three-layered feed-forward backpropagation ANN8.

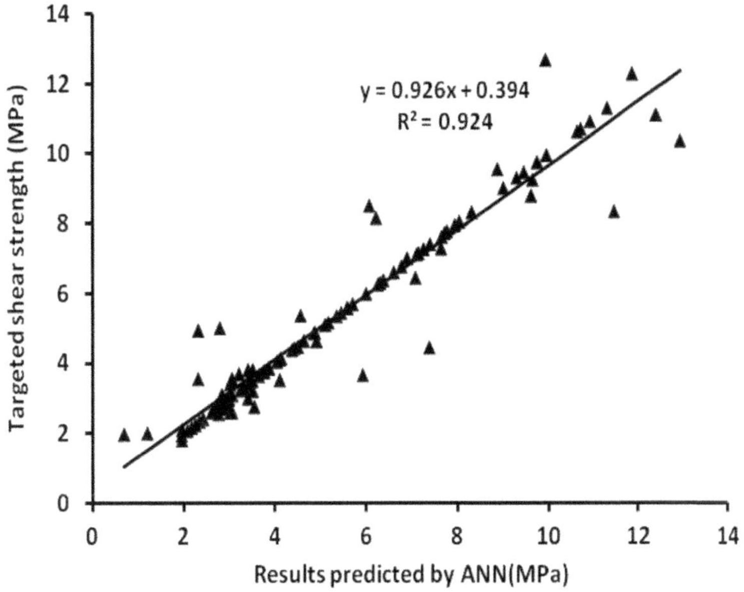

Figure 2 Graph of targeted shear strength versus predicted shear strength.

Parametric study

The parametric study is carried out with the developed artificial neural network to evaluate the influence of parameters like b, d, V_f, l_f/d_f, a/d, A_{st}, f_c, and clear L/D. The effect was studied by varying one of the input parameters with respect to varying span-to-depth ratio mainly and all other parameters are set to constant values. The values of L/D ratio and a/d ratio are assumed less than 2 to fit the beam in deep beam category.

The model is fed with some assumed data and the resulting output values of shear strength are studied. The range of parameters used in parametric study is shown in Table 2.

Results and discussion

The effect of each parameter on the shear strength of SFRC deep beam is studied by varying influencing parameters. The model is fed with the values of influencing parameters and resulting output values of shear strength are studied. The range of parameters used in parametric study is shown in Table 2. The results obtained by developed ANN are plotted against respective parameters.

Table 2 Values of trial data

Parameter	Range
Effective depth (d; mm)	300, 350, 400, 450, 500
Fiber content (V_f;%)	0, 0.5, 1, 1.25, 1.5, 1.75, 2
Fiber aspect ratio (l_f/d_f)	50, 60, 70, 80, 90
Shear span ratio (a/d)	0.5, 1, 1.5, 2
Longitudinal steel (%)	0, 0.5, 1, 1.5, 2, 2.5
Compressive strength (f_c; MPa)	50, 60, 70, 80
Span to depth ratio (L/D)	1, 1.25, 1.5, 1.75, 2

Effect of effective depth

Figure 3 shows the effect of increase in span-to-depth ratio on shear strength with increase in effective depth. It is found that for a constant volume fraction of 1% and fiber aspect ratio of 50, the shear strength of deep beam decreases as the depth of beam goes on increasing except at $d = 350$ mm ($d/b = 2.33$). The decrease in the shear strength varies from 18% to 26% for variation of depth from 300 to 500 mm. Similarly, it is noted that the shear strength decreases with increase in span-to-depth ratio (L/D) at different depths of high-strength SFRC deep beam shown in Figure 3. It resembles with the findings of Appa Rao (2007).

Effect of fiber volume fraction

The effect of fiber content on shear strength is observed from Figure 4. It shows the variation of shear strength with increase in fiber volume fraction for different values of L/D ratios. As the fiber volume increases, the shear strength is enhanced at all considered span-to-depth ratio. It is also noted that the shear strength decreases with the increase in L/D ratio for a volume fraction. For the assumed parameters shown in Figure 4, the decrease in shear strength with increase in L/D ratio at $V_f = 0.5\%$ is 25% and with the increase in volume fraction at $V_f = 2\%$, the decrease in shear strength is 6%. This predicts the effectiveness of steel fiber in enhancing the shear

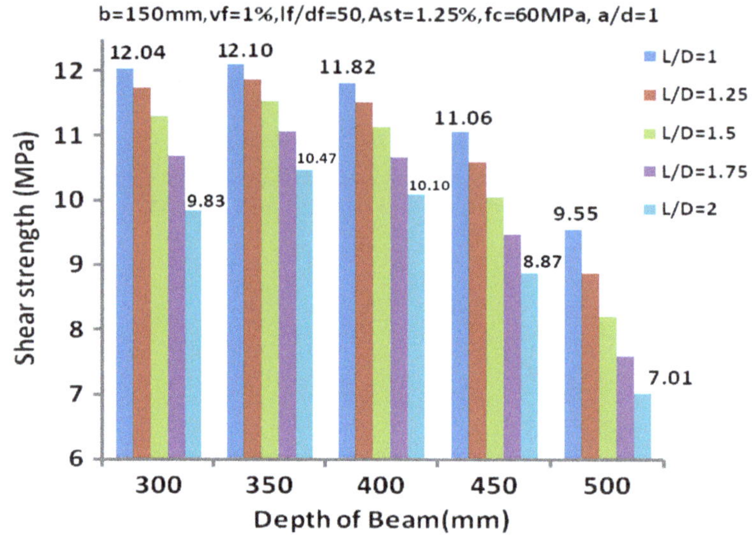

Figure 3 Effect of depth on shear strength of deep beam at different *L*/*D* ratios.

strength of high-strength deep beam. It can be seen that there is an almost linear increase in the shear strength of deep beams with increase in fiber contents as predicted by Remigijus and Gediminas (2007).

Effect of fiber aspect ratio

The fiber aspect ratio is related with the geometry of the fibers. The influence of fiber aspect ratio on shear strength for different volume fraction and *L*/*D* ratio (less than 2) is represented in Figure 5. Many researchers like Narayan and Darwish (1987, 1988), Manicka Selvem and Kuruvilla (1988), Remigijus S (2007), Khuntia et al. (1999) have combined the effect of fiber aspect ratio and volume fraction in their formulations.

Figure 5 represents the effect of fiber aspect ratio with varying *L*/*D* ratio from 1 to 2 and fiber volume fraction from 0.5% to 1.5% at the interval of 0.25. It predicts that for *L*/*D* ratio <1.25, the effect of fiber aspect ratio is negligible for considered values of parameters. However, the shear strength goes on decreasing

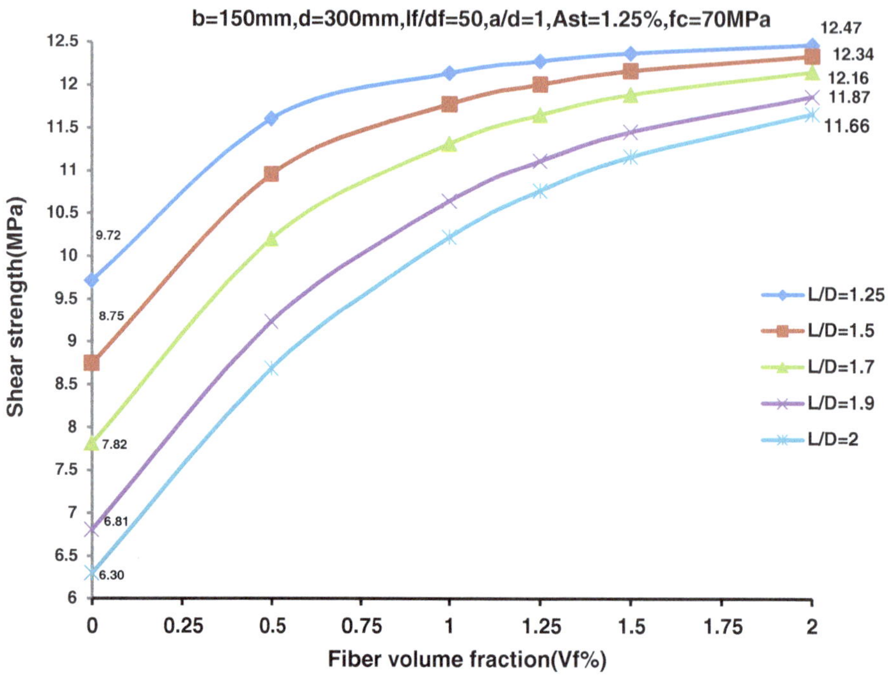

Figure 4 Effect of V_f percentage at different *L*/*D* ratios on shear strength deep beam.

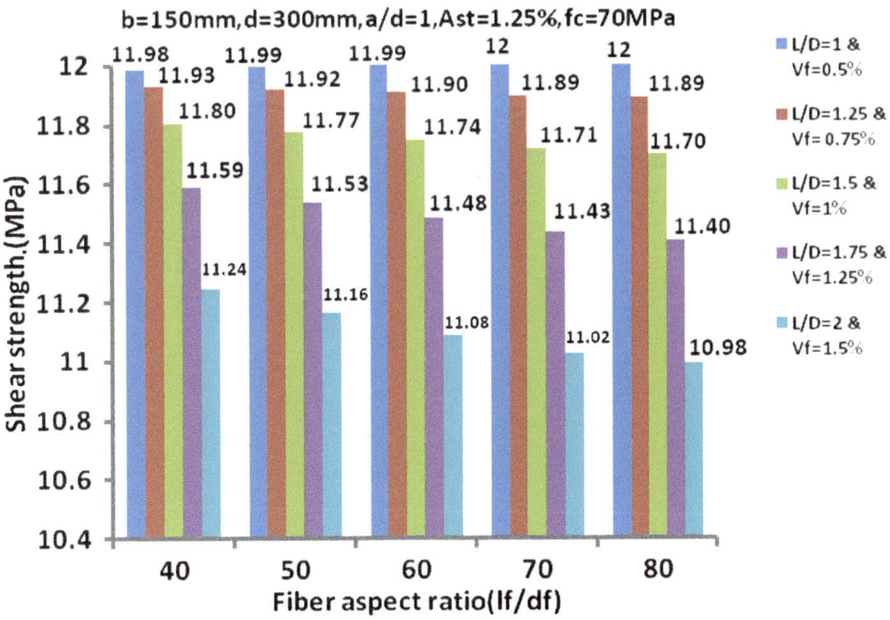

Figure 5 Effect of l_f/d_f at varying L/D and V_f percentage on shear strength of deep beam.

with increase in fiber aspect ratio though the volume fraction of fibers and L/D ratio is increased. The decrease in the shear strength by 0.8%, 1.6%, and 2.3% at $L/D = 1.5$, 1.75, and 2, respectively, is observed for change in fiber aspect ratio from 40 to 80 from Figure 5.

Effect of longitudinal steel

Figures 6 and 7 show the variation of shear strength with increase in longitudinal steel percentage. The relation is studied for the increase in fiber content as well as L/D ratio, respectively. Figure 6 obtained for variation of longitudinal steel along with volume fraction represents

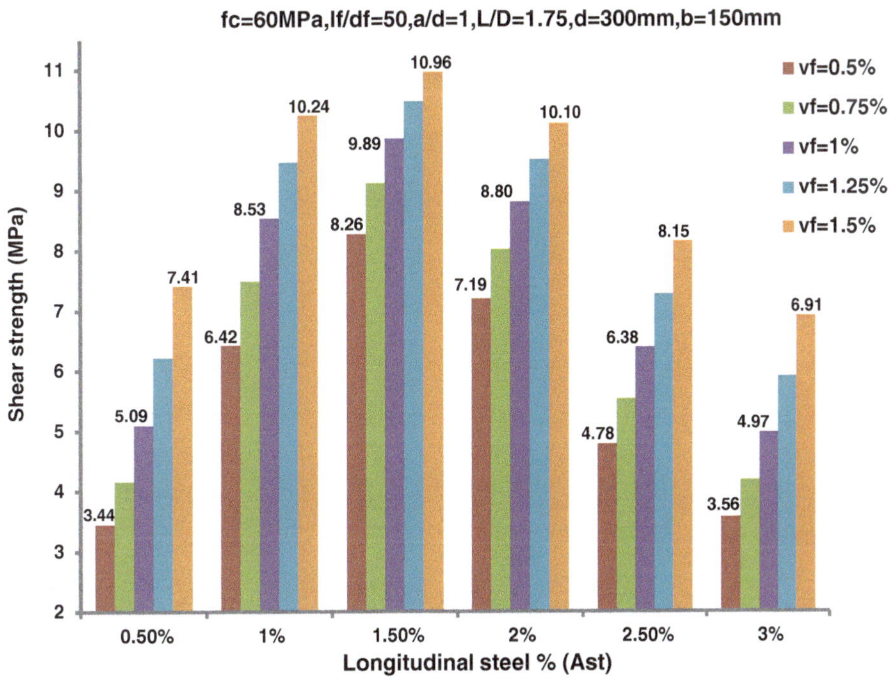

Figure 6 Effect of A_{st} percentage at different fiber volume fractions on shear strength.

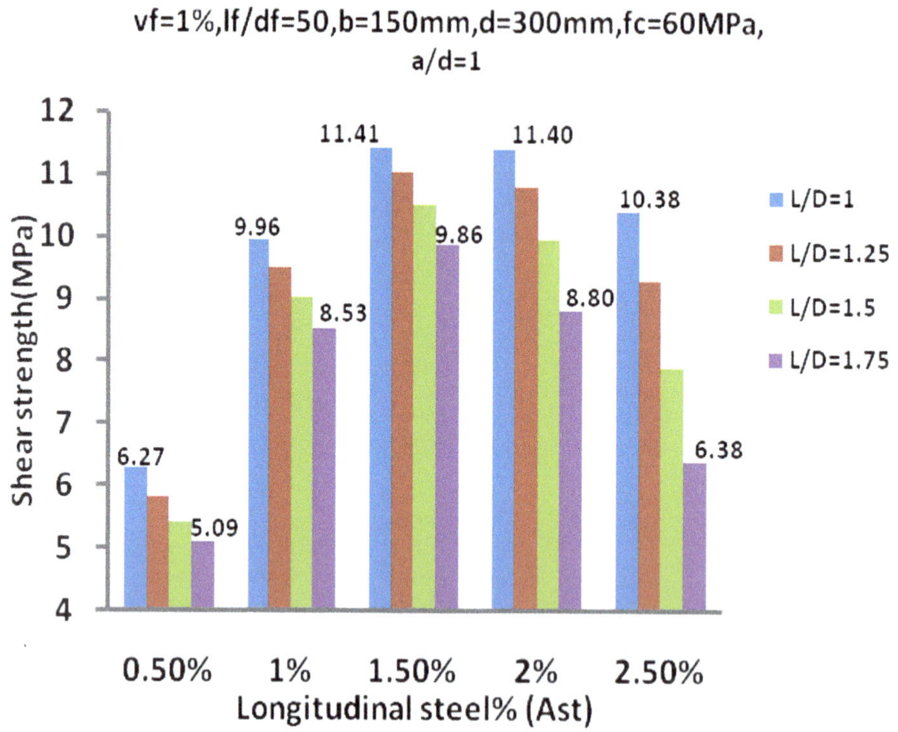

Figure 7 Effect of *L/D* at different percentages of longitudinal steel on shear strength.

increase in shear capacity of up to A_{st} =1.5% by 140%, 94%, and 48% for the range of V_f = 0.5%, 1%, and 1.5%, respectively. However, the shear strength decreases even with the increase in fiber dosage at A_{st} = 2%, 2.5%, and 3% at constant fiber aspect ratio of 50. The percentage of decrease in the shear strength at V_f = 0.5%, 1%, and 1.5% is found to be 57%, 50%, and 37%, respectively. It indicates that for lower values of longitudinal steel percentage, the addition of steel fibers contributes to the shear strength of high-strength deep beams Lee and Kim (2008), and longitudinal steel with 1.5% is the optimum value for the considered parameters in Figure 6. For further increase in A_{st} from 2% to 3%, the decrease in shear strength percentage at lower than

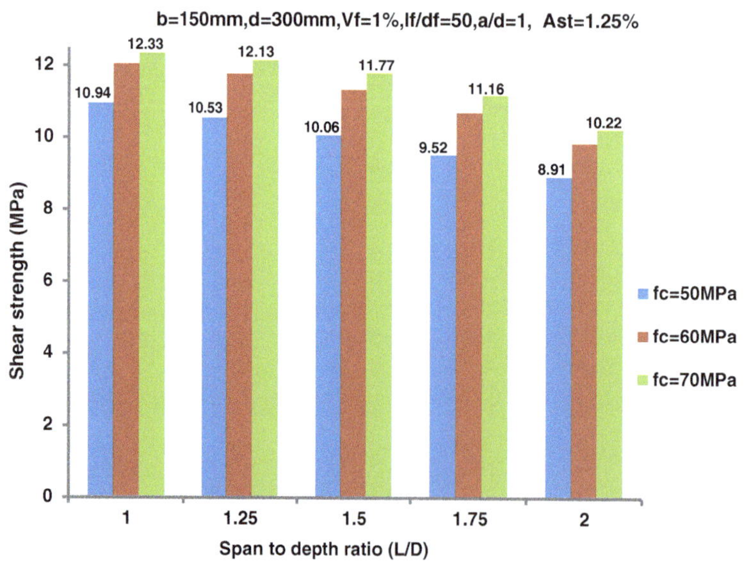

Figure 8 Effect of *L/D* at different compressive strength ratios on shear strength.

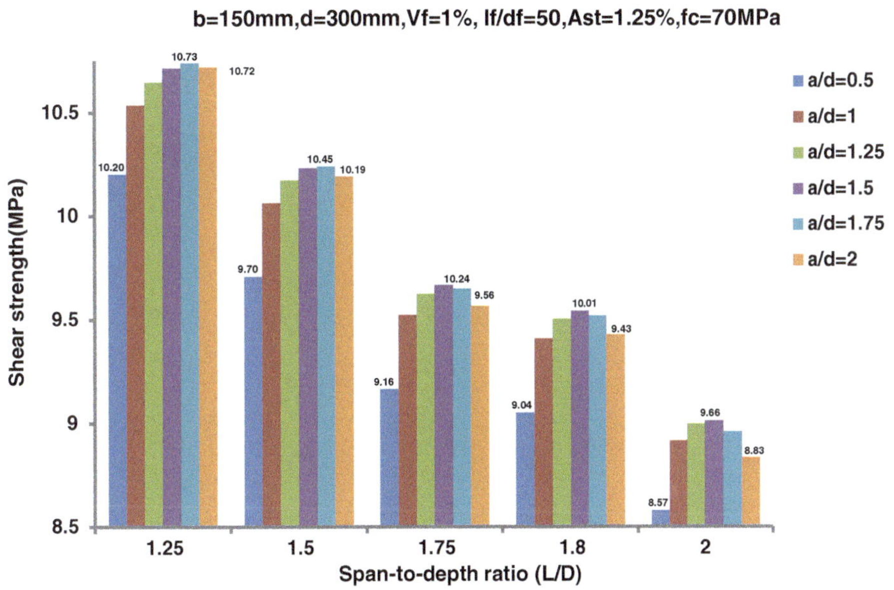

Figure 9 Effect of *L*/*D* at different *a*/*d* ratios on shear strength.

that of longitudinal steel and for the same fiber volume fraction is observed. This helps the optimum combination of longitudinal steel and volume fraction for high-strength SFRC deep beams.

The Figure 7 shows variation of the shear strength with increase in percentage of longitudinal steel at constant volume fraction (1%) for different $L/D < 2$. The increase in shear capacity is noted at $A_{st} = 1.5\%$ for all L/D ratios mentioned in Figure 7. The increase in strength by 82% and 91% is observed at $L/D = 1$ and 1.75, respectively.

It is observed that with the increase in span-to-depth ratio, the shear strength decreases at all values of longitudinal steel. The minimum decrease by 19% at $A_{st} = 0.5\%$ and maximum decrease by 39% at $A_{st} = 2.5\%$ is found for $L/D = 1$ to 1.75, respectively, as shown in Figure 7. The decrease in shear strength is enhanced with increase in the percentage in longitudinal steel and span-to-depth ratio.

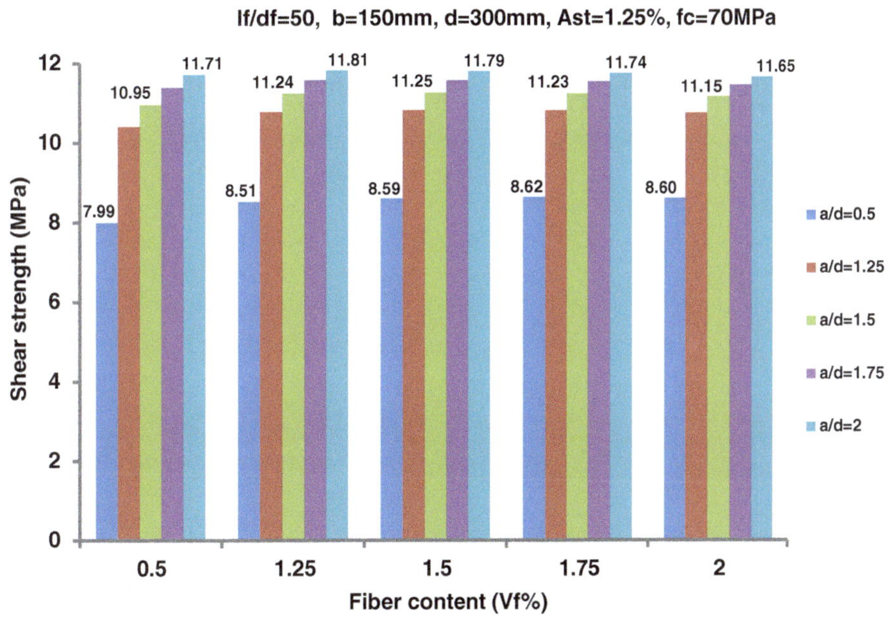

Figure 10 Effect of V_f percentage for different *a*/*d* ratios on shear strength of deep beam.

Table 3 Shear strength by ANN8, ANN7, and researchers for without longitudinal steel

Serial no.	Fiber volume fraction (%)	ANN8 results (MPa)	ANN7 results (MPa)	Narayanan and Darwish (1988)	Khuntia et al. (1999) (MPa)	Ashour et al. (1992)
1	0	2.66	2.93	2.82	3.49	2.5
2	0.5	4.18	4.21	3.64	4.01	3.41
3	1	6.47	5.04	4.23	4.54	3.77
4	1.5	8.83	5.58	4.79	5.06	4.41
5	2	10.42	5.6	5.32	5.6	5.05

Effect of compressive strength

The effect of compressive strength of concrete is plotted in Figure 8 for different L/D ratios; while other parameters are $V_f = 1\%$, $a/d = 1$, $l_f/d_f = 50$, $d = 300$ mm, and $b = 150$ mm.

It is found from Figure 8 that the shear strength of deep beam is enhanced by increase in compressive strength from 50 to 70 MPa for all L/D ratios but decreases with increase in L/D ratio at all considered compressive strengths. The depth of beam is constant. The result fairly agrees with the prediction of Appa Rao et al. (2007). The decrease in shear strength for variation in L/D from 1 to 2 is found to be 19% and 17% for compressive strength of concrete at 50 and 70 MPa, respectively, for other parameters shown in Figure 8.

Effect of shear span-to-depth ratio

The positions of load from support (shear span) affect the load-carrying mechanism of the flexure member. It is now well established that the point loads are directly transferred to the support through a strut, converting arch action into truss mechanism with increase in depth.

The shear span-to-depth ratio effect on deep beam for different L/D ratios is observed from Figure 9. The shear strength increases by 5%, 7%, 11%, 16%, and 12% at $L/D = 1.25$, 1.5, 1.75, 1.8, and 2, respectively, for variation in $a/d = 0.5$ to 1.5. The further increase in a/d decreases the shear strength. It is observed from Figure 9 that the shear strength is maximum at $a/d = 1.5$ for all L/D ratios considered. It also indicates that the shear strength decreases with the increase in L/D at a particular a/d ratio. The average decrease in the shear strength by 16% is found with increase in L/D ratio from 1.25 to 2 at each a/d ratio.

Figure 10 indicates the shear strength variation with fiber volume content at different shear span ratios. It represents that at a volume fraction $V_f = 1.5\%$ and $a/d = 1.5$, there is 1% increase and at $V_f = 2\%$ and $a/d = 1.5$, the shear capacity decreases by 2%. This indicates that addition of fiber volume fraction does not affect the load-carrying mechanism.

The results obtained from ANN8 are also compared with the results obtained from the equations (Slater et al. 2012) proposed by previous researchers Narayanan and Darwish (1988), Khuntia et al. (1999), Ashour et al. (1992), and the results of ANN7 model developed using same data with the same input parameters except L/D ratio. The equations proposed by Narayanan and Darwish (1988) and Ashour et al. (1992) do not include the effect of L/D ratio but consider the contribution of longitudinal steel to shear strength in their formulations. Khuntia et al. (1999) proposes equation based on contribution of concrete strength (f_c) and steel fiber content (V_f%) to the shear strength. The constant parameters assumed are $b = 150$ mm, $d = 300$ mm, $f_c = 70$ MPa, $a/d = 1$, $L/D = 1.9$ and crimped-type fibers. The comparison is made for variation in fiber volume as a key factor controlling the properties of SFRC concrete. Tables 3 and 4 show the results obtained for $A_{st} = 0\%$ and 1.5% at different volume fractions, respectively.

It can be observed from Table 3 that the shear strength at different V_f percentages by ANN7 is in good agreement with the results by other researchers where parameter L/D ratio is not considered. The results by ANN8 shows greater shear strength when the effect of

Table 4 Shear strength by ANN8, ANN7, and researchers for 1.5% longitudinal steel

Serial no.	Fiber volume fraction (%)	ANN8 results (MPa)	ANN7 results (MPa)	Narayanan and Darwish (1988)	Ashour et al. (1992)
1	0	7.10	5.91	5.71	6.11
2	0.5	8.51	7.48	7	7.87
3	1	10.02	7.47	7.6	9.62
4	1.5	10.99	7.34	8.15	11.36
5	2	11.53	7.24	8.68	13.12

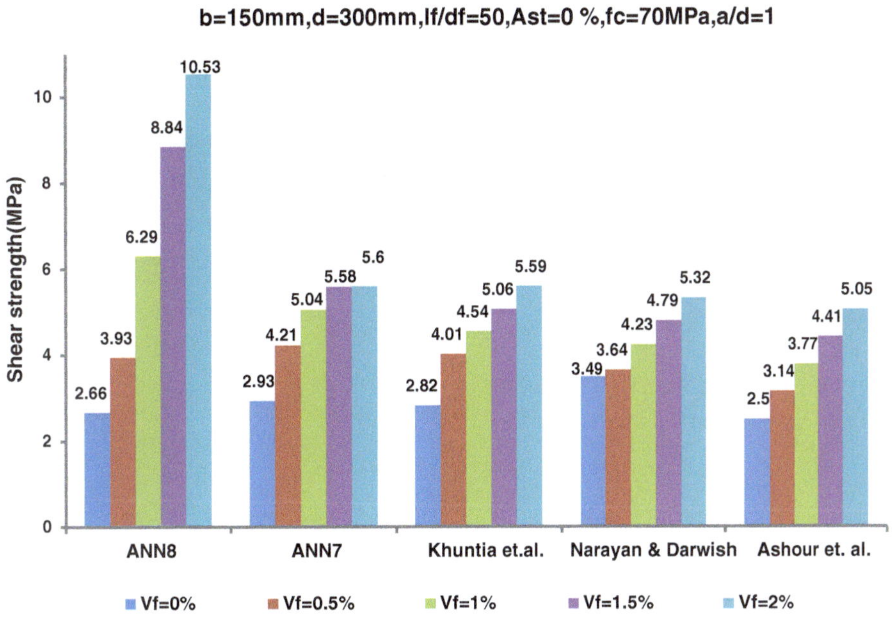

Figure 11 Comparison of ANN8, ANN7, and researchers for without longitudinal steel.

L/*D* ratio is incorporated in the analysis for conditions without longitudinal steel. The increase in shear strength for variation in fiber volume fraction from 0.5% to 2% is observed 28%, 33%, 58%, and 86% from the comparison of ANN8 and ANN7.

Table 4 for 1.5% longitudinal steel indicates good agreement between ANN7 and Narayanan and Darwish (1988) while there is also very good agreement between ANN8 and Ashour et al. (1992). The comparison of the results in Tables 3 and 4 shows that the equation proposed by Ashour et al. (1992) is more accurate to predict the results of high-strength SFRC deep beams for effect of *L*/*D* ratio.

The results are also plotted in Figures 11 and 12 which indicate that the variation in shear strength is linear for increase in fiber volume fraction as observed

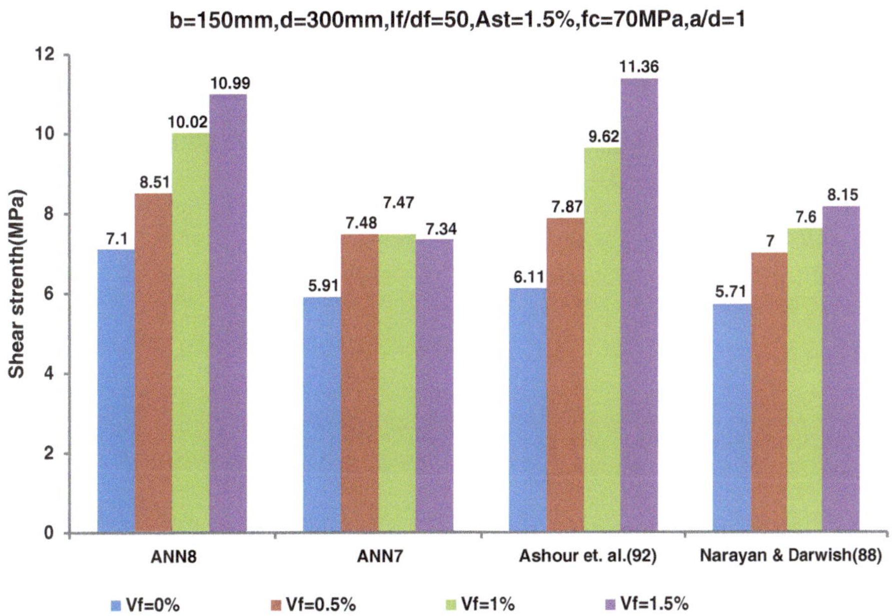

Figure 12 Comparison of ANN8, ANN7, and researchers with 1.5% longitudinal steel.

by ANN8. The equations of Narayanan and Darwish (1988) and Khuntia et al. (1999) do not consider the effect of L/D ratio (size effect) in their formulation and underestimate the shear strength of high-strength SFRC deep beams at various fiber volume fractions. The results by Ashour et al. (1992) show overestimation of the strength as fiber volume fraction increases for $L/D < 2$.

Conclusions

The artificial neural network (ANN8) is developed and utilized to predict the shear strength of steel fiber-reinforced deep beams which consists of input layer with eight neurons, hidden layer with twenty neurons, and output layer with one neuron.

The comparison between results of ANN8 with results of ANN7 and proposed equations by previous researchers reveals the reliability of neural network developed for prediction of shear strength of high-strength steel fiber-reinforced concrete deep beams.

The parametric study of variables affecting shear strength using the developed network shows identical variation of shear strength for a/d and L/D ratios of less than 2 as observed by various researchers through experimental research. The results of variation in span-to-depth ratio show decrease in shear strength for different influencing factors like fiber content, fiber aspect ratio, longitudinal steel, depth of beam, and shear span. This also shows that the size effect is predominant in decreasing the shear strength and must be taken into consideration as an important factor in analysis and design of high-strength SFRC deep beams. The result predicted by developed ANN8 resembles with the findings of previous researchers. This shows the validity and versatility of developed neural network for predicting the shear strength of high-strength SFRC deep beam using artificial neural network model.

The various parameters affecting the shear strength of SFRC deep beams make the analysis more complex and there is need to study the behavior of high-strength SFRC deep beams by suitable combinations of the parameters. The ANN proves to be the most powerful technique to analyze the complex behavior.

Additional file

Additional file 1: Statistics for input parameters for SFRC beams.

Abbreviations

a/d: shear span-to-effective depth ratio; ANN: artificial neural network; A_{st}: longitudinal steel area, mm(); B: width of the beam, mm; f_c: compressive strength of concrete; HSC: high-strength concrete; L/D: clear span-to-overall depth ratio; l_f: fiber length, mm; l_f/d_f: fiber aspect ratio; MPa: ; d: effective depth of beam; NSC: normal strength concrete; SFRC: steel fiber-reinforced concrete; V_f: fiber volume fraction, %.

Competing interests

The authors declared that they have no competing interests.

Authors' contributions

UN carried acquisition of the relevant data of results of experiments carried by various researchers from literatures, developed the neural network, analyzed and interpreted the results, and also drafted the manuscript. SK have critically revised the draft for intellectual contents and participated in interpreting the results. Both the authors read and approved the final manuscript.

Acknowledgements

The authors acknowledge the help provided by Department of Library of K.K. Wagh Institute of Engineering and Research Centre, Nasik and MCERC, Nashik for acquisition of data.

References

ACI committee 318 (2002) Building code requirements for structural concrete and commentary. American Concrete Institute, Detroit, USA
Adhikari BB, Mutsuyoshi H (2006) Prediction of shear strength of steel fiber RC beams using neural networks. Constr Build Mater 20:801–811
Al Ta'an SA, Al-Feel JR (1990) Evaluation of shear strength of fiber reinforced concrete beams. Cement and Conc Composites 12:87–94
Anant P, Modhera CD (2008) Micro mechanical crack and deformations study of SFRC deep beams. In: 33rd Conference on our world in concrete and structures. August 2008, Singapore, pp 25–27, August 2008
Appa Rao G, Injaganeri SS, Suresh P (2007) Investigation of size effect on shear strength of reinforced concrete beams. J of Struct Engg 33(06):499–504
Arafa M, Alquedra M, An-Najjar H (2011) Neural network models for predicting shear strength of reinforced normal and high strength concrete deep beams. J Appl Sci 11(2):266–274
Ashour SA, Hasanain GS, Wafa FF (1992) Shear behavior of high strength fiber reinforced beams. ACI Struct J 89(2):176–184
Aurello M, Ruig MF (2008) Shear strength of members without transverse reinforcement as a function of critical shear crack width. ACI Struct J 105:163–172
Choi K-K, Park H-G, Wight J (2007) Shear strength of steel fiber reinforced concrete without web reinforcement. ACI Structural J 104(S 58):12–20
Gohnert M (2006) Postulate on shear in reinforced concrete. J South African Inst Civil Engg 48(4):9–13
Gustavo J (2006) Shear strength of beams with deformed steel fibers. Concrete International, November, pp 57–66
IS: 456 (2000) Plain and reinforced concrete- code of practice. Bureau of Indian Standards, Manak Bhavan, New Delhi, India
Khuntia M, Stojadinovic B, Goel SC (1999) Shear strength of normal and high strength fiber reinforced concrete beams without stirrups. ACI Structural J 96(2):282–289
Kotsovos MD (1987) Shear failure of reinforced concrete beams. Engg Struct 9:32–38
Kumar S, Birari SV (2010) Neural Network modeling of shear strength of SFRC corbels without stirrups. App Soft Computing 10:135–148
Kwak Y-K (2007) Shear strength of steel fiber reinforced concrete beams without stirrups. ACI Struct J 104-S(58):420–429
Kwak Y, Eberhard MO, Kim W, KIM J (2002) Shear strength of steel fiber reinforced concrete beams without stirrups. ACI Struct J 99(4):530–538
Lee J-Y, Kim U-Y (2008) Effect of longitudinal tensile reinforcement ratio and shear span-depth ratio on minimum shear reinforcement in beams. ACI Struct J 105:134–144
Lim TY (1987) Bending behavior of steel fibre concrete beams. ACI Struct J 84:524–536
Madan SK, Rajesh K, Singh SP (2007) Steel fibers replacement of web reinforcement for RCC deep beams in shear. Asian J of Civil Engg (Building and Housing) 8(5):479–489
Manicka Selvem VK, Kuruvilla T (1988) An application of shear strength theories of concrete deep beams. J Struct Engg 15(2):69–75
Mau ST, Hsu TTC (1987) Shear strength prediction of deep beams with web reinforcement. ACI Struct J 84:513–523

Misrayah A, Banthia N (2002) Shear strength of steel fiber reinforced concrete. ACI Mater J 99(M (40)):473–479
Narayanan R, Darwish IYS (1987) Use of steel fibers as shear reinforcement. ACI Struct J 84:216–227
Narayanan R, Darwish IYS (1988) Fiber concrete deep beams in shear. ACI Struct J 85:141–149
Noghabhai K (2000) Beams of fibrous concrete in shear and bending: experimental and model. J Struct Engg-ASCE 126(2):243–251
Remigijus S, Gediminas M (2007) The influence of shear span ratio on load capacity of fiber reinforced concrete elements with various steel fiber volumes. Civil Engg and Manag XIII(3):209–215
Roberts TM, Ho NL (1982) Shear failure of deep fiber reinforced concrete beams. Int J Cement Conc Lightweight Conc 4(3):145–152
Sanand A, Saka MP (2001) Prediction of ultimate shear strength of reinforced concrete deep beams using neural networks. J Struct Engg 127:818–828
Shah RH, Mishra SV (2004) Crack and deformation characteristics of SFRC deep beam. Inst of Engg J CV (I) 85:44–48
Slater E, Moni M, Shahria Alam M (2012) Predicting the shear strength of steel fiber reinforced concrete beams. Const Build Mat 26:423–436

13

An improved CWT-based algorithm for the generation of spectrum-compatible records

Luis A Montejo[1*] and Luis E Suarez[2]

Abstract

The seismic design of most civil structures is usually accomplished using the response spectrum approach or simplified equivalent lateral force methods. However, some special tasks require the use of dynamic time history analyses. In the nuclear industry, for example, dynamic analyses are required in the design verification and seismic assessment of critical buildings and in the development of floor response spectra and free-field ground response spectra. The input motion for these analyses requires acceleration time series whose response spectrum matches a target design spectrum. This article revises the continuous wavelet transform (CWT) approach to generate spectrum-compatible records from the modification of acceleration time histories recorded in actual seismic events. The computational efficiency of the algorithm is increased greatly by performing the wavelet decomposition and details reconstruction via fast convolution using fast Fourier transforms. The new algorithm is evaluated using a typical design spectrum from the nuclear industry and different seed records.

Keywords: Wavelet transform; Fast convolution; Fast Fourier transform; Artificial earthquake records; Seismic design

Introduction

Independent of the approach selected, either displacement or force-based, most of the seismic designs of civil structures are performed using variations of the response spectrum method, where the ground motion is represented as a design response spectrum that is representative of the seismic hazard on the site. The (pseudo-accelerations or relative displacements) design response spectrum is used to estimate the force or deformation demand imposed on the structure. The main goal of this technique is to translate the ground motion into forces acting on the building. Nevertheless, for some special tasks or structures, dynamic time history analyses are required. In the standard American Society of Civil Engineers (ASCE) 7–10, for example, if the structure meets the characteristics necessary to be considered as irregular, dynamic analyses (either modal response spectrum or time history) are required to verify the design of the structure. In the nuclear industry, dynamic time history analyses are required in the design, verification, and seismic assessment of critical buildings. In addition, they are needed for the development of floor (or in-structure) response spectra and free-field ground response spectra. When dynamic time history analyses are to be performed, the seismic input needs to be defined as a time history of accelerations. To comply with code requirements, the acceleration series used should be compatible with the design spectrum, i.e., the response spectrum of the acceleration time history should match with the design spectrum in average.

To select these input records, one can look at historic records with their amplitudes scaled by a factor to optimize the fitting over the design spectrum (e.g., Bommer and Acevedo 2004). Alternatively, there are different methodologies available to generate 'synthetic' spectrum-compatible records. When using the first option, i.e., amplitude scaling, a larger number of records are required to obtain a reliable average of the system response due to the natural scatter of the records. On the other hand, if spectrum-compatible records are used, the number of analyses required to obtain a reliable estimate of the response is substantially reduced (e.g., Watson-Lamprey and Abrahamson 2006; Heo et al. 2011; Hancock et al. 2008).

* Correspondence: luis.montejo@upr.edu
[1]Department of Engineering Science and Materials, University of Puerto Rico at Mayaguez, Mayaguez, 00680, Puerto Rico
Full list of author information is available at the end of the article

There are several methodologies that have been proposed for the generation of spectrum-compatible records, like the adjustment of the power spectrum of a random process or the manipulation in the time or frequency domain of historic records. The criteria for the generation and evaluation of such records specified in standard seismic provisions (e.g., ASCE 7; American Society of Civil Engineers 2010) are rather slight when compared to the requirements in the nuclear regulatory guide RG 1.208 (US-NRC 2007). For example, RG 1.208 explicitly prohibits the use of synthetic motions that are not based on seed-recorded time histories so that only modified recorded ground motions can be used for site response analysis.

The available methodologies for obtaining spectrum-compatible accelerograms based on the modification of historic records can be classified into three groups: (1) those based on matching in the frequency domain by manipulation of the Fourier spectrum (e.g., Tsai 1972), (2) those grounded on wavelet adjustments to the recorded accelerograms at specific times (e.g., Lilhanand and Tseng 1988; Abrahamson 1992; Hancock et al. 2006; Al Atik and Abrahamson 2010), and (3) those centered on the manipulation of the wavelet coefficients obtained via continuous wavelet transform (e.g., Mukherjee and Gupta 2002; Suarez and Montejo 2003, 2005). Although an acceptable level of matching can be obtained by means of any of the available methodologies, the use of spectrum-compatible records is a matter of discussion in the earthquake engineering community, as the characteristics of the final compatible record may largely differ from those observed in real records e.g., (Naeim and Lew 1995; Bommer and Ruggeri 2002; Bommer and Acevedo 2004). Although these arguments may be valid, the discussion of this issue is beyond the scope of the paper. This article revises the continuous wavelet transform approach to generate spectrum-compatible records from the modification of actual registered seismic events and propose an alternative computation algorithm to reduce its computational cost.

Methods

CWT-based methodology for the generation of spectrum-compatible records

For completeness, this section briefly describes the methodology to manipulate historic earthquake records to obtain spectrum-compatible accelerograms based on the continuous wavelet transform (CWT). A comprehensive presentation of the CWT theory behind the methodology and the properties of the wavelet function used are available elsewhere (Suarez and Montejo 2005).

The CWT of a signal $f(t)$ can be defined as the sum over all times of the signal multiplied by scaled and shifted versions of a wavelet function Ψ as defined by Equation 1

$$C(s,p) = \int_{-\infty}^{+\infty} f(t)\Psi^*_{s,p}dt = \int_{-\infty}^{+\infty} f(t)\frac{1}{\sqrt{s}}\Psi^*\left(\frac{t-p}{s}\right)dt \tag{1}$$

The parameter s and p are used to scale and shift the wavelet, respectively. The asterisk * indicates complex conjugation and the normalizing factor $1/\sqrt{s}$ ensures that the energy is the same for all values of s. The result of the transform is a matrix of wavelet coefficients $C(s,p)$ that contain information about the signal at the scale s and time position p, that is, the CWT can be viewed as a two-dimensional transform that maps an original time signal $f(t)$ into a time-scale domain. Since the scale s can be related to frequency, the CWT is occasionally used in earthquake engineering to obtain simultaneous time-frequency representations of earthquake records (e.g., Montejo and Kowalsky 2008). Once the wavelet coefficients are calculated, the signal can be reconstructed using Equation 2:

$$\begin{aligned} f(t) &= \frac{1}{K_\psi}\int_{s=0}^{+\infty}\left(\int_{p=-\infty}^{\infty} C(s,p)\Psi_{s,p}(t)\frac{dp}{s^2}\right)ds \\ &= \frac{1}{K_\psi}\int_0^\infty D(s,t)ds \end{aligned} \tag{2}$$

where K_ψ is a constant that depends on the wavelet function selected for the analysis and the functions $D(s,t)$ are

$$D(s,t) = \int_{p=-\infty}^{\infty} \frac{1}{s^2} C(s,p)\Psi_{s,p}(t)dp. \tag{3}$$

They are referred to as the detail functions. They have a dominant frequency that depends on the type of wavelet. In practice, a set of n discrete values of the continuous scale s are used.

The selection of an appropriate wavelet function is crucial for the effective implementation of a given application. For strong motion data synthesis and analysis, the impulse response wavelet has been successfully used in the past. It is define as

$$\Psi(t) = e^{-\zeta\Omega|t|}\sin(\Omega t), \tag{4}$$

where ζ and Ω are the parameters that define the shape and central frequency of the wavelet, respectively. For the applications presented here, values of ζ = 0.05 and $\Omega = \pi$ are used. The mathematical properties and advantages of this wavelet in the analysis of earthquakes records are discussed in Suarez and Montejo (2005) and Hancock et al. (2006).

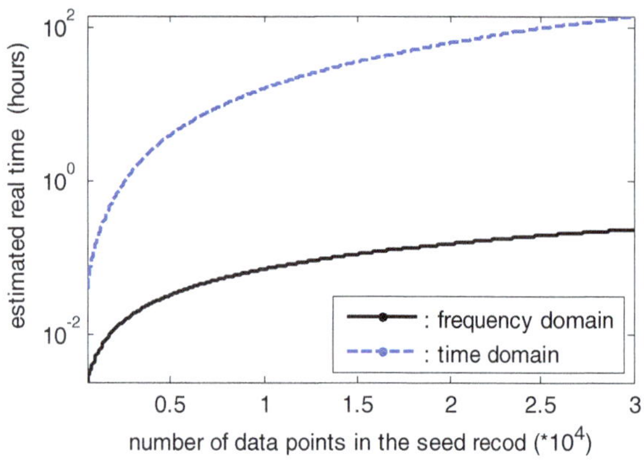

Figure 1 Time savings using the proposed algorithm implementation.

To obtain spectrum-compatible records one may proceed as follows:

1. Perform the CWT, using Equations 1 and 4, to obtain the wavelet coefficients $C(s,p)$ at selected values of s. The values of s are defined based on the periods or frequencies where a match with the target spectrum is desired. For the impulse response wavelet, the relation between s and the predominant frequency (f) in hertz and period (T) is given by Equation 5:

$$f = \frac{\Omega}{2\pi}\frac{1}{s} \qquad T = \frac{2\pi}{\Omega}s. \tag{5}$$

   In the applications presented in this work, a total of n = 100 frequency values uniformly spaced over the log frequency scale from 0.1 to 50 Hz is used as required in Appendix F of the Nuclear Regulatory Guide RG 1.208 (US-NRC 2007). Depending on the application, the user may decide to modify these values.
2. Once the wavelet coefficients $C(s,p)$ are calculated, the detail functions are constructed using Equation 3.
3. Calculate the ratios between the target spectrum $Sa_T(f_j)$ and the response spectrum of the historical record $Sa_R(f_j)$ at the frequencies defined in step 1:

$$R\left(f_j\right) = \frac{Sa_T\left(f_j\right)}{Sa_R\left(f_j\right)} \quad ; \quad j = 1, 2, \ldots, n. \tag{6}$$

4. Multiply each detail function $D(s_j,t)$ by the corresponding spectral ratio $R(f_j)$ and reconstruct the signal with Equation 2.

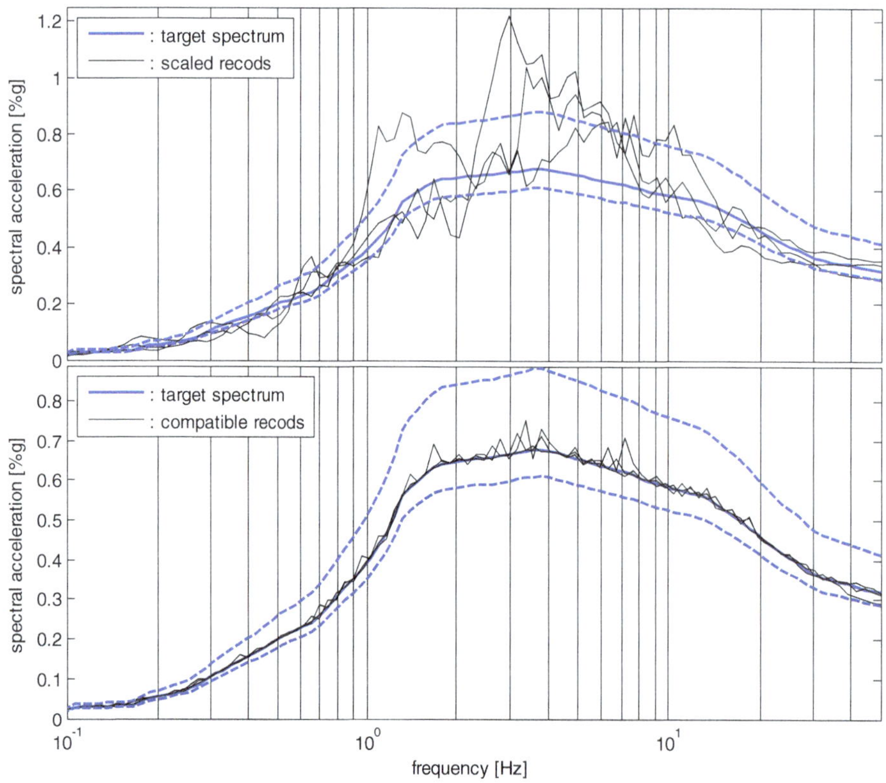

Figure 2 Target uniform hazard response spectra. Top: the target UHRS at the rock outcrop and the scaled response spectra for the three seed records. Bottom: the target UHRS and the response spectra of the resulting compatible records.

Table 1 Summary of the selected seed records

NGA number	Event	Station	Magnitude	Mechanism	R_{rupt} (km)	Vs30 (m/s)
1521	1999 Chi-Chi, Taiwan	TCU089	7.62	Reverse/ oblique	8.9	680
1508	1999 Chi-Chi, Taiwan	TCU072	7.62	Reverse/ oblique	7	468
284	1980 Irpinia, Italy	Auletta	6.90	Normal	9.6	1,000

5. Calculate and compare the response spectrum of the reconstructed signal $Sa_R(f_j)$ with the target spectrum and repeat steps 3, 4, and 5 until a desired level of matching is attained or a maximum number of iterations are reached.

Improved efficiency using fast convolution via FFT

The CWT approach previously described for generating spectrum-compatible records has been largely used in the past, offering stable solutions and good matching with different target spectra (e.g., Velez 2007; Priestley et al. 2007; Bahar and Taherpour 2008; Linzell and Nadakuditi 2011; Montejo et al. 2012). However, a common complaint is the amount of time that takes for the algorithm to run, especially when lengthy records are used as input. To improve the processing time, a new algorithm is proposed in this paper. It computes the wavelet decomposition, Equation 1, and constructs the detail functions, Equation 3, using a fast convolution implementation by means of the fast Fourier transform (FFT).

By looking at Equation 1, it can be said that the CWT for a fixed scaled s is the convolution of the signal $f(t)$ with the wavelet function scaled by s. The matrix of wavelet coefficients $C(s,p)$ is formed by changing the values of s and repeating the convolution operation. The original implementation of the algorithm computed these convolutions in the time domain. However, according to the convolution theorem, convolution in the time domain is equivalent to multiplication in the frequency domain, that is, if we have the two functions f and g in the time domain, its convolution can be calculated in the frequency domain using Equation 7:

$$f \otimes g = F^{-1}\{F\{f\} \bullet F\{g\}\} \tag{7}$$

where $\otimes$ denotes convolution, $\bullet$ denotes pointwise product, and F and F^{-1} denote the forward and inverse Fourier transform, respectively. Since we are dealing with discrete signals, transformations between the time and frequency domains call for the application of the discrete Fourier

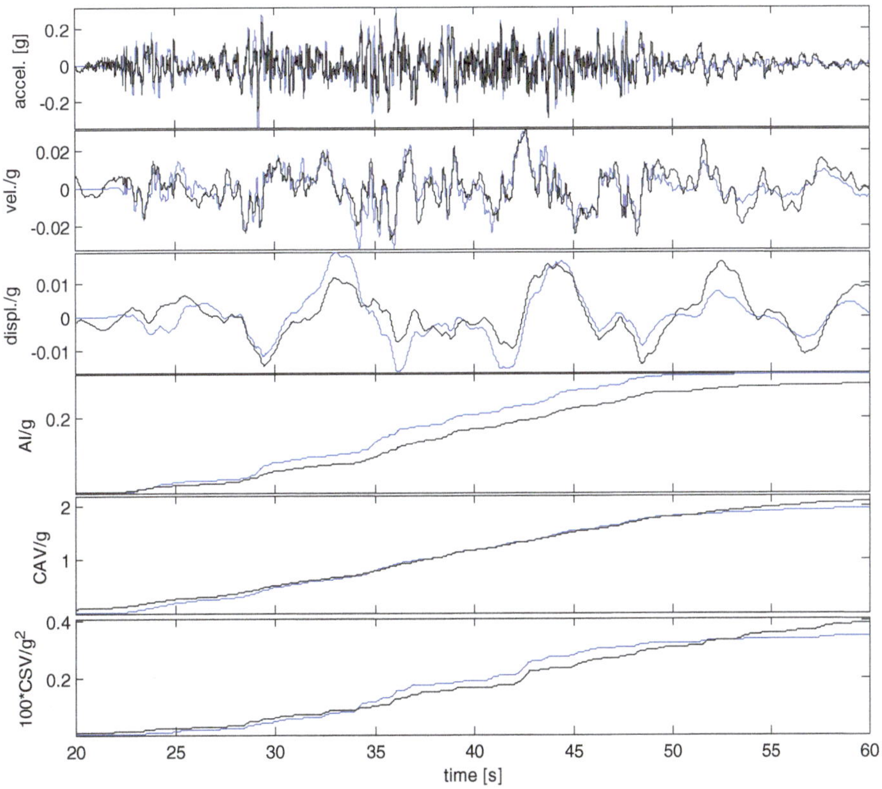

Figure 3 Characteristics of the seed motions corresponding to the TCU089 1999 Chi-Chi record. Top to bottom: time histories of acceleration, velocity, displacement, AI, CAV, and CSV for the scaled (blue line) and compatible records (black line) corresponding to the TCU089 1999 Chi-Chi record.

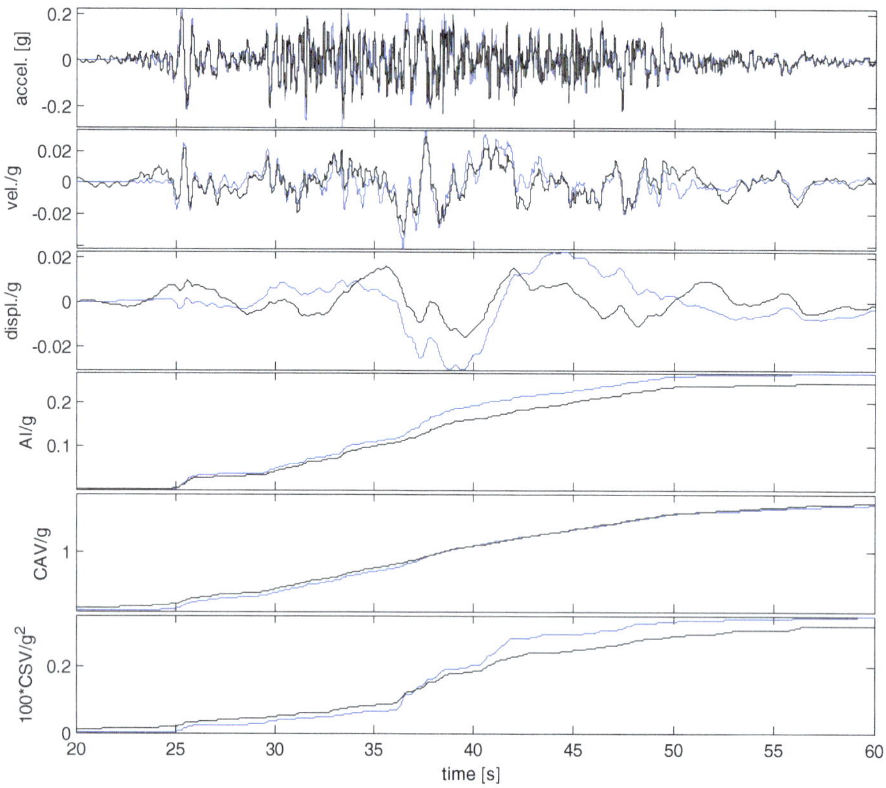

Figure 4 Characteristics of the seed motions corresponding to the TCU072 1999 Chi-Chi record. Top to bottom: time histories of acceleration, velocity, displacement, AI, CAV, and CSV for the scaled (blue line) and compatible records (black line) corresponding to the TCU072 1999 Chi-Chi record.

transforms (DFT). If the DFT is implemented using the FFT algorithm, the convolution via the frequency domain can be significantly faster than directly convolving the two time domain signals. While the computational cost for direct time domain convolution of two N-point sequences is of the order $O(N^2)$, the cost for frequency domain convolution using the FFT algorithm is $O(N \log N)$. Considering the typical length of the signals used in this application and the number of convolution operations required by the algorithm (one per each value of s), this translates into large savings in the computational cost. In the new implementation, the CWT is executed using Equation 8:

$$C(s,p) = f(t) \otimes \Psi_{s,p} = \text{IFFT}\{\text{FFT}\{f(t)\} \bullet \text{FFT}\{\Psi_{s,p}\}\} \tag{8}$$

where FFT and IFFT are, respectively, the direct and inverse discrete Fourier Transforms calculated with the FFT algorithm.

Note that in order to be able to compute the pointwise product (•), the lengths of the FFT for the record and for the wavelet must be the same. To obtain a faster and most precise implementation, the lengths of the wavelet and the record are set to the next power of 2 of the expected convolution length (2 * N − 1) (Smith 2002).

Finally, notice from Equation 3 that the detail functions can also be calculated using the FFT convolution between the wavelet coefficients for a fixed s and the wavelet function scaled by the same s value, as shown in Equation 9:

$$\begin{aligned} D(s,t) &= \frac{1}{s^2} C(s,p) \otimes \Psi_{s,p}(t) \\ &= \frac{1}{s^2} \text{IFFT}\{\text{FFT}\{C(s,p)\} \bullet \text{FFT}\{\Psi_{s,p}\}\}. \end{aligned} \tag{9}$$

Figure 1 shows the computational savings obtained using the proposed algorithm and assuming that the record is decomposed at 100 scale (frequency) values. It is seen that for a relative short record of 2,000 data points (e.g., 10 s sampled at 200 Hz), the estimated 'real time' that will take the new implementation of the algorithm to run is about 42 s, while the old implementation will take approximately 37 min. As the length of the seed record increases, the differences become exponentially larger (notice the logarithmic scale used in the y-axis). The estimated real times in Figure 1 were computed based on a laptop PC i7 at 2.90 GHz, 8 GB RAM, and 64-bit OS.

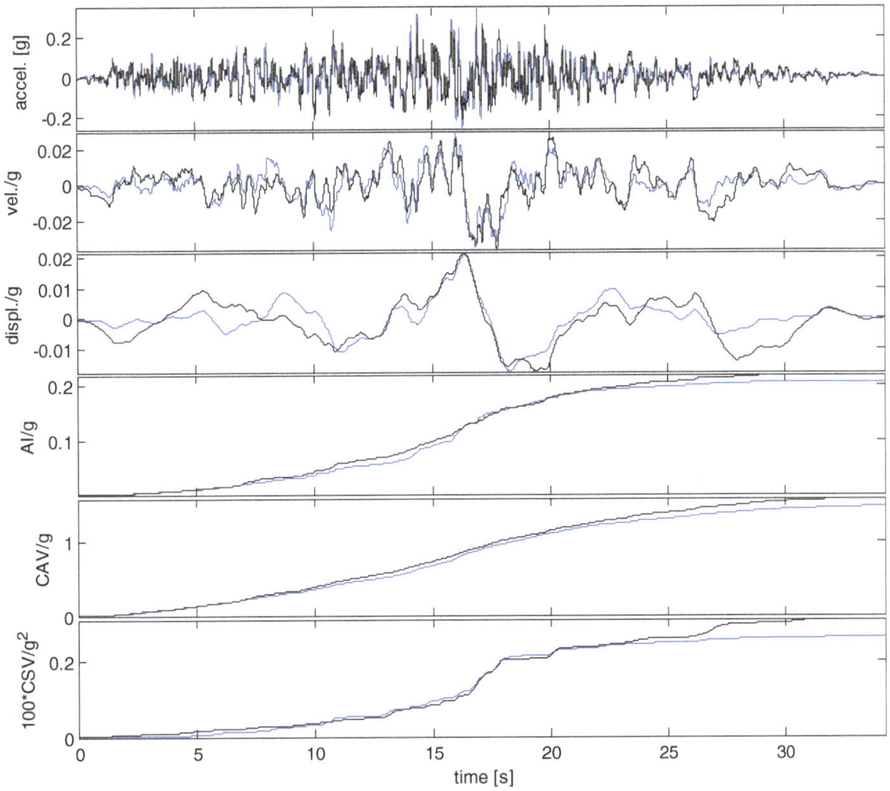

Figure 5 Characteristics of the seed motions corresponding to the Auletta 1980 Irpinia record. Top to bottom: time histories of acceleration, velocity, displacement, AI, CAV, and CSV for the scaled (blue line) and compatible records (black line) corresponding to the Auletta 1980 Irpinia record.

Results and discussion

The new implementation of the algorithm is evaluated using a design spectrum typical of the US nuclear industry. The target spectrum is extracted from Section 5 of RG 1.208 (US-NRC 2007) and represents a typical uniform hazard response spectrum (UHRS) at the outcrop rock (Figure 2). In practice, synthetic records compatible with this spectrum need to be generated to perform site analyses and develop a site-specific design spectrum.

Three different seed records to be used as input for the algorithm were selected from the PEER NGA database (PEER 2013) based on the similarity of the response spectrum shape of the records with the shape of the target spectrum. Table 1 summarizes the main characteristics for each of the records selected. Since the selection is based on spectral shape similarities, the amplitude of the records is scaled to provide a best fit. The scaled response spectra for the three records are displayed and compared with the target spectrum in Figure 2 (top).

Figure 2 (bottom) shows the response spectra for 5% damping of the resulting compatible records along with the target spectrum. The 90% and 130% design spectra are also displayed (dashed lines) as these are the spectral amplitude limits specified in RG 1.208. To prevent the spectra in large frequency windows from falling below the spectrum, RG 1.208 also requires that no more than nine adjacent frequency points fall below the target spectrum. It is seen that the compatible records generated satisfy these requirements.

Figures 3,4,5 display the time histories of acceleration, velocity, displacement, Arias intensity (AI), cumulative absolute velocity (CAV), and cumulative squared velocity (CSV) for the linearly scaled and compatible records generated. It is seen that the characteristics of the seed motions are, in general, retained in the compatible records.

Conclusions

An improved algorithm for faster generation of spectrum-compatible records was developed. The algorithm is based on the modification of the CWT coefficients of a historic record. The efficiency of the algorithm is improved by performing the required convolution operations in the frequency domain via fast Fourier transforms. It was shown that the algorithm can be used to obtain compatible records that satisfy current US NRC requirements. A Matlab implementation of the algorithm is available upon request from the corresponding author.

Competing interests
The authors declare that they have no competing interests.

Authors' contributions
Both authors contributed equally to this research work.

Acknowledgements
This work was performed at the University of Puerto Rico at Mayaguez under award NRC-HQ-12-G-38-0018 from the US Nuclear Regulatory Commission. The statements, findings, conclusions, and recommendations are those of the authors and do not necessarily reflect the view of the US Nuclear Regulatory Commission.

Author details
[1]Department of Engineering Science and Materials, University of Puerto Rico at Mayaguez, Mayaguez, 00680, Puerto Rico. [2]Department of Civil Engineering and Surveying, University of Puerto Rico at Mayaguez, Mayaguez, 00680, Puerto Rico.

References
Abrahamson NA (1992) Non-stationary spectral matching. Seismol Res Lett 63(1):30

Al Atik L, Abrahamson N (2010) An improved method for non-stationary spectral matching. Earthq Spectra 26(3):601–617

American Society of Civil Engineers (2010) ASCE 7 - minimum design loads for buildings and other structures. Reston, VA

Bahar O, Taherpour A (2008) Nonlinear dynamic behavior of RC buildings against accelerograms with partial compatible spectrum. In: 14th World Conference on Earthquake Engineering. , Beijing, China. 12–17 October 2008

Bommer JJ, Acevedo AB (2004) The use of real earthquake accelerograms as input to dynamic analysis. J Earthq Eng 8(1):43–91

Bommer JJ, Ruggeri C (2002) The specification of acceleration time-histories in seismic codes. Euro Earthquake Eng 16:3–16

Hancock J, Watson-Lamprey J, Abrahamson NA, Bommer JJ, Markatis A, McCoy E, Mendis R (2006) An improved method of matching response spectra of recorded earthquake ground motion using wavelets. J Earthq Eng 10(special issue 1):67–89

Hancock J, Bommer JJ, Stafford PJ (2008) Number of scaled and matched accelerograms required for inelastic dynamic analyses. Earthquake Eng Struct Dynam 37:1585–1607

Heo Y, Kunnath S, Abrahamson N (2011) Amplitude-scaled versus spectrum-matched ground motions for seismic performance assessment. J Struct Eng 137:278–288

Lilhanand K, Tseng WS (1988) Development and application of realistic earthquake time histories compatible with multiple damping design spectra. In: Proceedings of the 9th WCEE, vol 2. , Tokyo-Kyoto, Japan, pp 819–824

Linzell DG, Nadakuditi VP (2011) Parameters influencing seismic response of horizontally curved, steel, I-girder bridges. Steel Compos Struct 11–1:21–38

Montejo LA, González-Román LA, Kowalsky MJ (2012) Seismic performance evaluation of reinforced concrete-filled steel tube pile/column bridge bents. J Earthq Eng 16(3):401–424

Montejo LA, Kowalsky MJ (2008) Estimation of Frequency Dependent Strong Motion Duration via Wavelets and its Influence in Nonlinear Seismic Response. Computer Aided Civil and Infrastructure Engineering. 23(4):253–264.

Mukherjee S, Gupta VK (2002) Wavelet-based generation of spectrum compatible time-histories. Soil Dyn Earthq Eng 22(9–12):799–804

Naeim F, Lew M (1995) On the use of design spectrum compatible time histories. Earthq Spectra 11(1):111–127

PEER - Pacific Earthquake Engineering Research Center (2013) PEER NGA database. http://peer.berkeley.edu/peer_ground_motion_database. Accessed January 2013

Priestley MJN, Calvi GM, Kowalsky MJ (2007) Direct displacement based seismic design of structures. IUSS Press, Pavia, Italy

Smith S (2002) Digital signal processing: a practical guide for engineers and scientists. California Technical Publishing, San Diego, CA

Suarez LE, Montejo LA (2003) Generación de registros artificiales compatibles con un espectro de respuesta mediante la transformada wavelet. In: Proc. del II Congreso Nacional de Ingeniería Sísmica. , Medellín, Colombia November 2003

Suarez LE, Montejo LA (2005) Generation of artificial earthquakes via the wavelet transform. Int J Solids Struct 42:5905–5919

Tsai NC (1972) Spectrum-compatible motions for design purposes. J Eng Mech Div ASCE 98:345–356

US-NRC (2007) Regulatory guide 1.208 - a performance-based approach to define the site-specific earthquake ground motion. http://www.nrc.gov/reading-rm/doc-collections/reg-guides/power-reactors/rg/01-208/01-208.pdf. Accessed January 25, 2013

Velez E (2007) Experimental cyclic behavior of reinforced concrete wall houses loaded in their weak direction. Thesis. Department of Civil Engineering, University of Puerto Rico at Mayaguez

Watson-Lamprey JA, Abrahamson NA (2006) Bias caused by use of spectrum compatible motions. In: Proceedings of the 8th US National Conference on Earthquake Engineering. , San Francisco, California. 18–22 April 2006

14

Pull-out behavior of galvanized steel strip in foam concrete

Maziar Ramezani*, Juan Vilches and Thomas Neitzert

Abstract

The interface and bond between concrete and reinforcing steel are the most fundamental problems of reinforced concrete structures. In this paper, the pull-out strength of galvanized steel strips with different geometries and hole patterns in foam concrete blocks are investigated experimentally and numerically. Foam concrete mixtures of 1,200-kg/m^3 density were obtained by mixing cement and water in a mortar mixer together with ultrafoam as the foaming agent and Quick-Gel as the viscosifier. A theoretical model is developed to predict the bond-slip relationship between the strip and the concrete. This model is further implemented in a finite element simulation of the pull-out tests through the ABAQUS user subroutine. The results show good correlation between experimental, theoretical, and finite element simulation analyses. The influence of the steel strip geometries on the maximum pull-out force is also studied, and it was found that the strips with the greatest hole area, hole diameter, and circumference areas have higher pull-out forces and the increase is nearly linear.

Keywords: Bond stress-versus-slip relationship; Foam concrete; Pull-out test; Steel strip

Introduction

In the industrialized world, new and modern techniques and materials, such as composites, have become common in today's construction industry. These new materials are lightweight, energy efficient, aesthetically attractive, and efficiently handled and erected. Composite structural assemblies (CSAs) are products with performance superior to existing building elements, based on combinations of materials such as embedded light gauge steel components, settable fillers, and coatings of sheet materials to provide a variety of finishes (Akin et al. 2005). The materials are combined or assembled in various ways to provide improved performance in strength and stiffness and associated properties of acoustic filtering, thermal energy conservation, vibration resistance, moisture barriers or absorption, and fire behavior. The structural strength of these products is determined by the grade and thickness of the steel, the configuration of internal panel components, and by the features of infill material (Balendran et al. 2002).

The current paper focuses on novel materials for application in CSAs and the bonding performance between steel and infill materials. The bond between concrete and steel elements is one of the most important properties contributing to the successful functioning of a composite panel (Kayali 2004). The main contribution to bond strength comes from the chemical adhesion and the friction resistance occurring between the steel and the concrete as a result of the surface effects. Composite structures made of various materials can fail in several ways depending on material properties of components, design methods, and loading cases, but the most important is the cooperation of all elements (Alterman et al. 2011). The lack of compatibility between elements leads to bond failure, sliding of reinforcement bars and/or steel strips, local deformations, and finally cracking. Destructive measurements of shear strength through pull-out and push-in tests are commonly used methods to assess the quality of a connection between steel elements and fillers, e.g., concrete.

There are only few studies available in the literature about the structural bond properties between lightweight concrete and steel bars with pull-out test (see e.g., Cao and Chung 2001; De Lorenzis et al. 2002; Chang 2003; Al-mahmoud et al. 2007). Ben Romdhane and Ulm

* Correspondence: maziar.ramezani@aut.ac.nz
Centre for Advanced Manufacturing Technology (CAMTEC), School of Engineering, Auckland University of Technology, Auckland 1010
New Zealand

(2002) investigated the steel-concrete interface with computational mechanics. They modelled micro-to-macro mechanisms of the interface and studied the numerical implementation of this model in combination with a discrete crack model for concrete. Few finite element (FE) simulations of pull-out tests are also available in the literature (e.g., Wu et al. 2009; Wei et al. 2012), yet without accurate modelling of the interface and bond behavior. Casanova et al. (2012) presented a new finite element approach to model the steel-concrete bond effects. Their model proposed to relate steel, represented by truss elements, with the surrounding concrete in the case where the two meshes are not necessarily coincident. Most of the previous investigations about bonding behavior focused mainly on experimental and theoretical studies of shear stress between reinforcing bars and concrete specimens (e.g., Bouazaoui and Li, 2008; Fang et al. 2006; Khandaker and Hossain 2008), and plastic bars and concrete samples (e.g. Won et al. 2008), but not directly on steel strips, which are applicable to this project. This study investigates bonding performance and analyzes it through pull-out tests to provide data for the theoretical and FE analysis of CSAs. Because bonding is a general problem for wall panel design and other uses of concrete that employ steel to concrete contact zone, this research undertakes an analysis of the influence of various steel strip designs to understand bonding behavior. With the obtained results, the research enables understanding on the main features of metal face materials commonly used in wall panels.

Methods

Theoretical model

To have a better understanding of the pull-out test, a theoretical model is developed in this section which is suitable for pull-out test of a steel strip in concrete. The shear bond properties between the two materials can be determined by this model in the form of the bond stress-versus-slip relation. The model is based on the previous work of Banholzer et al. (2005), where the model is modified to fit the steel strip as the reinforcing element. Considering Figure 1, if an axial load P is applied at the end of the steel strip ($x = L$), the change in load dF_s over a distance dx along the strip is a function of the shear stress in terms of x and the circumference of the strip in contact with the concrete:

$$\frac{dF_s}{dx} = 2(a+b)\tau \tag{1}$$

where F_s is the force in the strip, a and b are the length and width of the steel strip cross section, and τ is the shear stress in the strip. By assuming that the interface is sheared due to an applied pull-out force and

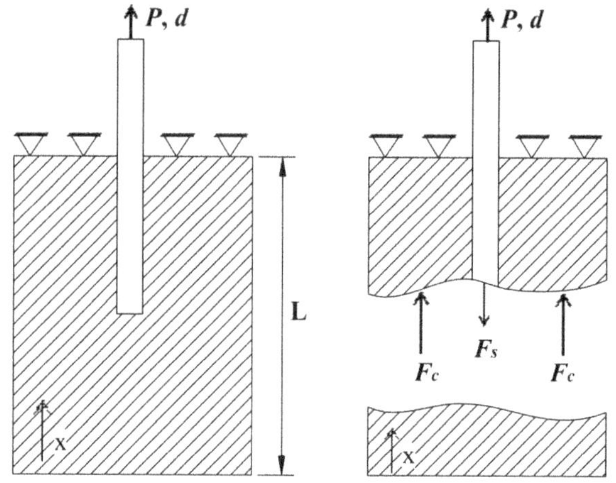

Figure 1 Pull-out test configuration and equilibrium of forces.

considering the absolute displacement of the concrete at a point x in relation to its origin as u_c and that of the strip as u_s, the slip s can be defined as the difference of u_c and u_s, as follows:

$$s = u_s - u_c. \tag{2}$$

Differentiating Equation 2 with respect to x yields

$$\frac{ds}{dx} = \varepsilon_s - \varepsilon_c \tag{3}$$

where ε_s and ε_c are the strain of the strip and the concrete, respectively. Using Hooke's law leads to

$$\frac{ds}{dx} = \frac{F_s}{E_s A_s} - \frac{F_c}{E_c A_c} \tag{4}$$

where E_s, A_s, E_c, and A_c are the modulus of elasticity of the strip in tension, the cross-sectional area of the strip, modulus of elasticity of the concrete in compression, and the load-carrying area of the concrete, respectively. As illustrated in Figure 1, the load application is idealized as a single pull-out load P at the top of the strip ($x = L$) with a corresponding displacement d applied in x direction. The global equilibrium of forces in an arbitrary chosen cross section yields

$$F_s + F_c = 0. \tag{5}$$

For demonstration purposes (see Figure 2), the relationship between τ and s is given as $\tau = \kappa\, s$, where κ is constant. It should be pointed out that a linear dependence between the shear stress and the slip is not

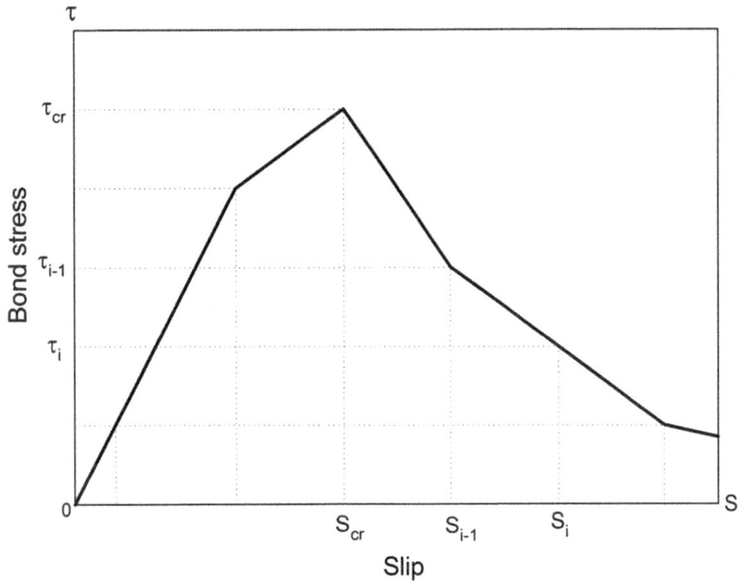

Figure 2 ***N*** **piecewise linear relationship of bond stress versus slip.**

assumed throughout the contact zone. The theoretical model uses a piecewise method where the nonlinear behavior of the contact interface is simplified by several linear pieces as illustrated in Figure 2. Combining these linear pieces gives a good approximation of the general nonlinear behavior. The pull-out problem can then be represented by a second-order differential equation, as follows:

$$F''_s - \beta^2 F_s = 0; \qquad F_s(L) = P, \quad F_s(0) = 0$$
$$\beta^2 = 2(a+b)\left(\frac{1}{E_s A_s} + \frac{1}{E_c A_c}\right).\kappa \tag{6}$$

where β is a material parameter which includes the bond stress-versus-slip relation. Another possible approach is to model the experimental situation in respect to the local slip s. After substituting Hooke's law for the strip and the concrete respectively into Equation 3, differentiating the results with respect to x and using Equations 1 and 5, the following second-order differential equation in terms of s can be obtained:

$$\begin{aligned} s'' &= 2(a+b)\left(\frac{1}{E_s A_s} + \frac{1}{E_c A_c}\right)\tau(s) = T(s) \\ T(s) &= 2(a+b)\beta\tau(s) \\ \beta &= \frac{1}{E_s A_s} + \frac{1}{E_c A_c} \\ s' &= \beta F_s \end{aligned} \tag{7}$$

Equation 7 represents the basic relationships between the second-order derivative of the local slip s and the local bond stress $\tau(s)$. Knowing that at $x = 0$ the force in the strip is zero ($s'(0) = 0$), assuming the slip s at the loaded strip end is d ($s(L) = d$), and denoting the

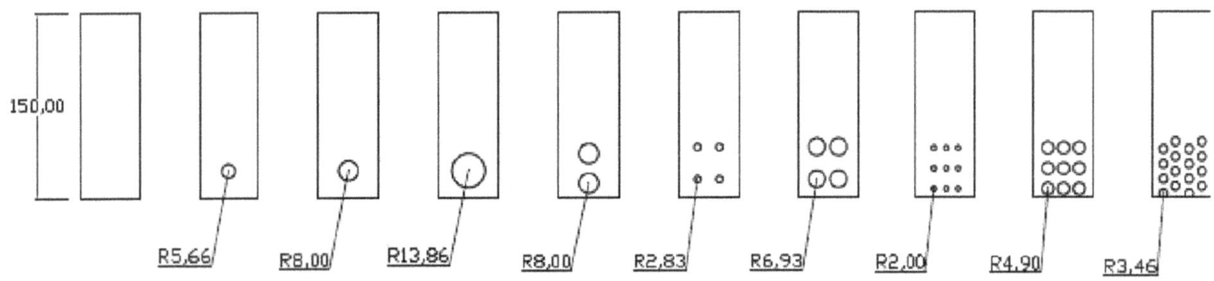

Figure 3 **Patterns of holes in the steel strips (all sizes in mm).**

coordinate of the strip by $0 \le x \le L$, the mathematical representation of the pull-out test can be expressed either as a boundary value problem

$$s'' = T(s), \quad s'(0) = 0, \quad s(L) = d \tag{8}$$

or as a corresponding initial value problem for a strip end slip δ at $x = 0$

$$s'' = T(s), \quad s'(0) = 0, \quad s(0) = \delta. \tag{9}$$

Banholzer et al. (2005) assumed an N piecewise linear bond law with no limitation of N (Figure 2) to achieve a general mathematical description of the bond law. The function $T(s) = 2\,(a + b)\,\beta\,\tau(s)$ can be expressed for an interval i ($s_{i-1} \le s \le s_i$) as

$$T(s) = m_i(s - s_{i-1}) + T_{i-1}, \tag{10}$$

$$m_i = \frac{T_i - T_{i-1}}{s_i - s_{i-1}}, \quad T_0 = s_0 = 0.$$

Consequently, the force F_s in the strip is normalized as well:

$$q(x) = \beta F_s(x) = s'(x). \tag{11}$$

The force in the strip at $x = L$ corresponds to the pull-out load P, and therefore, we have $s'(L) = q(L) = \beta P$. Now, the initial value problem (Equation 9) can be easily solved in an iterative process for any given $T(s)$ and δ, with Runge-Kutta numerical integration procedure, and the normalized pull-out force can be determined for a given displacement d, such that the boundary condition at the bottom of the strip $s'(0) = 0$ is satisfied.

Experimental analysis

Foam concrete mixtures were prepared at 1,200-kg/m^3 density by using stable foam concrete compositions, which were designed based on the investigation of Kearsley and Mostert (2005). Properties of the foam concrete were analyzed for density and compressive strength based on the standard testing methods for physical and mechanical properties (ASTM C495 and ASTM C39). Experiments showed that by using the theoretical design method of Kearsley and Mostert (2005), it is possible to predict the compressive strength and density for the concrete mixtures. Ordinary Portland Cement was used for the foam concrete mixtures. To prepare the foam, ultrafoam was used as the foaming agent and Quick-Gel (Baroid, Houston, TX, USA) as the viscosifier. These were mixed with water in the foam generator until the foam bubble size was uniform and stable (usually 2 min). Then, the foam was added to the cement matrix while stirring, with continual mixing for 1 to 2 min. When the mixture of foam and cement matrix was uniform, the prepared foam concrete was poured into the test mold with slight vibrations to fill up the mold completely. Specimens were demolded after 24 h and then moist cured in a standard curing room for a further 28 days in order to test the samples with a standard testing machine.

Table 1 Relationships of hole patterns

Hole pattern	Number of holes	Radius of holes (mm)	Area of holes (mm^2)
S0	0	0	0
S1-6	1	5.66	100.6
S4-3	4	2.83	100.6
S9-2	9	2	113.1
S1-8	1	8	201.1
S2-8	2	8	402.1
S14-3	14	3.46	526.5
S1-14	1	13.86	603.5
S4-7	4	6.93	603.5
S9-5	9	4.9	678.9

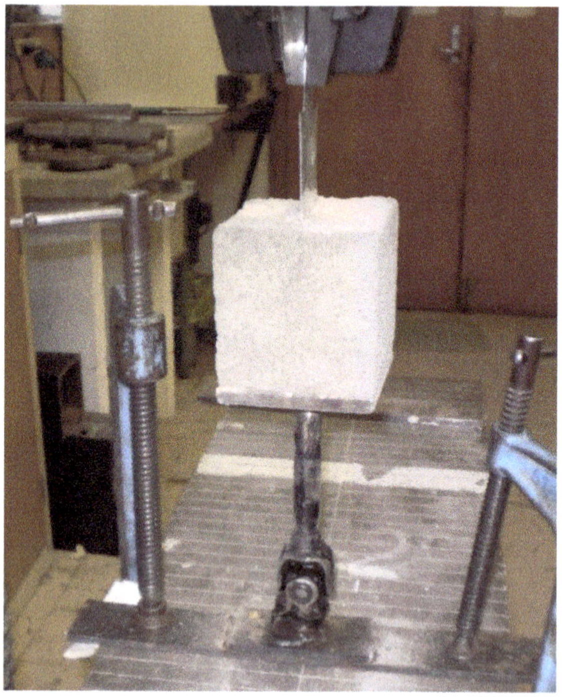

Figure 4 Testing mechanism with a freely adjustable ball joint and a plate with embedded bolts within foam concrete samples.

Standard cylinders of 100-mm diameter were used for the compressive strength tests, which were carried out in a testing machine of 100-kN capacity at a loading displacement rate of 0.1 mm/s. Three cylinders were tested for compressive strength after 28 days according to the ASTM C39 standard test method for compressive strength of cylindrical concrete specimens, and the mean value was calculated. The average compressive strength of the foam concrete was 8.8 MPa.

The pull-out test method was used to evaluate the shear bonding strength between the steel and the foam concrete. All the pull-out tests followed the ASTM C900 standard. The ram velocity for all the tests was set at 0.2 mm/s to allow slow deformation of the concrete and the steel strips. The pull-out test samples include a foam concrete cube (100 mm × 100 mm × 100 mm) with galvanized steel sheet (50 mm × 150 mm) standing in the middle of the concrete cube. A hot-dip galvanized steel sheet, 0.75-mm thick, grade G250 from New Zealand Steel (Private Bag, New Zealand) was used. Steel sheets were pre-cleaned and degreased by using soapy water without damaging the coating before getting it in contact with the fresh concrete.

The influence of the steel strip geometries was studied as well. The major purpose of the pull-out test experiments was to investigate the effect of ten different hole patterns (Figure 3) in the steel strips on the bonding behavior. Ten different hole patterns were prepared including different numbers of holes (0, 1, 2, 4, 9, and 14), radius, and distribution (Table 1). The reason for using different sizes and numbers of holes was to investigate the influence of contact cross section and hole patterns of the strip on the pull-out strength. These holes were filled with the concrete, and the effect of mechanical interlock produced by the concrete inside the holes will be studied. In codifying the hole pattern produced, steel plates were denominated as S0, S1-6, S4-3, S9-2, S1-8, S2-8, S14-3, S1-14, S4-7, and S9-5, for each number and approximate radius of holes.

Three specimens for each single set of strips were prepared and tested to determine the load-versus-displacement curves. Therefore, 30 tests had been carried out in total for all hole patterns. The primary pull-out tests showed that the results do not have significant variations, and therefore, it was decided to use three repetitions for each set of strip geometries. The pull-out tests were carried out using a universal testing machine equipped with a frame, which holds cubic samples and with a freely adjustable ball joint for pull-out tests as shown in Figure 4. Four M10 bolts were embedded through the cubic foam concrete samples and fixed to a plate at the bottom of the foam concrete cube to hold the sample in place. The adjustable ball joint for pull-out tests assures that the pull-out forces were centric to the strips during loading. The load was applied to the metal strip

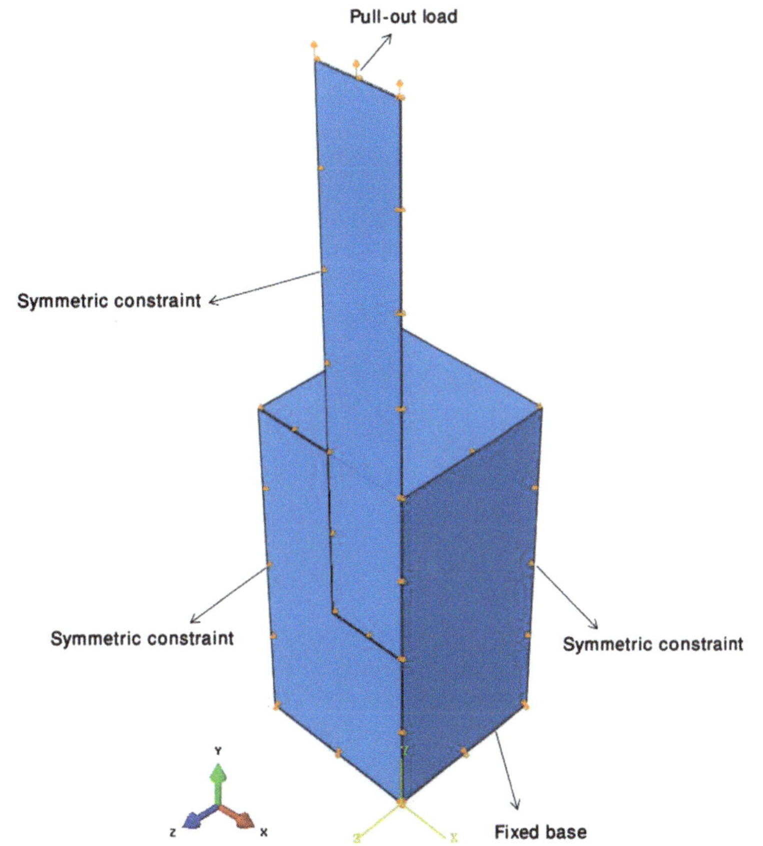

Figure 5 Load and boundary conditions for the FE model.

through a mechanical joint and evenly increased while controlling the displacement at a rate of 0.05 mm/s. Both the load applied and the displacement of the steel strip were measured and recorded on a computer until the strip was extracted from the cubic sample. Thus, a load-deformation curve was obtained from each experiment. It is worth pointing out that the slippage is not equal to the grip displacement throughout the contact zone. In the development of the theoretical model, it is only assumed that the slip s at the loaded strip end is equal to the grip displacement d ($s(L) = d$). The measured displacement is then related to the slip of the contact zone using the equations developed in the 'Theoretical model' section (see Equation 8).

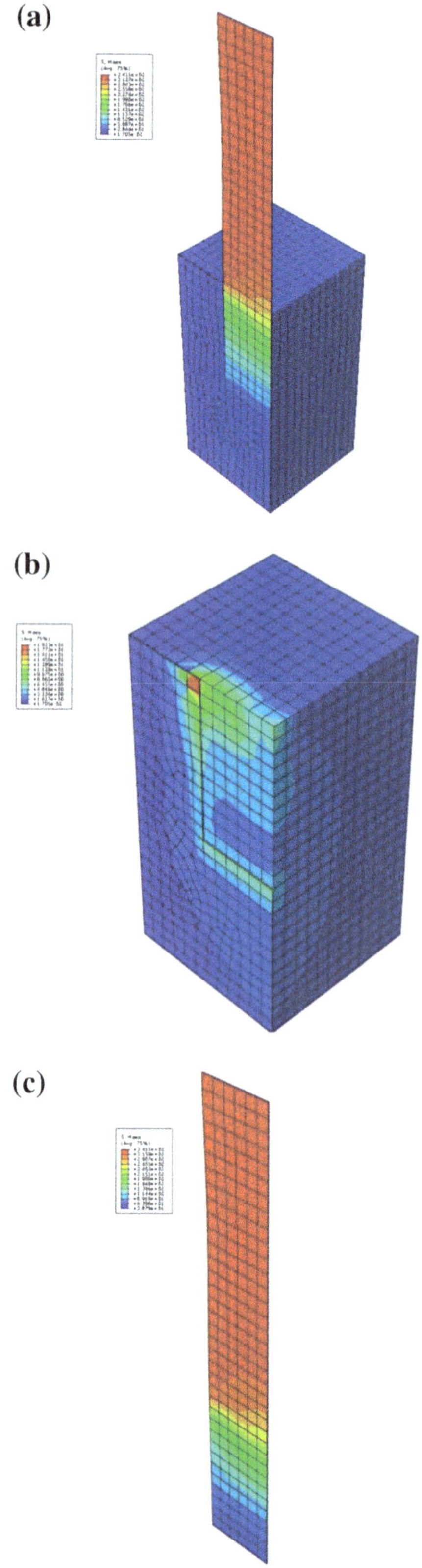

Figure 6 Distribution of von Mises stress in pull-out test of the steel strip S0 and foam concrete at d = 6 mm. (a) Whole structure, **(b)** concrete block, and **(c)** steel strip.

Finite element simulations

The pull-out test was further studied by finite element simulations. A three-dimensional finite element model was developed using ABAQUS release 6.12 (Simulia, Providence, RI, USA) for this purpose. This is an implicit analysis code suitable for modelling both material and geometric nonlinearities and has good contact modelling. The FE analysis comprised an elastic-plastic simulation of the pull-out process and the determination of the load-displacement curve in the steel strip. The FE simulations were conducted for all the hole patterns depicted in Figure 3, and the FE results were compared with experimental and theoretical results. The material data for FE simulations were obtained from compression test of foam concrete and tensile test of steel strip. The galvanized steel sheet had the elastic modulus of 200 GPa and Poisson's ratio of 0.3 with the ultimate yield stress of 357 GPa. The foam concrete had a density of 1,200 kg/m^3, Poisson's ratio of 0.2, and compressive strength of 8.8 MPa. The elastic modulus of the foam concrete was 5,535 MPa.

In the FE simulations, the bond stress-slip ($\tau - s$) relations obtained from experiments and theoretical modelling were used as the input data for the FE model. Bond stress-slip relations are considered as the constitutive relations for the behavior of the contact interface. The pull-out tests were then simulated by applying the shear constitutive behavior to the FE model by a user-defined subroutine. UINTER user-defined subroutine was used to define the interfacial constitutive behavior which proposes a new numerical model for simulating the bond-slip behavior of the composite structure. User subroutine UINTER is called at points on the slave surface of a contact pair with a user-defined constitutive model defining the interaction between the surfaces. The subroutine can be used to define the mechanical (normal and shear) interactions between surfaces. In this way, any special or proprietary interfacial constitutive behavior could be defined through the user subroutine UINTER written in FORTRAN in ABAQUS/Standard, which may include

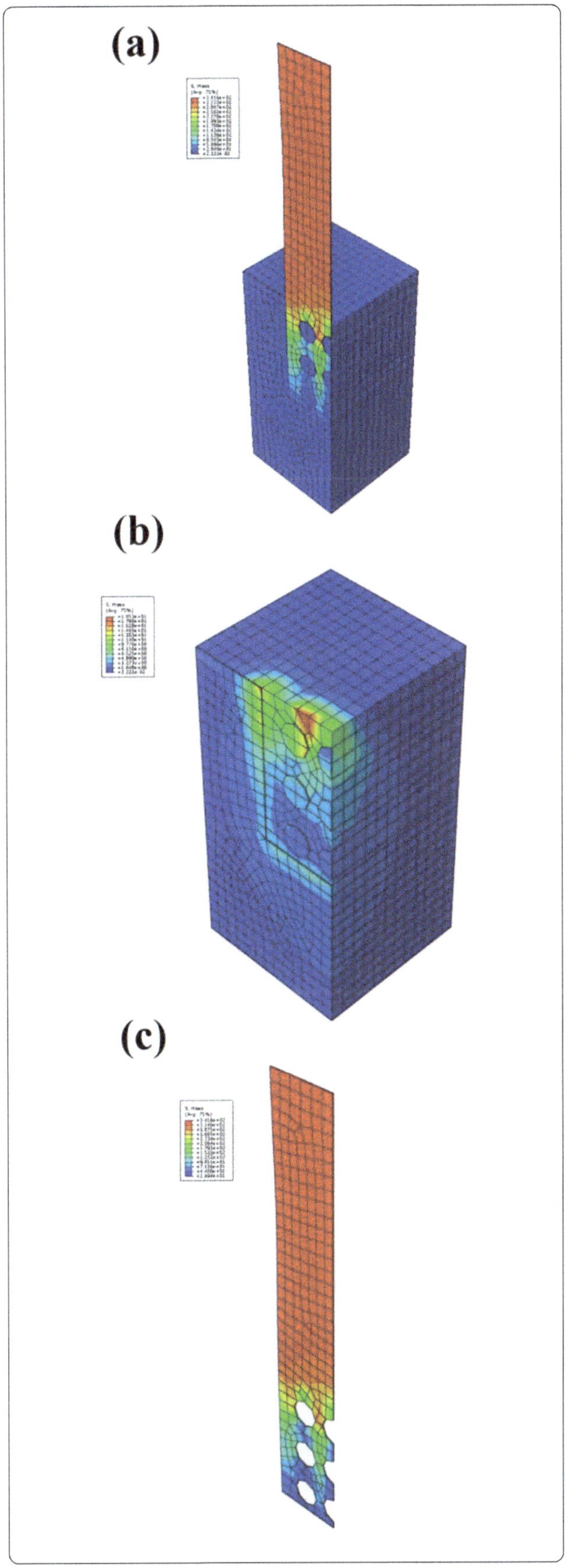

Figure 7 Distribution of von Mises stress in pull-out test of the steel strip S9-5 and foam concrete at *d* = 6 mm. (a) Whole structure, **(b)** concrete block, and **(c)** steel strip.

both the normal and tangential behaviors. This feature provides a strong tool for simulating the bond-slip behavior. User subroutine UINTER was called for each contact constraint location of affected contact pairs in every increment of the ABAQUS/Standard analysis. The input to this user subroutine includes the current relative position of a particular constraint point on the slave surface with respect to the corresponding closest point on the master surface, as well as the incremental relative motion between these two points. The normal direction constitutive behavior was modelled by using the penalty approach once the interaction surfaces are in contact; otherwise the normal strength is zero.

To minimize computation time, a quarter of the model was simulated due to symmetry and appropriate constraints were imposed on the symmetry plan. The load and boundary conditions are illustrated in Figure 5. The base of the concrete was fixed and the symmetry planes were constrained to model the actual behavior of the whole model. The pull-out load was applied to the top of the steel strip. The FE mesh was uniform and generated using C3D8R element which is an eight-node linear brick with reduced integration and hourglass control. Hourglass control formulation was used to control the distortion of solid elements during the final stages of the pull-out tests. The FE simulations were repeated without hourglass control, and the results clearly showed that the hourglass control formulation increases the accuracy of the model. In order to gauge how reasonable and accurate the results are on a given mesh, FE mesh convergence study was carried out. The results showed that the mesh size presented in Figure 6 produces a satisfactory balance of accuracy and computation time. The results showed that the finer mesh produced typically the same result; however, as the mesh was made finer, the computation time increased significantly.

The concrete damaged plasticity model was assumed for the concrete. The concrete damaged plasticity model provides a general capability for modelling concrete and other quasi-brittle materials in all types of structures. This model uses concepts of isotropic damaged elasticity in combination with isotropic tensile and compressive plasticity to represent the inelastic behavior of concrete. The concrete damaged plasticity model is based on the assumption of scalar (isotropic) damage and is designed for applications in which the concrete is subjected to arbitrary loading conditions. The model takes into consideration the degradation of elastic stiffness induced by plastic straining both in tension and

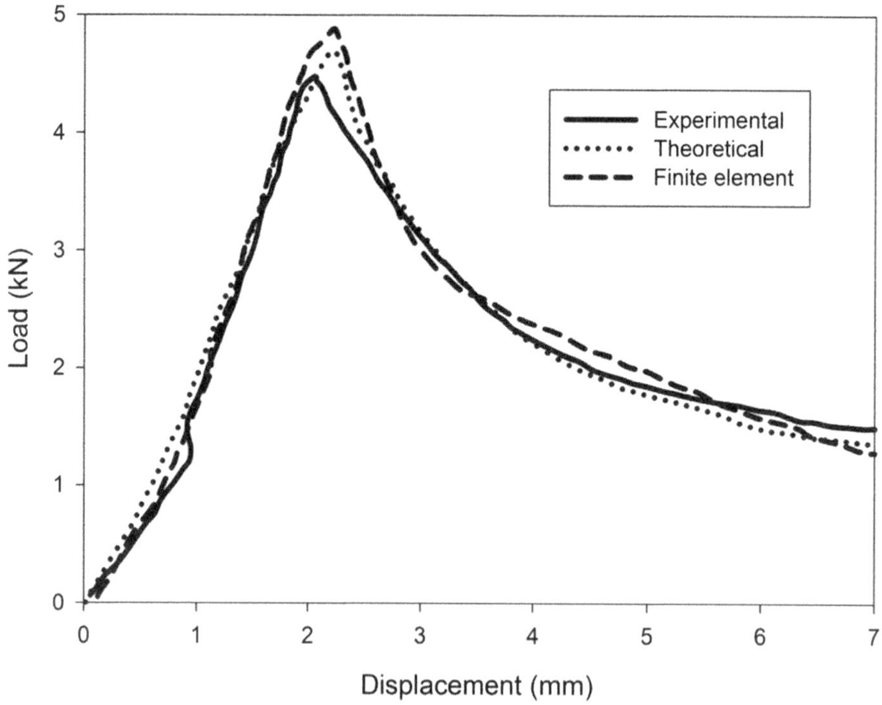

Figure 8 Load-displacement curves for the pull-out test of the steel strip S0.

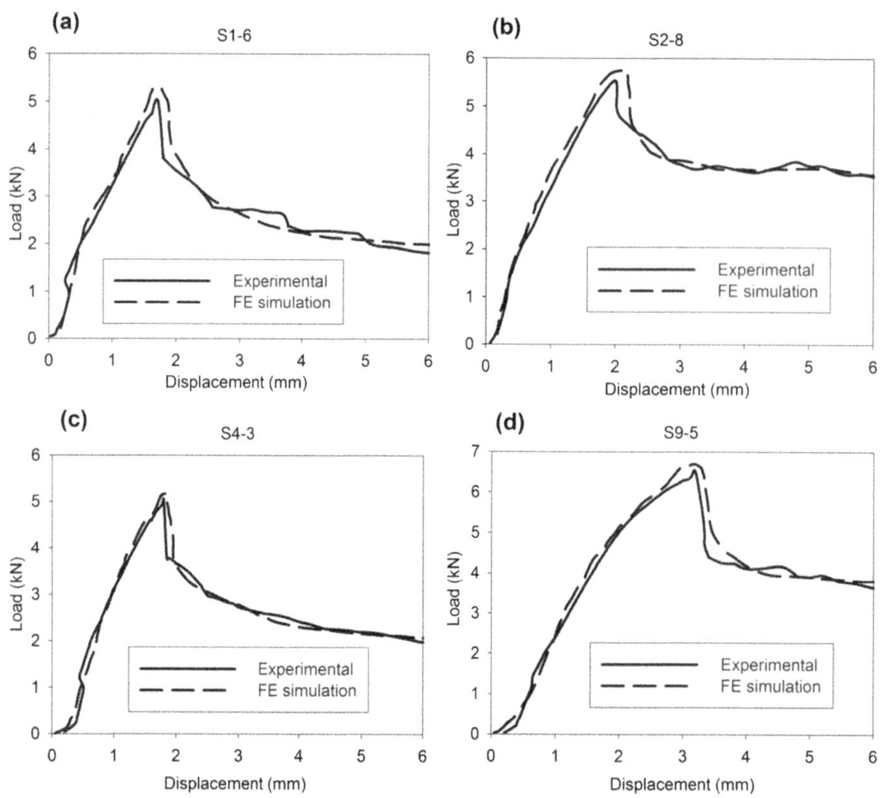

Figure 9 Load-displacement curves for the pull-out tests of the steel strips. **(a)** S1-6, **(b)** S2-8, **(c)** S4-3, and **(d)** S9-5.

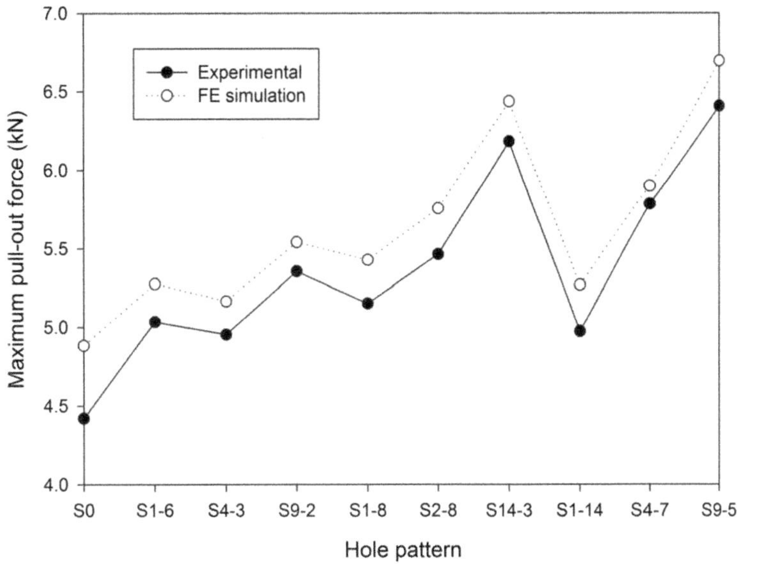

Figure 10 Comparison of experimental and FE simulation pull-out forces for different strip geometries.

compression. It also accounts for stiffness recovery effects under cyclic loading.

The node-to-surface contact between the strip and the concrete was modelled through the master-slave algorithm. General contact automatically assigns master and slave roles for contact interactions, and ABAQUS calculates an overclosure tolerance based on the size of the underlying element facets on a slave surface. Slave surfaces (or nodes) in a particular interaction are repositioned onto the associated master surface (or nodes) if the two surfaces are initially overclosed by a distance smaller than the calculated tolerance. Initial gaps between surfaces remain unchanged by default adjustments. If a portion of a slave surface is initially overclosed by a distance greater than the calculated tolerance, ABAQUS automatically generates a contact exclusion for this surface portion and its associated master surface. Therefore, general contact does not create interactions between surfaces (or portions of surfaces) that are severely overclosed in the initial configuration of the model, and these surfaces can freely penetrate each other throughout the analysis.

A cohesive behavior was introduced between the contact surfaces. Cohesive behavior was defined as part of the surface interaction properties that are assigned to a

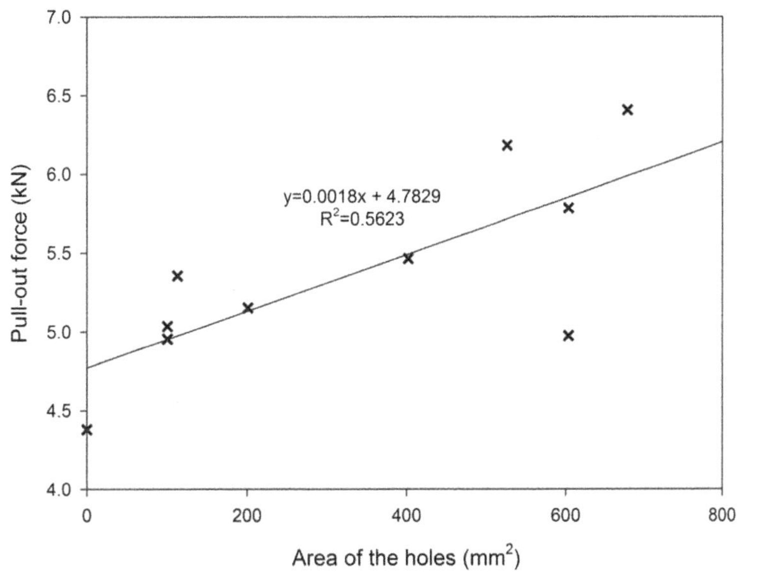

Figure 11 The effect of hole area on the pull-out force.

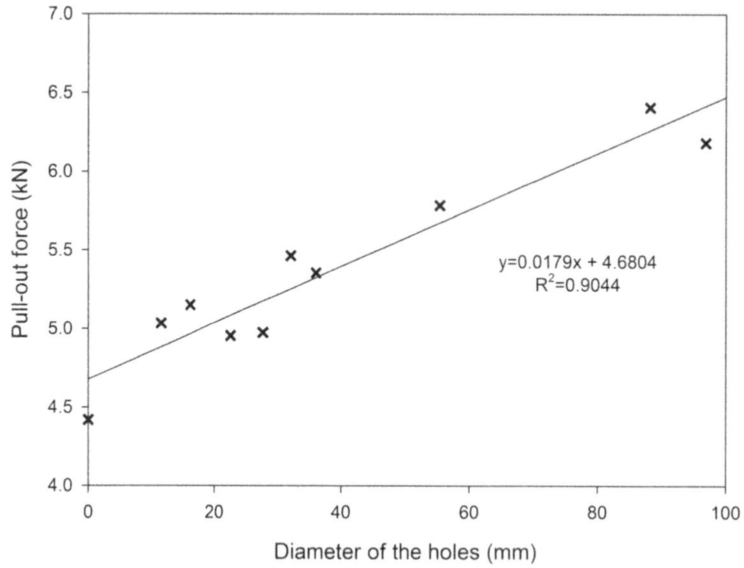

Figure 12 The effect of hole total diameter on the pull-out force.

contact pair. Surface-based cohesive behavior was defined as a surface interaction property and was used to model the surface interaction between the foam concrete and the strip. The cohesive behavior assumes that failure of the cohesive bond is characterized by progressive degradation of the cohesive stiffness, which is driven by a damage process. It also allows the failed nodes to reenter contact with post-failure cohesive behavior. The nonlinear geometry NLGEOM option in ABAQUS was used throughout the analysis. Figures 6 and 7 show the distributions of the von Mises stress in the strips and the concrete blocks in the pull-out tests of the steel strips S0 and S9-5, respectively.

Results and discussions

The results of the pull-out tests between the steel strips and the aerated concrete, and theoretical and FE analyses are presented and discussed here. The main purpose of the pull-out experiments was to find relationships between ten-hole patterns and mechanical properties of aerated concrete in order to evaluate and predict the bonding

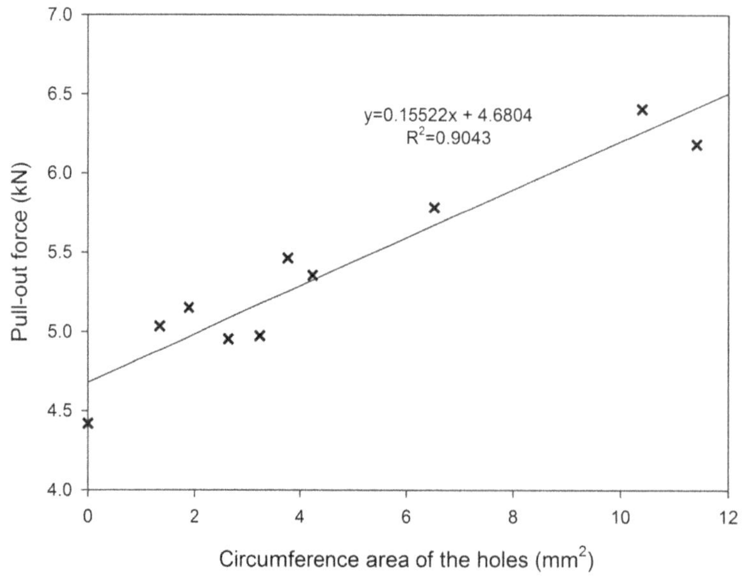

Figure 13 The effect of hole circumference area on the pull-out force.

between steel and lightweight concrete through a numerical simulation.

Three specimens for each single set (three identical samples) of strip design were prepared and tested to determine the bonding strength during pull-out tests. The reference hole area was 100.6 mm^2 with a 11.32-mm diameter. The reference area was increased gradually in order to verify the effect of the anchorage of the concrete embedded into holes in various designs.

The load-displacement curves for the original pull-out test of the steel strip without hole is depicted in Figure 8 for the experimental, theoretical, and FE simulation analyses. It can be seen from the figure that the results show very good correlation, with the FE model predicting slightly higher pull-out force. It is worth pointing out that the theoretical model developed in the section 'Theoretical model' is only valid for the pull-out test of the steel strip without any hole in terms of predicting the load-displacement curve. However, the bond-slip relationship obtained from this model can be further used in FE simulations of all other strip geometries through the user-defined subroutine UINTER. The load-displacement curves obtained from the experiments and FE simulations for all other strip geometries were also compared and show very good agreement. For brevity, only the results for four cases, S1-6, S2-8, S4-3, and S9-5, are presented in Figure 9; however, all other cases show the same trend. It can be seen from the figure that the pull-out load drops suddenly after the maximum value is achieved and then starts to decrease slowly, which indicates that the loading capacity still exists even after debonding. It is found from pull-out experiments that a long softening stage (>20 d_1) exists after the maximum load, where d_1 is the displacement corresponding to the maximum pull-out load. The experimentally and FE simulation predicted maximum pull-out forces were measured and depicted in Figure 10 for all the hole patterns. In general, the FE simulations tend to predict slightly higher pull-out force as shown in Figure 10 with the differences less than 10%.

Figures 11, 12 and 13 illustrate the maximum pull-out force for all pull-out results of the ten series of samples with respect to the total hole area, the total diameter of the holes, and the total circumference area of the holes, respectively. According to Figure 11, a correlation coefficient of R^2 = 0.5623 was achieved, which shows a weak relationship of data for different hole patterns. However, it can be seen that by discarding the S1-14 sample, the correlation improves greatly. Figures 12 and 13 show strong correlations of R^2 = 0.9044 and R^2 = 0.9043, respectively. In general, the figures reveal that there is a nearly linear increase of the pull-out forces on steel strips with holes with respect to specimens without holes if the hole sizes are similar. In general terms, pull-out force increases when all the three parameters (total hole area, diameter, and circumference area) increase. The main reason for the different behavior of S1-14 is that the area of chemical adhesion was reduced due to the holes cut into the steel strips, and the mechanical interlock introduced by the concrete inside the holes is not enough to compensate for the loss of adhesion in weaker matrix.

Conclusions

The goal of this paper was to describe bonding processes between steel and concrete specimens experimentally and by applying analytical and FE methods. The results of investigations on the bonding behavior between lightweight concrete and steel strips enable a better understanding on the use of lightweight concrete in wall panels. The results show that the experimental, theoretical, and FE simulation results match very well. Three geometrical parameters affecting the pull-out force were also investigated in this paper, namely the total area of the holes, the diameter of the holes, and the circumference area of the holes. The diameter of the holes or circumference area of the holes was found to give stronger correlation with the pull-out forces; however, weaker correlations exist between the pull-out forces and the total area of the holes. The strips with the greatest hole area, hole diameter, and circumference areas had higher pull-out forces, and the increase is nearly linear.

Competing interests

The authors declare that they have no competing interests.

Authors' contributions

MR carried out the theoretical and finite element simulation analyses. JV conducted the experimental tests and assisted the simulations. TN participated in the study design and team supervision. All authors contributed to and approved the final manuscript.

Acknowledgements

The authors like to thank the Foundation of Research, Science and Technology (FRST) of New Zealand for the generous financial support received during the investigation.

References

Akin O, Gül R, Cüneyt A (2005) Effect of steel fibers on the mechanical properties of natural lightweight aggregate concrete. Mater Lett 59:3357–3363

Al-mahmoud F, Castel A, François R, Tourneur C (2007) Effect of surface pre-conditioning on bond of carbon fibre reinforced polymer rods to concrete. Cement Concrete Comp 29:677–689

Alterman D, Vilches J, Neitzert T (2011) An analysis of the bonding energy through pull-out tests for aerated concrete with various steel strip geometries. Adv Mater Res 275:55–58

Balendran RV, Zhou F, Nadeem A, Leung A (2002) Influence of steel fibres on strength and ductility of normal and lightweight high strength concrete. Build Environ 37:1361–1367

Banholzer B, Brameshuber W, Jung W (2005) Analytical simulation of pull-out tests—the direct problem. Cement Concrete Comp 27:93–101

Ben Romdhane MR, Ulm FJ (2002) Computational mechanics of the steel-concrete interface. Int J Numerical Anal Methods Geomechanics 26(2):99–120

Bouazaoui L, Li A (2008) Analysis of steel/concrete interfacial shear stress by means of pull out test. Int J Adh & Adhes 28:101–108

Cao J, Chung D (2001) Degradation of the bond between concrete and steel under cyclic shear loading, monitored by contact electrical resistance measurement. Cem Concr Res 31:669–671

Casanova A, Jason L, Davenne L (2012) Bond slip model for the simulation of reinforced concrete structures. Eng Struct 39:66–78

Chang J (2003) Bond degradation due to the desalination process. Construct Build Mater 17:281–287

De Lorenzis L, Rizzo A, La Tegola A (2002) A modified pull-out test for bond of near-surface mounted FRP rods in concrete. Composites: Part B 33:589–603

Fang C, Lundgren K, Plos M, Gylltoft K (2006) Bond behaviour of corroded reinforcing steel bars in concrete. Cem Concr Res 36:1931–1938

Kayali O (2004) Bond of steel in concrete and the effect of galvanizing. Galv Steel Reinf Concr 8:229–270

Kearsley EP, Mostert HF (2005) Designing mix composition of foam concrete with high fly ash contents. Use of foamed concrete in construction. ICE, London, In, pp 29–36

Khandaker M, Hossain A (2008) Bond characteristics of plain and deformed bars in lightweight pumice concrete. Construct Build Mater 22:1491–1499

Wei G, Liu G, Xu C, Sun X (2012) Finite element simulation of perfect bonding for single fiber pull-out test. Adv Mater Res 418–420:509–512

Won J, Park C, Kim H, Lee S, Jang C (2008) Effect of fibers on the bonds between FRP reinforcing bars and high-strength concrete. Composites: Part B 39:747–755

Wu JJ, Cheng XH, Qi Y (2009) Simulation and analysis of single fibre pull-out test with FEM. Shanghai Jiaotong Daxue Xuebao/Journal of Shanghai Jiaotong University 43(10):1601–1604. 1608

15

Development of an integrated remote monitoring technique and its application to para-stressing bridge system

Ayaho Miyamoto[1*], Minoru Motoshita[2] and Joan R Casas[3]

Abstract

Bridge monitoring system via information technology is capable of providing more accurate knowledge of bridge performance characteristics than traditional strategies. This paper describes not only an integrated Internet monitoring system that consists of a stand-alone monitoring system (SMS) and a Web-based Internet monitoring system (IMS) for bridge maintenance but also its application to para-stressing bridge system as an intelligent structure. IMS, as a Web-based system, is capable of addressing the remote monitoring by introducing measuring information derived from SMS into the system through Internet or intranet connected by either PHS or LAN. Moreover, the key functions of IMS such as data management system, condition assessment, and decision making with the proposed system are also introduced in this paper. Another goal of this study is to establish the framework of a para-stressing bridge system which is an intelligent bridge by integrating the bridge monitoring information into the system to control the bridge performance automatically.

Keywords: Bridge monitoring; Para-stressing system; Stand-alone monitoring system; Web-based Internet monitoring system; Structural performance

Introduction

Japan's investments in social infrastructure systems have steadily increased since the high economic growth period that started in the late 1960s. Social infrastructure systems have been reinforced in a short period of time. The public has been provided with stable daily social lives. With the maturing of society, however, shrinking public finance offers only limited funds for infrastructure development. Conventional maintenance practice has difficulty attending to the deterioration of existing infrastructure systems that has been occurring intensively (Miyamoto and Isoda 2012). Future challenges include not only the maintenance of excellent performance of infrastructure with limited funds to hand it over to the next generations but also the enhancement of services that infrastructure renders.

Bridges have recently been lacking required performance due to the damage ascribable to the increases of vehicle size and traffic volume and seeing their performance deteriorate as they have been in service over a long time under aggressive environments. Deterioration of bridges in particular has been increasing the need of investigations and inspections, and placing heavy burdens on bridge administrators. To minimize bridge upgrade and maintenance costs while funds are limited, lengthen service lives, reduce life cycle costs, and maintain the level of services provided by bridges, early investigations and remedial actions are required in the course of shift from reactive to proactive response, and appropriate identification and evaluation of bridge conditions and prediction of deterioration are demanded. Few engineers are equipped with bridge investigation, inspection, or diagnostic skills. Frequent investigations, inspections, or diagnoses are therefore difficult to make. Under such demanding conditions, a conceivable highly accurate and efficient inspection technique is a remote monitoring system for structures via communication networks (Miyamoto and Motoshita 2004).

At present, bridges are designed based on the use of a live load model present in the respective codes, uniformly

* Correspondence: miya818@yamaguchi-u.ac.jp
[1]Graduate School of Science & Engineering, Yamaguchi University, 2-16-1, Tokiwadai, Ube, Yamaguchi 755-8611, Japan
Full list of author information is available at the end of the article

applied to all bridges. Therefore, some bridges carrying a relatively small volume of traffic are designed to carry the loads of heavy vehicles that are unlikely to act during their service life, resulting in uneconomical design. In other cases, non-standard bridges, as long-span bridges (cable-stayed, arches,...), may be subject to very high levels of live load that the bridge most probably will never experience. This is because the live load model in the code is normally calibrated for short- and medium-span bridges, and the direct application to very long spans derives on the over-estimation of the loading. If it was possible to design a bridge so that it can carry the design load equivalent to the loads of vehicles that usually travel and control bridge performance to structurally offset the loads of heavy vehicles that may be hazardous to the bridge, safety and reliability of the bridge may be ensured against such unexpected heavy loads (Kim et al. 2000). Reducing the design live load may also lead to the reduction of the volume of materials used. This approach should therefore be taken actively also from a viewpoint of resource saving (Holnicki-Szulc and Rodellar 1998).

Unexpected heavy loads, such as big earthquake and strong typhoon, may be handled by a para-stressing system that exerts required counterforce as required (Montes 1996). The para-stressing system activates control devices including actuators to offset stresses induced by the strain or deformation of the bridge to restore the bridge to the original condition. The para-stressing system is composed of detection and sensing, decision-making and instruction, and control functions. The detection and sensing function involves the sensing of loading and the identification of conditions of structures (Yan et al. 2010, 2011). The decision-making and instruction function helps determine whether control measures are required or not and send instructions to the control function whenever necessary. The control function is aimed at operating actuators or other devices to apply required forces to the structure. In the papers of Magaña et al. (1998) and Casas (2000), the active control of cable-stayed bridges subject to accidental loads due to earthquake is presented. There, only a theoretical approach was made, using a finite element model of a cable-stayed bridge as the prototype where the control algorithms were applied. The research wanted to demonstrate the feasibility and workability of using the control algorithms in the controllability of such a para-stressing system. The results showed that the control algorithms were very effective in reducing the magnitude of the deflections of the deck caused by the vertical component of an earthquake. In the present paper, the control is applied to a physical and not only mathematical model, and the exceptional load is due to the traffic load instead of earthquake. The experience is a new step forward to the implementation in full-scale structures. In this study, an Internet monitoring system (IMS) via communication networks was built for a cable-stayed bridge model and then a para-stressing system was developed using the IMS.

Methods

Outline of integrated remote monitoring system

Monitoring is an important technique for evaluating the soundness and diagnosing bridge performance based on the real-time measurements of strain, displacement, vibration, and other parameters. The goal should be not only simply to measure the deformations of bridges but also to obtain useful knowledge for bridge management by efficiently recording, processing, and using measurement results. Conventional monitoring relied on manual measurement. Increasing the monitoring efficiency requires the realization of unmanned automated real-time measurement and the development of an environment for providing numerous bridge-associated people with

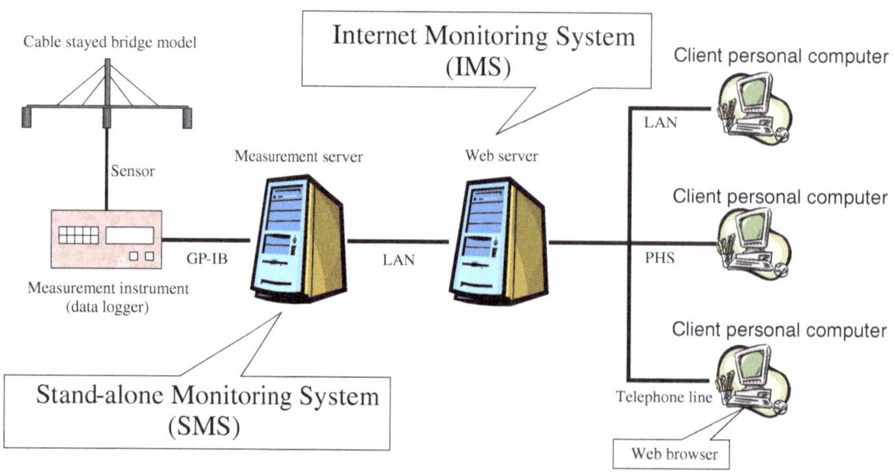

Figure 1 Composition of integrated remote monitoring system.

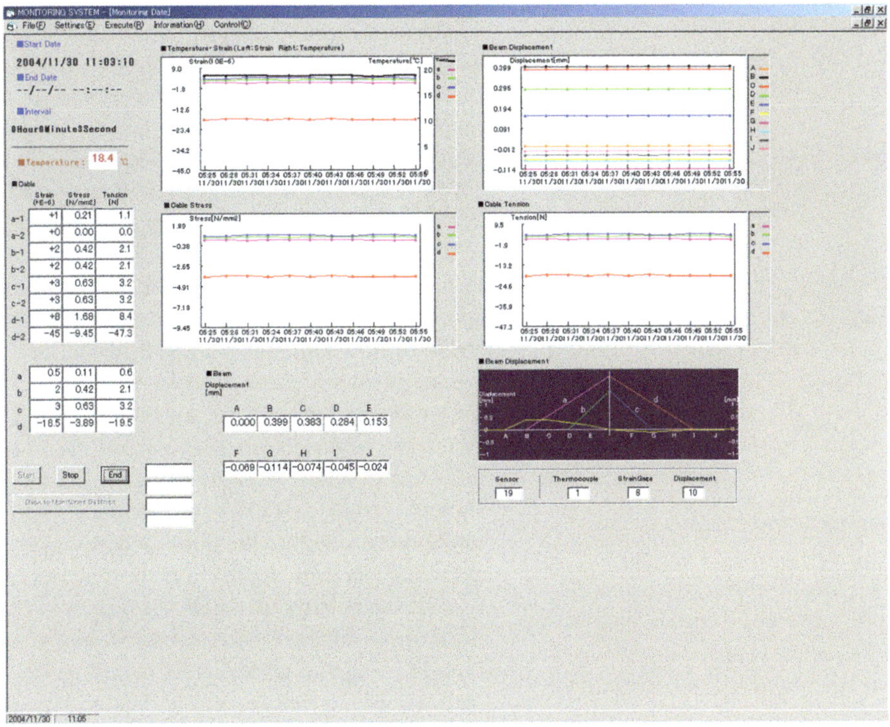

Figure 2 Screen output from SMS.

access to necessary data from the place where they need the data. The systems built in this study are the stand-alone monitoring system (SMS), which is installed mainly in the field to make real-time measurement and visualize measurement results, IMS, which enables remote measurement and data collection via communication networks, and another monitoring system, which integrates the two systems. An outline of the integrated remote monitoring system is given in Figure 1. The integrated remote monitoring system is used as a sensing tool for the para-stressing system to identify the present condition of the bridge in terms of deformation, stress, and other parameters, and to confirm post-control conditions of the bridge.

Outline of stand-alone monitoring system

The SMS works on a measurement computer installed at the bridge site where field monitoring takes place. SMS measures external forces acting on the bridge and the behavior of the bridge subjected to external forces. SMS periodically measures the stresses and displacements of bridge members and temperature around the bridge using sensors installed in the bridge, visualizes the collected data in graphs and charts, and stores them in a measurement server.

SMS that monitors a cable-stayed bridge model is composed of a measurement instrument (data logger) that monitors the behavior of the model and a measurement

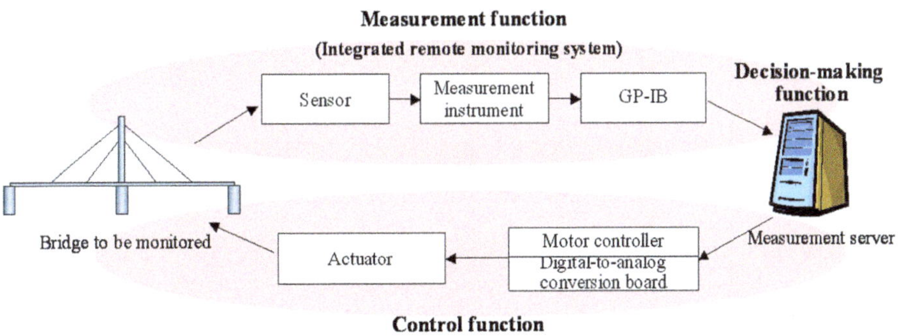

Figure 3 Composition of para-stressing system.

Figure 4 Cable-stayed bridge model.

server that sends measurement instructions to the measurement instrument and records data (Figure 1). The measurement server is equipped with measurement control and data storage programs and records monitoring data and displays them in graphs. Figure 2 shows a screen output by SMS for a two-span cable-stayed bridge model.

Outline of Internet monitoring system

To increase measurement and data collection efficiency of SMS at the bridge site, an IMS using communication networks was built by incorporating a Web server. Using IMS, which is capable of transmitting monitoring data at real time, enables quick identification of bridge performance and response, and efficient management and use of collected measurement data regardless of time or place. With IMS, early discovery of problems by field monitoring, simultaneous monitoring of multiple bridges, data use for other purposes than bridge management, guarantee of data uniformity, and reduction of inspection and other maintenance work also become possible.

For practical IMS implementation, however, preventing illegal attacks of Internet invaders such as system destruction, data manipulation, and eavesdropping is very important. In IMS, the range of authority of the user to access the system and data is classified into the manipulation of measurement results, data use, or data retrieval. Some users are authorized to access the common file server in the Web server, but others not. Thus, restrictions are imposed on the system use. The system is intended to allow as many bridge-related people as possible to retrieve and use data. Users are classified into administrator, member, or guest according to the range of authority (security depth) to restrict access to the system. A multi-level authentication is adopted in which authentication is done at each security depth. Wielding the broadest range of authority of the administrator requires authentication at three security depths. This system adopts server authentication, client authentication, and password authentication based on security socket layer (SSL) that encrypts data.

Outline of para-stressing system

Concept of a para-stressing system

Para-stressing is an intelligent technology based on a new concept of response to external forces acting on structures by controlling structural members including material

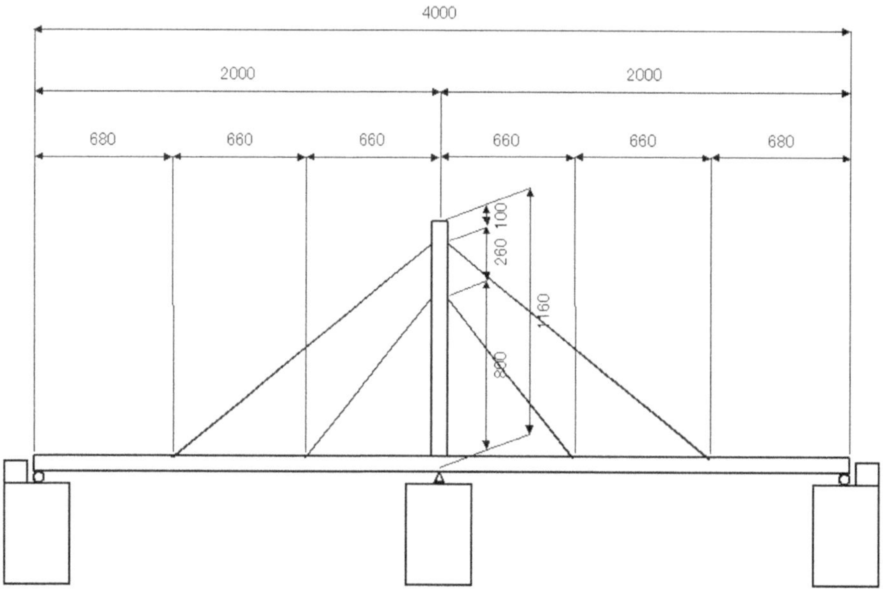

Figure 5 Dimensions of the cable-stayed bridge model (in mm).

Table 1 Measurement items

Sensor	Measurement item	Number of measurement points
Thermopile	Temperature	1
Strain gauge	Cable tension and stress	8
Displacement gauge	Main girder displacement	10

properties in real time while regarding the entire structure as a self-organized system (Montes 1996). Shown in Figure 3 is a self-organized structure that senses, determines, and controls external forces by itself. The system senses the conditions of the bridge based on the measurements obtained by the integrated remote monitoring system, determines the present serviceability and safety condition of the bridge using the measurement server, and sends control instructions to the actuator whenever necessary. A system was constructed for automatically carrying out a series of these jobs.

Development of para-stressing system

Described below are the three functions of the para-stressing system shown in Figure 3: sensing, decision-making, and control/actuation functions.

Sensing function The sensing function is intended to accurately identify the conditions of the bridge. The integrated remote monitoring system is used to perform this function. With the para-stressing system, displacements of main girders are controlled by varying the cable tension of the cable-stayed bridge model. The sensing function is therefore responsible for accurately grasping cable tension before and after the control measure is taken.

Decision-making function The decision-making function calculates the tension that should be applied to cables to reduce the main girder displacement or vibration due to large vehicle loads to the level under normal loads (optimal counterforce for controlling displacement). The cable tension measured by the sensing function is compared with the calculated counterforce. Control instructions are continually sent to the control function until the cable tension becomes identical to the optimal counterforce.

Control function The control function varies cable tension by stressing or relaxing the tension using the actuator on the bridge to offset additional loads of large vehicles.

Results and discussion

Example of application

Description of the structure

To check the effectiveness of the proposed method, a para-stressing system was developed for a cable-stayed bridge model as shown in Figure 4. The dimensions of the bridge model are shown in Figure 5. The two spans have a span length of 2.0 m. The cross-sectional dimensions of the deck which is a box girder with five rooms are 31.0 cm in width and 7.0 cm in height, and for the pylon, the main cross-sectional dimensions are 8.0 cm in width and 5.0 cm in thickness.

Description of the monitoring system

The measurement items used for the para-stressing system are listed in Table 1. The location of sensors and main girder displacement measurement points are shown in Figure 6.

Description of the test

The objective function during the tests is to control the maximum deflection in the deck due to the crossing of vehicles. To this end, a para-stressing system is developed, and the following test procedures were also developed.

When a large vehicle reaches the bridge, its weight is measured with a load cell at the starting point of the bridge. When the measured force is uploaded into the system, a program is activated to calculate the counterforce required for controlling the deck deflection. The

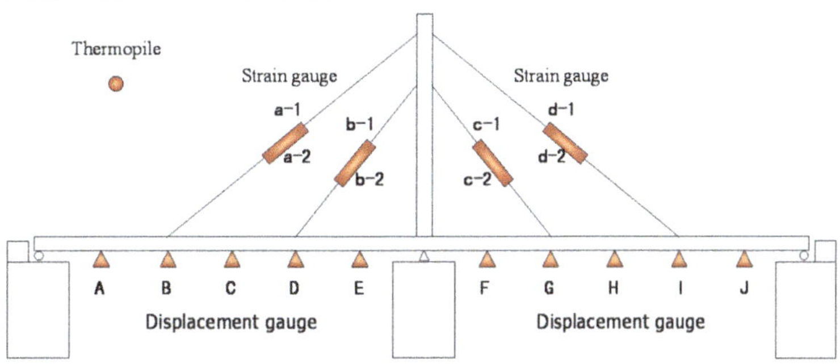

Figure 6 Sensor positions and measurement points.

system calculates the cable tension (optimal counterforce) required to hold the main girder displacement under normal loads (allowable displacement) when the heavy vehicle reaches the first measurement point (A). Once the optimal counterforce is determined, instructions are sent to the actuator to start operation, stressing/releasing of stay cables starts and, at the same time, the monitoring system begins to measure cable tension. The decision-making function compares the measured cable tension with the optimal counterforce and repeatedly stresses and relaxes cable tension until an agreement is reached. When the adjustment of tension is finished, the load is moved to measurement point A where a displacement gauge is in place as shown in Figure 6. When the large vehicle reaches measurement point A, the above process is repeated for the next measurement point B. Operation is repeated from the time a large vehicle reaches the starting point of the bridge model until the vehicle reaches the end, and the deck deflection is measured each time the control measure is taken.

The types of loads used in the test are listed in Table 2. The allowable loads represent the weight of vehicles that normally travel on the bridge, which are lower than the present design standards. The large vehicle loads are loads applied by vehicles that rarely travel on the bridge and are likely to be hazardous to the bridge. In the test, the main girder displacement was measured under large vehicle loads while no control measures were taken and when displacement was reduced to the level under allowable loads. The objective of the test was to confirm the restoration of the displacement, thanks to control measures. Tests were conducted in four cases with varying combinations of large vehicle and allowable loads as shown in Table 3.

In test case (i) in Table 3, the exceptional live load was 40 N and the allowable live load was 10 N. First, the main girder displacement was monitored while an exceptional live load of 40 N was crossing the bridge. The obtained deflection in each measuring point (A to J) was used as the data in the case without control measures (δ_1). Afterwards, the main girder displacement was measured when the main girder displacement under the load of 40 N was intended to be reduced to that under an allowable live load of 10 N. The measured result was used as data in the case where control measures were taken (δ_2). The difference was calculated between displacements in the cases with and without control measures and compared to the deflection obtained when the normal load of 10 N is applied (δ_3). In this way, the efficiency of the para-stressing system can be defined. The efficiency of the system is measured in terms of the so-called *rate of restoration R*, defined as follows:

$$R(\%) = \frac{\delta_1 - \delta_2}{\delta_1 - \delta_3} \times 100$$

In cases (ii) to (iv) in Table 3, similar tests were conducted with varying combinations of exceptional and allowable live loads.

Table 2 Types of loads

Allowable live load	Large live load
Below allowable design load	Exceeding allowable design load
10 N, 30 N	40 N, 50 N

Table 3 Test cases

	Large live load (N)	Allowable live load (N)
Case (i)	40	10
Case (ii)	40	30
Case (iii)	50	10
Case (iv)	50	30

Test results

The results of test case (i) in Table 3 are discussed here. Figure 7 compares the test result when the large vehicle load was applied at loading point B with finite element method (FEM) analysis result.

Rates of restoration at respective measurement points when the load was controlled at points A to E are listed in Figure 8.

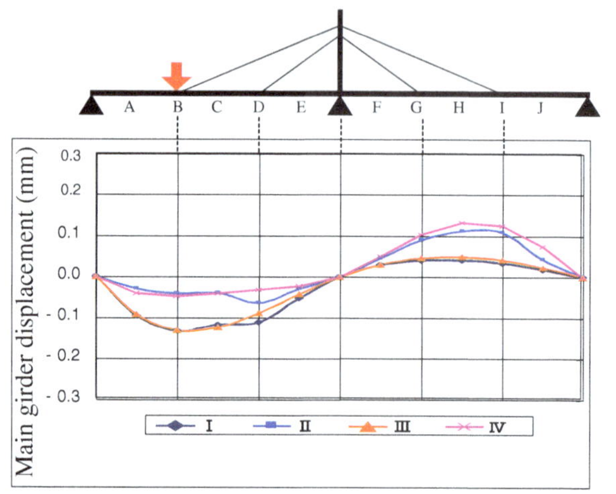

Figure 7 Comparison of results in case (i) at loading point B. I, main girder displacement under exceptional vehicle load without any control measures (test result (δ_1)); II, main girder displacement under exceptional vehicle load when cable tension was controlled with the objective of reducing the deflection to that corresponding to the normal vehicle (test result (δ_2)); III, main girder displacement under exceptional vehicle load without any control measures (FEM result); IV, main girder displacement under exceptional vehicle load when cable tension was controlled (FEM result).

40→10N		Loading point A	B	C	D	E	Mean value
Measurement point	Edge						
	A	128	109	95	86	90	102
	B	110	98	88	82	68	89
	C	95	85	78	76	62	79
	D	83	86	79	75	54	75
	E	117	104	105	113	74	103
	Pylon						
	F	76	85	82	79	136	
	G	86	93	90	83	114	
	H	86	90	86	81	109	
	I	89	92	86	79	98	
	J	49	16	4	- 38	- 139	
	Edge						
Mean Value		107	96	89	86	70	90

Figure 8 List of measurements in test case (i) (rate of restoration in %).

Discussion

Figure 7 shows a good approximation between experimental and FEM results. This suggests the validity of the adopted FEM model for the development of the para-stressing system. According to Figure 8, the rate of restoration is nearly 80% despite variation according to the measurement point. Thus, the para-stressing system was effective for restoring the bridge after displacement.

As is obvious from Figure 7, large negative displacement occurred in the unloaded span. This phenomenon is attributable to the balance of stiffness between the main girders and pylon of the cable-stayed bridge model. The stiffness of the main girder is much higher than that of the pylon.

Figure 8 shows that the rate of restoration was higher near the end of the bridge than near the pylon. This is attributable to the structural properties of the cable-stayed bridge model dependent on the stiffness balance among the main girders, pylon, and cables.

Conclusions

This paper introduces an IT-based bridge remote monitoring system which enables real-time monitoring and control for unexpected heavy loads such as big earthquake and strong typhoon. The integrated remote monitoring system is also composed of not only the Internet but also other types of information technology such as the latest information processing and soft computing technologies.

The major results of this study and the tasks for the future are the following:

1. The monitoring system built in this study is effective not only for investigating or inspecting structures but also for building a para-stressing system to control the bridge performance under exceptional loads.
2. The para-stressing system developed in this study achieved a rate of restoration of 80% related to the vertical deflection of main girders. Although this is a quite effective value, further investigation on the structural properties of the bridge to increase the rate of restoration is needed.
3. Both the monitoring and para-stressing systems are dependent on communication networks. The following requirements should therefore be satisfied.
   (a) As a security measure, SSL that encrypts data has been used, and a multi-level authentication has been adopted according to security depths. Efforts are required to enhance the security a level further.
   (b) Simultaneously monitoring multiple bridges by a single bridge management system or using para-stressing systems requires efficient data processing, guarantee of data reliability and homogeneousness, and early detection of and recovery from system malfunctions.

Competing interests
The authors declare that they have no competing interests.

Authors' contributions
AM and MM established an idea of the integrated remote monitoring system for existing bridges and also of the theoretical background. They also carried out many kinds of experimental works as para-stressing system. JRC contributed numerous discussions and advice. All authors read and approved the final manuscript.

Acknowledgements
This research was supported by JSPS KAKENHI (Grant-in-Aid for Scientific Research (S)), grant no. 16106007. The authors would like to thank the students and staffs of Yamaguchi University, Japan for their great help and a lot of advice.

Author details
[1]Graduate School of Science & Engineering, Yamaguchi University, 2-16-1, Tokiwadai, Ube, Yamaguchi 755-8611, Japan. [2]Kyowa Sekkei CO., LTD., Ushitora, Osaka, Ibaraki 567-0877, Japan. [3]Department of Construction Engineering, School of Civil Engineering, Technical University of Catalunya, Campus Nord, c/Jordi Girona 1-3, Modulo C-1, Barcelona 08034, Spain.

References

Casas JR (2000) Structural control of bridges: recent experiences in cable-stayed bridges. In: Casciati F, Magonette G (eds) Structural control for civil and infrastructure engineering, Proceedings of the 3rd International Workshop on Structural Control, Paris, July 2000. World Scientific, New Jersey, pp 85–92

Holnicki-Szulc J, Rodellar J (1998) Requirements and potential applications in mechanical and civil engineering, vol 65, Smart structures, NATO Science Series. Kluwer, Boston

Kim S, Yoon C, Kim B (2000) Structural monitoring system based on sensitivity analysis and a neural network. Comput-Aided Civ Infstr Eng 15(4):309–318

Magaña ME, Rodellar J, Casas JR, Mas J (1998) Active control of cable-stayed bridges. Smart structures, NATO Science Series, vol 65. Kluwer, Boston, p 65

Miyamoto A, Isoda S (2012) Sensitivity analysis of mechanical behaviors for bridge damage assessment. Struct Eng Mech 41(4):539–558

Miyamoto A, Motoshita M (2004) An integrated internet monitoring system for bridge maintenance. In: Proceedings of the second international workshop on structural health monitoring of innovative civil engineering structures. ISIS Canada Research Network, Manitoba, pp 573–584

Montes S (1996) A global concept for 21st century bridge: para-stressing. In: Proceedings of FIP symposium on post-tensioned concrete structures., London, pp 739–744

Yan Z, Miyamoto A, Jiang Z, Liu X (2010) An overall theoretical description of frequency slice wavelet transform. Mech Systs Signal Process 24(2):491–507

Yan Z, Miyamoto A, Jiang Z (2011) Frequency slice algorithm for modal signal separation and damping identification. Comput Struct 89(1–2):14–26

16

Structural health monitoring of a cantilever beam using support vector machine

Satish B Satpal[1], Yogesh Khandare[1], Anirban Guha[1*] and Sauvik Banerjee[2]

Abstract

In this article, the effectiveness of support vector machine (SVM) is examined for health monitoring of beam-like structures using vibration-induced modal displacement data. The SVM is used to predict the intensity or location of damage in a simulated cantilever beam from displacements of the first mode shape. Twelve levels of damage intensities have been simulated at 12 locations, and six levels of white Gaussian noise have been added, thereby obtaining 1,008 simulations. About 90% of these are used for training the SVM, and the remaining are used for testing. The trained SVM is able to predict damage intensity and location of all the training set data with nearly 100% accuracy. The test set data reveal that SVM is able to predict damage intensity and damage location with errors varying from 0.28% to 4.57% and 0% to 20.3%, respectively, when there is no noise in the data. Addition of noise degrades the performance of SVM, the degradation being significant for intensity prediction and less for damage location prediction. The results demonstrate the use of SVM as a powerful tool for structural health monitoring without using the data of healthy state.

Keywords: Support vector machine, Structural health monitoring, FE simulation, noisy data

Introduction

The process of implementing a damage identification strategy for aerospace, civil, and mechanical engineering infrastructure is referred to as structural health monitoring (SHM) (Farrar and Worden 2007). Any changes in the material and/or geometric properties of these structures are considered as damage. The modal parameters (natural frequencies, mode shapes, and modal damping) are functions of the physical parameters (mass, stiffness, and damping), and therefore, it is reasonable to assume that the existence of damage leads to changes in the modal properties of the structure.

Several researchers have studied vibration-based methods and a detailed literature review has been reported by Zou et al. (1999). Though the changes in natural frequencies give useful information regarding the structure's status, i.e., whether it is damaged or not, that is not sufficient to locate the damage. As a result, methods based on mode shape data and frequency response function of the structure have been subject of recent studies (Mal et al. 2005). It has been demonstrated that the approximate location and severity of an unknown defect can be determined using a damage index, which is calculated by comparing the changes in extracted signal features at the healthy and damaged states. Both built-in piezoelectric patches and the laser vibrometer have been used as response sensors in an effort to examine their sensitivity for health monitoring of metallic and composite structures (Banerjee et al. 2009).

Traditional neural network approaches have suffered difficulties with generalization, producing models that over-fit the data. This is a consequence of the optimization algorithms used for parameter selection and the statistical measures used to select the 'best' model (Gunn 1998). On the other hand, support vector machine (SVM) is a novel machine learning method based on statistical learning theory introduced by Vapnik (1999) and is gaining popularity due to many features. SVM requires few training data and avoids over-fitting of data. Shan et al. (2012) used the multi-class pattern classification algorithm of C-support vector machine and the regression algorithm of ε-support vector machine to identify the damage location and damage extent, respectively, in a railway continuous girder bridge. SHM of aerospace structures based on dynamic strain

* Correspondence: anirbanguha1@gmail.com
[1]Department of Mechanical Engineering, Indian Institute of Technology Bombay, Powai, Mumbai 400076, India
Full list of author information is available at the end of the article

measurements using SVM classification has been studied by Loutas et al. (2012). The use of SVM for prediction of fault in power systems has been demonstrated by Kumar et al. (2011). They used support vector classification to predict the damage location.

Bulut et al. (2005) demonstrate the damage detection in civil structure using SVM classifier and wavelets. They found that the SVM was a robust classifier in the presence of noise, whereas wavelet-based compression gracefully degrades its classification accuracy. Kurek and Osowski (2010) used SVM classifier for fault diagnosis of the broken rotor bars of squirrel-cage induction motor. A new monitoring and diagnostic method using support vector data description (SVDD) was proposed by Chendong et al. (2012), which only needs samples under healthy condition. They selected several nodes of the monitored structure and decomposed the signals from these nodes with wavelet packet transform. To monitor the structural health efficiently, they assembled a combined feature using wavelet packet energy distributions of these nodes. The feature was then applied as the input to a SVDD classifier.

In this work, we have considered normalized displacement data of the first mode of a cantilever beam for training ε-SVM in regression problem. This approach needs only damaged state data of the structure (beam). Finite element (FE) analysis of the beam is carried out in the FE package ABAQUS® (ABAQUS Inc., Providence, RI, USA). The damages are simulated at 12 different damage locations, and 12 damage intensities at each location are considered in this study. The efficiency of SVM is examined by adding noise levels of 30 to 80 dB in steps of 10 dB to the original data.

Methods

Finite element modeling and analysis

A cantilever beam of length 450 mm, width 40 mm, and thickness 3mm with Young's modulus 207 GPa, density 7,850 kg/m^3, and Poisson's ratio 0.33 is simulated. A 4-node rectangular shell element of size 5 × 10 mm is used for meshing. The damage at a given location is simulated by reducing the thickness from 15% to 70% in steps of 5% over a length of 10 mm. The schematic of the damage locations starting from 100 mm, separated by 25 mm up to 375 mm from fixed end, is shown in Figure 1a. First modal displacement values are obtained at 91 points along the center line of the beam. There are 12 damage locations and 12 damage intensities at each location, giving a total of 144 cases for simulations, and data (displacement of the 91 points at first mode shape) obtained from each case is used in SVM. Mode shape data is obtained from the points highlighted in Figure 1b along the length of beam at the spacing of 5 mm.

Overview of support vector machine

SVMs are a set of related supervised learning methods used for classification and regression (Gunn 1998) which use machine learning theory to maximize predictive accuracy while automatically avoiding over-fit to the data. In this section, we present a brief theoretical formulation of SVM for linear regression analysis.

SVM was initially developed for solving classification problems and successfully applied in regression problems. The general formulation of regression learning is carried out as follows (Gunn 1998).

Consider the problem of approximating the set of data,

$$(y_1, x_1), \ldots, (y_1, x_1), x \in R^n, y \in R, \tag{1}$$

with a linear function

$$f(x) = (w \cdot x) + b, \tag{2}$$

where

$$w = \sum_{i=1}^{l} (a_i - a_i^*) x_i \tag{3}$$

$$b = -\frac{1}{2} w \; [x_r + x_s] \tag{4}$$

The optimal regression function is given by the minimum of the functional,

$$\phi(w, \xi, \xi^*) = \frac{1}{2} \|w\|^2 + C \;\; \sum_{i=1}^{l} \xi_i + \sum_{i=1}^{l} \xi_i^* \Bigg) \tag{5}$$

where C is a pre-specified value, and ξ^*, ξ are slack variables representing upper and lower constraints on the outputs of the system. Using an ε-insensitive loss function,

$$L_\varepsilon(y) = \begin{cases} 0 for |f(x) - y| < \varepsilon \\ |f(x) - y| - \varepsilon \; otherwise \end{cases} \tag{6}$$

The solution is given by

$$\max_{\alpha, \alpha^*} W(\alpha, \alpha^*) = \max_{\alpha, \alpha^*} \left\{ \begin{array}{l} -\frac{1}{2} \sum_{i=1}^{l} (\alpha_i - \alpha_i^*) \left(\alpha_j - \alpha_j^*\right) (x_i \cdot x_j) \\ + \sum_{i=1}^{l} \alpha_i (y_i - \varepsilon) - \alpha_i^* (y_i + \varepsilon) \end{array} \right\} \tag{7}$$

Alternatively,

$$\bar{\alpha}, \bar{\alpha}^* = \underset{\alpha, \alpha^*}{\arg\min} \left\{ \begin{array}{l} -\frac{1}{2} \sum_{i=1}^{l} (\alpha_i - \alpha_i^*) \left(\alpha_j - \alpha_j^*\right) (x_i \cdot x_j) \\ -\sum_{i=1}^{l} y_i (\alpha_i - \alpha_i^*) + \sum_{i=1}^{l} (\alpha_i - \alpha_i^*) \varepsilon \end{array} \right\} \tag{8}$$

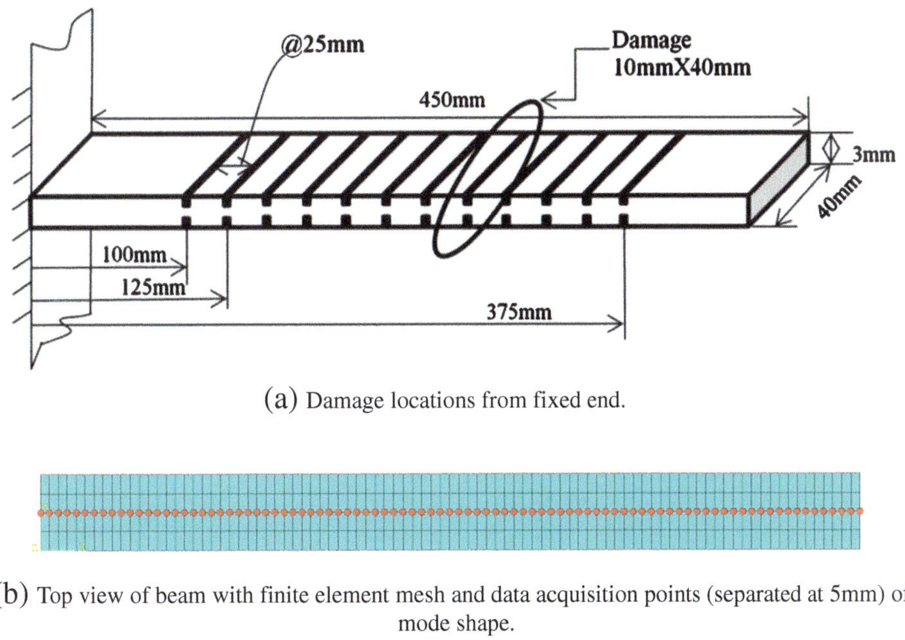

Figure 1 Schematic of the damage locations and mode shape data. Damage locations from fixed end (**a**). Top view of the beam with finite element mesh and data acquisition points (**b**).

with the constraints,

$$0 \le \alpha_i \le C, i = 1, \dots, l \tag{9}$$

$$0 \le \alpha_i^* \le C, i = 1, \dots, l \tag{10}$$

$$\sum_{i=1}^{l} \left(\alpha_i - \alpha_i^*\right) = 0 \tag{11}$$

Solving Equation 7 along with constraints Equations 9 and 10 determines the Lagrange multipliers (α_i, α_i^*), and the regression function is given by Equation 1.

The Karush-Kuhn-Tucker conditions that are satisfied by the solution are as follows:

$$\bar{\alpha}_i \bar{\alpha}_i^* = 0, i = 1, \dots, l \tag{12}$$

Therefore, the support vectors are points where exactly one of the Lagrange multipliers is greater than zero. When $\varepsilon = 0$, we get the L_1loss function, and the optimization problem is simplified to

$$\min_{\beta} \frac{1}{2} \sum_{i=1}^{l} \sum_{j=1}^{l} \beta_i \beta_j \left(x_i \cdot x_j\right) - \sum_{i=1}^{l} \beta_i y_i \tag{13}$$

where β is the difference in Lagrange multipliers with constraints

$$-C \le \beta_i \le C, i = 1, \dots, l \tag{14}$$

and

$$\sum_{i=1}^{l} \beta_i = 0 \tag{15}$$

A Matlab® toolbox (MathWorks, Natick, MA, USA) developed by Gunn (1998) is used for support vector regression analysis. The code requires 'training' and 'testing'. In our case, displacement at 91 points along the beam in the first mode shape obtained from a vibration simulation is used as the 'input', and the location or intensity of damage is the output to the SVM in the training stage. An unknown set of displacements at those 91 points is then used as the input to the trained SVM, and the damage intensity (or location) 'predicted' by the SVM is compared with the known damage intensity (or location) which would have caused those displacements. This is known as testing.

Performance of the SVM model depends on the proper selection of the SVM parameters (C, ε). Hence, it is one of the critical tasks in SVM analysis for the given data set. There is no universal rule to select the best parameters for SVM analysis. These need to be determined by trial and error. However, Cherkassky and Ma (2004) proposed a simple yet practical analytical approach to SVM regression parameter setting directly from the training data. However, in the present paper, the parameters are selected by trial and error.

In order to simulate experimental uncertainties, white Gaussian noise is added in the original mode shape data using Matlab® function, *awgn(x, snr)*. This function generates a vector of white Gaussian noise of mentioned signal-to-noise ratio (*snr*) value in decibel scale and adds to the signal vector x.

Results and discussion

Case 1: prediction of damage intensity

Twelve damage intensities have been simulated at each of the 12 locations using ABAQUS®, thus giving 144 simulations. The vibration-induced displacements of the beam at 91 locations (evaluation points) for each of these simulations are used as input to the SVM, and the damage intensity is used as the output. The format for the training and test set data used to predict the damage intensity using the SVM is shown in the Figure 2. Damages of 12 different intensities are simulated at the location L_1. The displacements of the beam at the first mode shape for all the damage intensities except 15% are fed to the SVM as training input and the corresponding damage intensities as training output. After training, the mode shape displacements at 15% damage are fed to the trained SVM, and the output (Damage $\text{intensity}_{\text{predicted}}$) is compared to the value 15 (Damage $\text{intensity}_{\text{actual}}$). The error is evaluated according to the formula given in Equation 16. The parameters for SVM are taken as follows: C = infinity, e = 0, exponential radial basis function (ERBF) kernel, and ε-insensitive loss function. The procedure is repeated for all the damage locations (L_1 to L_{12}). To find the damage location, mode shape data for all the locations (at particular damage intensity) are used. Mode shapes at one of the locations is used as test input data set whose location is to be predicted and the remaining are used for training.

$$\%\ \text{error} = \left(\frac{|\text{Damage intensity}_{\text{predicted}} - \text{Damage intensity}_{\text{actual}}|}{\text{Damage intensity}_{\text{actual}}}\right) \times 100 \quad (16)$$

The SVM analysis is carried out using the mode shape data considering no noise at each location for all the damage intensities. The results obtained in damage intensity prediction in terms of percentage error are plotted in Figure 3. Thereafter, noise levels of 30 to 80 dB are added in steps of 10 dB to the original mode shape data, and similar exercises are carried out for each noise level. The errors have been averaged over damage locations and plotted in Figures 4 and 5. Inspection of these figures shows that error in the damage intensity prediction by SVM is high for low damage intensity (15% to 25%) at increased noise level. If we accept that SVM should only be used to predict the damage intensity above 25%, then the error in the prediction for no noise case varies from 1.55% to 2.45%.

The plot for 80-dB noise level is seen to be nearly overlapping with no noise case, which indicates that the effect of this level of noise is very low. Noise in the level of 30 to 50 dB however causes significant differences in the results.

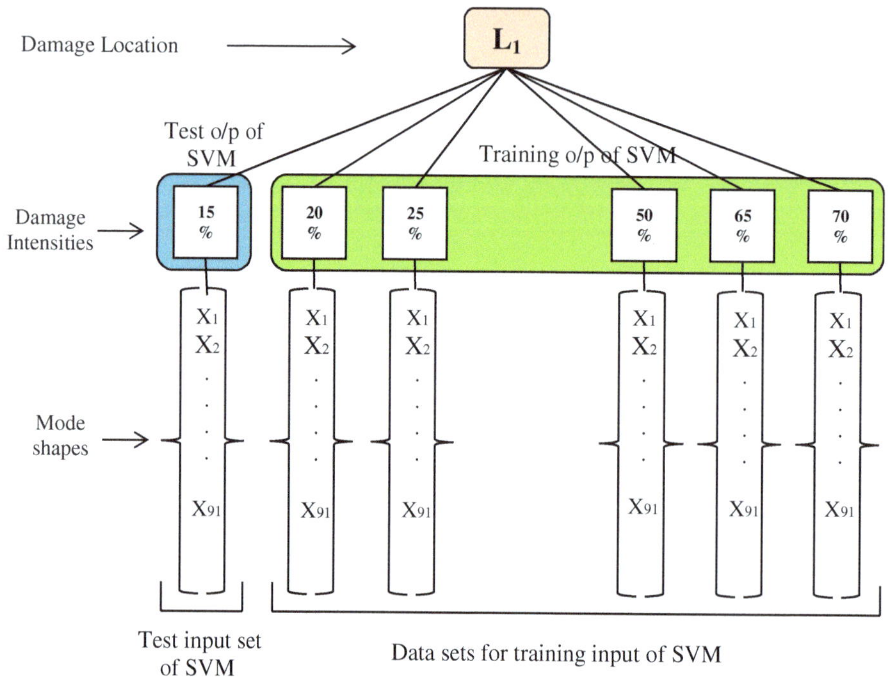

Figure 2 Format of data used for SVM analysis for prediction of intensity for a fixed location.

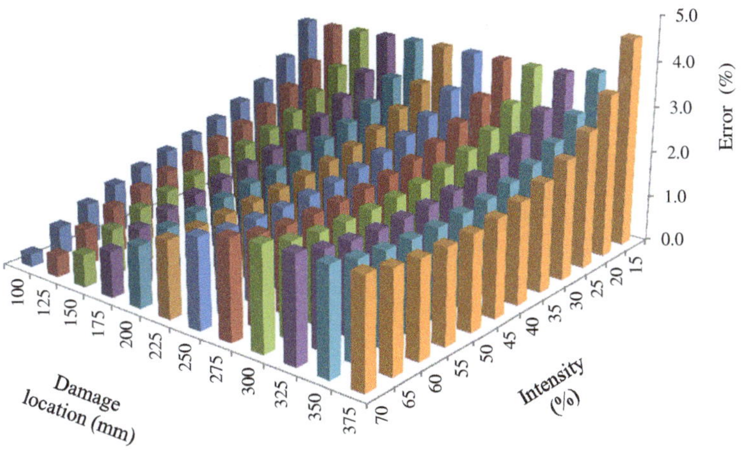

Figure 3 **Error in damage intensity prediction for noise-free data.**

The capability of SVM to predict damage intensity is good at high damage intensity and deteriorates with increase in noise. At low level of noise (80 dB), it is able to predict even low levels of damage. At the higher levels of noise (70 to 30 dB), the prediction of high damage (beyond 25%) remains acceptable, but prediction of low damage intensity deteriorates significantly. At the highest level of noise (30 dB) and lowest level of damage (15%), the error in prediction is as bad as 95.59%. The overall trend of the plots indicates that high level of damage is better predicted than low damage, the only significant deviation being the range of 30% to 50% damage for 30-dB noise.

Case 2: prediction of damage location

An exercise similar to case 1 is carried out except that in this case, for a given damage intensity, the damage location is unknown. The error in location prediction is calculated using Equation 17 at each location considering all damage intensities, and it is plotted in Figure 6. The percentage error in the location prediction is significant only near the fixed end (location, 100 mm) where it is 20%. For the locations in the middle span of the beam (125 to 350 mm), the prediction of damage location is quite accurate, varying from 1.46% to 4.93%. The training parameters for SVM are taken as follows: C = infinity, $e = 0$, ERBF kernel, and ε-insensitive loss function.

The percentage error is calculated as follows:

$$\%\ \text{error} = \left(\frac{\left|\text{Damage location}_{\text{predicted}} - \text{Damage location}_{\text{actual}}\right|}{\text{Length of beam}}\right) \times 100 \tag{17}$$

Thereafter, a similar analysis is carried out by adding white Gaussian noise (30 to 80 dB) in steps of 10 dB in the mode shape data. The error in location prediction for all damage intensities with noise added in the data is calculated. The error for all damage intensities is averaged over each location for different noise levels and plotted in Figures 7 and 8. The error in the prediction of damage

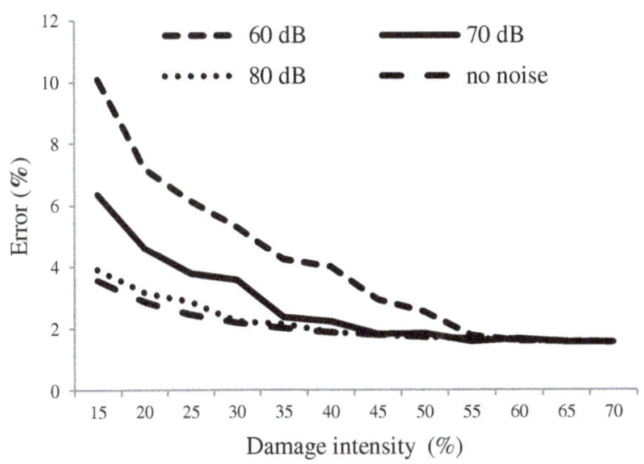

Figure 4 **Variation of percentage error with noise levels 30, 40, and 50 dB.** To find damage intensity averaged over damage location.

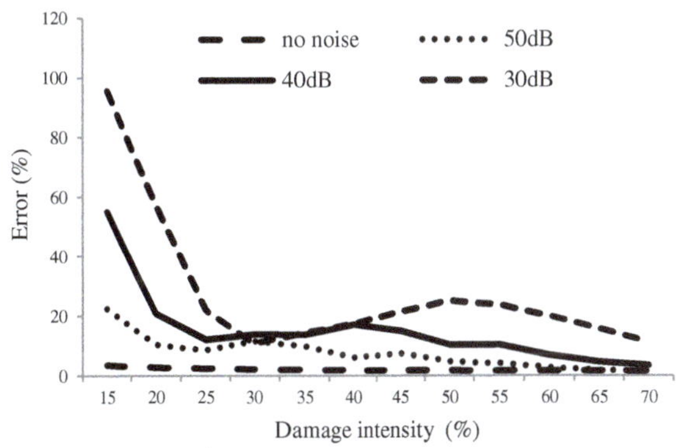

Figure 5 **Variation of percentage error with noise levels 60, 70, and 80 dB.** To find damage intensity averaged over damage location.

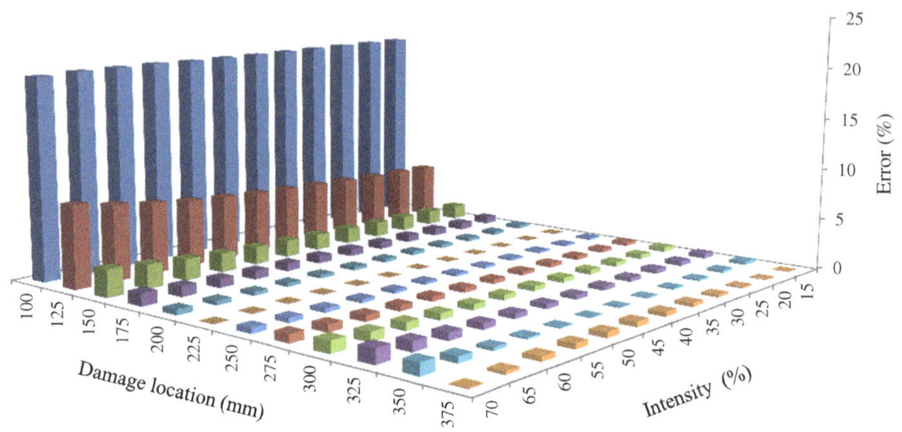

Figure 6 Error in damage location prediction for noise-free data.

locations near both fixed and free ends of the beam is high compared to those in the middle span of the beam.

Noise levels of 80, 70, and 60 dB produce almost the same trend as that of no noise. These error graphs nearly overlap with that of no noise except near the free end (where the error increases up to 5%). Thus, noise up to 60 dB has very little effect on the capability of SVM to predict damage location. As noise level increases from 50 to 30 dB, the error increases considerably. However, all of them show the same trend of low error (below 6%) in the middle span and high error at near both fixed and free ends. The increase in error with increase in noise is very significant near the free end (0.31% to 30%), but not so significant near the fixed end (20% to 31%). Thus, SVM predicts damage locations in a cantilever beam quite well if the damage is in the middle span of the beam. Damage locations near the free end continue to be predicted accurately if the noise level is reasonably low. However, damage locations near the fixed end are difficult to predict even with no noise.

Conclusions

This work attempts to evaluate the capability of support vector machines for predicting the intensity or location of damage for health monitoring of a cantilever beam. The 1,008 simulations have been studied using 12 levels of damage intensities, 12 locations, and 7 noise levels (including no noise). The SVM has been trained with vibration-induced displacements collected at 91 points for the first mode shape as input and damage intensity or location as output using only 90% of the simulations, the rest being kept aside for testing. After training, the SVM is able to predict any damage intensity or location of the training set data with almost zero error. The errors in the test set without noise are less than 2.5% for damage intensity prediction and below 6% for damage location prediction in the central section of the beam. Location prediction degrades drastically near the fixed end of the beam for any level of noise and is inaccurate near the free end only for high levels of noise. On the contrary, damage intensity prediction is significantly

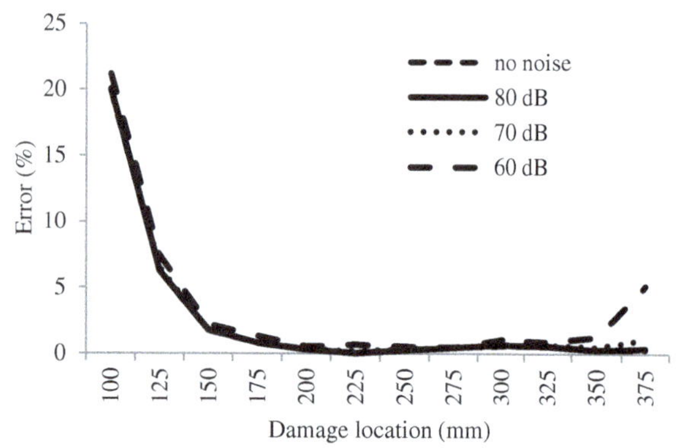

Figure 7 Noise effect on percentage error averaged over intensity (no noise, 50, 40, and 30 dB).

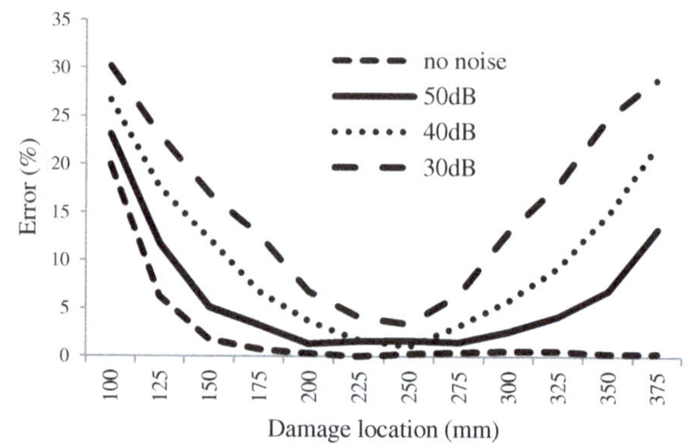

Figure 8 Noise effect on percentage error averaged over intensity (no noise, 80, 70, and 60 dB).

affected by noise for low damage intensity. High noise causes significant deterioration in performance at damage intensities lower than 25%. The worst level of noise (30 dB) results in 11% to 22% error for damage intensities higher than 25% and a much higher error at lower damage intensities.

Competing interests

The authors declare that they have no competing interests.

Authors' contributions

SBS carried out the finite element simulations. YK performed the support vector machine training. AG optimized the performance of support vector machine. SB prepared the finite element model. All authors read and approved the final manuscript.

Author details

[1]Department of Mechanical Engineering, Indian Institute of Technology Bombay, Powai, Mumbai 400076, India. [2]Department of Civil Engineering, Indian Institute of Technology Bombay, Powai, Mumbai 400076, India.

References

Banerjee S, Ricci F, Monaco E, Mal A (2009) A wave propagation and vibration-based approach for damage identification in structural components. Journal of Sound and Vibration 322:167–183

Bulut A, Sing AK, Shin P, Fountain T, Jasso H, Yan L, Elgamal A (2005) Real-time nondestructive structural health monitoring using support vector machines and wavelets. In: SPIE 5770, Advanced Sensor Technologies for Nondestructive Evaluation and Structural Health Monitoring, 180. UC Santa Barbara, Santa Barbara, 13 May 2005

Chendong D, Iyan L, Qiang G (2012) Structure health monitoring using support vector data description and wavelet packet energy distributions. Applied Mechanics and Materials 135–136:930–937

Cherkassky V, Ma V (2004) Practical selection of SVM parameters and noise estimation for SVM regression. Neural Netw 17:113–126

Farrar CR, Worden K (2007) An introduction to structural health monitoring. Phil Trans R Soc A 365:303–315

Gunn S (1998) Support vector machines for classification and regression. ISIS technical report. Image Speech & Intelligent Systems Group. University of Southampton

Kumar SK, Jayabarathi T, Naveen S (2011) Fault identification and location in distribution systems using support vector machines. European Journal Scientific Research 51:53–60

Kurek J, Osowski K (2010) Support vector machine for fault diagnosis of the broken rotor bars of squirrel-cage induction motor. Neural Computation & Application 19:557–564

Loutas TH, Panopoulou A, Roulias D, Kostopoulos V (2012) Intelligent health monitoring of aerospace composite structures based on dynamic strain measurements. Expert Systems with Applications 39:8412–8422

Mal A, Ricci F, Banerjee S, Shih F (2005) A conceptual structural health monitoring system based on vibration and wave propagation. Structural Health Monitoring 4(3):283–293

Shan D, Fu C, Li C (2012) Experimental investigation of damage identification for continuous railway bridges. Journal of Modern Transportation 20:1–9

Vapnik VN (1999) An overview of statistical learning theory. IEEE Trans Neural Networks 10:988–999

Zou Y, Tong L, Steven GP (1999) Vibration-based model-dependent damage (delamination) identification and health monitoring for composite structures—a review. Journal of Sound and vibration 230(2):357–378

Geometrical metallic shell behavior study under compression

Ram Ranjan Sahu[1,3*] and Pramod Kumar Gupta[2]

Abstract

The loading response of shell structures depends on their shapes. The individual shape behaves uniquely to the loading, and different shapes give different load displacement graphs. If these geometries are combined, then their load displacement graphs can be different. It may happen that their combined behavior can be harnessed to be used as better energy absorber in a controlled manner, which they cannot if they are used individually. Experiments were conducted on two separate geometries as well by joining them with weld. The first geometry was the top cylindrical and frusta bottom. The second geometry was the shape of inverted bell crater and half-spherical shells. Combining these two shapes produced the third geometrical shape. The large deformation was obtained by crushing these geometrical shells between two flat plates. The finite element analysis was used to simulate the crush phenomenon. The behavior of the geometrical shell to large deformation was understood with load displacement graph for different samples. The energy absorbed by the different samples was calculated and compared. The parameters, like material and thickness, were varied for the samples to see their effect on the large deformation behavior. Moreover, friction role is discussed on the crush phenomenon. This paper also gives an idea on the different parameters which can affect the energy-absorbing capacity of the combined geometrical shells.

Keywords: Geometrical shell, Crushing, Large deformations, Energy absorption, Load-deflection, Finite element analysis

Introduction

Shell structures are good energy absorbers and have been attracting researchers to study their response for large deformation. In day-to-day life, many types of shell structures are being used in construction. These shell structures could be a combination of many shell geometries. It can be a combination of cylinder, frusta, spherical shells, etc. Many researchers tried to understand these shells under different loading conditions. The collapse behavior of thin spherical shells under quasi-static loading was studied by Gupta et al. (2008). Three-dimensional numerical simulations were carried out for all the specimens tested under quasi-static loading using ANSYS (ANSYS, Inc., Canonsburg, PA, USA). All the stages of collapse of the shell including nonsymmetrical lobe formation were simulated. Material, geometric, and contact nonlinearities were incorporated in the analysis. The obtained results were compared with the experiments. They investigated that the relatively thick shells deform axi symmetrically and the major load is absorbed by the rolling plastic hinges. When the thickness is reduced considerably, the inward dimpling is followed by multiple numbers of nonsymmetrical lobes which are caused by the formation of stationary hinges. Deformations and energy absorption capacity of thin-walled tubes with various section shapes (circular, square, rectangular, hexagonal, triangular, pyramidal, and conical) were investigated both experimentally and numerically by Nia and Hamedani (2010). The tubes with same volume, height, average section area, thickness, and material were subjected under axial quasi-static loading. They found that the section geometry has considerable effect on the energy absorption. The larger the number of section edges, the greater is the energy absorption capacity. This is due to an increase of the number of folds and plastic hinges in sections with larger number of edges. For their test, the absorbed energy per

* Correspondence: RamRanjan.Sahu@tatatechnologies.com
[1]Civil Engineering Department, Indian Institute of Technology Roorkee, Uttarakhand Pin 247667, India
[3]Engineering Research Centre, Tata Motors-Pune, India
Full list of author information is available at the end of the article

unit mass was maximal for cylindrical tubes. Postlethwaite and Mills (1970) studied the frustum and proposed the mean crushing force calculation for external collapse. The failure modes of frusta with experiments were studied by Mamalis and Johnson (1983). They found that the load-deflection curves of the frusta were more regular than those of the cylinders. It was also observed that with the increase in wall thickness, the post-buckling load increased in a parabolic manner. They also observed that the post-buckling load decreased with an increase in semi-apical angle. It was also observed that the thin frusta deformed in diamond shape while the thick ones deformed into axisymmetric rings. Alghamdi (1991) worked on direct inversion and outward flattening for frusta. Abaqus finite element program (Dassault Sytemes, Veilizy-Villacoublay, France) was used by Aljawi and Alghamdi (1999) to simulate the frusta inversion process. They found a good agreement between the finite element (FE) and experimental results. The collapse studies on varying wall thicknesses of the metallic frusta subjected to axial compression were done by Gupta (2008). The mode of collapse forms by the development of one concertina fold followed by the plastic zone. During the development of mode of collapse, some portions of the frusta move radially inward and some radially outward. Experimental investigation for axial compression of thin-walled circular tubes was done by Guillow et al. (2001). Alghamadi (2001) paper reviews the common shapes of collapsible energy absorbers and their different modes of deformations. The typical geometrical shell studied for large deformation behavior is shown in Figure 1.

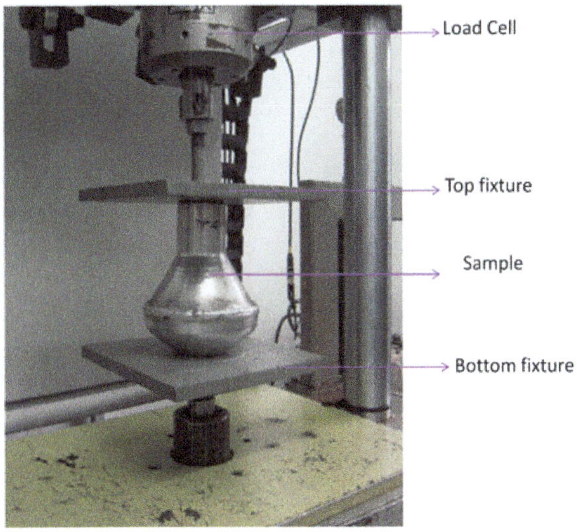

Figure 2 Experimental setup.

Experiments

Experimental setup

The experimental setup consists of rigid top and bottom fixtures and aluminum samples as shown in Figure 2. Aluminum samples are compressed between these two rigid fixtures through Instron Universal Machine (Instron, Norwood, MA, USA).

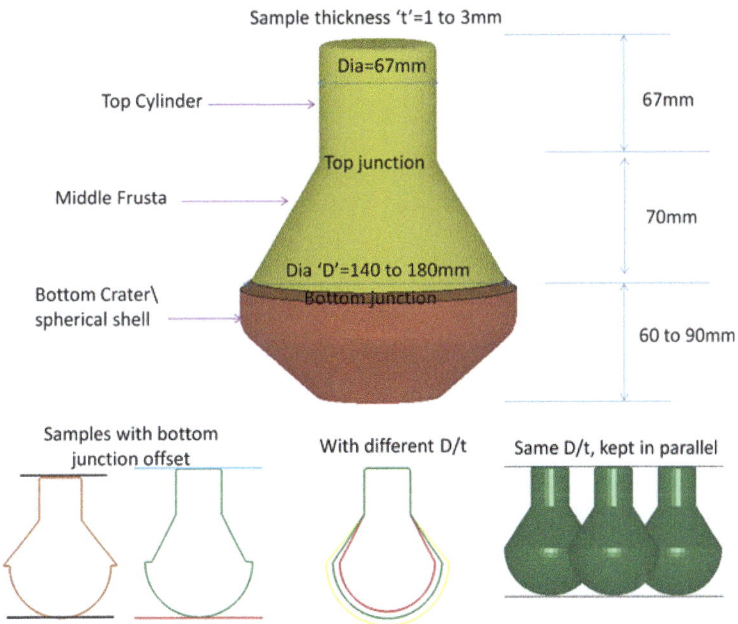

Figure 1 Test samples shape.

Figure 3 Experimental samples.

The aluminum samples were made by spinning process. The top half and bottom half were made separately. The both the halves were joined by aluminum weld. Top and bottom samples are shown in Figure 3 before and after joining.

The typical shape of the top half is a combination of cylinder and frusta. The bottom half is of inverted bell crater shape, and few samples were with spherical shapes. After joining the two halves, junctions are formed, namely, top and bottom junction. The typical dimensions and parametric models are shown in Figure 1.

The top fixture is fixed to load cell of the Instron machine and is stationary. The bottom fixture is attached to moving ram of the machine. The sample is held between these two fixtures. The Instron machine has a maximum capacity up to 4 ton. The load cell is kept below the top stud. The bottom ram moves up and down with the maximum ram stroke up to ± 125 mm. The system is hydraulically operated and can operate at quasi-static and transient load conditions. The ram on which tube is kept was moved up with speed of 5 mm/min to ensure the quasi-static condition. The alignment of the sample and its fixture is assured with the machine axis. The testing was done to the maximum displacement up to 150 mm.

Material properties

The tensile testing of the material was done by preparing test specimens as per ASTM E8 (American Society for Testing and Materials, West Conshohocken, PA, USA) specifications, as shown in Figure 4a. These test specimens were cut from the samples. These test pieces were tested with standard tensile testing machine at room temperature. A specimen after test is shown in Figure 4b. With repeated test, the average value of stress strain is shown in Figure 4c. From the graph,

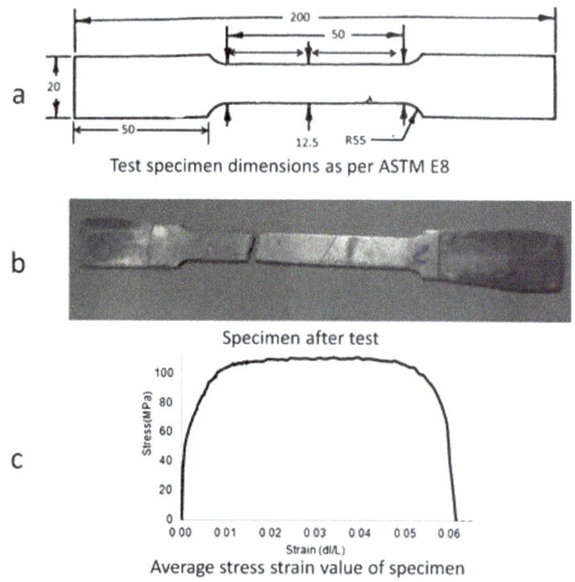

Figure 4 Material testing. (**a**) Test specimen dimensions as per ASTM E8. (**b**) Specimen after test. (**c**) Average stress strain value of specimen.

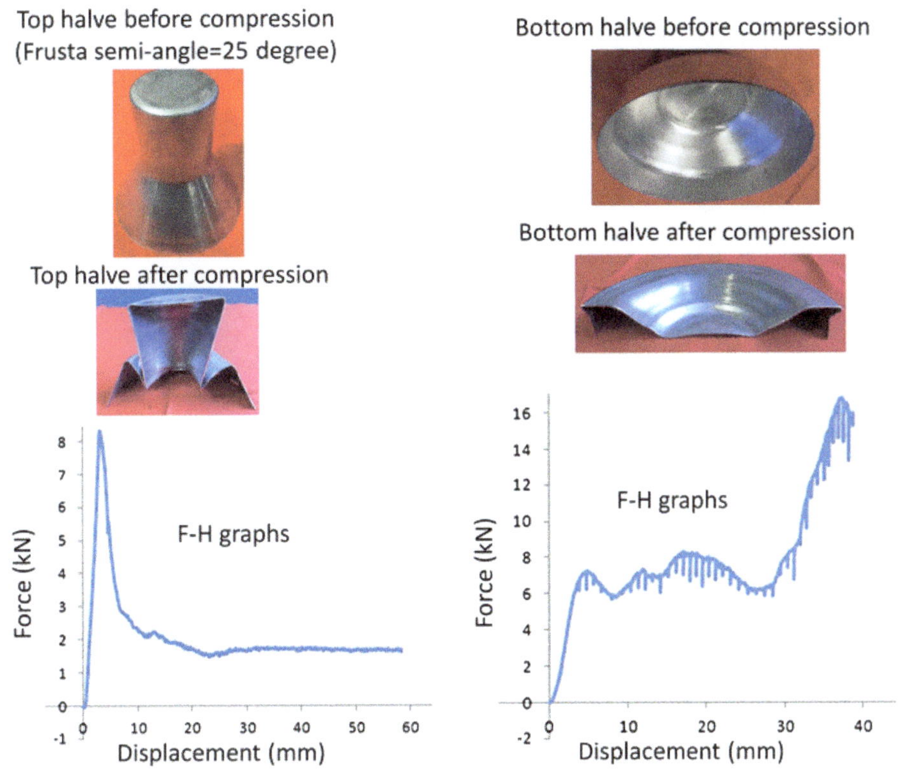

Figure 5 Deflected shape and *F-H* graph of individual halves.

material properties (tensile strength = 108 MPa at 4.6% strain and yield strength = 65 MPa) were extracted as per ASTM E8 standards.

The fixtures were made of steel having 0.2% proof stress of 800 MPa and an elastic modulus of 210 GPa. Since proof stress of fixture is much greater than that of the samples, it can be regarded as rigid.

Experimental results

The experiments were conducted on individual halves as well on combined geometry samples. Figure 5 shows the shape before and after compression and their respective force-displacement (*F-H*) graph. The unit followed for force was in kilonewton (1,000 N) in *y*-axis and displacement in millimeters at *x*-axis. The top half of the sample mainly shows the inward inversion of frusta portion, while inward dimpling is shown for bottom half of the crater. For the energy calculation for the bottom half, the portion with sudden rise in the *F-H* graph is not taken.

The typical behavior of the combined geometrical shell can be explained by Figure 6, where deflected shapes are also shown in different stages of the graph. The *F-H* graph starts from point 1 till it achieves the maximum force to point 2. This is the force requirement for the geometry to start the process of plasticity. This force starts the inward inversion of frusta portion of top half and inward dimpling of bottom half. Once this process starts, the graph drops down and stabilizes and goes up to point 3. After this point, the inverted frusta and bottom dimpled halves contact each other, which resulted into increases in force which is denoted by point 4.

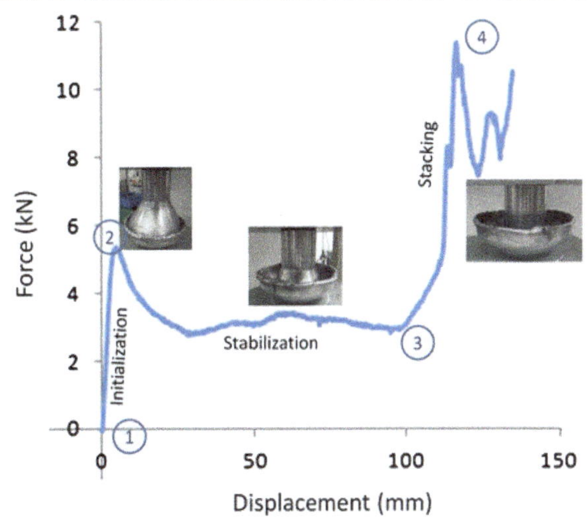

Figure 6 Deflected shape and *F-H* graph of combined geometry.

Table 1 Sample categories and their results

Category	Sample geometry	Sample description	Sample number	Parameter	Stroke (H) (mm)	Energy (kJ)	Weight of the sample (kg)	ω = (kJ/kg)	Average F (kN)
Experimental	Separate geometry	Bottom half only	S1	Shape, bell crater	38.78	0.196	0.092	2.13	5.05
		Top half only	S2	Frusta semi-angle = 25°	58.28	0.123	0.0825	1.49	2.11
	Combined geometry	Frusta semi-angle = 30° (sample 30_1)	S3	Crater bottom half, bottom junction mean diameter = 154 mm	144.59	0.543	0.198	2.74	3.76
		Frusta semi-angle = 30° (sample 30_2)	S4	Crater bottom half, bottom junction mean diameter = 156 mm	140.68	0.699	0.198	3.53	4.97
		Frusta semi-angle = 30° (sample 30_3)	S5	Crater bottom half, bottom junction mean diameter = 155 mm	148.50	0.71	0.198	3.59	4.78
		Frusta semi-angle = 25° (sample 25_1)	S6	Crater bottom half, bottom junction mean diameter = 147 mm	114.50	0.2088	0.191	1.09	1.82
		Frusta semi-angle = 25° (sample 25_2)	S7	Spherical bottom half, bottom junction mean diameter = 140 mm	157.00	0.641	0.202	3.17	4.08
Theoretical (base sample with bottom junction diameter = 160° and t = 1 mm)	Combined geometry	Different offset	S8	Offset_out10	150.00	0.61	0.263	2.32	4.07
			S9	Offset_in10	150.00	0.57	0.236	2.42	3.80
		Different 'D / t'	S10	$D / t = 140$	150.00	0.879	0.198	4.44	5.86
			S11	$D / t = 160$	150.00	0.742	0.235	3.16	4.95
			S12	$D / t = 180$	150.00	0.626	0.277	2.26	4.17
		Different 'μ'	S13	$\mu = 0.1$	150.00	0.742	0.235	3.16	4.95
			S14	$\mu = 0.2$	150.00	0.638	0.235	2.71	4.25
			S15	$\mu = 0.3$	150.00	0.583	0.235	2.48	3.89
		Different 't'	S16	T = 1 mm	150.00	0.742	0.235	3.16	4.95
			S17	T = 2 mm	150.00	2.18	0.47	4.64	14.53
			S18	T = 3 mm	150.00	3.22	0.706	4.56	21.47
		Different 'material'	S19	Tested	150.00	0.742	0.235	3.16	4.95
			S20	Mat 1050 (m1)	150.00	0.483	0.235	2.06	3.22
			S21	Mat 6082 (m2)	150.00	2.11	0.235	8.98	14.07
			S22	Mat 7020 (m3)	150.00	3.01	0.235	12.8	20.07
		Different lattice	S23	Single sample	150.00	0.596	0.194	3.07	3.97
			S24	Three samples in parallel	150.00	1.78	0.584	3.05	11.87

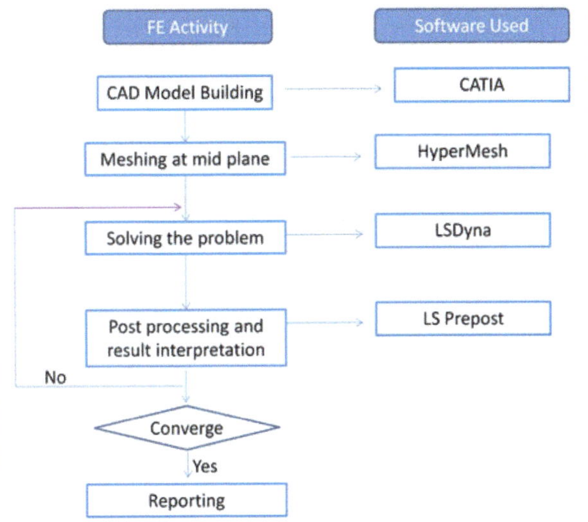

Figure 7 Step for analytical analysis.

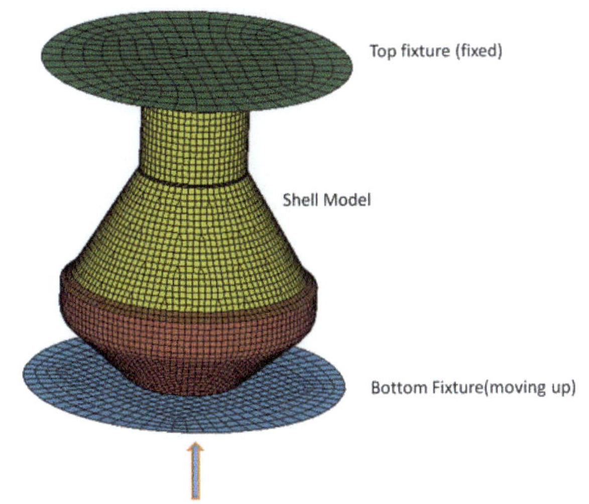

Figure 8 FE model of test simulation.

Table 1 contains the geometrical parameter details, energy contents, specific energy, average force for the sample tested along with the analytical sample results. From the table, it is evident that the specific energy (ω) reduces with the offset and higher D / t ratio at bottom junction. It also reduces with higher coefficient of friction while increases in higher strength material. The average force is observed to be maximal for higher-thickness and higher-strength materials.

Finite element simulation

The analytical approach for test simulation is discussed below. The step followed for analytical analysis is shown in Figure 7.

The sample computer-aided design (CAD) models were made in CATIA software (Dassault Sytemes, Veilizy-Villacoublay, France). Once the CAD models are ready, their middle surface can be extracted for FE model building. HyperMesh (2011) was used for FE model building. Aluminum samples were represented with shell elements at their middle surface. Triad and quad shell elements were used for FE model building. The element sizes chosen were fine enough to fully represent the details of geometry, like small curved fillets. The top and bottom fixtures were represented with rigid elements. The LsDyna (Livermore Software Technology Corporation 2003) explicit solver was used to solve the problem. The material models *MAT_RIGID and *MAT_PIECEWISE_LINEAR_PLASTICITY were used for fixtures and samples, respectively. The contact type *CONTACT_AUTOMATIC_SURFACE_TO_SURFACE was used to define contacts between disjoint parts. Also, for self-contact, *CONTACT_AUTOMATIC_SINGLE_SURFACE was used. Coulomb friction type was used to define the coefficient of friction between samples and fixtures. Shell element formulation chosen was Belytschko-Tsay because of its less computation cost with good accuracy. FE model fully represents the sample and simulates the compression process. Result interpretation was done through LS-PREPOST (Livermore Software Technology Corporation 2007). Figure 8 shows the FE shell model of test simulation.

Theoretical analysis confidence on correct simulation to test was established and with this confidence, the various parametric theoretical samples were prepared. The base sample diameter at bottom junction for theoretical models is kept at 160 mm for all types of parametric models except for samples in which the D / t values are 140 and 160. The parameters could be offset at the bottom junction, different D / t, different coefficients of friction, different thicknesses, different materials, samples lattice, etc. These theoretical samples are enlisted in Table 1 and their results are discussed.

The tested sample S3 was simulated with FE as shown in Figure 9. The deflected shape and their F-H graph compared well. This shows the FE assumptions and material test data taken in analysis are correct.

The energy balance graph of FE simulation for sample S3 is shown in Figure 10. The undesired energies like kinetic energy, hourglass energy, and sliding energy are minimal (less than 10% of the total energy) and hence, the total energy is contributed only through sample internal energy which is caused by the deformation. The graph represents a perfect energy balance of simulation and hence ensures the correctness of simulation and convergence. Also, this

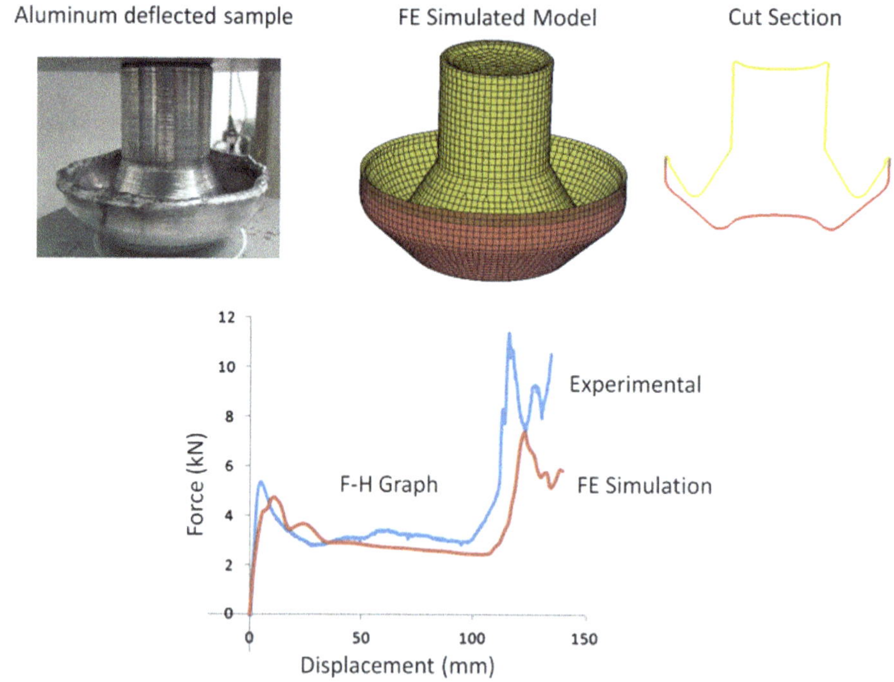

Figure 9 FE and test result comparison.

energy balance graph gives confidence on the element type chosen and on its formulation.

Results and discussion

The FE-simulated *F-H* graph and von Mises stress contour for sample S3 at different deformation stages are shown in Figure 11. The stress contour in red colored which indicates the plasticity of metal, is depicted for rolling hinges formed due to inward inversion of frusta portion at *H* = 70 mm. Contours at the last stage of deformation is also shown where the top cylindrical portion has undergone plasticity. Hence, by plotting stress contour at different deformation stages, the relative plasticity of the different portions of the sample can be assessed.

The *F-H* graph and cut section of sample S8 (outward 10 mm offset) is shown in Figure 12. The initial rising portion of the graph is due to the inward dimpling of bottom spherical shell which continues up to *H* = 45 mm. Further enforced displacement starts necking, and sudden drop in *F-H* graph is noticed. This necking initiates the inward inversion of conical portion until it reaches *H* = 125 mm. Afterwards, the bottom spherical and inward inverted portions touch each other, leading to a sudden jump in effort. This shows that the outward offset at bottom junction

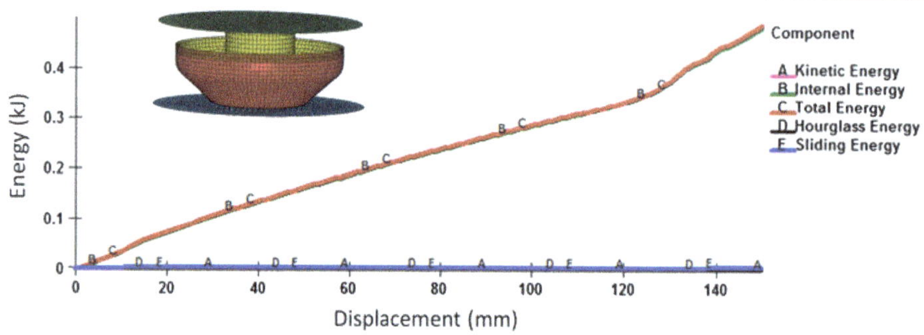

Figure 10 Energy balance of FE simulation.

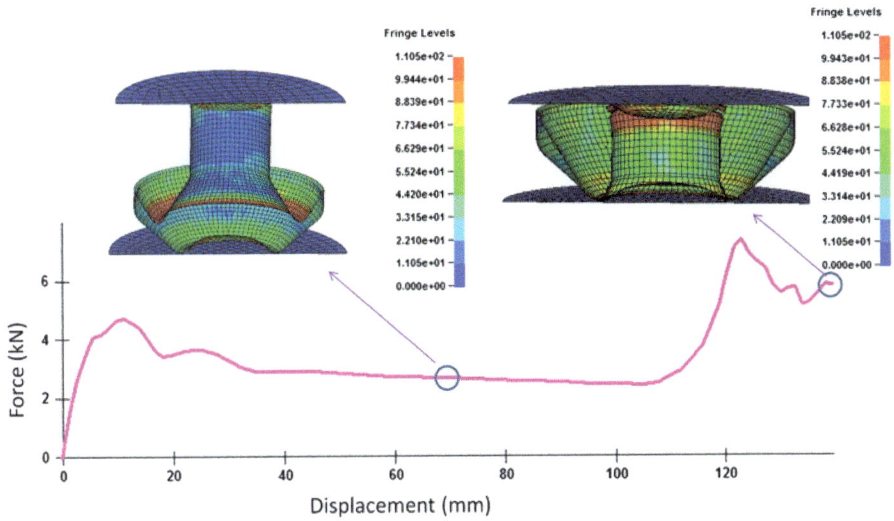

Figure 11 FE-simulated *F-H* graph and stress contour.

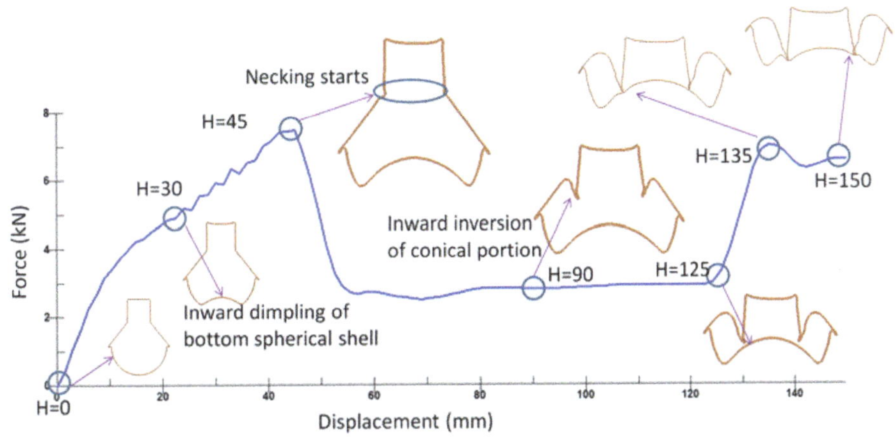

Figure 12 *F-H* graph and deflected cut section for sample S8.

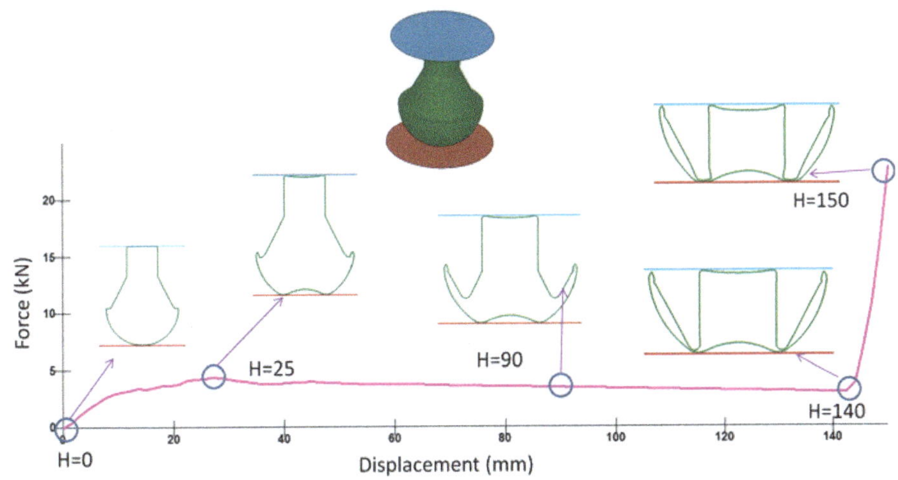

Figure 13 *F-H* graph and deflected cut section for sample S9.

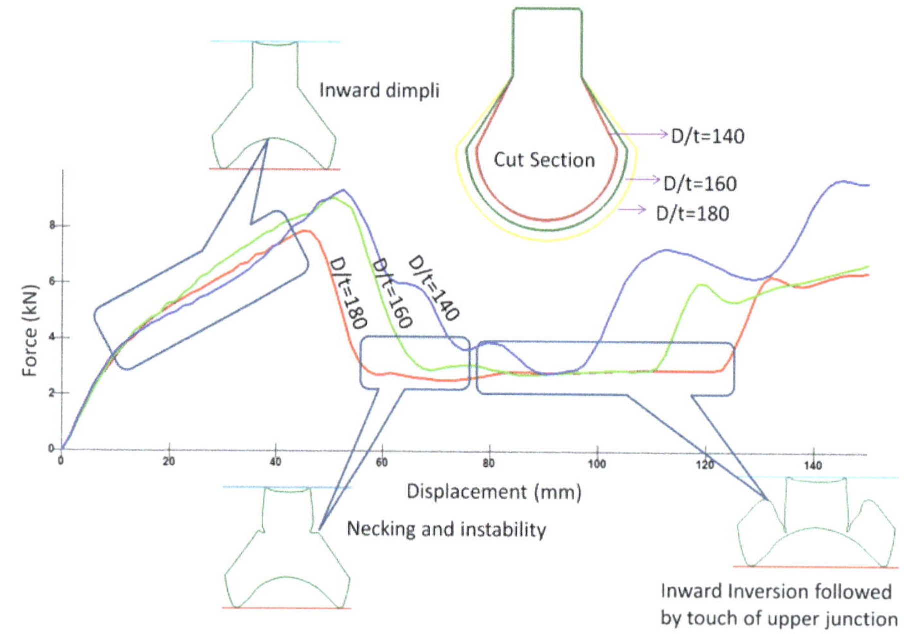

Figure 14 *F-H* graph for different *D* / *t* samples.

leads to inward dimpling first, followed by necking and inward inversion of conical portion.

The *F-H* graph and deflected cut section at different stages of deflection for sample S9 (inward 10 mm offset) are shown in Figure 13. The initial rising portion of the graph is due to the inward dimpling of bottom spherical shell and setting plasticity at the offset which happens at H = 25 mm. Afterwards, the conical portion is inverted from its bottom to the upper portion. In this process the rolling hinges volume decreases and hence the *F-H* graph trend declined. This inversion continues until point reaches H = 140 mm. Afterwards, the bottom of the top cylindrical portion touches to dimpled spherical portion, causing rise in the effort. In this whole process, the bottom spherical and top cylindrical portions are not utilized effectively. The inward dimpling at the bottom junction provokes the inversion, and this featured phenomenon could be used in large deformation applications.

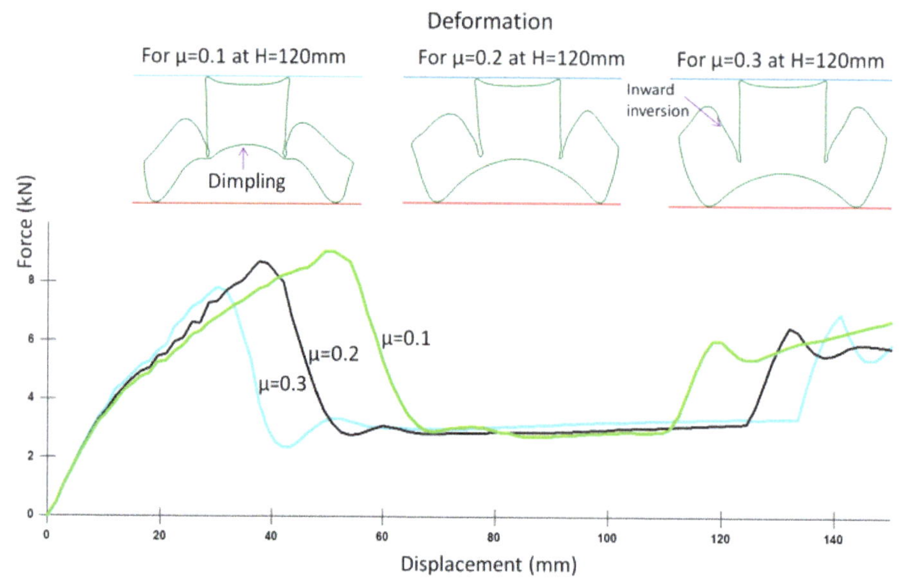

Figure 15 Effect of coefficient of friction on *F-H* graph.

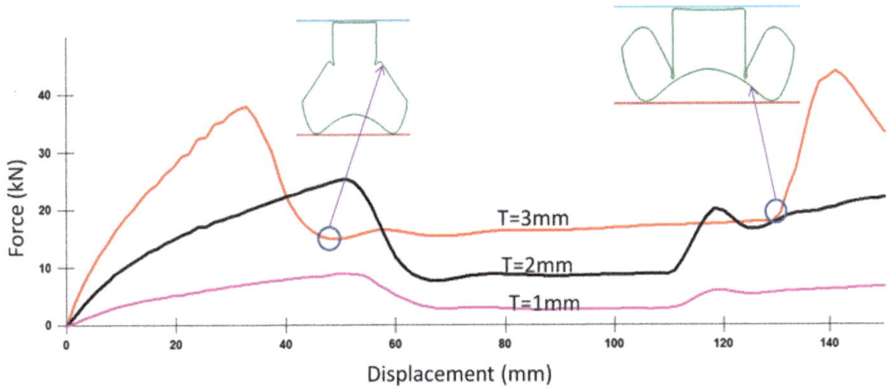

Figure 16 Thickness effect on *F-H* graph.

Three samples (S10, S11, and S12) were made with $D\ /\ t$ ratios at bottom junction as 140, 160, and 180. The superimposed graphs of these three samples are shown in Figure 14. The graph for all three samples rises to approximately 50 mm of stroke, which is mainly due to inward dimpling of bottom spherical shell. After further application of the enforced bottom plate motion, the necking starts at the top junction (also some instability occurs at the bottom junction for $D\ /\ t = 140$) which leads to a sudden drop in *F-H* graph. Afterwards, the inward inversion of the middle frusta starts. On further displacement of the bottom plate, the top junction touches to the dimpled bottom spherical shell and, hence, the force suddenly increases. The bottom junction continues to press the bottom spherical shell until it reaches $H = 150$ mm. The wide stretch in the graph is noticed for inward inversion of middle frusta for $D\ /\ t = 180$ while it is less for $D\ /\ t = 140$. The higher $D\ /\ t$ ratio renders more inward inversion of the middle frusta while the lesser D / t ratio in the sample does not; however, instability is noticed in the frusta portion.

The effect of coefficient of friction μ between contacts was studied on the samples S13, S14, and S15 who have the same shape with $D\ /\ t = 160$ at the bottom junction. Figure 15 shows superimposed *F-H* graph for samples of $\mu = 0.1$, 0.2, and 0.3. It is clear from the figure that at higher coefficient of friction, the inward dimpling is reduced due to the higher friction at the bottom, which invokes the early start of inward inversion of the middle frusta. With lesser value of coefficient of friction, it is reversed where inward dimpling is allowed more than the frusta inward inversion.

The effect of thickness was studied on samples S16, S17, and S18, and the *F-H* graph of these samples is shown in Figure 16. The samples were with mean diameter of 160° at the bottom junction and their thicknesses were varied from 1 to 3 mm. The larger thickness has difficulty in inward dimpling, and hence early necking followed by inversion starts. Therefore, the higher thickness sample shows large stretch for inward inversion of the frusta portion. Also, higher thickness yields more energy (see Table 1).

Three types of aluminum alloy along with tested materials have been used, i.e., high, medium, and low strengths to see the material effect on the deformation pattern. The properties are enlisted in Table 2.

The examples of high strength alloy are 7xxx, e.g., 7020 and 7075, and they are comparable to high-strength steel. The medium strength alloy is 6xxx series; for example, 6082. Their strength is comparable to mild steel. The low-strength aluminum alloy could be in the pure form of aluminum, for example, 1050A and 1050-O. The geometrical sample for this study was taken as $D\ /\ t = 160$ at bottom junction and these were assigned with four types of material properties. The samples are S19, S20, S21, and S22 as shown in Table 1. The shape of *F-H* graph along with the deformation pattern at 50% and 100% of compression is shown in Figure 17. The material having lower strength facilitates the inward inversion of frusta portion, while it shows instability for

Table 2 Material properties

Material name	Properties		
	Tensile strength (MPa)	Yield strength (MPa)	Elongation (%)
Tested	108	65	4.6
m1 (1050)	130	60	20
m2 (6082)	340	310	11
m3 (7020)	430	420	11

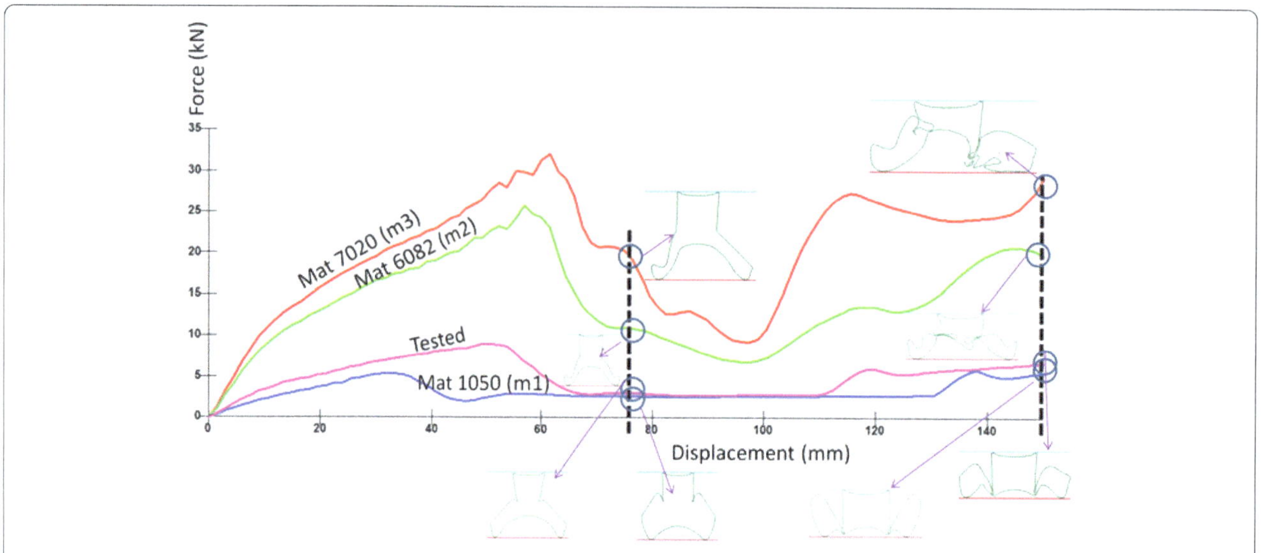

Figure 17 Effect of material on *F-H* graph.

samples with higher strength materials. Also looking at the last stage deformations, the pattern is regular for samples of lower strength materials, while it is of irregular pattern for samples with strength materials.

The energy (kJ) bar chart for samples of different materials are plotted in Figure 18, for 7020, 6082, tested, and 1050 materials whose yield values are highest to lowest, respectively. The 7020 alloy has the maximum yield value as compared to all and which gives the highest value of energy in its sample, while 1050 gives the least. Hence, it can be said that the material having higher yield value will have more deformation energy keeping other the same condition.

The samples S23 and S24 were prepared to study the effect of samples lattice on load displacement pattern. The sample S23 was a single sample (D / t = 160), while S24 has three numbers of S23 kept in parallel.

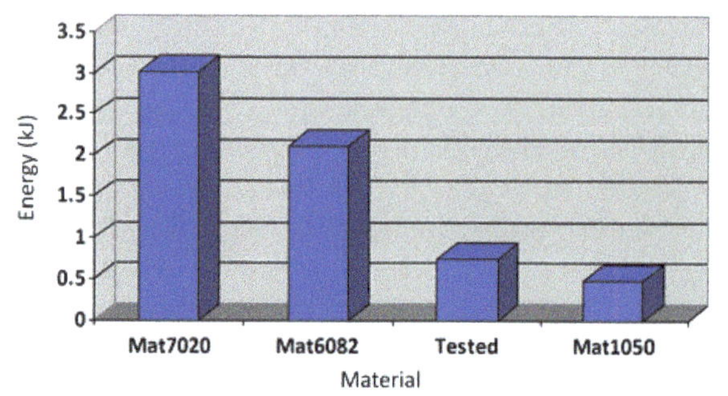

Figure 18 Energy comparison of the samples with different materials.

The effect could be clearly seen in Figure 19, where the effort is linear multiplication of single sample, for higher numbers of samples kept in parallel. Hence, to absorb more energy, one can keep the samples in parallel lattice in desired numbers. The number can be decided by total energy absorption divided by the energy absorption of a single sample.

Conclusion

The detailed experimental and computational studies on the metallic geometrical shape were done. It is learnt that a geometrical shape made up of a combination of the different basic shapes can be utilized for controlled energy absorption. The geometrical parameters of this shape affect their response to loading and their deformation pattern. For example, the less D / t ratio at bottom junction may invoke instability (Figure 14). Moreover, the coefficient of friction plays an important role in this kind of deformation phenomenon. The higher coefficient of friction can be used for the early start of inward inversion in the frusta portion and can prevent inward dimpling of the bottom spherical portion. With higher thickness, the overall energy increases. In addition, the higher thickness invokes early inversion of frusta and stretches this zone (Figure 16).

The simulation was perfected by getting the perfect energy balance equation. The flexibility to model various types of parametric samples with simulation software was utilized. The effect of geometrical, materials, thickness, and other parameters could be studied with these analytical models.

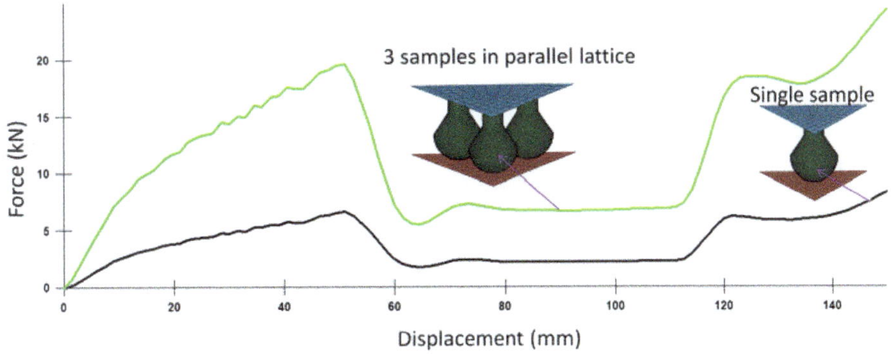

Figure 19 Energy comparison of samples of different lattices.

Competing interests
The authors declare that they have no competing interests.

Acknowledgments
The authors gratefully acknowledge the support of Engineering Research Centre of Tata Motors at Pune-India, for providing the necessary support for simulation and solving the problems. Author also thank the craftsmen of proto shop of Tata Motors for shaping and welding such complicated job. Also thanks to the CAD team for making the asked geometry in CATIA software from where the FE model could be built.

Author details
[1]Civil Engineering Department, Indian Institute of Technology Roorkee, Uttarakhand Pin 247667, India. [2]Department of Civil Engineering, Indian Institute of Technology Roorkee, Uttarakhand Pin 247667, India. [3]Engineering Research Centre, Tata Motors-Pune, India.

References
Alghamdi AAA (1991) Design of simple collapsible energy absorber. King Abdulaziz University, Thesis

Alghamdi AAA (2001) Collapsible impact energy absorbers: an overview. Thin-Walled Struc 3(2):189–213

Aljawi AAN, Alghamdi AAA (1999) Investigation of axially compressed frusta as impact energy absorbers. In: Gaul L, Brebbia AA (eds) Computational methods in contact mechanics IV. WIT, Southampton, pp 431–43

Guillow SR, Lu G, Grzebieta RH (2001) Quasi-static axial compression of thin-walled circular aluminium tubes. Int J Mech Sci 43(9):2103–2123

Gupta NK, Mohamed SN, Velmurugan R (2008) Experimental and theoretical studies on buckling of thin spherical shells under axial loads. Int J Mech Sci 50:422–432

Gupta PK (2008) A study on mode of collapse of varying wall thickness metallic frusta subjected to axial compression. Thin-Walled Struc 46:561–571

HyperMesh11 software (2011) A product of Altair Engineering HyperWorks, for fininite element pre-processing. www.altairhyperworks.com, 1820 E Big beavers, Troy, MI 48083

Livermore Software Technology Corporation (2003) LsDyna A program for nonlinear dynamic analysis of structures in three dimension, software and its user manual. Livermore Software Technology Corporation pp 94550–1740: V970

Livermore Software Technology Corporation (2007) LS-PrePost - an advanced pre/post processor for LS-Dyna. Livermore Software Technology Corporation, Livermore, California, pp 94550–1740, V2.1

Mamalis W, Johnson W (1983) The quasi-static crumpling of thin-walled circular cylinders and frusta under axial compression. Int J Mech Sci 25(9/10):713–32

Nia AA, Hamedani JH (2010) Comparative analysis of energy absorption and deformations of thin walled tubes with various section geometries. Thin-Walled Struc 48(12):946–954

Postlethwaite HE, Mills B (1970) Use of collapsible structural elements as impact isolators, with special reference to automotive applications. J Strain Anal 5(1):58–73

18

Minimum depth of soil cover above long-span soil-steel railway bridges

Morteza Esmaeili[1*], Jabbar Ali Zakeri[1] and Parisa Haji Abdulrazagh[2]

Abstract

Recently, soil-steel bridges have become more commonly used as railway-highway crossings because of their economical advantages and short construction period compared with traditional bridges. The currently developed formula for determining the minimum depth of covers by existing codes is typically based on vehicle loads and non-stiffened panels and takes into consideration the geometrical shape of the metal structure to avoid the failure of soil cover above a soil-steel bridge. The effects of spans larger than 8 m or more stiffened panels due to railway loads that maintain a safe railway track have not been accounted for in the minimum cover formulas and are the subject of this paper. For this study, two-dimensional finite element (FE) analyses of four low-profile arches and four box culverts with spans larger than 8 m were performed to develop new patterns for the minimum depth of soil cover by considering the serviceability criterion of the railway track. Using the least-squares method, new formulas were then developed for low-profile arches and box culverts and were compared with Canadian Highway Bridge Design Code formulas. Finally, a series of three-dimensional (3D) finite element FE analyses were carried out to control the out-of-plane buckling in the steel plates due to the 3D pattern of train loads. The results show that the out-of-plane bending does not control the buckling behavior of the steel plates, so the proposed equations for minimum depth of cover can be appropriately used for practical purposes.

Keywords: Soil-steel bridge, Railway bridges, Large span, Finite element analysis, Minimum cover depth, Serviceability criterion

Introduction

Soil-steel bridges are categorized as composite structures and are composed of corrugated steel plates buried in engineered soil materials. One of the failure modes for these bridges can be initiated by shear or tension failure in the soil cover and can result in structural buckling due to an inadequate depth of soil above the metal structure. This mode of failure is often avoided in the design codes by specifying a minimum depth of soil cover.

The minimum depth of soil cover that has been used for soil-steel railway bridges by various codes such as the California Transportation System (*S*/5) CALTRANS (2000), British Design manual (*S*/5) DMRB (2001), American Iron and Steel Institute (*S*/4) CSPI and AISI (2002), and American Association of State Highway and Transportation Officials (*S*/5) (AASHTO Highway Bridges 2000) was originally empirical and defined by a fraction of the bridge span (*S*). Additionally, ASTM A796 introduced the minimum depth of cover, $H_{\min}$, as $H_{\min} = 0.55S\sqrt{(\mathrm{AL})d/EI}$, which depends on the *S*, axle load (AL), eccentric distance (*d*), elastic modulus (*E*), and the moment of inertia (*I*) of the corrugated steel plates (ASTM 1982). The Australian Rail Track Corporation (ARTC 2005) also recommended specific values for the minimum depth of cover for different classes of railway tracks.

However, in the OHBDC (1992), the empirical formulas for the depth of cover were modified based on the results of a finite element (FE) analysis that considered the geometric shape of the metal structure and the axle load of a truck (Hafez and Abdel-Sayed 1983). Furthermore, the minimum required depth of cover $H_{\min}$ in the third edition of the OHBDC (1992) and in the current Canadian Highway Bridge Design Code (CHBDC 2006) has been specified as the greater of 0.6 m or $D_h/6\ ((D_h/D_v)^{0.5})$ with a maximum of 1.5 m, where D_h and D_v are

* Correspondence: m_esmaeili@iust.ac.ir
[1]School of Railway Engineering, Iran University of Science and Technology, University Street, Hengam Aveneu, Narmak, Tehran 1684613114, Iran
Full list of author information is available at the end of the article

the effective span and rise of the metal structure, respectively CSPI and AISI (2002; CHBDC 2006). This empirical-based formula is commonly used to calculate the depth of cover for highway and railway bridges (Abdel-Sayed and Salib 2002; Hafez 1981). An analytical formula has not yet been established specifically for the minimum cover depth of railway bridges although several structural studies have been done on dynamic and static behaviors of soil-steel structures (Flener et al. 2005; Flener and Karoumi 2009; Manko and Beben 2005; Manko and Beben 2008).

The minimum depth of cover criterion was fundamentally developed to avoid problems associated with soil failure above the crown of a soil-steel bridge (Mohammed et al. 2002). Although this criterion may be sufficient for the serviceability of highway bridges, some revisions may be essential for railway bridges (due to heavier axle loads and relevant dynamic effects), especially for long-span conditions (Peck and Peck 1984). On the other hand, the serviceability and the riding comfort criteria for railway bridges have remarkable importance compared with highway bridges. From the structural design aspect of long-span railway bridges, the critical sections must be stiffened either by stiffening metal sheets or by utilizing sandwich panels as concrete-filled metal sections. The serviceability criteria of the soil-steel railway bridges with spans larger than 8 m that use stiffened panels have not been previously examined and are the main subject of this paper. Therefore, for this study, the CHBDC (2006) equations for minimum cover depth were modified to define the minimum required cover depth for railway bridges based on the two-dimensional (2D) and three-dimensional (3D) FE analysis results using the well-known PLAXIS codes for the aforementioned cases.

Regarding the special serviceability criteria for railway bridges, for computing the minimum depth of cover, the permissible settlement of the railway track and the buckling of the conduit walls in different sections along the longitudinal axis of the bridges in 3D analyses were controlled. For these cases, the spans from 8.07 up to 13.46 m of box culverts and 14.13 up to 23.40 m of low-profile arches using stiffened and non-stiffened deep corrugated panels have been considered.

This study assumes that all the metal structures are buried in well-graded gravel (GW) as engineered backfill material. The material nonlinearity of the soil and metal structure as well as the stage construction effects were accounted for in the numerical analyses, and the railway load model LM71 CEN (2002) has been applied.

The purpose of this study is to reexamine the equations of minimum cover depth of soil-steel railway bridges based on best-fitted curves of the numerical results of the 2D FE analyses. Thereafter, the results are compared with the values obtained from the CHBDC (2006) method.

Methods

The primary subject of the current study is to introduce a new set of minimum soil cover equations for long-span railway bridges based on the numerical interpolation of the results of the 2D FE analyses. These equations relate the minimum depth of cover to the relative stiffness of backfill and the culvert structures and effective span of the railway bridges.

To determine the variation trend of the minimum depth of cover along with its governing parameters (geometry, length of span, and the panel stiffness) through numerical analyses, the permissible settlement of the track, metal structure buckling, and soil body failure criteria have been checked initially for each bridge structure for a 0.6-m depth of cover (the minimum limit of cover depth specified by CHBDC). When all of the defined criteria have not been fulfilled simultaneously, the depth of soil cover above the crown was increased, and the analyses were then restarted for a new depth of cover. The minimum depth of cover in which all of the criteria were simultaneously fulfilled was chosen as the *minimum depth of soil cover* for a specific bridge structure.

In this manner, the results of the 2D FE analyses present specific patterns for the calculation of the minimum depth of cover for box culverts and low-profile arches. In order to check the applicability of the proposed equations for minimum depth of cover in practical problems, a series of 3D finite element analyses with more realistic idealization of the railway superstructure components and the lateral slope of bridge embankment were carried out to control the out-of-plane buckling in the steel plates.

The concept of using 2D and 3D FE analyses to determine the pattern of the minimum depth of cover for soil-steel railway bridges is shown in the flowchart of Figure 1.

Procedure of the numerical modeling

In this section, the numerical modeling specifications and the assumptions used in the analyses will be explained in detail. This explanation details the various aspects of numerical modeling such as the model geometry and boundary conditions, material properties, loading pattern, and monitoring of the desired results.

Model geometry and boundary conditions

Because the box and low-profile arch geometries of soil-steel railway bridges are commonly used for road or highway underpasses Planning and Budget Organization (1985), for this study, the geometry with four various spans were selected for the 2D and 3D analyses. The goal was to model the actual shape of the soil-steel railway bridges. Detailed dimensions of the bridges are

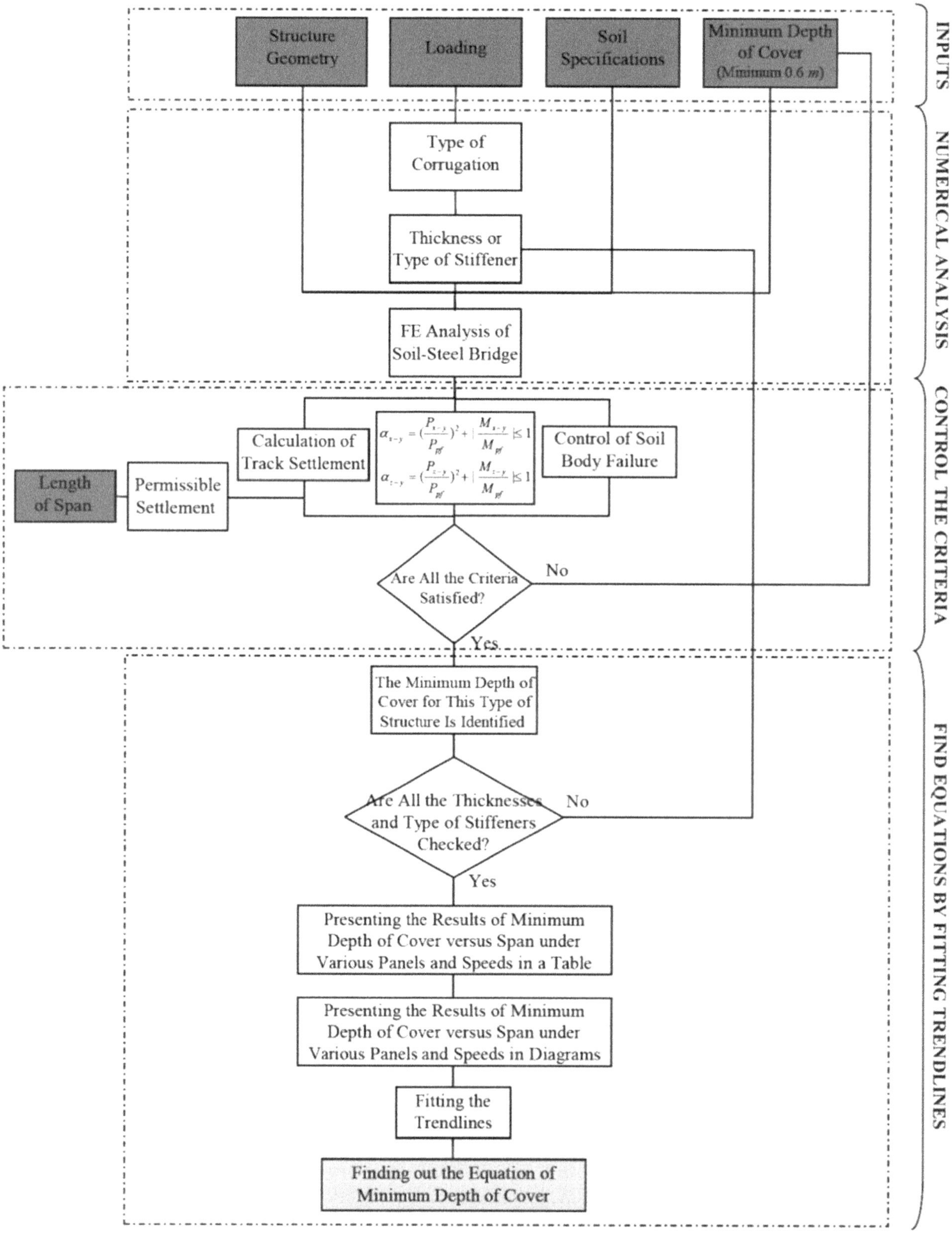

Figure 1 The procedure of finding the pattern of minimum covers depth for soil-steel railway bridges.

shown in Table 1 and Figure 2 Planning and Budget Organization (1985).

Two-dimensional idealization

Two-dimensional idealization was initially used wherein a transverse slice of unit length of the structure was assumed to be in a state of plane strain. In a plane-strain idealization, deformations perpendicular to the plane of the idealized structure are assumed to be zero.

The continuum media around the structure were extended to a triple distance of the outer side of the foundations in width for each side of the structure and four times of the total height of the structure by considering the minimum depth of cover beneath the structure. Standard fixities were applied as boundaries to the outer surfaces of the continuum media (see Figure 3). The vertical displacements are not fixed in vertical boundaries, but all displacements and rotations are fixed in the bottom boundary. These boundaries were specified after several analyses to minimize the effects of the boundaries on the results and geostatic pressures.

Because a large number of calculations (considering geometry and panel types, different depth of cover, axle loads, and loading patterns) were required to determine the preliminary new pattern of the minimum depth of cover, a 2D idealization was initially used to reduce the computing power and time required to perform the analyses. A 3D idealization was then used to control the out-of-plane buckling in the corrugated steel plates, to check the applicability of the 2D-based equations of minimum cover in practical 3D problems.

Three-dimensional idealization

As in practice, the train loading pattern has a 3D nature, so the derived values of minimum depth of soil cover based on 2D finite element analyses should be rechecked to consider the relevant 3D effects. These effects remarkably relates to the realistic load distribution pattern in railway superstructure as well as substructure components and the out-of-plane buckling modes of the steel plates. Consequently, the 3D numerical modeling of the selected structures was taken into consideration.

To reduce the computational costs in the 3D idealization, only half of the structure was modeled, which takes advantage of the symmetry of the railway cross section and loading pattern. Considering the assumed depth of the soil cover (1.5 m of covering material including the ballast and embankment layers) and a slope of 1 vertically to 1.5 horizontally of the ballast shoulders, the longitudinal dimension of each bridge was modeled. The 3D mesh was created using railway cross-sectional segments along the longitudinal axis of each structure.

For the 3D model, the boundaries were defined as 60 m in width from each side of the centerline of the structure and 60 m in depth from the underneath of foundation (see Figure 3). As in the 2D idealization, standard fixtures were applied as boundaries to the outer surfaces of the continuum media, and the boundaries were specified after several analyses to minimize the effects of the boundaries on the results and geostatic pressures.

The sequence of placing the engineering fill layers was presented in the 2D and 3D models. Therefore, each layer of the FE mesh was constructed sequentially. Because the actual number of layers employed in the placement of the backfill in a typical soil-steel bridge is too large to be conveniently used in the FE analyses, a smaller number of layers were used in the simulation of the construction process Abdel-Sayed et al. (1993).

Types of elements

To analyze the soil-steel railway bridge as a soil-structure interaction problem, two types of elements were defined in the solution process. The first group that deals with the structural elements such as the conduit

Table 1 Categories of the dimensions and use of the clearance boxes under low-profile arches and box culverts

	Box culvert	Low-profile arch
Span × Rise (Use of the clearance box under the structure)	8.07 m × 3.06 m	14.13 m × 5.37 m
	(Common for two-line local roads with minimum width)	(Common for two-line roads without considering the future development with ideal width)
	10.51 m × 3.55 m	16.52 m × 5.33 m
	(Common for two-line local roads with ideal width)	(Common for two-line main roads with considering the future development with minimum width)
	11.02 m × 3.24 m	20.95 m × 6.64 m
	(Common for two-line roads without considering the future development with minimum width)	(Common for developed four-line main road with minimum width)
	13.46 m × 3.49 m	23.40 m × 6.44 m
	(Common for two-line roads without considering the future development with ideal width)	(Common for developed four-line main road with ideal width)

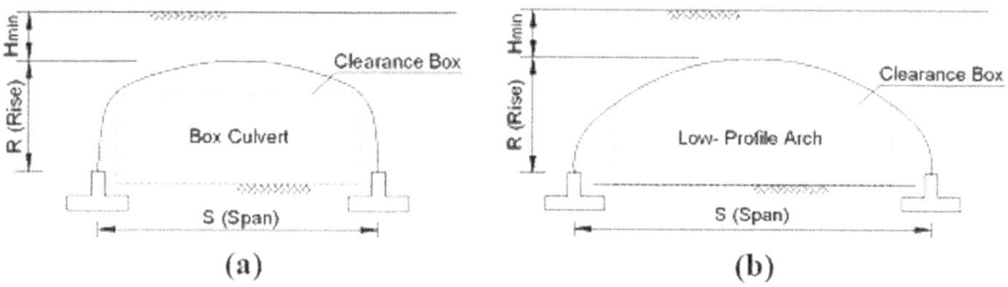

Figure 2 Clearance boxes under (a) box culverts and (b) low-profile arches.

walls and foundation system was introduced as beam elements and plate elements for the 2D and 3D FE modeling. The rails, which were idealized only in the 3D FE analysis, were modeled using plate elements.

For the second group that was used to idealize the engineered backfill, natural ground, and ballast material, 15-node triangular isotropic elements were used in the 2D analysis to have more accuracy results in the evaluation of stress and strain. On the other hand, for the 3D analysis, 15-node wedged elements were used to model the soil environment as well as the ballast and sleepers.

Full-bonded conditions are considered between materials in 2D and 3D FE analyses which cause more conservative results due to greater stresses induced in metal structures. On the other hand, the studies performed by Duncan (1979), MacDonald (2010), and Peterson et al. (2010) demonstrated that, in most cases, the effects of slip between the metal structure and the backfill is ignorable.

Material properties

Because of the serviceability and the riding comfort criteria for railway bridges, the critical sections are often stiffened either by stiffening metal sheets or by utilizing sandwich panels as concrete-filled metal sections. Because of obvious difference between the geometrical shape of the box and low-profile arch (LPA) types of the bridges, the initial analysis results show that for box and LPA, the critical sections are in the haunches and in the midspan, respectively.

The rigidity effects of the panels on the minimum depth of cover were examined using non-stiffened and stiffened panels of deep corrugations (Figure 4) for each of the structures. The panels were idealized as elasto-plastic beams or plates for the 2D and 3D FE idealizations. For the special case of the composite panel type VI, the cracked moment of inertia of the section was considered in the numerical calculations. The equivalent parameters of the beam which represents the corrugated profile are calculated based on the model of El-Sawy (2003). A summary of the parameters such as the axial stiffness (*EA*), the flexural rigidity (*EI*), the Poisson's ratio (ν), the plastic moment capacity (M_p), and the compressive strength (P_p) for various types of panels are shown in Table 2 (Abdulrazagh 2009).

The behavior of the soil-steel bridge foundation and pedestals was idealized as elastic beams for the 2D analyses and as elastic plates for the 3D FE analyses, which

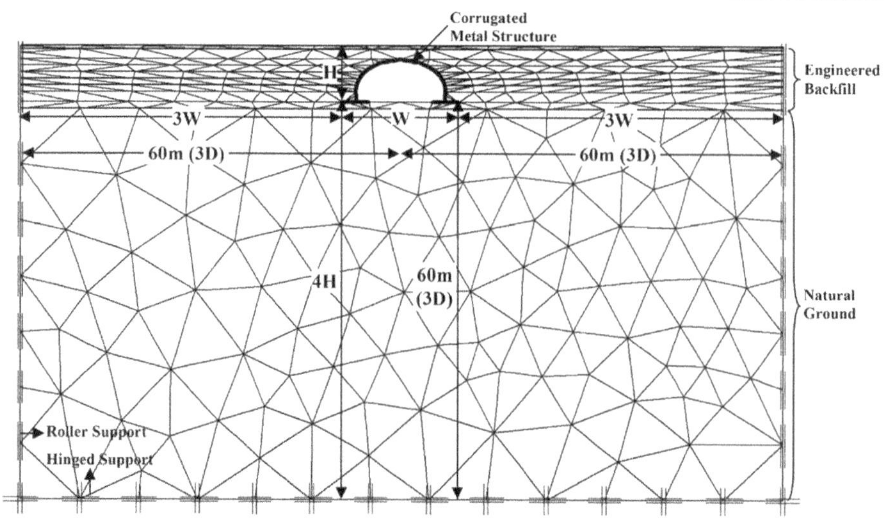

Figure 3 Finite element models (2D and 3D).

was used to maintain the elastic characteristics under load. This was assumed because the initial studies on the behavior of the foundation system demonstrated that, for the design specifications of the foundation system and under the conditions used for this study, the foundation and pedestals did not exhibit a considerable amount of nonlinear behavior. Therefore, to reduce the computational time, the foundation system was assumed to have an elastic characteristic. The parameters used to idealize this type of material model such as the concrete compressive strength (f_c), axial stiffness (*EA*), flexural rigidity (*EI*), unit weight of concrete, and Poisson's ratio (ν) per unit length of the foundation system are shown in Table 3 (Abdulrazagh 2009).

To eliminate the effects of subgrade soil types and the slope of the trench walls on the analytical results of the minimum depth of cover, GW with a minor cohesion was selected for the engineered backfill from the soil group I of the CHBDC soil classifications CHBDC (2006) as a continuum media around the metal structures. A 0.3 m thick layer of ballast was also considered for both the 2D and 3D models. Therefore, in this study, the definition of *minimum depth of cover* also includes the depth of the ballast layer. In other words, the vertical distance between the lower level of sleepers and the crown is defined as the *depth of cover*. The behavior of the soil and ballast layers was defined as a nonlinear material based on the Mohr-Coulomb criterion. The soil and ballast specifications such as the type of material behavior, unit weight of soil (γ), Young's modulus (E_{ref} 2.07×10^4 to approximately 3.00×10^4 kN/m^2), Poisson's ratio, friction angle (ϕ), and cohesion (c) are shown in Table 4 (Abdulrazagh 2009).

The railway superstructure (rails and sleepers) was not idealized for the 2D analysis to simplify the calculations; however, it was idealized in the 3D analysis using PLAXIS 3D code because of its particular capabilities for modeling the longitudinal dimension of the structures and to obtain more accurate results by taking the lateral stiffness of the superstructure into account. To idealize the rail and sleepers in the 3D analyses, the properties of the UIC 60 rail type and the mono-block concrete sleepers were specified for the plate and soil type elements of the railway superstructure model, respectively (see Tables 5 and 6; Abdulrazagh (2009)). The rails and sleepers were modeled as linear elastic materials to eliminate the undesired behavior of the superstructure for

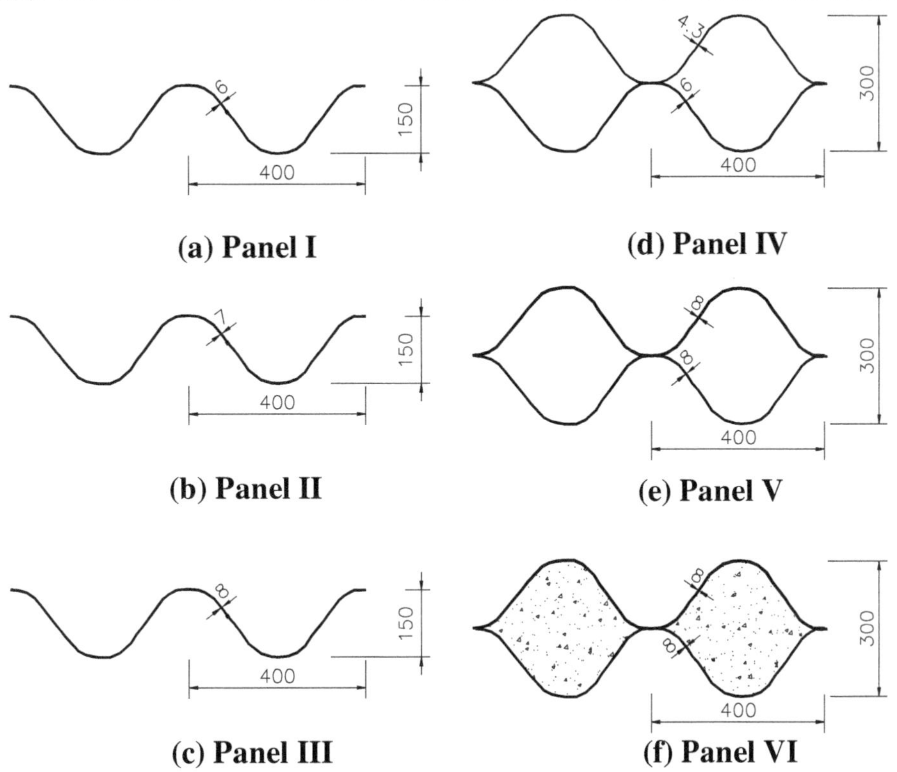

Figure 4 Types of conduit wall panels with dimensions in millimeters. (**a**) Standard 400 mm × 150 mm corrugation with 6-mm thickness, panel I. (**b**) Standard 400 mm × 150 mm corrugation with 7-mm thickness, panel II. (**c**) Standard 400 mm × 150 mm corrugation with 8-mm thickness, panel III. (**d**) Stiffened corrugated panel with 4.3-mm corrugated plate, panel IV. (**e**) Stiffened corrugated panel with 8-mm corrugated plate, panel V. (**f**) Stiffened corrugated panel with concrete filling, panel VI.

Table 2 Element specifications of conduit wall panels (corrugated steel plates)

Material model	Panel type	*EA* $\left(\frac{kN}{m}\right)$	*EI* $\left(\frac{kN.m^2}{m}\right)$	Poisson's ratio	M_p $\left(\frac{kNm}{m}\right)$	P_p $\left(\frac{kN}{m}\right)$
Elasto-plastic	I	1.65×10^6	4,630.80	0.3	117.30	2,478.00
Elasto-plastic	II	1.93×10^6	5,414.20	0.3	137.07	2,892.00
Elasto-plastic	III	2.18×10^6	6,151.80	0.3	155.67	3,270.00
Elasto-plastic	IV	2.81×10^6	20,846.29	0.3	226.71	4,218.00
Elasto-plastic	V	4.36×10^6	33,061.56	0.3	330.61	6,540.00
Elasto-plastic	VI	1.85×10^9	125,864.65	0.3	330.61	6,540.00

the determination of the minimum depth of cover for each analysis.

Loading model and loading pattern

The railway standard static loading pattern LM71 CEN (2002) that was used in the 2D and 3D FE analyses is shown in Figure 5. This loading pattern is also commonly used to design traditional railway bridges.

The factored loading diagram, which was directly applied to the 2D and 3D models, included the half magnitudes of those shown in Figure 5a,b (as a wheel load quantity). The longitudinal distributed loading model of LM71 (see Figure 5a) was selected for the 2D analysis to simplify the investigation of the moving effects of the railway load on the structure through a large amount of calculations in the numerical procedure. As an example, the factored strip loads of the 2D models are shown by AA and BB in Figure 6. The BB strip loads in Figure 6 for the 2D analyses were extended to boundaries.

Because the location of the railway moving load varies along the bridge, it is important to investigate the effects of the maximum lateral pressures, centric and eccentric loads on the foundation system, and crown and conduit walls. Therefore, the defined loading patterns in the 2D and 3D FE analyses were located in three critical locations as follows: (1) the right corner of the factored AA strip load for the 2D analysis (or the first right factored concentrated load for the 3D analysis) was located in line with the left foundation wall to study the stability of the structure against overturn moments and conduit wall buckling (Figure 6a, location 1); (2) the factored railway loading pattern was located symmetrically at the midspan to control the creation of a plastic hinge at the crown and at the conjunction points of the crown and haunches (Figure 6b, location 2); and (3) the right corner of the factored AA strip load in the 2D analysis (or the first right factored concentrated load in the 3D analysis) was located in line with the right foundation wall to investigate the eccentric loading effects on the analytical results (Figure 6c, location 3).

The initial measurements of the 2D and 3D idealization show that the railway location at the midspan defined by location 2 produced the maximum thrust in box and low-profile bridges. These measurements also show that the extension of the strip loads to the four sleepers adjacent to the outer sides of the spring line for location 2 in the 3D analysis results in a reasonable effective length for the longitudinal idealization of the railway track and consideration of the load effects (see Figure 7).

Definition of minimum cover governing criteria

For this study, the minimum cover for all of the cases in the numerical analysis is known as the cover depth, which creates three major modes of serviceability failure. First, by setting the minimum cover criterion, excessive uneven settlements in the railway bridge, which control the passenger comfort criteria, will be prevented (Neidhart 2005). Second, it prevents buckling in the various structural components. Once the settlement of the railway track does not exceed the permissible vertical deflection of the railway bridge, a buckling of the conduit wall, which results in soil tensile stresses, will not occur; however, the latter criterion was also checked by assigning nonlinear material properties to the wall conduits (Abdel-Sayed and Salib 2002). The third mode of failure, which controls the bearing capacity of the system as a type of foundation, is the soil local collapse. The mechanism of control for this type of failure was included in the numerical analysis by

Table 3 Element specifications of foundations and pedestals

Material	Material model	f_c $\left(\frac{kN}{m^2}\right)$	*EA* $\left(\frac{kN}{m}\right)$	*EI* $\left(\frac{kN.m^2}{m}\right)$	Unit weight $\left(\frac{kN}{m^3}\right)$	Poisson's ratio
Pedestal	Elastic	35,000	2.24×10^6	119,466.67	25	0.1
Foundation	Elastic	35,000	1.12×10^6	14.93×10^6	25	0.1

Table 4 Element specifications of soil and ballast

Material	Soil model	γ $\left(\frac{kN}{m^3}\right)$	E_{ref} $\left(\frac{kN}{m^2}\right)$	Poisson's ratio	φ	C_{ref} $\left(\frac{kN}{m^2}\right)$
Soil	Mohr-Coulomb	22	Increases 0.5×10^5 with every 1-m increase of depth	0.2	40°	7
Ballast	Mohr-Coulomb	20	2.07×10^4	0.2	40°	0

assigning the Mohr-Coulomb criterion to the soil material. These criteria will be discussed in more detail in the following subsections.

Permissible settlement of the railway track

For a soil-steel railway bridge, the minimum depth of cover must be defined in such a way to prevent the uneven settlements. In railway bridges, the criterion which restricts the settlement is related to speed of passing trains. According to CEN (2001) and the studies by Neidhart (2005), the settlement of the railway track δ satisfies the comfort criteria as:

$$\delta \le 0.625 \frac{l^2}{V_{tr}^2}, \tag{1}$$

where V_{tr} is the running speed of the trains (in meters per second), and l is the distance between two points along the railway track where the differential settlement is desired (in meters). This equation results in a maximum allowable settlement (for a passing speed of 120 km/h) of about 62 mm at a distance equal to l (span/2) = 10.473 m. The permissible settlement for each of the predefined structures at a speed of 120 km/h (33.33 m/s) is shown in Table 7 (CEN 2001).

For this study and for the case of the railway bridge, the relative settlements between the spring line and midspan were checked for various depths of cover at different sections along the longitudinal axis of the bridge in the 3D FE analysis (at the sections of the embankment heel δ_{AB}, ballast heel δ_{CD}, sleeper edge δ_{EF}, outer edge of rail δ_{GH}, and axis of the railway track δ_{IJ} for the 3D analyses (see Figure 8) was compared to Equation 1. A similar approach was used in the 2D FE analysis to check the relative settlements between the spring line and midspan at the x and y plane (δ_{AB}).

To satisfy the serviceability of the railway bridge, the vertical deformation of analytical bridge should be less than the permissible settlement defined in Table 7. In other words, if a specific soil cover depth satisfies all of the aforementioned relative settlements at the same time, then it is defined as the *minimum depth of soil cover* for a specific structure. This depth was checked for each of the structures.

Buckling control in the conduit walls

The strength required for the service life of the soil-steel bridge to avoid the buckling of the conduit wall is controlled by the combined bending moment and axial thrust as follows (CSPI and AISI 2002):

$$\alpha = \left(\frac{P}{P_{pf}}\right)^2 + \left|\frac{M}{M_{pf}}\right| \le 1, \tag{2}$$

where P and M are the axial thrust and bending moments due to the dead load and railway load in conduit walls, and they are calculated using the 2D or 3D FE analyses. P_p and M_p are the compressive strength and plastic moment capacity of the predefined conduit wall sections, respectively, and are shown in Table 2 and Figure 4.

This criterion was automatically controlled by defining an elasto-plastic behavior for the conduit walls as beam elements for the 2D FE analysis. As the dead and live loads in the 3D FE analysis cause axial thrusts and bending moments at both the x-y plane (P_{11} and M_{11}) and the x-z plane (P_{22} and M_{22}) so Equation 2 should be controlled in both directions. More details of the buckling control in both planes are presented in the following sections. However, the initial investigations of the different 3D models revealed that the axial thrusts and bending moments at the x-y plane (cross section) are dominant to the x-z plane values, so they control the buckling of the structures.

Control of the soil body failure

For the soil-steel bridges, the soil above the conduit is prone to shear failure if the embankment is subjected to loads that are eccentric with respect to the conduit axis. An insufficient depth of cover results in a local soil

Table 5 Element specifications of rails

Material	Material model	E $\left(\frac{kN}{m^2}\right)$	EA $\left(\frac{kN}{m}\right)$	EI $\left(\frac{kN.m^2}{m}\right)$	Unit weight $\left(\frac{kN}{m}\right)$	Poisson's ratio
Rail	Elastic	2.0×10^8	1,517,400	6,110	0.6034	0.25

Table 6 Element specifications of concrete sleepers

Material	Material model	γ $\left(\frac{kN}{m^3}\right)$	E_{ref} $\left(\frac{kN}{m^2}\right)$	K_0	Poisson's Ratio
Sleeper	Elastic	25	2.8×10^7	-	0.1

failure, which fundamentally arises in the presence of tensile or shear stresses over the maximum allowable stress defined by the Mohr-Coulomb criterion (Abdel-Sayed and Salib 2002; Lee 2002; Beben 2009; Flener 2010). This type of failure was a 'non-converged deformation' during the reconstruction of elasto-plastic stiffness matrix which was automatically controlled by the software as the soil body failure during the setting of the acceptable soil cover on the structure in the numerical analysis.

Results and discussion

Numerical results

In this section, the 2D and 3D FE numerical results are discussed, and the developmental procedure of the interpolation equations is described.

Results of the 2D finite element analyses

A comparison between the results of the 2D analysis for box and low-profile arches is shown in Figure 9.

As a result of the various numerical analyses and satisfying the aforementioned criteria, the minimum depth of cover was evaluated against the span of railway box and low-profile arch bridges using various panel types (stiffness) and with respect to the permissible settlement criterion. A brief review of these data demonstrates that the trends of the minimum depth of cover for box bridges differ from that of the low-profile arches. This is due to the different structural geometry of the boxes which results in a different mechanism of behavior under load. Thus, to ease the process of finding the new pattern of minimum depth of cover, the data have been separately categorized for boxes and low-profile arches in the following step.

Regarding the power trend type of the CHBDC formula of minimum depth of cover $D_h/6((D_h/D_v)^{0.5})$ and the trend of changes in the cover depth for the boxes and low-profile arches versus the span, exponential, polynomial, or power trend lines were assumed to properly fit to the data of 2D results. Therefore, two types of exponential, one type of polynomial, and one type of power trend line were examined as follows: (1) exponential trend line as a function of the effective span (D_h); (2) power trend line as a function of the effective span (D_h); (3) exponential trend line as a function of D_h/D_v (D_v effective rise of the structure); and (4) polynomial trend line as a function of D_h/D_v. A comparison of the R^2 values of the trend lines shows that the exponential and power functions of D_h represent a more accurate estimation of the data. However, due to the dimensions of the formulas for calculating the minimum depth of cover that must be in meters, an exponential function (D_h) was chosen to describe the change in the minimum depth of cover versus the effective span.

In order to account for moment of inertia and length of span which are the representative of relative stiffness of backfill and culvert structure and the geometry of metal culver structure, the basic form of the minimum depth of cover resulted from 2D analysis for the railway boxes and high-profile arches is introduced as follows:

$$H_{\min} = \alpha\left(N_f\right)^{\beta} e^{\mu D_h}, \tag{3}$$

where the dimensionless ratio N_f is defined by the following:

$$N_f = \frac{E_s D_h{}^3}{EI}, \tag{4}$$

in which E_s is the secant modulus of the backfill, which depends upon the quality and compaction of backfill and the depth of cover; E is the modulus of elasticity of the corrugated plates; and I is the moment of inertia of

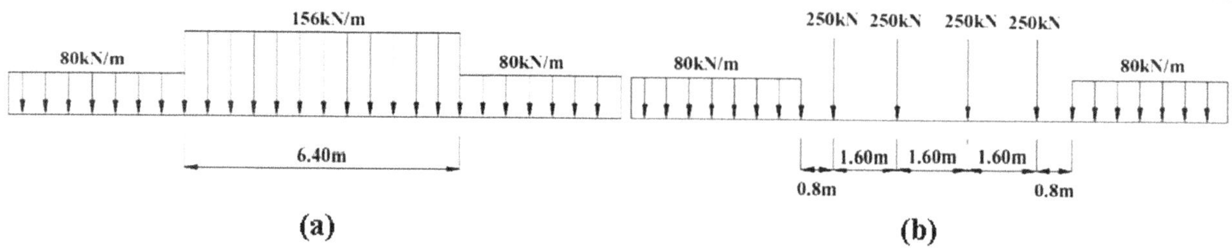

Figure 5 Eurocode model LM71 railway bridge loading diagram. (**a**) Longitudinal distribution of the unfactored axle loads used in 2D FE models. (**b**) Unfactored loading diagram used in 3D FE models (CEN 2002).

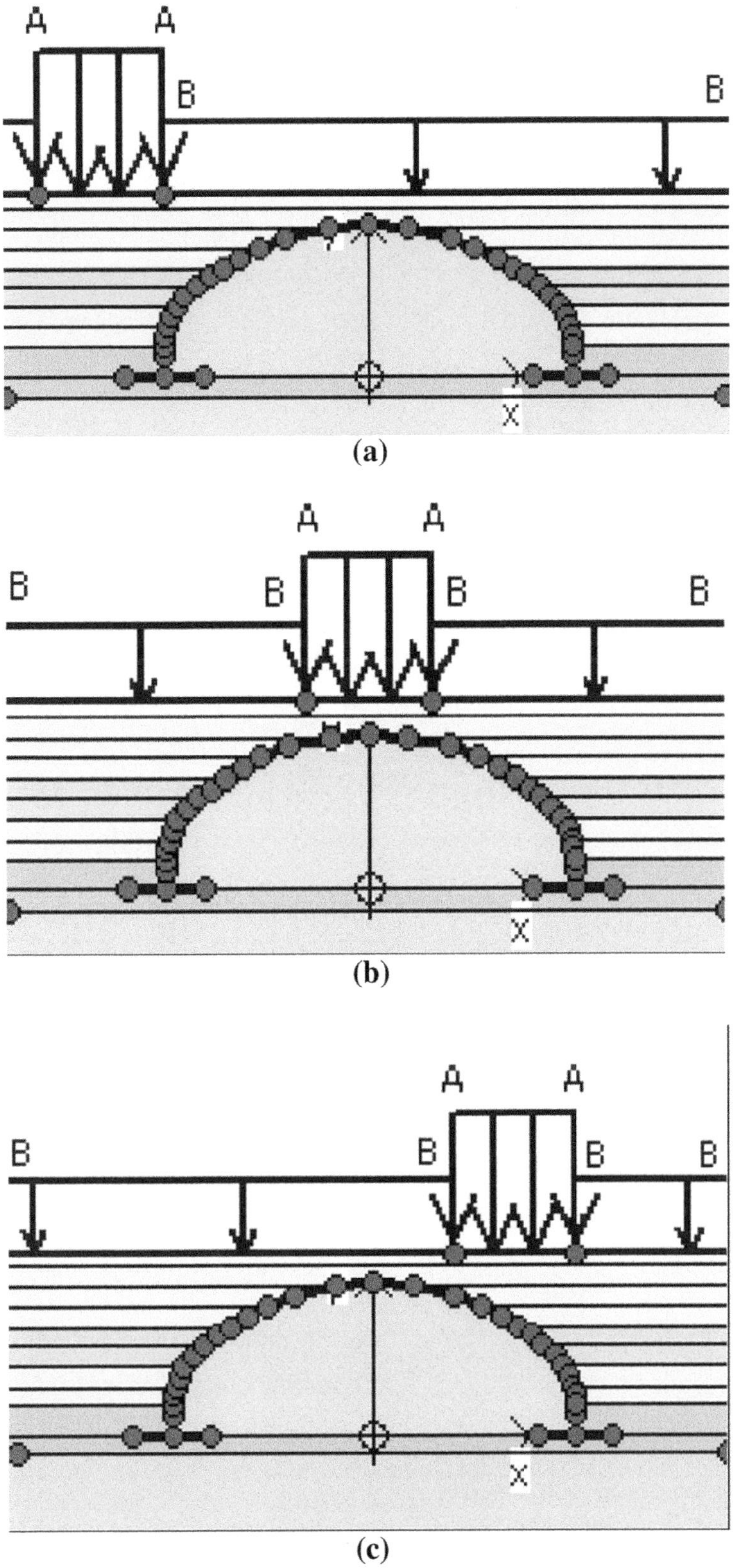

Figure 6 Location of factored railway load. (**a**) The right corner of factored AA strip load in 2D analysis located in line with the left foundation wall (location 1). (**b**) The factored railway loading pattern located symmetrically on midspan (location 2). (**c**) The right corner of factored AA strip load in 2D analysis located in line with right foundation wall (location 3).

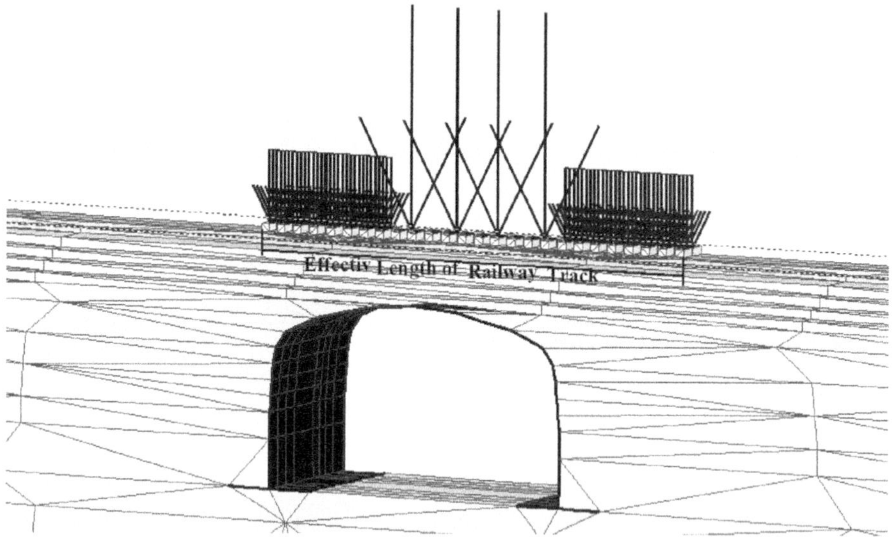

Figure 7 Effective length for longitudinal idealization of railway track in 3D FE analyses.

the conduit wall per unit length of the bridge. Variables α, β, and μ in Equation 3 are unknown constants. These constants were calculated separately for the boxes and low-profile arches using the least-squares method to determine the function of best fit. The final set of parameters is listed below:

$$\begin{cases} \alpha = 0.0139 \\ \beta = 0.25 \\ \mu = \frac{1}{3} \end{cases} \text{BOX} \qquad \begin{cases} \alpha = 0.0139 \\ \beta = 0.25 \\ \mu = \frac{1}{3} \end{cases} \text{LPA.}$$

By substituting the above parameters into Equation 3, the equations for the minimum depth of cover are given by Equations 5 and 6, assuming that all the metal structures are buried in GW with 24 MPa of secant elasticity modulus as an average value in the mid-height of the structure:

$$H_{\min} = 0.0139 N_f^{0.25} e^{D_h/3} \quad R^2 = 0.90 \text{ for boxes,} \tag{5}$$

$$H_{\min} = 0.0100 N_f^{0.25} e^{D_h/7} \quad R^2 = 0.87 \text{ for low-profile arches.} \tag{6}$$

As shown in the above equations, the R^2 values are close to 1.0, which represents the considerable accuracy of the interpolation function with minimal divergence from the 2D FE numerical results. Regarding the maintenance operations, buckling criterion, and AASHTO and CHBDC limits, a recommended minimum of 0.6 m and maximum of 1.5 m must be maintained for the depth of cover using the following expressions:

$$H_{\min} = \min\begin{cases} 1.5\, m \\ \max\begin{cases} 0.6\, m \\ H = 0.0139\left(\dfrac{E_s D_h^3}{EI}\right)^{0.25} e^{D_h/3} \end{cases} \end{cases} \text{for boxes,} \tag{7}$$

$$H_{\min} = \min\begin{cases} 1.5m \\ \max\begin{cases} 0.6m \\ H = 0.0100\left(\dfrac{E_s D_h^3}{EI}\right)^{0.25} e^{D_h/7} \end{cases} \\ \text{for low-profile arches,} \end{cases} \tag{8}$$

The results of the 3D finite element analysis for control of buckling

Checking the applicapability of the 2D-based equations of minimum cover in practical three-dimensional problems, a series of three-dimensional FE analyses were carried out to control the out-of-plane buckling in the steel plates due to three-dimensional pattern of the train

Table 7 Permissible settlement at speed 120 km/h

Structure type	Span (m)	*l* (m)	$\delta_{\text{permissible}}$ (mm)
Box culvert	8.074	4.035	9.00
	10.514	5.255	15.00
	11.023	5.511	17.00
	13.456	6.728	25.00
Low-profile arch	14.130	7.065	28.00
	16.518	8.259	38.00
	20.946	10.473	62.00
	23.405	11.702	77.00

Figure 8 Points to be studied in tension of spring line and mid-span. It is for the measurement of relative settlements in box bridges: (**a**) at *x-y* plane and (**b**) along the longitudinal dimension of the bridge.

loads. In this regard, the combination of thrusts and bending moments are checked using Equation 2 at two different sections (under rail and in the middle of track) in the midspan (points G and I in Figure 8b) and at in tension of spring line (points H and J in Figure 8b). The results of buckling control (α) for panel VI are illustrated in Figure 10. This figure shows that the combination of thrusts and bending moments in both planes ($\alpha_{x\text{-}y}$ and $\alpha_{y\text{-}z}$) are quite smaller than the maximum limitation of $\alpha_{\text{allowable}} = 1.0$. Also, this figure shows that $\alpha_{x\text{-}y}$ is always greater than $\alpha_{y\text{-}z}$ which means that the out-of-plane bending does not control the buckling behavior of the steel plates. The comparison between Figure 10a,b shows that the buckling under the rail and in the middle of track has almost similar values.

As can be seen in Table 8 for the structures with wall stiffness less than those of panel VI, in order to withstand under railway loads, the height of cover has significantly increased. To make a decision on the best alternative for bridge construction from the design aspect, a technical and economical comparison should be made between two different cases: the bridges with the small stiffness and high depth of cover and the bridges with the high stiffness and shallow depth of cover considering the limitation of the vertical project level (for optimizing the earth work in the project). However, none of the cases show an $\underline{\alpha}$ value greater than 1.0.

In summary, the results of Figure 10 and Table 8 show that the out-of-plane bending ($\alpha_{y\text{-}z}$) does not control the buckling behavior of the steel plates. Thus, the proposed equations for minimum depth of cover can be appropriately used for practical purposes.

Discussion

The results of the 2D FE analyses are compared to the standard codes, and the newly proposed equations as

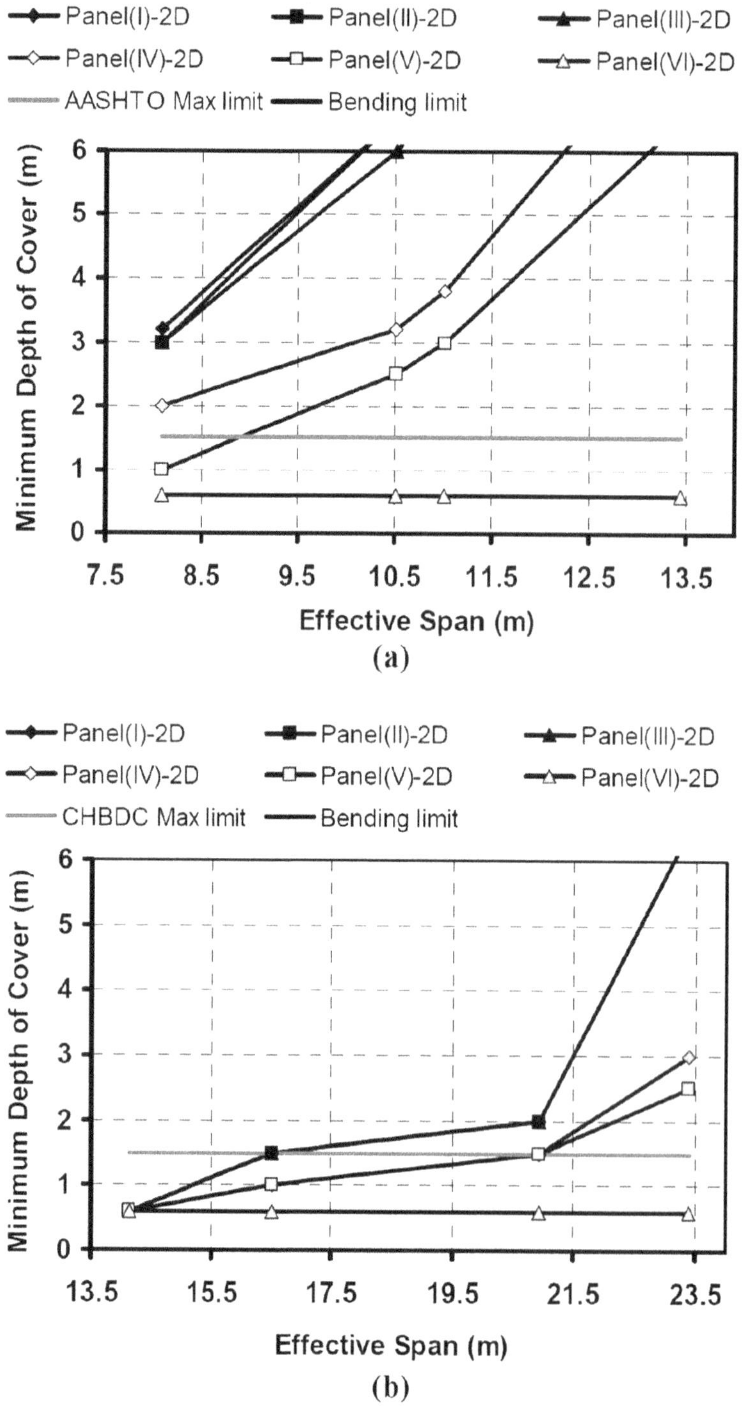

Figure 9 **The comparison of 2D results for (a) box bridges and (b) low-profile arches.**

well as the validity limit of these formulas are discussed in the following subsections.

Validity range of the derived equations

The results of the 2D FE analyses are compared to the values of the proposed box equations (referred to as the *formula* in the legends) and the *AASHTO limit* are shown in Figure 11. There is good agreement between the 2D FE results and the values of Equation 5. The conformity of Equation 5 with the 2D results was evaluated using an R^2 value that was calculated using $R^2 = 1 - SSE/SST$, where $SSE = \Sigma\left(Y_i - \hat{Y}_i\right)^2$ and $SST = (\Sigma Y_i^2) - ((\Sigma Y_i)^2)/n$. The R^2

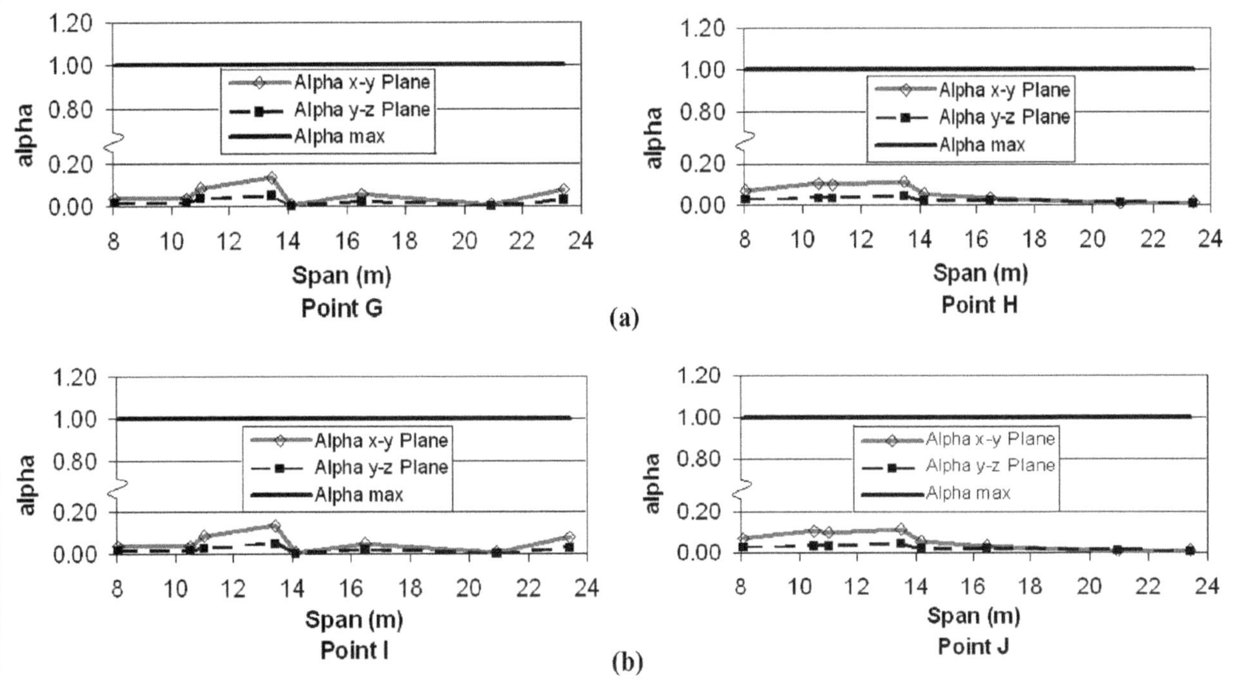

Figure 10 Control of buckling for structures with panel VI: (a) Under the rail. (b) In the middle of track.

value for this equation was calculated to be 0.90, which demonstrates a reasonable agreement with the 2D results.

It should be noted that an equation is not given in the CHBDC and AASHTO for the minimum depth of cover of box culverts. Only a minimum of 0.3 m and a maximum of 1.5 m are specified in the CHBDC (2006) and AASHTO Highway Bridges (2000), respectively. By accounting for the buckling criterion of the railway box bridges, a minimum of 0.6 m (called the *bending limit*) was obtained for the depth of cover, which is indicated by the dashed lines in Figure 11.

Figure 11a shows that panels IV and V with a span of 8.07 m and panel VI for all of the spans follow the

Table 8 Control of buckling for the structures with profiles other than panel VI

Span (mm)	Points G and I				Points H and J			
	Panel I	Panel III	Panel IV	Panel V	Panel I	Panel III	Panel IV	Panel V
Box 8.07	6.0 m[a]				6.0 m[a]			
	0.157[b] 0.153[c]				0.689[b] 0.687[c]			
10.51	-	6.5 m[a]				6.5 m[a]		
		0.297[b] 0.294[c]				0.730[b] 0.719[c]		
11.02	-		6.5 m[a]				6.5 m[a]	
			1.118[b] 0.105[c]				0.873[b] 0.857[c]	
13.46	-			6.5 m[a]				6.5 m[a]
				0.134[b] 0.131[c]				0.984[b] 0.976[c]
LPA 14.13	0.6 m[a]				0.6 m[a]			
	0.543[b] 0.540[c]				0.490[b] 0.485[c]			
16.52	1.5 m[a]				1.5 m[a]			
	0.333[b] 0.279[c]				0.829[b] 0.821[c]			
20.95	2.0 m[a]				2.0 m[a]			
	0.490[b] 0.434[c]				0.552[b] 0.506[c]			
23.40	6.5 m[a]				6.5 m[a]			
	0.494[b] 0.491[c]				0.903[b] 0.887[c]			

[a] Depth of cover; [b] a_{x-y}; [c] a_{y-z}.

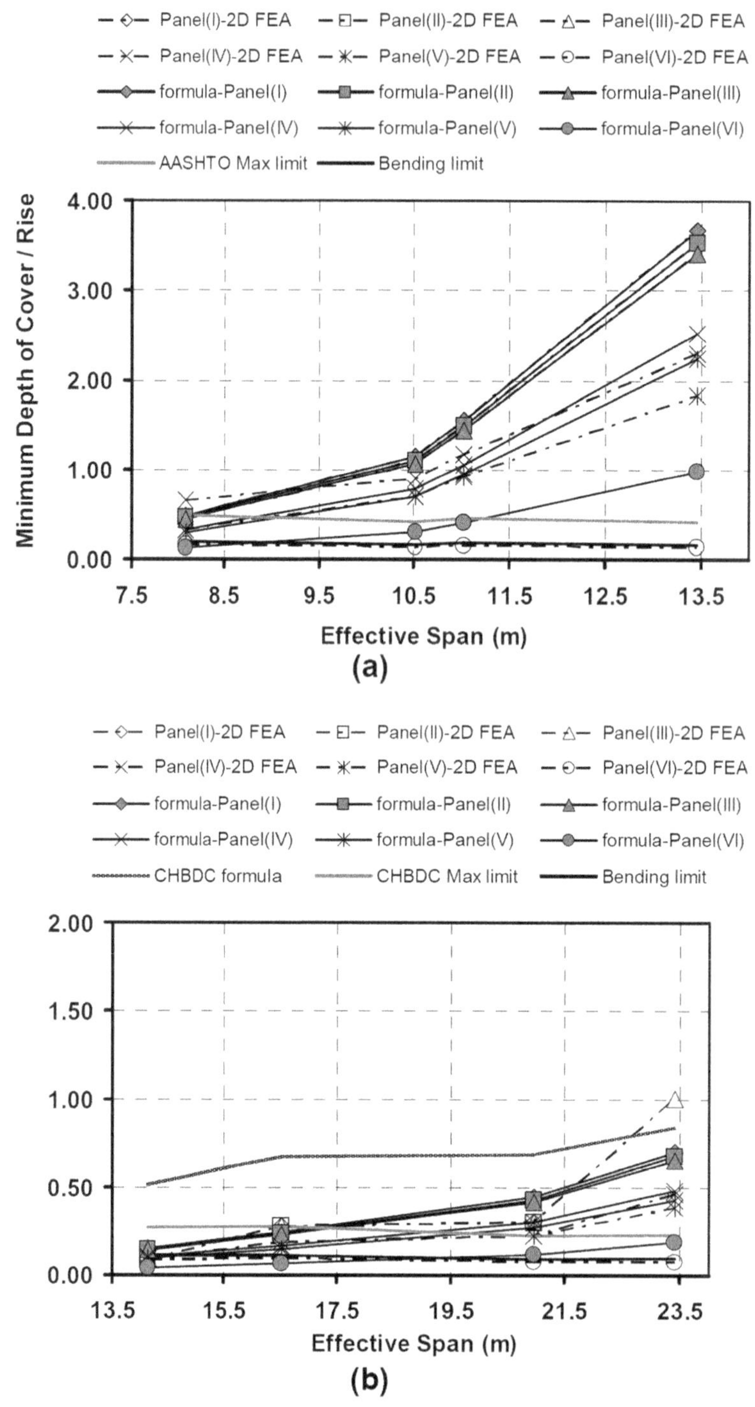

Figure 11 Results of 2D FE analyses in comparison with two values. (**a**) Box formula and the AASHTO limit. (**b**) LPA formula and the CHBDC limit.

AASHTO max limit. The figure also presents that the agreement between the 2D FE results and the values of Equation 5 is good for all panels (R^2 = 0.90) specially panels IV to VI. The results of 2D FE analysis for panels I to III are not shown in this figure because of the high value of minimum depth of cover. In these cases, usually stiffer panels are used to reduce the required minimum cover depth.

The results of the 2D FE analysis are compared to the values of the proposed low-profile arch equation (referred to as the *formula* in the legends), and the *CHBDC max limit* and the *bending limit* (0.6 m for the low-

Table 9 The R^2 values for 2D equations for all spans

Panel type	Box	LPA
I	0.91	0.83
II	0.92	0.87
III	0.93	0.88
IV	0.98	0.97
V	0.96	0.98
VI	-	-

profile arches) are illustrated in Figure 11b. The R^2 value of 0.87 for Equation 6 demonstrates a reasonable agreement with the 2D FE results.

The values of the minimum depth of cover specified by the CHBDC are also shown in Figure 11b. The CHBDC recommends using the greater of $D_h/6((D_h/D_v)^{0.5})$ and 1.5 m for non-stiffened panels Hafez and Abdel-Sayed (1983). A comparison between the *CHBDC formula* and the resulting data shows that the minimum depth of cover specified by the CHBDC is much higher than those obtained from 2D FE analyses (without considering the 0.6-m limit). This difference may refer to the recommendation of more conservative values by the standards, so the recommended values should be applicable for all practical conditions. Figure 11b shows that the panel IV with the span of 16.52 m, panel V with the span of 16.520 and 20.95 m, and panel VI for all of the spans follow the *CHBDC max limit.*

The overall conformity of the newly established equations obtained from the 2D FE analysis was discussed in the previous section. However, to investigate the validity of the newly established formulas with respect to the panel types, the R^2 value was calculated separately for each type of panel. The R^2 values are shown in Table 9.

From the results shown in the table, the equations are applicable for panels I to V for boxes and low-profile arches because the R^2 values are greater than 0.85. It is recommended to use the equations conservatively for panel VI or to use the minimum value of 0.6 m for the depth of cover. The results of the 2D FE analysis for panel VI represented a depth of cover less than 0.6 m. A minimum of 0.6 m must be maintained for the *CHBDC minimum (bending) limit.*

Conclusions

The minimum depth of cover requirements given by different codes are typically based on vehicle loads, non-stiffened panels, and only the geometrical shape of the metal structure to avoid the failure of soil cover above a soil-steel bridge. In this paper, the effects of spans larger than 8 m (using stiffened panels under railway loads) are investigated using an FE analysis. For this study, 2D and 3D FE analyses of four low-profile arches and four box culverts with spans larger than 8 m were performed to develop new patterns for the minimum depth of soil cover. Using the least-squares method to adopt the best-fit equation of the numerical data, two new sets of formulas were recommended. Based on the numerical results, the primary research findings are summarized as follows:

1. The minimum depth of cover increases exponentially along with an increase in the span of boxes and low-profile arches.
2. The increase of rigidity of the wall panels significantly decreases the depth of cover required above the bridges. The efficiency of the stiffened panels is more pronounced for large spans.
3. Different trends of the minimum depth of cover were determined for box bridges and low-profile arches. This difference is due to the various structural geometries of the boxes that resulted in a different mechanism of behavior under load.
4. Exponential forms of the minimum depth of cover for railway boxes and low-profile arches were developed as a function of the span and relative stiffness of the backfill and the culvert structure, which exhibited good conformity with the 2D FE analysis results ($R^2 > 0.85$).
5. A survey on the validity limit of the new formulas established from the 2D FE analyses demonstrated that the equations are appropriate to calculate the minimum depth of cover for all of the boxes and low-profile arches with spans greater than 8 m with stiffened and non-stiffened panels. However, a minimum depth of cover of 0.6 m must always be maintained.

Competing interests
The authors declare that they have no competing interests.

Authors' contributions
As corresponding author, ME defined the main theme and objective of the study for PHA, explained the research methodology, processed the obtained numerical results, and guided the calculation of closed form equations for minimum soil cover on selected bridges. ME was involved in language editing and finalizing the manuscript writing. JAZ edited the first draft of the paper and controlled the obtained numerical results. PHA accomplished the whole of the numerical simulations and prepared the first draft of the paper. All authors read and approved the final manuscript.

Authors' information
ME is an assistant professor at the School of Railway Engineering in IUST. JAZ is an associate professor at the School of Railway Engineering in IUST. PHA is a PhD candidate at the University of Tehran.

Author details
[1]School of Railway Engineering, Iran University of Science and Technology, University Street, Hengam Aveneu, Narmak, Tehran 1684613114, Iran. [2]School of Civil Engineering, University of Tehran, Tehran 111554563, Iran.

References

Abdel-Sayed G, Salib SR (2002) Minimum depth of soil cover above soil-steel bridges. J Geotech Geoenviron Eng ASCE 128(8):672–681

Abdel-Sayed G, Bakht B, Jaeger LG (1993) Soil-steel Bridges, Design and Construction. McGraw-Hill Pub, New York

Abdulrazagh PH (2009) 2D and 3D Finite element analysis of railway soil-steel bridges with attitudes towards minimum depth of soil cover. Iran University of Science and Technology, M.Sc. Thesis

ARTC (2005) Buried Corrugated Metal Structures. Australian Rail Track Corporation, Adelaide

ASTM (1982) A796/A796M: Standard Practice for Structural Design of Corrugated Steel Pipe, Pipe-Arches, and Arches for Storm and Sanitary Sewers and Other Buried Applications. ASTM International, West Conshohocken, PA

Beben D (2009) Numerical analysis of soil-steel bridge structure. Baltic J Road & Bridge Eng 4(1):13–21

CALTRANS (2000) Bridge Design Specifications. California Transportation Board, Sacramento, CA

CEN (2001) EN1990 Eurocode: Basis of Structural Design. Draft Annex A2: Application for Bridges (Normative). European Committee for Standardization, Brussels

CEN (2002) Eurocode 1: Actions on Structures Part 2 - Traffic Loads on Bridges (prEN 1991–2). European Committee for Standardization, Brussels

CHBDC (2006) CAN/CSA-S6-06 Section 7: Buried Structures. National Standard of Canada, CSA International, Ontario

CSPI and AISI (2002) Handbook of Steel Drainage & Highway Construction Products, 2nd edn. American Iron and Steel Institute, Washington DC

DMRB (2001) Design of corrugated steel buried structures with spans greater than 0.9 meters and up to 8.0 meters: BD12/01, vol. 2, section 2, part 6. British Design Manual for Roads and Bridges, London

Duncan JM (1979) Behavior and design of long-span metal culverts. J Geotech Eng Division ASCE 105(3):399–418

El-Sawy KM (2003) Three-dimensional modeling of soil-steel culverts under the effect of truckloads. J Thin-Walled Struct 41:747–768

Flener EB (2010) Testing the response of box-type soil-steel structures under static service loads. J Bridge Eng ASCE 15(1):90–97

Flener EB, Karoumi R (2009) Dynamic testing of a soil-steel composite railway bridge. J Eng Struct 13(12):2803–2811

Flener EB, Karoumi R, Soundquist H (2005) Field testing of a long-span arch steel culvert during backfilling and service. J Struct Infrastruct Eng 1(3):181–188

Hafez H (1981) Soil-steel structures under shallow cover. PhD dissertation. University of Windsor

Hafez H, Abdel-Sayed G (1983) Soil failure in shallow covers above flexible conduits. Can J Civil Eng 10:654–661

Highway Bridges AASHTO (2000) Soil-Corrugated Metal Structure Interaction Systems, 2nd edn. American Association of States Highway and Transportation, Washington, DC

Lee J (2002) The performance of soil-steel bridges with shallow soil cover under vehicle load. Proceedings 5th Int Conf on Bridge Maintenance. Saf Manage 1:42–50

MacDonald L (2010) Numerical modelling of vehicle loads on buried orthotropic steel shell structures. Dalhousie University, M.Sc Thesis

Manko ZZ, Beben D (2005) Research on steel shell of a road bridge made of corrugated plates during backfilling. J Bridge Eng ASCE 10(5):592–603

Manko ZZ, Beben D (2008) Dynamic testing of a corrugated steel arch bridge. Can J Civil Eng 35(3):246–257

Mohammed H, Kennedy JB, Smith P (2002) Improving the response of soil-metal structures during construction. J Bridge Eng ASCE 7(1):6–13

Neidhart TH (2005) True-to-scale *in situ* tests determining dynamic performance of earthworks under high speed train loading. Geotechnics Roads 1:27–34

OHBDC [Ontario Highway Bridge Design Code] (1992) Buried Structures, 3rd edn. Ontario Ministry of Transportation, Toronto

Peck OK, Peck RB (1984) Experience with flexible culverts through railroad embankments. Proc 2nd Int Conf on Soil Mechanic Found Eng 2:89–98

Peterson DL, Nelson CR, Li G, McGrath TJ, Kitane Y (2010) Recommended design specified for live load distribution to buried structures. NCHRP Report 647. Transportation Research Board, Washington

Planning and Budget Organization (1985) Code 88 Criteria of Geometrical Design of Roads and Crossings. Planning and Budget Organization, Tehran

Evaluating the displacement amplification factors of concentrically braced steel frames

Mussa Mahmoudi* and Mahdi Zaree

Abstract

According to seismic design codes, nonlinear performance of structures is considered during strong earthquakes. Seismic design provisions estimate the maximum roof and story drifts occurring during major earthquakes by amplifying the drifts computed from elastic analysis at the prescribed seismic force level with a displacement amplification factor. The present study tries to evaluate the displacement amplification factors of conventional concentric braced frames (CBFs) and buckling restrained braced frames (BRBFs). As such, static nonlinear (pushover) analysis and nonlinear dynamic time history analysis have been performed on the model buildings with single and double bracing bays, and different stories and brace configurations (chevron V, invert V, and X bracing). It is observed that the displacement amplification factors for BRBFs are higher than that of CBFs. Also, the number of bracing bays and height of buildings have a profound effect on the displacement amplification factors. The evaluated ratios between displacement amplification factors and response modification factors are from 1 to 1.12 for CBFs and from 1 to 1.4 for BRBFs.

Keywords: Buckling restrained braced frame, Concentrically steel braced frame, Displacement amplification factor, Ductility factor, Overstrength factor

Introduction

It is well recognized that most disasters due to moderate or severe earthquake ground motions are caused by the failure of civil engineering facilities, many of which were presumed to have been designed and constructed to provide protection against natural hazards. Much of the damages and collapse of structures during severe earthquakes primarily occurred due to excessive displacement in stories. In force-based seismic design, the force demand is generally determined on the basis of the structural linear response. Studies show that structures designed by modern seismic code procedures are likely to undergo large cyclic deformations in the inelastic range when subjected to a severe tremor. In current seismic codes, design base shears are calculated by reducing the elastic to the inelastic strength demands using the response modification factor (R) (Lee et al. 2004). Similarly, the displacement demand of a structure is estimated by multiplying a linear displacement response by the displacement amplification factor (C_d). The displacement amplification factor is the structural response parameter most widely employed for evaluating the inelastic performance of structures. C_d is also the parameter explicitly or implicitly used in most common design procedures.

There are several systems that can be used effectively to provide resistance to seismic lateral forces. Conventional concentric braced frame (CBF) and buckling restrained braced frame (BRBF) are the most efficient and common structural systems in steel construction to resist lateral forces, especially for structures in highly seismic regions. The use of concentric braces in framed structures offers an attractive system for seismic resistance, primarily due to their efficiency in providing lateral stiffness, hence limiting inter-story as well as overall lateral deformations (Goggins et al. 2006). In other words, because of complete truss action, steel braces improve the lateral strength and stiffness of the structural system and participate in seismic energy dissipation by deforming inelastically during an earthquake (Davaran and Hoveidae 2009).

Lateral displacements on structural buildings have been of great concern for engineers. Several researchers

* Correspondence: m.mahmoudi@srttu.edu
Faculty of Civil Engineering, Shahid Rajaee Teacher Training University, Tehran, Iran

have tried to investigate the displacement amplification factors of structural systems. Uang and Maarouf (1994) have discussed the effects of building and predominant earthquake ratios, types of yield mechanisms, and structural overstrength on the displacement amplification factor. Kim and Choi (2004) showed that the structural displacements decrease with the increase in BRB stiffness. They also found that the story-wise distribution of BRB, in proportion to the story drifts and story shears, results in better structural performance. Mahmoudi (2004) evaluated the displacement amplification factor and proposed a value to estimate the maximum lateral structural displacement, without using the nonlinear analysis. He also calculated the ratio of displacement amplification factor and response modification factor. In their study, Kiggins and Uang (2006) found that the use of BRB with steel moment frames will reduce residual story drifts and permanent deformations which can conversely lead to obtaining larger value of response modification factor.

Studies show various behavior factors for concentric braced frames (Mahmoudi and Zaree 2010). Furthermore, based on the previous design provisions, codes give constant value of displacement amplification factor for conventional CBFs and BRBFs, which do not consider the structure characteristic (number of stories and bracing bays). To overcome this inadequacy, the present paper has also focused on the evaluation of displacement amplification factor and its relation to the response modification factor of both CBFs and BRBFs. Here, the nonlinear static pushover analysis and nonlinear dynamic time history analysis were conducted by considering the behavior of members in life safety structural performance level as suggested by the Federal Emergency Management Agency (FEMA)-356 (2000).

Concentric braced frames

Bracings as lateral load-resistant system are one of the most commonly used methods to resist lateral loads such as earthquake. The braced frame response to earthquake loading depends mainly on the asymmetric axial resistance of the bracing members (Broderick et al. 2008). Conventional steel bracings dissipate considerable energy yielded under tension, but buckle without much energy dissipation in the compression range of cyclic loading (Kumar et al. 2007). If buckling of steel brace is restrained and the same strength is ensured both in tension and compression, the energy absorption of the brace will be markedly increased, and the hysteretic property will be simplified. Considering limited ductility and energy dissipation capacity of conventional CBF systems, efforts were made to develop new systems with stable hysteretic behavior, significant ductility, and large energy dissipation capacity. One such system with an improved seismic behavior is the BRBF. A typical BRB consists of a yielding steel core encased in a mortar-filled steel hollow section to restrain buckling, non-yielding and buckling restrained transition segments, and non-yielding and unrestrained end zones (Figure 1) (Sahoo and Chao 2010). Axial forces in BRBs are primarily resisted by steel cores which are laterally braced continuously by the surrounding mortar and steel encasement to avoid their buckling under compressive loads. This allows the steel core to yield in tension and compression, thereby significantly increasing the energy dissipation capacities of BRBs as compared to conventional steel braces (Figure 2) (Sahoo and Chao 2010).

Methods

Structural model and design

To evaluate the displacement amplification factor and its relation to response modification factor, 30 conventional CBFs and 20 BRBFs with 3, 5, 7, 10, and 12 stories as well as a bay length of 5 m were designed. Three different bracing types (i.e., X, chevron V, and chevron inverted V) for conventional CBFs and two bracing types (chevron V and chevron inverted V) for BRBFs were considered. The height of every model structure was fixed at 3.2 m. Figure 3 shows the plan of the model structures with the braces located in single and double bays. The model buildings were designed to take into account the Iranian Earthquake Resistance Design Code (Standard No. 2800) (BHRC 2005) and Iranian National Building Code, part 10, steel structures design (MHUD 2009). The buckling restrained brace members were designed according to the seismic provision of AISC (2005). The beam-column connections were assumed to be pinned so that the seismic load was resisted mainly by braces.

For brace designs, the double channel sections and the plate sections were used for CBFs and BRBFs, respectively. The effective length factor (K) considered for brace design is 0.5 for X braces, 1 for V and inverted V conventional concentric braces, and 0 for buckling restrained braces. Meanwhile, the IPB sections were used for the column in 3- and 5-story buildings, and the box sections were preferred in 7-, 10-, and 12-story buildings. Table 1 presents details about the structural members selected for the seven-story model frame with inverted V braces.

Displacement amplification factor

Both structural and nonstructural damages observed during earthquake ground motions are primarily produced by lateral displacements. Thus, the estimation of lateral displacement demands is of significant importance in performance-based design methods (Hajirasouliha and Doostan 2010). According to modern seismic design provisions, building structures undergo inelastic deformation during severe earthquakes. Therefore, these provisions

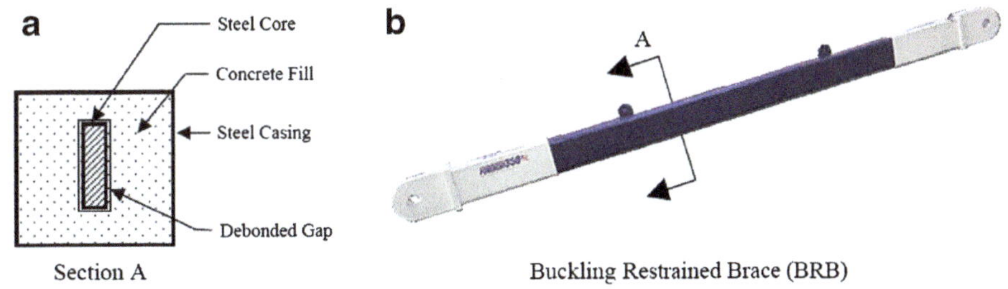

Figure 1 Typical BRB element. From Hussain et al. (2006).

permit a designer to reduce the elastic seismic force demand through a response modification factor. The response modification factor is the ratio required to maintain the structure from elastic to the inelastic design strength. Since reduced seismic force is used in the design, computed displacements from an elastic analysis are amplified in order to estimate the actual deformations following a severe earthquake. The displacement (or drifts), calculated through structural analysis, is not the real one; rather, it is less than the maximum structural displacement during strong tremors. The seismic design provisions estimate the maximum roof displacement and story drifts by augmenting the elastic analysis of displacement amplification factor (C_d) (Uang and Maarouf 1994):

$$\Delta_{\max} = \Delta_W \times C_d \tag{1}$$

where Δ_{max} is the maximum inelastic displacement (roof or story drifts), Δ_W is the displacement calculated by elastic analysis, and C_d is the displacement amplification factor.

Figure 4 represents the structural relations of a base shear and roof displacement, which can be developed by

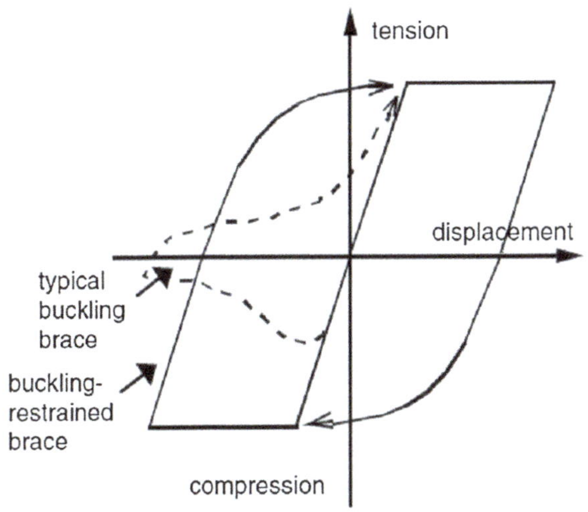

Figure 2 Axial force displacement behavior of braces. From AISC (2005).

a nonlinear analysis. In this figure, real nonlinear behavior is idealized by a bilinear elasto-plastic relation. Displacement amplification factor (C_d) and response modification factor (R) are determined as follows (Uang and Maarouf 1994):

$$C_d = \mu \times R_S \tag{2}$$

$$R = R_\mu \times R_S \tag{3}$$

where R_μ is a reduction factor due to ductility, R_S is the overstrength factor, and μ is the structural ductility factor defined as follows:

$$\mu = \frac{\Delta_{\max}}{\Delta_y} \tag{4}$$

where, according to Figure 4, Δ_{max} is the maximum displacement for the first life safety performance in the structure, and Δ_y is the yield displacement observed in the structure.

Several formulas of reduction factor due to ductility have been proposed by previous researchers such as Riddell (1989), Nassar and Krawinkler (1992), Miranda (1993), and Fajfar (2002). In the simple version of the N2 method proposed by Fajfar (2002), R_μ is written as follows:

$$\begin{aligned} R_\mu &= (\mu-1)\frac{T}{T_C} + 1 (T < T_C) \\ R_\mu &= \mu (T \geq T_C) \end{aligned} \tag{5}$$

where T_C is the characteristic period of the ground motion, and T is the fundamental period.

The strength revealed during the formation of plastic hinges is called overstrength, which is one of the important parameters in seismic design of structures. The overstrength factor R_S is defined by Mahmoudi and Zaree (2011) as follows:

$$R_S = \frac{V_u}{V_W} \times R_1 \times R_2 \tag{6}$$

According to Figure 4, V_W is the design base shear of the building, and V_u is the base shear with relevance to

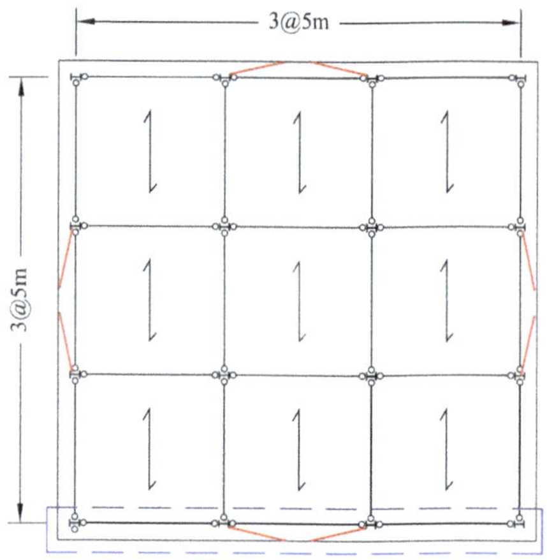

Figure 3 Plan views of buildings and brace locations in the frames.

Table 1 Sectional properties of seven-story model structures with inverted V braces

Braced type	Number of story	Interior column	Exterior column	Braces	Beam
Conventional CBFs					
Single bay brace frame	1	Box 150 × 150 × 10	Box150×150×10	2UNP120	IPE360
	2	Box 150 × 150 × 10	Box150×150×10	2UNP160	IPE360
	3	Box 150 × 150 × 10	Box 150 × 150 × 10	2UNP180	IPE360
	4	Box 250 × 250 × 15	Box 150 × 150 × 10	2UNP200	IPE360
	5	Box 250 × 250 × 15	Box 150 × 150 × 10	2UNP200	IPE360
	6	Box 300 × 300 × 20	Box 150 × 150 × 10	2UNP220	IPE360
	7	Box 300 × 300 × 20	Box 150 × 150 × 10	2UNP220	IPE360
Double bays brace frame	1	Box 150 × 150 × 10	Box 150 × 150 × 10	2UNP100	IPE360
	2	Box 150 × 150 × 10	Box 150 × 150 × 10	2UNP120	IPE360
	3	Box150 × 150 × 10	Box 150 × 150 × 10	2UNP120	IPE360
	4	Box150 × 150 × 10	Box 150 × 150 × 10	2UNP140	IPE360
	5	Box 200 × 200 × 15	Box 200 × 200 × 15	2UNP160	IPE360
	6	Box 200 × 200 × 15	Box 200 × 200 × 15	2UNP160	IPE360
	7	Box 250 × 250 × 15	Box 250 × 250 × 15	2UNP160	IPE360
BRBFs					
Single bay brace frame	1	Box 150 × 150 × 10	Box 150 × 150 × 10	PL50×15	IPE360
	2	Box 150 × 150 × 10	Box 150 × 150 × 10	PL50×15	IPE360
	3	Box 150 × 150 × 10	Box 150 × 150 × 10	PL50×18	IPE360
	4	Box 150 × 150 × 10	Box 150 × 150 × 10	PL50×20	IPE360
	5	Box 200 × 200 × 15	Box 150 × 150 × 10	PL50×20	IPE360
	6	Box 200 × 200 × 15	Box 150 × 150 × 10	PL50×20	IPE360
	7	Box 200 × 200 × 15	Box 150 × 150 × 10	PL60×20	IPE360
Double bays brace frame	1	Box 150 × 150 × 10	Box 150 × 150 × 10	PL50×10	IPE360
	2	Box 150 × 150 × 10	Box 150 × 150 × 10	PL50×10	IPE360
	3	Box 150 × 150 × 10	Box 150 × 150 × 10	PL50×10	IPE360
	4	Box 150 × 150 × 10	Box 150 × 150 × 10	PL50×10	IPE360
	5	Box 150 × 150 × 10	Box150 × 150 × 10	PL50×12	IPE360
	6	Box 200 × 200 × 15	Box 200 × 200 × 15	PL50×12	IPE360
	7	Box 200 × 200 × 15	Box 200 × 200 × 15	PL50×12	IPE360

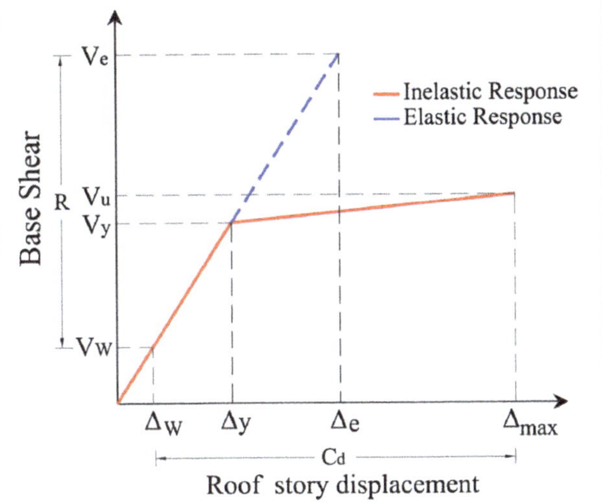

Figure 4 General structure response.

the first life safety performance of the structural members. R_1 accounts the difference between actual and nominal static yield strengths. For structural steel, a statistical study shows that the value of R_1 may be taken as 1.05 (Schmidt and Bartlett 2002). The parameter R_2 may be used to consider an increase in yield stress as a result of strain rate effect during an earthquake. For the strain rate effect, a value of 1.10 or a 10% increase could be used (Uang 1991).

To confirm displacement amplification factor (C_d) obtained from pushover analysis, Equation 7 was used for dynamic analysis:

$$C_d = \frac{\Delta_{max}}{\Delta_W} \times R_1 \times R_2 \times R_{SP} \tag{7}$$

where Δ_W is the design displacement, and R_{SP} is the post-buckling overstrength factor for CBFs. The post-buckling factors for CBFs in type V, inverted V, and X with single and two bracing bays are 1.11, 1.08, and 1.28, respectively (Mahmoudi and Zaree 2011).

Considering Equations 2 and 3, the ratio of C_d and R is thus:

$$\frac{C_d}{R} = \frac{\mu \times R_S}{R_\mu \times R_S} = \frac{\mu}{R_\mu} \tag{8}$$

According to Equation 8, it is feasible to evaluate the ratio of μ/R_μ instead of C_d/R. For building frame systems, various codes present the numerical values of the ratio between C_d and R. For instance, this ratio for concentrically steel braced frames is from 0.5 to 1 in NEHRP (1994) and IBC (2000) and is equal to 0.7 in the Iranian Earthquake Resistance Design Code (Standard No. 2800) (BHRC 2005).

Table 2 Characteristics of earthquake ground motions

Record	Year	Peak ground acceleration(g)	Duration(s)
El Centro	1940	0.348	53.7
Naghan	1977	0.723	5
Tabas	1978	0.915	25

Nonlinear analysis

Most structures experience inelastic deformations when subjected to severe earthquake ground motions. Therefore, the nonlinear behavior of structures should be taken into account to have accurate estimation of deformation demands. The extensive set of nonlinear analyses of the model buildings presented opportunities for the investigation of a large number of different response characteristics. Different performance criteria were also defined to verify structural or nonstructural elements under various performance levels. Structural performance level life safety (LS) is considered for the C_d assessment carried out in the present study. In the LS performance, the structure, or any part of it, does not collapse, retaining integrity and residual load capacity after the earthquake. The structure is significantly damaged and may have moderate permanent drifts but retains its full vertical load bearing capacity and sufficient residual lateral strength and stiffness to protect life even during strong aftershocks. Thus, nonlinear static (pushover) analysis and nonlinear time history analysis were used at life safety structural performance level. To do so, the SNAP-2DX (Rai et al. 1996) program was used.

Nonlinear static (pushover) analysis

Pushover analyses provide information on many response characteristics that cannot be obtained from an elastic static or elastic dynamic analysis. Furthermore, the ability of nonlinear static procedures to predict the maximum roof displacement caused by ground

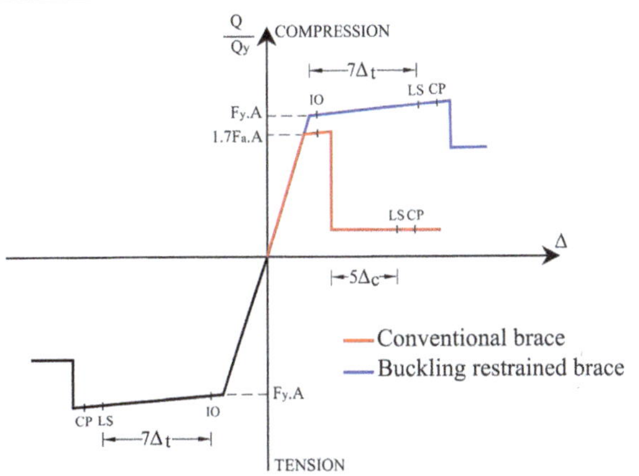

Figure 5 **Generalized force-deformation relation for steel brace elements (FEMA-356).** From Hajirasouliha and Doostan (2010).

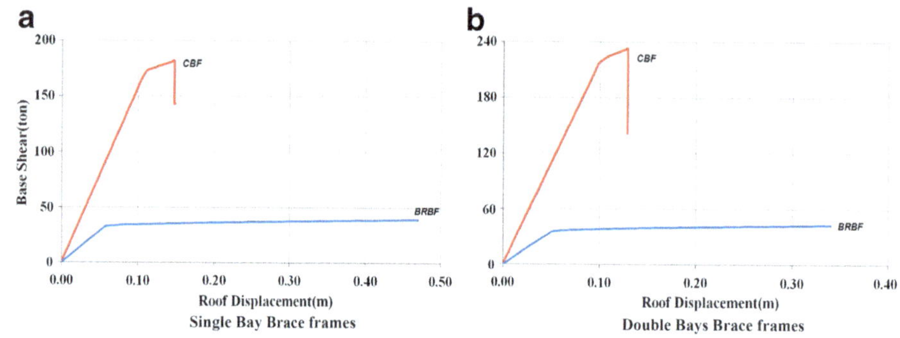

Figure 6 Roof displacement-base shear pushover curve for a seven-story inverted V braced frames. (**a**) Single bay brace frames. (**b**) Double bay brace frames.

Table 3 Displacement amplification factors for concentric braced frames result from pushover analysis

Brace type	Number of story	Single bay brace frame						Double bays brace frame					
		μ	R_S	R_μ	C_d	R	C_d/R	μ	R_S	R_μ	C_d	R	C_d/R
Conventional CBFs													
Chevron inverted V	3	1.45	4.24	1.31	6.15	5.55	1.11	1.53	6.09	1.35	9.31	8.22	1.13
	5	1.44	3.75	1.44	5.40	5.40	1.00	1.26	5.08	1.26	6.40	6.40	1.00
	7	1.39	3.72	1.39	5.17	5.17	1.00	1.25	4.74	1.25	5.92	5.92	1.00
	10	1.30	3.51	1.30	4.56	4.56	1.00	1.19	4.70	1.19	5.59	5.59	1.00
	12	1.29	3.50	1.29	4.51	4.51	1.00	1.18	4.29	1.18	5.06	5.06	1.00
Chevron V	3	1.48	3.82	1.32	5.65	5.04	1.12	1.39	5.22	1.26	7.25	6.58	1.11
	5	1.38	2.98	1.38	4.11	4.11	1.00	1.34	4.05	1.34	5.43	5.43	1.00
	7	1.42	2.88	1.42	4.08	4.08	1.00	1.28	3.67	1.28	4.69	4.69	1.00
	10	1.35	2.80	1.35	3.78	3.78	1.00	1.26	3.41	1.26	4.29	4.29	1.00
	12	1.32	2.78	1.32	3.66	3.66	1.00	1.26	3.27	1.26	4.05	4.05	1.00
X brace	3	1.68	3.61	1.49	6.06	5.38	1.13	1.50	5.86	1.34	8.79	7.85	1.12
	5	1.51	3.38	1.51	5.10	5.10	1.00	1.44	4.46	1.44	6.28	6.28	1.00
	7	1.52	3.05	1.52	4.64	4.64	1.00	1.45	4.16	1.45	6.03	6.03	1.00
	10	1.48	2.92	1.48	4.32	4.32	1.00	1.37	3.96	1.37	5.42	5.42	1.00
	12	1.42	2.86	1.42	4.06	4.06	1.00	1.37	3.67	1.37	5.03	5.03	1.00
BRBFs													
Chevron inverted V	3	9.47	2.41	6.76	22.82	16.30	1.40	9.33	3.41	6.68	31.80	22.85	1.39
	5	8.94	1.78	8.94	15.90	15.90	1.00	7.31	2.60	7.31	19.00	19.00	1.00
	7	7.54	1.74	7.54	13.12	13.12	1.00	6.72	2.36	6.72	5.89	5.89	1.00
	10	5.91	1.57	5.91	9.27	9.27	1.00	6.25	2.00	6.25	2.51	2.51	1.00
	12	5.02	1.40	5.02	7.02	7.02	1.00	5.25	1.78	5.25	9.34	9.34	1.00
Chevron V	3	8.77	2.53	6.30	22.18	15.93	1.39	8.89	3.39	6.38	30.13	21.62	1.39
	5	8.07	1.85	8.07	14.96	14.96	1.00	7.09	2.43	7.09	17.23	17.23	1.00
	7	7.25	1.84	7.25	13.34	13.34	1.00	6.91	2.19	6.91	15.13	15.13	1.00
	10	5.49	1.70	5.49	9.33	9.33	1.00	6.24	1.76	6.24	10.98	10.98	1.00
	12	4.75	1.58	4.75	7.50	7.50	1.00	5.11	1.48	5.11	7.56	7.56	1.00

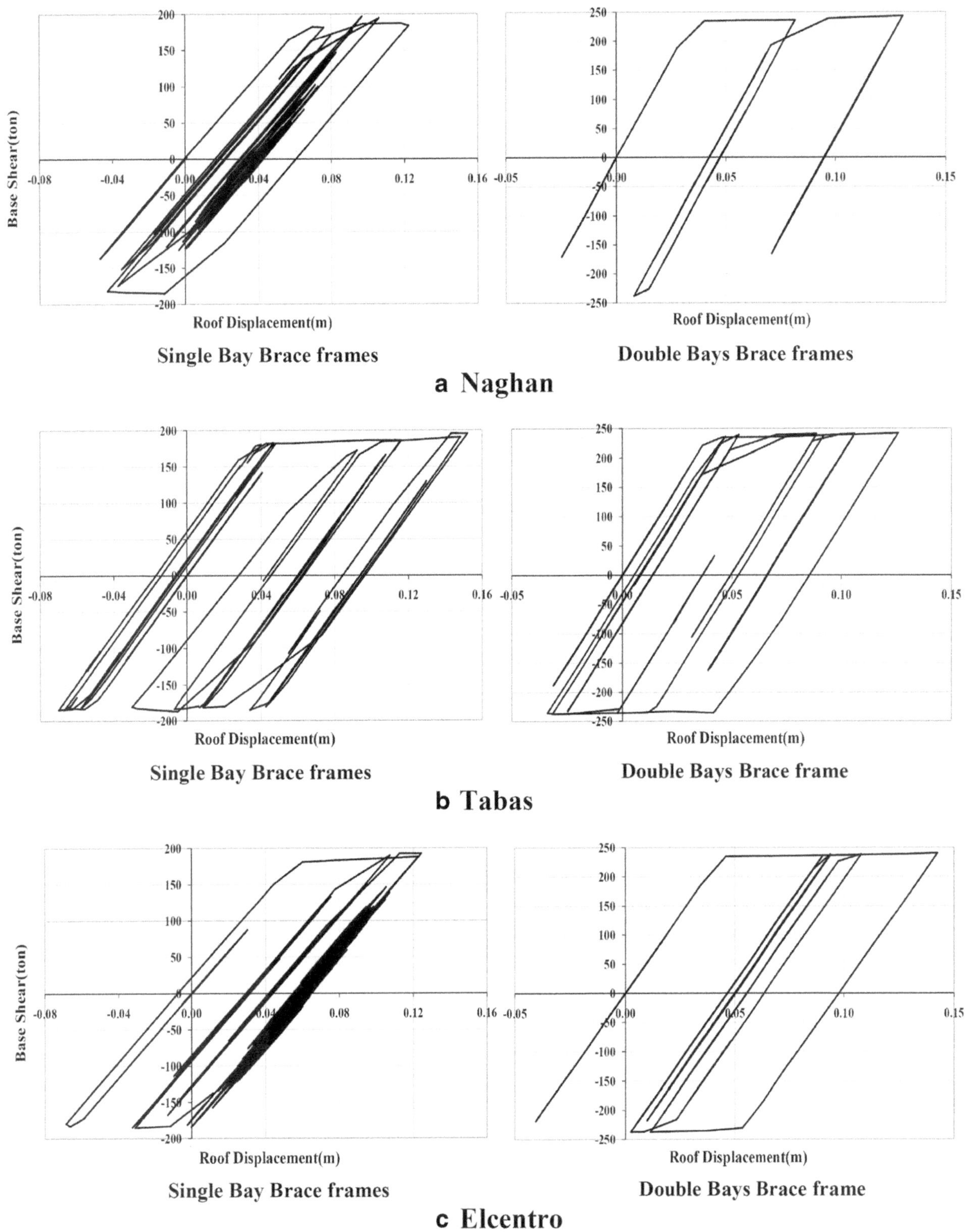

Figure 7 Nonlinear time history curves for seven-story conventional inverted V CBFs. (**a**) Naghan, (**b**) Tabas, and (**c**) El Centro earthquakes.

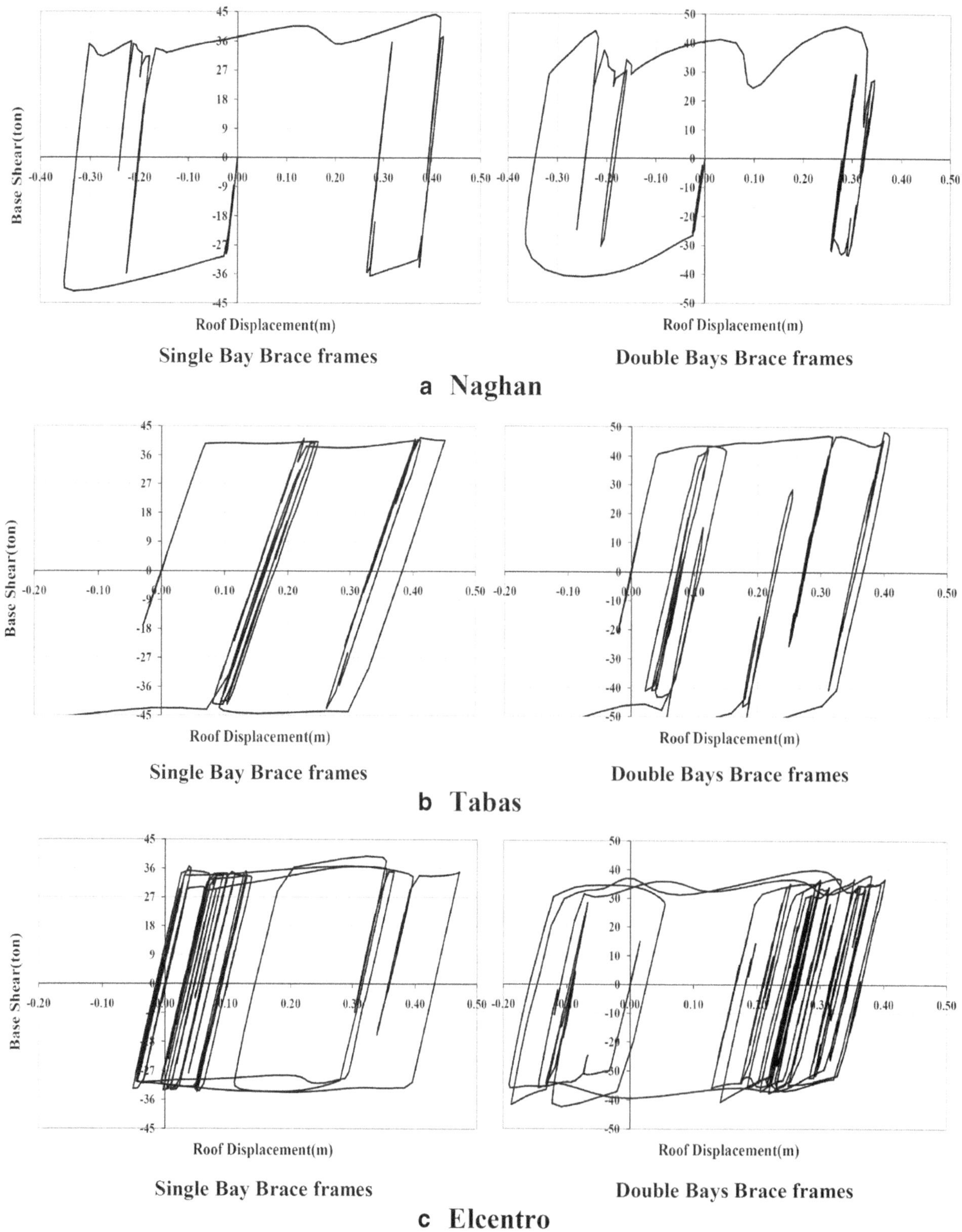

Figure 8 Nonlinear time history curves for seven-story inverted V BRBFs. (**a**) Naghan, (**b**) Tabas, and (**c**) El Centro earthquakes.

Table 4 Displacement amplification factors for concentric braced frame results from dynamic analysis

Brace type	Number of story	Single bay brace frame			Double bay brace frame		
		Δ_W(mm)	Δ_{max}(mm)	C_d	Δ_W(mm)	Δ_{max}(mm)	C_d
Conventional CBFs							
Chevron inverted V	3	10	48	6.03	4	31	9.27
	5	20	82	5.24	14	72	6.55
	7	36	134	4.84	27	127	6.03
	10	77	279	4.64	60	247	5.29
	12	113	370	4.21	89	330	4.75
Chevron V	3	11	47	5.45	6	35	7.32
	5	25	87	4.26	19	78	5.21
	7	45	139	3.87	32	123	4.86
	10	86	255	3.70	73	231	3.97
	12	125	346	3.46	101	314	3.86
X brace	3	8	34	6.21	6	34	8.95
	5	20	75	5.54	15	67	6.60
	7	37	121	4.83	28	117	6.18
	10	77	230	4.39	62	216	5.17
	12	125	333	3.93	90	298	4.90
BRBFs							
Chevron inverted V	3	13	248	21.87	7	199	32.36
	5	31	420	15.40	19	340	20.14
	7	42	460	12.57	32	406	14.69
	10	91	708	8.99	71	691	11.26
	12	154	956	7.15	117	915	9.06
Chevron V	3	13	229	20.98	8	208	29.3
	5	33	407	14.34	21	327	17.96
	7	42	506	13.98	40	491	15.13
	10	91	991	8.82	74	725	10.28
	12	152	943	7.14	128	840	7.56

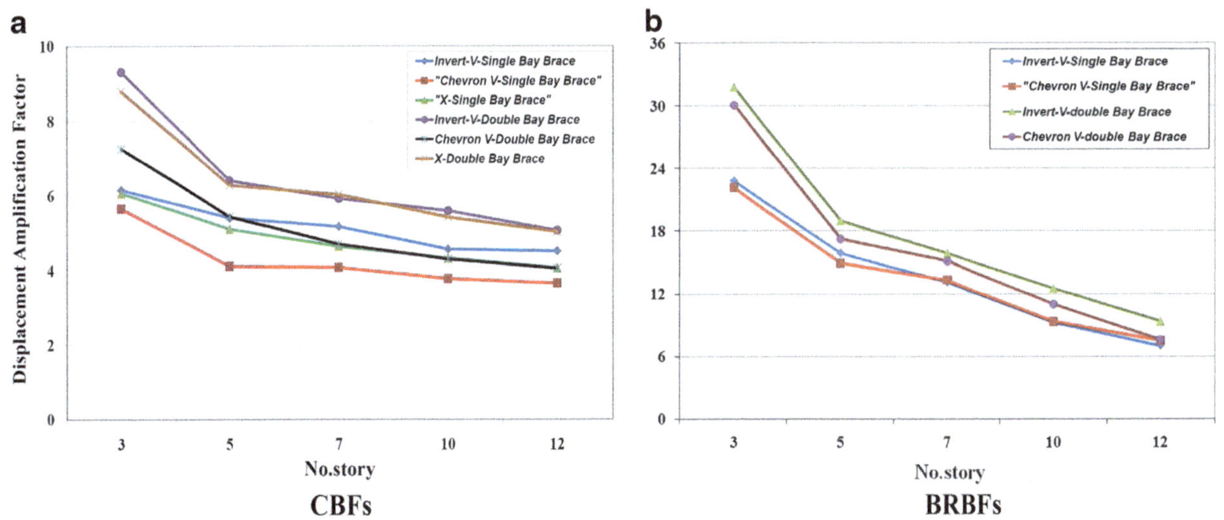

Figure 9 Displacement amplification factors for concentric braced frames. (**a**) CBFs and (**b**) BRBFs.

motion for CBFs was emphasized (Moghaddam and Hajirasouliha 2006). To achieve overstrength factor (R_S), ductility factor (μ), reduction factor due to ductility (R_μ), and finally displacement amplification factor (C_d) and its relation to response modification factor (R), the pushover analysis was carried out by subjecting a structure to monotonically increasing lateral forces with an invariant height-wise distribution.

Nonlinear time history analysis

Nonlinear time history analysis of a detailed analytical model is perhaps the best option to estimate deformation demands (Hajirasouliha and Doostan 2010). Although pushover analysis has advantages over elastic analysis procedures, underlying assumptions, the accuracy of pushover predictions, and limitations of current pushover procedures must be identified. Thus, nonlinear time history analyses have been performed to confirm the adequacy of static (pushover) analyses. Nonlinear dynamic analyses were carried out by employing suites of time history of the El Centro, Naghan, and Tabas earthquake matching with the design spectrum. The properties of the records used for this study are summarized in Table 2.

Modeling nonlinear behavior of braces, a generalized force-deformation relation was used for steel brace element as suggested by FEMA-356 (2000) (Figure 5). For buckling restrained braces, the model presented in Tables five, six, and seven in FEMA-356 (2000) were considered for both tension and compression behavior (Figure 5). The post-yield stiffness of beams, columns, and braces was initially assumed to be 2%. In Figure 5, Q, Q_y, and Δ are the generalized component load, expected strength, and component displacement, respectively. For conventional brace in compression, the residual strength after degradation is 20% of the buckling strength, and life safety plastic

Table 5 Comparison of displacement amplification factors for concentric braced frames

Brace type	Number of story	Single bay brace frame		Double bay brace frame	
		Pushover analysis	Time history analysis	Pushover analysis	Time history analysis
Conventional CBFs					
Chevron inverted V	3	6.15	6.03	9.31	9.27
	5	5.40	5.24	6.40	6.55
	7	5.17	4.84	5.92	6.03
	10	4.56	4.64	5.59	5.29
	12	4.51	4.21	5.06	4.75
Chevron V	3	5.65	5.45	7.25	7.32
	5	4.11	4.26	5.43	5.21
	7	4.08	3.87	4.69	4.86
	10	3.78	3.70	4.29	3.97
	12	3.66	3.46	4.05	3.86
X brace	3	6.06	6.21	8.79	8.95
	5	5.10	5.54	6.28	6.60
	7	4.64	4.83	6.03	6.18
	10	4.32	4.39	5.42	5.17
	12	4.06	3.93	5.03	4.90
BRBFs					
Chevron inverted V	3	22.82	21.87	31.80	32.36
	5	15.90	15.40	19.00	20.14
	7	13.12	12.57	5.89	14.69
	10	9.27	8.99	2.51	11.26
	12	7.02	7.15	9.34	9.06
Chevron V	3	22.18	20.98	30.13	29.3
	5	14.96	14.34	17.23	17.96
	7	13.34	13.98	15.13	15.13
	10	9.33	8.82	10.98	10.28
	12	7.50	7.14	7.56	7.56

deformation Δ_{LS} is equal to $5\Delta_C$ (Δ_C is the axial deformation at expected buckling load). On the other hand, for conventional brace in tension and buckling restrained brace, the life safety plastic deformation Δ_{LS} is equal to $7\Delta_T$ (Δ_T is the axial deformation at expected tensile yielding load). Based on earlier tests, the compression strength of BRB was assumed to be 10% larger than the strength in tension (Clark 2000). In this paper, the compression strength for BRB is considered equal to the tension strength. To capture the greatest demands on braces and beams, flexible beams were used.

Results and discussion

Base shear vs. maximum roof displacement nonlinear pushover analysis for a seven-story inverted V CBF and BRBF with single and double bracing bays are shown in Figure 6. As it is shown, the stiffness of CBF is higher than BRBF, whereas the ductility for BRBF is higher compared to CBF. Table 3 shows the displacement amplification factors (C_d) and its relations to response modification factors (R) for both CBFs and BRBFs. Figures 7 and 8 show nonlinear dynamic time history analysis results of the El Centro, Naghan, and Tabas earthquakes for a seven-story inverted V CBF and BRBF with single and double bracing bays. Table 4 shows the displacement amplification factors (C_d) for both CBFs and BRBFs assessed from nonlinear dynamic time history analysis.

The ductility in CBFs has lower values due to deterioration in strength and degradation of stiffness due to brace buckling in cyclic loading. Thus, it can be said that the overstrength factors have main effects on displacement amplification factors for CBFs. The number of bracing bays and structure height have an effect on overstrength and thus on C_d. On the other hand, these factors as such have no obvious result on ductility, so changing the structure height and the number of bracing bay has no effect on the ratio C_d/R.

In BRBFs, because of brace energy dissipation capacity in tension and compression, the ductility has high values and becomes the main parameter to determine displacement amplification factors. Also, structure height has a profound effect on ductility, so for BRBFs, variation in the number of stories has obvious impression on displacement amplification factors and the ratio C_d/R. On the other hand, structure characteristic cause little variation in overstrength factors and so in C_d and the ratio C_d/R.

Variation in displacement amplification factors for different types of concentric braced frames are shown in Figure 9. The comparison of pushover and nonlinear dynamic analysis displacement amplification factors are shown in Table 5. According to the results, pushover analysis provides good predictions of seismic demands for concentrically braced steel frames.

Conclusions

This paper assesses the displacement amplification factor (C_d) and the ratio between C_d and R factor of 30 conventional CBFs and 20 BRBFs in life safety structural performance level. For this purpose, the nonlinear static (pushover) analysis and nonlinear dynamic time history analysis have been performed on the buildings with single and two bracing bays, various stories, and different buckling restrained brace and conventional brace configurations. The beam-column connections were assumed to be pinned so that the seismic load was resisted mainly by braces. The results of this study can be summarized as follows:

- The displacement amplification factors increase with the decrease of structure height and the increase in the number of bracing bays. However, the number of bracing bays has no effect on the ratio C_d/R in both CBFs and BRBFs. The ratio between C_d and R factor is from 1 to 1.12 and 1 to 1.4 for CBFs and BRBFs, respectively.
- The displacement amplification factors for CBFs in type V, inverted V, and X are evaluated as 4.40, 5.20, and 4.90 for single bracing bay and as 5.40, 6.80, and 6.60, respectively, for double bracing bays.
- The obtained displacement amplification factors for different types of BRBFs with single bracing bay vary from 7 to 22.50, and for double bracing bays, these are from 8 to 31.
- The structure height in CBFs has no effect on ratio C_d/R, but in BRBFs, this has an effect on the ratio C_d/R because of ductility variation. Thus, in ductile brace frame systems (high ductility) and stiff buildings (low fundamental period), the ratio C_d/R is higher than 1 when the fundamental period (T) is lower than the period of the ground motion (T_C).

Competing interests

The authors declare that they have no competing interests.

Authors' contributions

MM proposed the subject of the research and participated in the sequence alignment. He also controlled the results. ZM carried out the structural analysis (linear and nonlinear) and data processing and participated in the sequence alignment. All authors read and approved the final manuscript.

Acknowledgment

This work was supported financially by the Shahid Rajaee Teacher Training University. We would like to thank for this support.

References

AISC (2005) Seismic provisions for structural steel buildings. American Institute of Steel Constructions, Illinois

BHRC (2005) Iranian code of practice for seismic resistance design of buildings: standard no.2800, 3rd edn. Building and Housing Research Center, Tehran

Broderick BM, Elghazouli AY, Goggins JM (2008) Earthquake testing and response analysis of concentrically-braced sub-frames. J Constructional Steel Res 64 (9):997–1007

Clark P (2000) Evaluation of design methodologies for structures incorporating steel unbonded braces for energy dissipation. In: 12th World Conference on Earthquake Engineering Proceedings. New Zealand Society for Earthquake Engineering, paper no. 2240. NZSEE, Wellington
Davaran A, Hoveidae N (2009) Effect of mid-connection detail on the behavior of X-bracing systems. J Constructional Steel Res 65(4):985–990
Fajfar P (2002) Structural analysis in earthquake engineering—a breakthrough of simplified nonlinear methods. 12th European Conference on Earthquake Engineering, London, 9–13 September 2002
FEMA (2000) Prestandard and commentary for the seismic rehabilitation of building, FEMA-356. Federal Emergency Management Agency, Washington, DC
Goggins JM, Broderick BM, Elghazouli AY, Lucas AS (2006) Behaviour of tubular steel members under cyclic axial loading. J Constructional Steel Res 62(2):121–131
Hajirasouliha I, Doostan A (2010) A simplified model for seismic response prediction of concentrically braced frames. Adv Eng Softw 41(3):497–505
Hussain S, Benschoten PV, Satari MA, Lin S (2006) Buckling restrained braced frame (BRBF) structures: analysis, design and approvals issues. Proceedings of the 75th SEAOC Annual Convention, Long Beach, 13–16 September 2006
IBC (2000) International building code. International Code Council, Washington, DC
Kiggins S, Uang CM (2006) Reducing residual drift of buckling-restrained braced frames as a dual system. J Eng Struct 28(11):1525–1532
Kim J, Choi H (2004) Behavior and design of structures with buckling-restrained braces. J Eng Struct 26(6):693–706
Kumar GR, Kumar SRS, Kalyanaraman V (2007) Behaviour of frames with non-buckling bracings under earthquake loading. J Constructional Steel Res 63(2):254–262
Lee SS, Goel SC, Chao SH (2004) Performance-based seismic design of steel moment frames using target drift and yield mechanism. 13th World Conference on Earthquake Engineering, Vancouver, B.C. Canada, 1–6 August 2004. Paper No 266
Mahmoudi M (2004) The ratio of displacement amplification factor to force reduction factor. 13th World Conference on Earthquake Engineering, Vancouver, B.C. Canada, 1–6 August 2004. Paper No. 1917
Mahmoudi M, Zaree M (2010) Evaluating response modification factors of concentrically braced steel frames. J Constructional Steel Res 66(10):1196–1204
Mahmoudi M, Zaree M (2011) Evaluating the overstrength of concentrically braced steel frame systems considering members post-buckling strength. J Constructional Steel Res 9(1):57–62
MHUD (2009) Iranian national building code (part 10): steel structure design. Ministry of Housing and Urban Development, Tehran
Miranda E (1993) Site-dependent strength-reduction factors. Struct Eng 119 (12):3503–3519
Moghaddam H, Hajirasouliha I (2006) An investigation on the accuracy of pushover analysis for estimating the seismic deformation of braced steel frames. J Constructional Steel Res 62(4):343–351
Nassar A, Osteraas J, Krawinkler H (1992) Seismic design based on strength and ductility demands. In: Proceedings of the Earthquake Engineering Tenth World Conference. Balkema, Rotterdam, pp 5861–5866
NEHRP (1994) Recommended provisions for the development of seismic regulations for new building. Seismic Safety Council, Washington, DC
Rai DC, Goel SC, Firmansjah G (1996) User's guide: structural nonlinear analysis program (SNAP-2DX). , Department of Civil and Environmental Engineering, University of Michigan, Ann Arbor
Riddell R, Hidalgo P, Cruz E (1989) Response modification factors for earthquake resistant design of short period buildings. Earthquake Spectra 5(3):571–589
Sahoo DR, Chao SH (2010) Performance-based plastic design method for buckling restrained braced frames. J Eng Struct 32(9):2950–2958
Schmidt BJ, Bartlett FM (2002) Review of resistance factor for steel: resistance distributions and resistance factor calibration. Can J Civ Eng 29:109–118
Uang CM (1991) Establishing R (or Rw) and C_d factor for building seismic provision. J Struct Eng 117(1):19–28
Uang CM, Maarouf A (1994) Deflection amplification factor for seismic design provisions. J Struct Eng 120(8):2423–243

20

Structural damage identification using signal processing method

Mallikarjuna D Reddy[1*] and Seetharaman Swarnamani[2]

Abstract

The objective of the current work is to show the effectiveness of using wavelet transform for detection and localization of small damages. The spatial data used here are the rotational mode shapes of the damaged and undamaged plate-like structures. The continuous wavelet transform using complex Gaussian wavelet is used to get the spatially distributed wavelet coefficients so as to identify the damage position on a square plate. The rotational mode shape data of the square plate with damage of different sizes are obtained using ANSYS 9.0. Damage identification for different boundary conditions is studied.

Keywords: Damage identification, Gaussian wavelet, Structural damage, Mode shapes, Wavelets

Introduction

Damage in a mechanical (or) structural system may be contributed by various factors, such as excessive response, accumulative crack growth, wear and tear of working parts, and impact by a foreign object. Structural health monitoring has emerged as a reliable, efficient, and economical approach to monitor system performance, detect such damage, asses/diagnose structural health condition, and make corresponding maintenance decisions; consequently, structural safety and functionality will be significantly improved and a condition-based maintenance procedure can be developed. Due to localization of damage in structure techniques using global averaging procedures, changes in eigen frequencies are less sensitive to initial or small changes. Hence, techniques that process the local information based on wavelets have emerged recently. An application of spatial wavelet theory to damage identification in structures was proposed by Liew and Wang (1998). They calculated the wavelet coefficients along the length of the beam based on the numerical solution for the deflection of the beam; the damage location was then indicated by a peak in the coefficients of the wavelets along the length of the beam. Wang and Deng (1999) described a method for detecting the location of localized defects. Quek et al. (2001) also used wavelet analysis for crack identification in beams under both simply supported and fixed-fixed boundary conditions. Hong et al. (2002) used the Lipschitz exponent for the detection of singularities in beam modal data. The Mexican hat wavelet was used, and the damage extent has been related to different values of the exponent. The correlation, however, of the damage extent to the Lipschitz exponent is sensitive to both sampling distance and noise, resulting in limited accuracy of the prediction. Recently, an interesting comparison between a frequency-based and mode shape-based method for damage identification in beam-like structure has been published by Kim et al. (2003). Later, Chang and Chen (2004) used spatially distributed signals by Gabor wavelet transform so that the distributions of wavelet coefficients could identify the damage position of a rectangular plate by showing a peak at the position of the damage which was very sensitive to the damage size. Also, Abdo and Hori (2001) made numerical study of the relation between damage characteristics, and changes in the dynamic properties are presented. It was found that the rotation of mode shape has the characteristic of localization at the damaged region even though the displacement modes are not localized.

Despite the extensive studies on vibration analysis of damaged plates, only few effective and practical techniques are found for very small damage identification. In this paper a study is carried out to investigate the influence of using the rotation of mode shapes as an input

* Correspondence: dmreddy@revainstitution.org
[1]Reva Institute of Technology and Management, Bangalore Karnataka 560064, India
Full list of author information is available at the end of the article

for the wavelets on damage identification for different scenarios of damage for different boundaries.

Methods

Using rotational mode shapes

In this section, further study is carried out to investigate the influence of using the rotation of mode shapes (Abdo and Hori 2001) as an input for the wavelets on damage identification for different scenarios of damage for different boundaries. The plate, whose length and width is equally divided into 25 elements so the simulated model has a total of 625 elements, is used for analysis. Damage is simulated by reducing the thickness of one element at different locations. Four-noded shell element with six degrees of freedom per node (three translations and three rotations (U_x, U_y, U_z, θ_x, θ_y, θ_z)) was used. Since the rotation around the Z-axis (θ_z) is very small for thin plates, only rotations around the X-axis (θ_x) and Y-axis (θ_y) are taken into consideration. These two rotations are normalized with respect to the square root of the sum of squares. The maximum slope at any point of a thin plate can be calculated as (Liew and Wang 1998)

$$Maximum \;\; slope = \sqrt{(\theta_x)^2 + (\theta_y)^2}. \quad (1)$$

Thus, for the sake of generality in this study, the following invariant is used for rotation at the jth node of the ith mode shape.

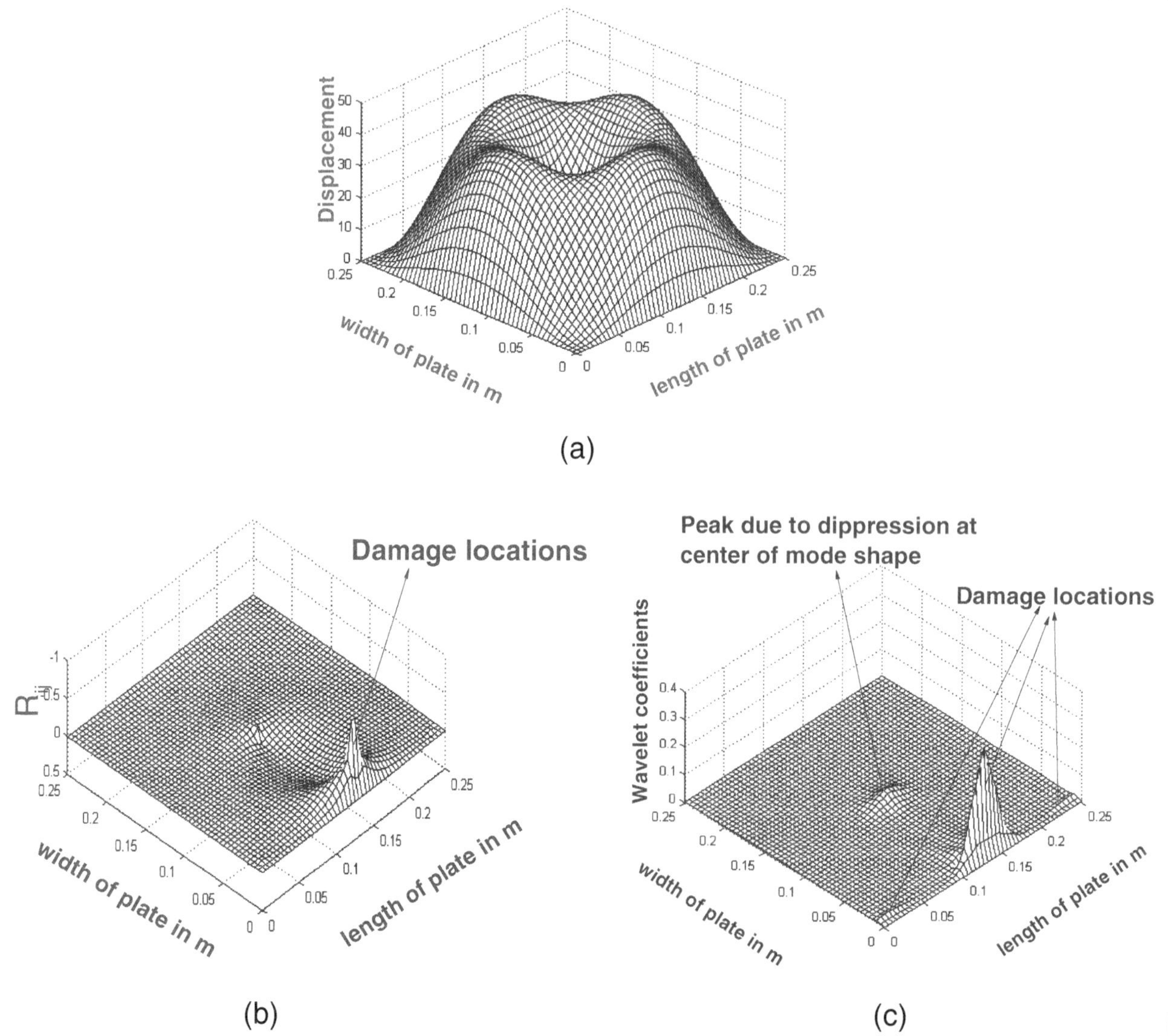

Figure 1 Rotation mode shape of the damaged plate. (**a**) The first rotation mode shape of the (33.33%) damaged plate (boundary conditions are fixed at four edges). (**b**) Absolute difference of the first rotation of mode shapes between the intact and damaged plate. (**c**) The distributions of wavelet coefficients for scale (s = 2) based on the absolute difference of the first rotation of mode shape.

$$R_{ij} = \sqrt{(\theta_x)_{ij}^2 + (\theta_y)_{ij}^2} \quad (2)$$

The procedure of the damage detection is as follows:

(a) Find the analytical mode shapes of the structure.
(b) Calculate the spatial wavelet coefficients of the mode shapes.
(c) Plot the value of wavelet coefficients in the full region for each scale of wavelets.
(d) Examine the distributions of wavelet coefficients at each scale. A sudden change in the distributions of the wavelet coefficients identifies the damage position.

Results and discussion

All-edges-fixed boundary conditions

In this particular case with a fixed plate, 33.33% severity is introduced (Figure 1a shows the first rotational mode shape of a fixed plate with three damages). Two damages are at the corner of plate, and one is at the center of a particular edge. It is observed from Figure 1b that the damage at the center of the edge is clearly located by high amplitude of absolute difference of the rotation mode shape. The corner damages are not highlighted, as seen in Figure 1b. To increase the sensitivity of the method to identify corner damages, the absolute difference in rotation mode shape is given as input to wavelet transform. The corresponding wavelet coefficient distribution is as shown in Figure 1c. It is clearly observed in

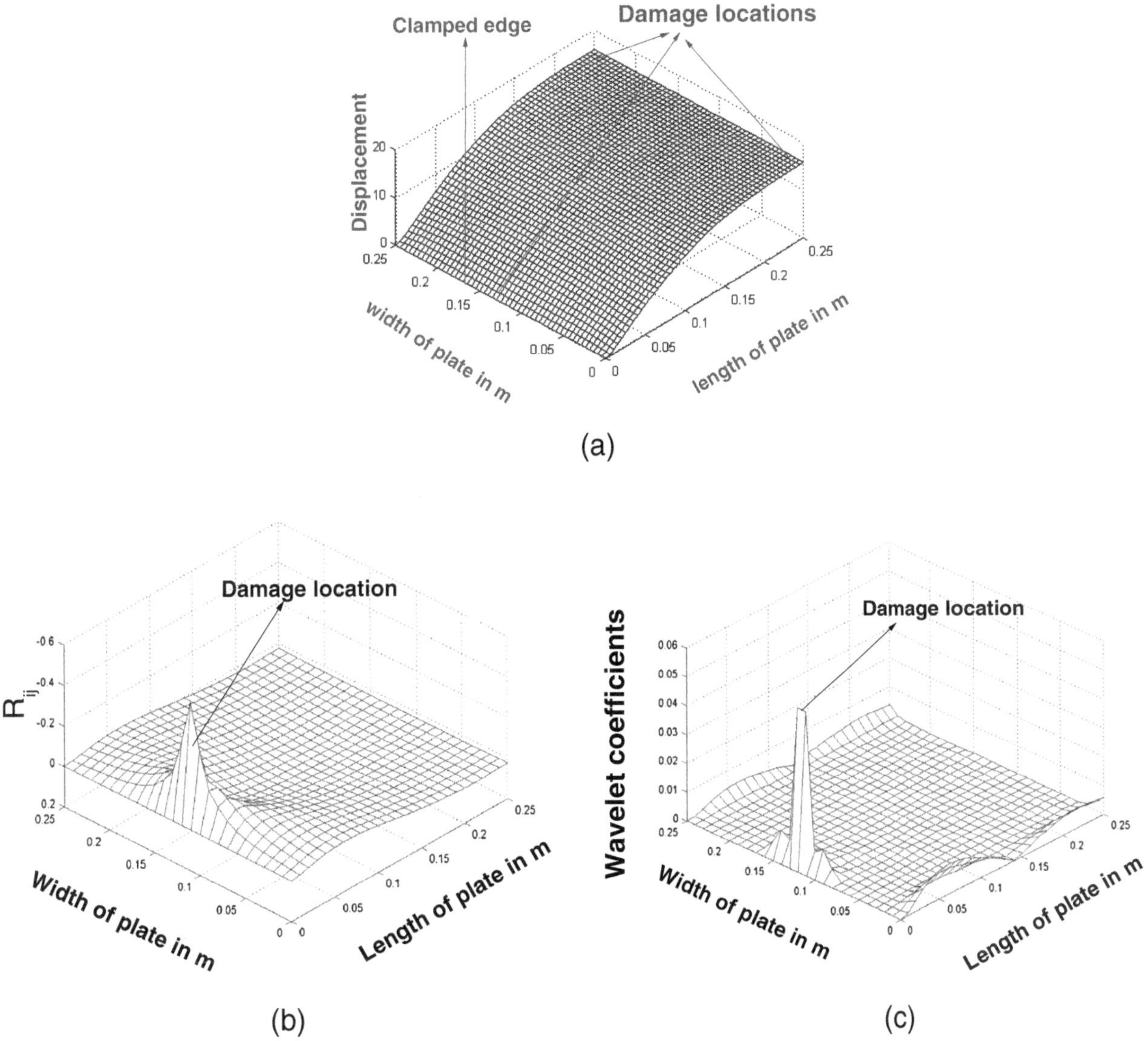

Figure 2 Rotation mode shape of the damaged cantilever plate. (**a**) The first rotation mode shape of the (33.33%) damaged cantilever plate. (**b**) Absolute difference of the first rotation of mode shapes between the intact and damaged plate. (**c**) The distributions of wavelet coefficients for scale (s = 2) based on the absolute difference of the first rotation of mode shape.

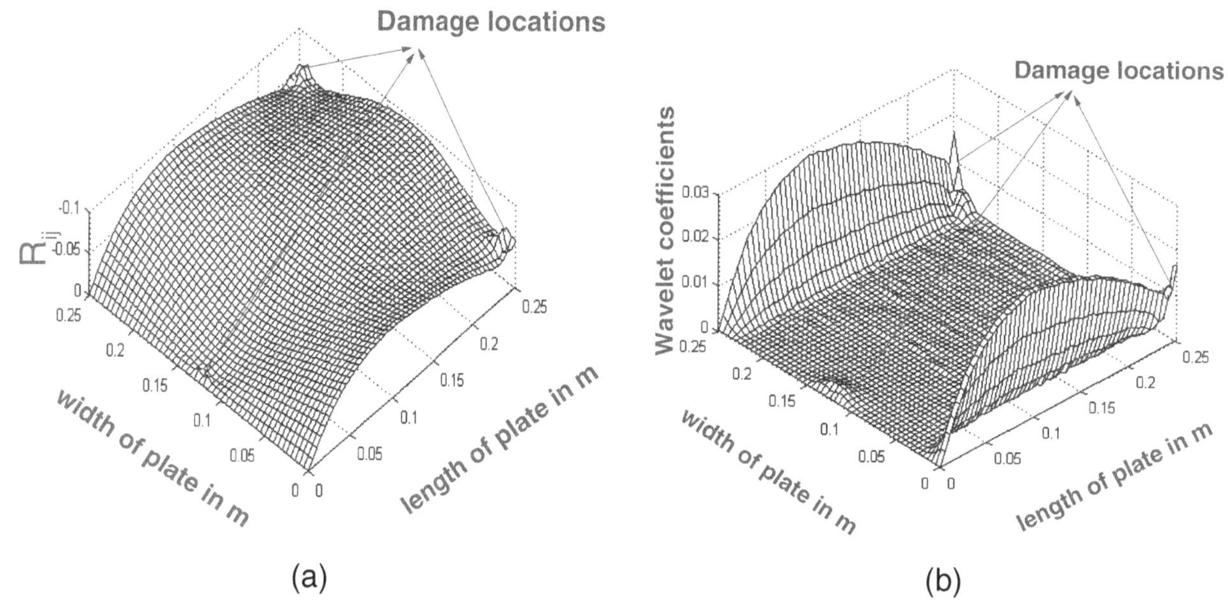

Figure 3 Second rotation of mode shapes. (**a**) Absolute difference of the second rotation of mode shapes between the intact and damaged plates. (**b**) The distributions of wavelet coefficients for scale (s = 2) based on the absolute difference of the second rotation of mode shape.

Figure 1c that the damage at the corner is identified by high coefficients at that location as indicated. Hence, wavelet-transformed rotation mode shape has the potential to identify damages at any locations, which would not be possible by using only rotation mode shape alone. Another observation in Figure 1b,c is that there is a center peak exactly at the center of the plate, which is due to the maximum slope (depression) of the rotation mode shape at the center.

One-edge-fixed boundary conditions (cantilever plate)

The same plate with 33% damage condition is analyzed for a mode shape as a cantilever, as shown in Figure 2a. Two damages are at the free-edge corner of the plate, and one is at the center of plate's clamped edge.

It was observed from Figure 2b that the damage at the center of the clamped edge is clearly located by high amplitude of absolute difference of the first rotation mode shape. However, the two damages at the free-edge corner are not highlighted, as shown in Figure 2b. Further, to increase the sensitivity of the method to identify corner damages, the absolute difference of the first rotation mode shape was inputted to the wavelets. The wavelet coefficient plot as shown in Figure 2c clearly shows which is not sensitive to damages at the free end of the plate.

To find out damages at the free end, absolute difference of the second rotation mode shape is used, as shown in Figure 3a which clearly indicates all three damages. To increase the sensitivity of the method, respective rotation mode shape was inputted to the wavelet. The corresponding wavelet coefficient distribution is as shown in Figure 3b. This clearly identifies the peaks with high wavelet coefficients at damage locations.

Conclusions

The objective of this paper is to apply spatial wavelet transform to highlight the sensitivity for detection and localization of damage in a plate structure with all boundaries fixed, using rotational mode shape as input. It was observed that by using modal data as input, damage can be identified exactly if the reduction in thickness is more than 10%.

Changes in rotations of mode shapes due to the presence of structural damage, represented here in a numerical finite element model, have been investigated. The results of the aluminum plate model demonstrate the usefulness of the changes in the rotation of mode shape as a diagnostic parameter in detecting and locating damage with regard to the plate with different boundary conditions. For identifying damages at a boundary, the method of using absolute difference rotational mode shape data is not sufficient. Thus, boundary damages are best identified when the absolute difference rotation mode shape is given as input to wavelet transform and the coefficients plotted in a length-width plane. It is observed that by using rotation modal data as input, damage can be identified exactly if the reduction in thickness is more than 5%. Below 5% reduction in thickness, absolute difference rotation mode shape data input to wavelet is more sensitive.

Competing interests
The authors declare that they have no competing interests.

Authors' contributions
MDR carried out the modeling and wavelet analysis part in Matlab and in CAE softwares. SS carried out the crack detection method for different boundary conditions by taking difference in the healthy and damaged mode shape plate. MDR and SS together analyzed the sensitivity of wavelets to crack detection at different levels of scales. Both authors read and approved the final manuscript.

Acknowledgments
The authors thank the Department of Mechanical Engineering of the Indian Institute of Technology Madras, Chennai, for the modeling and simulation support. The authors are immensely thankful for the support rendered by the editor of this journal for editing the manuscript.

Author details
[1]Reva Institute of Technology and Management, Bangalore Karnataka 560064, India. [2]Indian Institute of Technology Madras, Chennai, Tamilnadu 600036, India.

References
Abdo MAB, Hori M (2001) A numerical study of structural damage detection using changes in the rotation of mode shapes. J Sound Vib 251(2):227–239

Chang CC, Chen LW (2004) Damage detection of a rectangular plate by spatial wavelet approach. Int J Appl Acoust 65:819–832

Hong JC, Kim YY, Lee HC, Lee YW (2002) Damage detection using Lipschtiz exponent estimated by the wavelet transform: applications to vibration modes of a beam. Int J Solids Struct 39:1803–1816

Kim TJ, Ryu YS, Cho HM, Stubbs N (2003) Damage identification in beam type structures: frequency based method vs. mode shape based method. Eng Struct 25:57–67

Liew KW, Wang Q (1998) Application of wavelet theory for crack identification in structures. J Eng Mech 124:152–157

Quek S, Wang Q, Zhang L, Ang K (2001) Sensitivity analysis of crack detection in beams by wavelet techniques. Int J Mech Sci 43:2899–2910

Wang Q, Deng X (1999) Damage detection with spatial wavelets. Int J Solids Struct 36:3443–3468

Deformation capacity of unreinforced masonry walls subjected to in-plane loading: a state-of-the-art review

Amir Hosein Salmanpour[1*], Nebojsa Mojsilovic[1] and Joseph Schwartz[2]

Abstract

A research project on the deformation capacity of unreinforced masonry structures is underway at the Institute of Structural Engineering of ETH Zurich. The development of the basic building blocks for the displacement-based design of unreinforced masonry structures is the objective of the present research project, which should be seen as a first step in an initiative to investigate the limits of the deformation capacity of unreinforced masonry walls. This paper presents a summary review of previous experimental and analytical studies on the deformation capacity of unreinforced masonry walls subjected to in-plane loading. This review is the first phase of the aforementioned research program. A summary of 71 shear tests on unreinforced masonry walls is presented in the form of a database, along with the statistical analysis and discussion of the tests results. Furthermore, three different computational approaches for structural masonry, i.e. micro-modelling, macro-modelling and macro-element discretization, are discussed, and a review of macro-elements for the in-plane response of unreinforced masonry walls is presented. The reviewed models are discussed and a set of conclusions is given. Special attention is devoted to the deformation capacity parameter throughout the paper. Finally, the paper shows the limitations of our current state of knowledge of the deformation capacity of structural masonry.

Keywords: Deformation capacity; Experimental research; Shear test; Analytical research; In-plane response; Macro-element; Structural masonry; URM

Introduction

Masonry structures and materials represent one of the oldest building concepts available. Masonry construction is a traditional, widely used, extremely flexible and economical construction method with considerable potential for future developments. However, possibly due to the substantial empirical knowledge collected over several centuries of utilization of masonry as a structural material, the need for establishing a more modern basis for the design of masonry structures has not been appreciated in the same manner as for concrete structures. As a result, conventional masonry design practice is overly conservative, particularly in regard to the assessment of seismic resistance. Hence, the potential of masonry has not yet been fully exploited and there is a clear need for better utilization. For example, while current codes of practice severely limit the use of unreinforced masonry (URM) in construction, mainly because of the requirement of over-conservative values for the force-reduction factor (q-factor in Europe or R-factor in the US), recent studies show that the performance of structurally-designed low-rise URM buildings should be considered adequate for the category of ordinary buildings even in regions with appreciable seismic hazard. Furthermore, these studies also show that unreinforced masonry is still a very competitive choice for two- or three-story residential buildings (Magenes et. al 2009 and Lourenco et al. 2009).

Based on the positive experience gained in recent years in developing the basis for the displacement-based seismic design of concrete structures, it appears that the most feasible approach to enhance the rationality for the design of masonry structures is to apply the same fundamentals

* Correspondence: salmanpour@ibk.baug.ethz.ch
[1]Institute of Structural Engineering, ETH Zurich, CH-8093, Zurich, Switzerland
Full list of author information is available at the end of the article

in the analysis of masonry structures. A more consistent representation of the material resistance as well as of the (seismic) loads leads to more economical (or at least more reliable) designs in general, and especially for masonry structures where the safety margin, i.e. safety factor at present, is mostly based on experience rather than on quantified engineering modelling. Obviously, the first step towards the development of such an approach for masonry structures is to investigate the limits of the deformation capacity of masonry structures.

Given the above, a research project on the deformation capacity of unreinforced masonry structures has been started at the Institute of Structural Engineering of ETH Zurich. The objective of the present research project is the development of the basic building blocks for the displacement-based design of unreinforced masonry structures, which should be seen as the first step in an initiative to investigate the limits of the deformation capacity of unreinforced masonry walls. The research project includes a thorough survey and assessment of existing experimental and theoretical research in the area of the deformation capacity of structural masonry, an intensive testing campaign as well as developing and introducing new sophisticated mechanical models for structural masonry. This paper presents a summary review of previous experimental and theoretical studies on the deformation capacity of unreinforced masonry walls subjected to in-plane loading. This review is the first phase of the above-mentioned research program. It is worth noting here that even the conventional force-based seismic design approach is strongly connected to the deformation capacity parameter through the force-reduction factor which is used in the estimation of design force of structures in the force-based seismic design approach. Furthermore, the deformation capacity plays a crucial role in the seismic assessment and retrofitting of existing structures, which has become one of the main research topics in structural engineering.

Deformation capacity of structural masonry

In masonry structures subjected to seismic action, if local brittle failure modes, e.g., out-of-plane failure, are prevented by providing proper connections between intersecting walls and also between walls and diaphragms, a rather ductile global behaviour governed by the in-plane response of walls can develop. Hence, the investigation of the deformation capacity of masonry structures should start by studying the in-plane behaviour of masonry walls and their constitutive elements, i.e. piers and spandrels (in the case of walls with openings). Even though spandrels have a significant effect on the in-plane response of URM masonry walls, it seems, however, that the deformation capacity of masonry walls is mainly identified with the deformation capacity of piers. This is because experimental studies on spandrel and pier elements have shown that for most wall geometries the spandrel deformation capacity is larger than the pier deformation capacity; see, e.g., (Beyer and Dazio 2012).

The in-plane behaviour of masonry walls subjected to horizontal and vertical forces has been investigated in various test programs. As indicated by experiments, the in-plane response of masonry walls depends mainly on their failure mechanism. In the case of low vertical load and/or poor quality mortar, seismic loads cause shearing of the wall in two parts and sliding of the upper part on the other part. The mechanism is called sliding shear failure. The in-plane response of masonry walls failing in sliding shear mode is very stable and close to an elastic, perfectly plastic response with high energy dissipation and displacement capacity. In the case of sliding shear failure mode, the displacement capacity is very large. However, for practical applications it should be limited since the shear walls normally interact with other building elements.

The diagonal shear mode (simple shear mode) occurs where the principal tensile stress exceeds the in-plane tensile strength of the masonry. Peak resistance is governed by the formation and development of diagonal cracks. In the case of the diagonal shear mode, the typical response of masonry walls is characterized by rapid strength and stiffness degradation, moderate energy dissipation and limited displacement capacity. Although the diagonal shear mechanism, which often governs the in-plane response of masonry walls subjected to seismic loads, has limited deformation capacity, a classification of such a mechanism as simply brittle would lead to significant underestimation of the seismic capacity of masonry buildings. Hence, a moderate ductility, or better, a non-negligible nonlinear behaviour and deformation capacity has to be recognised for the diagonal shear failure mode (Magenes and Penna 2011).

Rocking-flexural, or simply flexural, failure usually takes place in the case of a high moment/shear ratio, i.e. in slender walls. As the horizontal load increases, bed joints crack in tension and shear is carried by the compressed masonry. The final failure is obtained by crushing of the compressed corner. In general, the in-plane response of masonry walls failing in the rocking-flexural mode is almost nonlinear elastic with very moderate hysteretic energy dissipation and negligible strength degradation. Regarding the displacement capacity, very large displacements can be obtained, especially when the axial load is low compared to the compressive strength of masonry. Actually, if no other failure mechanisms occur, the displacement, which can be attained in a rocking response, can be limited only by second order (P-Δ) effects associated with overturning (Magenes and Calvi 1997).

A substantial amount of experimental and theoretical work has been carried out to investigate the in-plane seismic response of masonry piers. This paper presents a summary review of these studies, with special attention given to the ultimate deformation capacity of masonry piers. It should be noted that very little experimental and theoretical research has been conducted so far on the response of masonry spandrels and only recently they have been subjected to full-scale in-plane testing (Magenes and Penna Magenes and Penna 2011). Clearly, to better understand the deformation capacity of structural masonry, there is a need for a thorough investigation of the in-plane behaviour of masonry spandrels.

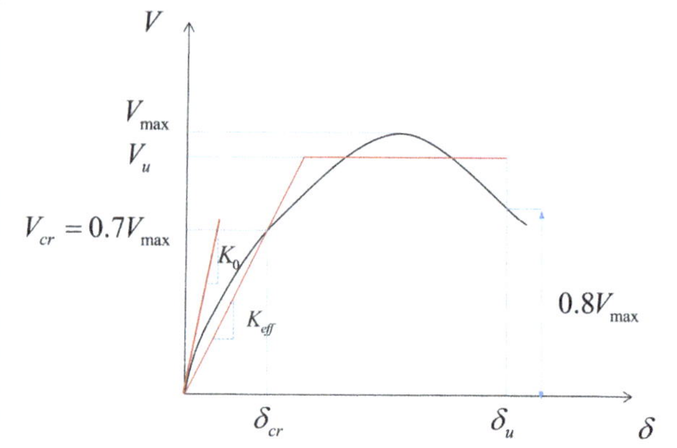

Figure 1 Bilinear idealization of the hysteretic envelope (Tomazevic 1999).

Experimental research on the deformation capacity of unreinforced masonry walls

Experimental research database

As the initial phase of the research project, a review of the technical literature on experimental research on the deformation capacity of structural masonry was conducted. The main objective of the literature review was to provide a comprehensive database of available test results for statistical analysis of the deformation capacity of structural masonry as well as for identifying the decisive parameters for our own experimental work.

Table S1 presents a summary of 71 tests that have been reviewed so far; see Additional file 1: Table S1. It gives information regarding material properties of the constituents: unit dimensions and compressive strengths of units, f_b, mortar, f_m and masonry (perpendicular to the bed joints), f_x; specimen geometry: wall length, l_w, height, h_w and thickness, t_w; boundary conditions and applied vertical pre-compression, σ_0. The database is limited to the shear tests conducted on full-scale unreinforced, unconfined masonry shear walls made of clay bricks and bed joints with general purpose mortar, whose joints had a nominal thickness of 10 mm. Different types of head joints including fully mortared (F), unfilled (U), mortar pocket (MP) and tongue and groove (TG) have been considered. All tests were static-cyclic, except for test MI1m, in which the lateral displacement was applied monotonically. Where the composition of the mortar has not been specified, the mortar is a cement-lime mortar. Further, it should be noted that a considerable number of reviewed shear tests on structural masonry have not been considered in the database since they did not provide useful information on the deformation capacity.

Additional file 1: Table S1 also reports the results of the reviewed tests in terms of failure mechanism and parameters of the idealized bilinear envelope; see Figure 1. Regarding the failure mechanism, the data was classified into four categories: shear-dominated (SH), flexure-dominated (F), sliding (SL) and hybrid (H), i.e. combined shear and flexural, failure modes. The classification was based on the shape of the reported hysteretic loops and available photos, sketches and description of the damage propagation and failure modes given in the reviewed reports. Among the analysed 71 tests, 24 tests (34%) were characterized by the shear-dominated failure mode, 22 tests (31%) by the flexure-dominated failure mode and 21 tests (29%) by the hybrid failure mode. Only 4 tests (6%) failed by sliding.

In order to ensure consistency throughout the database, the actual hysteretic behaviour of the walls was approximated by the linear elastic, ideal plastic bilinear envelope illustrated in Figure 1 (Tomazevic 1999). In order to determine the idealized bilinear envelope curve, after constructing the hysteretic envelope, three parameters had to be identified: the effective stiffness (K_{eff}), the ultimate displacement capacity (δ_u) and the ultimate shear strength (V_u). The effective stiffness is calculated from the secant of the cyclic envelope at $0.7V_{max}$, where V_{max} is the maximal lateral load obtained from the test. The ultimate displacement capacity is the displacement corresponding to a strength degradation of 20%. The definition of the ultimate displacement capacity is to some extent subjective. However, the above criterion has been widely used for the definition of the ultimate deformation capacity by a majority of researchers and been adopted by most of the current structural codes. The ultimate shear strength is obtained by equating the areas under the experimental and bilinear envelopes.

Statistical analysis of the experimental data

Figure 2 illustrates the ultimate drift capacity (the ultimate displacement capacity divided by the height of the specimen, δ_u/h_w) for each failure mode except for the sliding failure mode, where the tests were interrupted before reaching the ultimate displacement capacity of the specimens. As mentioned before, the displacement capacity of these walls is very large.

As indicated by Figure 2, the mean values of the ultimate drift capacity for walls failing in flexural, hybrid and shear failure modes were 1.21%, 0.96% and 0.40%, respectively. The deformation capacity of walls that have undergone a

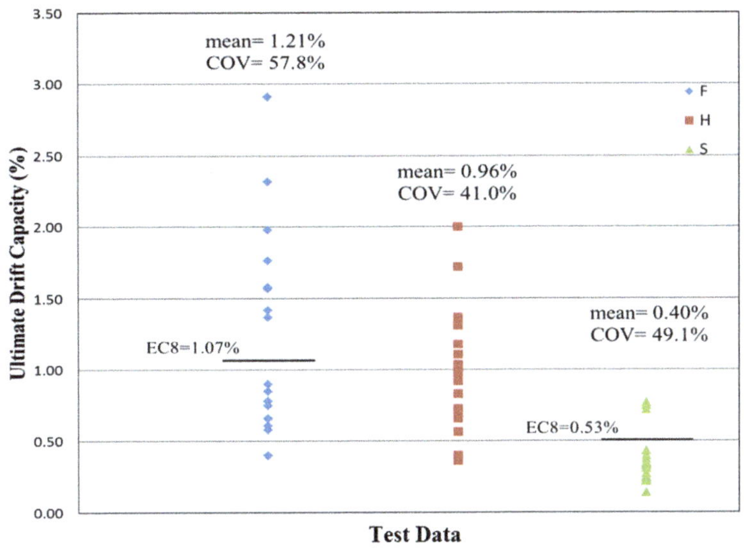

Figure 2 Ultimate drift capacity of unreinforced masonry walls obtained from tests.

shear-dominated failure mode is much lower than that of walls failing by hybrid or flexure-dominated modes. Thus, it was concluded that the global displacement capacity of unreinforced masonry structures is mostly controlled by the displacement capacity of walls failing in shear-dominated modes. In the following, we will discuss the ultimate drift capacity of walls with shear-dominated failure modes in more detail.

Figure 3 illustrates the ultimate drift capacity of walls that failed in the shear-dominated failure mode. The hatched and solid bars correspond to the specimens with cantilever and fixed ends boundary conditions, respectively. As shown, the ultimate drift capacity of walls with shear failure modes ranges between 0.14% and 0.78% with a mean value of 0.40% and a coefficient of variation (COV) of 49.1%. It should be mentioned that tests that did not reach the ultimate limit state have been excluded from the analysis and from Figures 2 and 3 (4 tests in the case of shear-dominated failure). The minimum value of the drift capacity corresponded to wall 16_2. It is important to point out that in this case, such a rather limited ultimate drift capacity of 0.14% is basically due to the

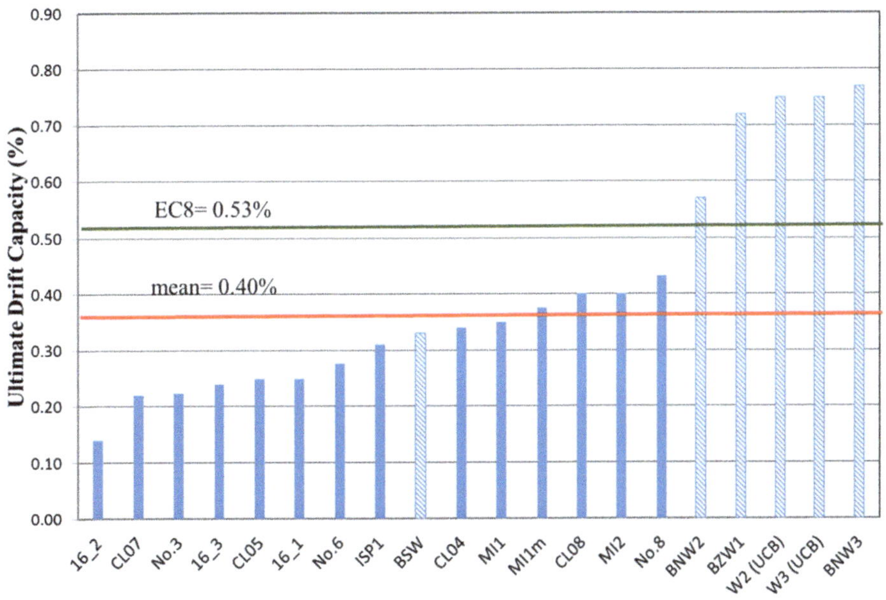

Figure 3 Ultimate drift capacity of walls with shear-dominated failure mode.

definition of the ultimate state as the state corresponding to a strength degradation of 20%. As given in (Frumento et al. 2009) the hysteretic envelope of wall 16_2 exhibits sudden strength degradation after reaching the maximum shear strength, but after that the wall exhibits further deformation capacity before the collapse. A similar behaviour was observed for walls 16_1 and 16_3 from the same reference. Hence, in order to take advantage of the complete deformation capacity of masonry structures, it seems necessary to develop more consistent criteria for the ultimate deformation capacity.

The maximum value of the ultimate drift capacity was reached for specimen BNW3, which was tested under cantilever boundary conditions. For this wall, the early occurrence of shear cracks and the large difference between the cracking load and the maximum shear load were typical (Frumento et al. 2009).

From Figure 3, it can be seen that the specimens with fixed ends boundary conditions exhibited lower drift capacity than those with cantilever boundary conditions. Further inspection of the comparable data revealed that the ultimate drift capacity decreases as the vertical precompression increases or as the aspect ratio of the specimen, i.e. h_w/l_w, decreases. The influence of the other factors, e.g., head joint type and size effect (specimens with the same aspect ratio), could not be investigated because of the inhomogeneity of the available experimental data.

It is clear from Figure 2 that the ultimate drift capacity exhibited rather large scatter. The corresponding (large) values of COV for walls with flexure-dominated, hybrid and shear-dominated failure modes were 57.8%, 41.0% and 49.1%, respectively. Due to this scatter, it is not easy to identify a rational value for the ultimate displacement capacity of unreinforced masonry shear walls based only on the available experimental data. However, some guidance for practicing engineers must be provided and such values are given by structural codes. In Europe, the ultimate drift capacity provided by Annex 3 of Eurocode 8 (CEN-EN 1998-3 2005) is 0.53% for unreinforced masonry walls in the shear failure mode and 1.07% for walls failing in the flexural mode for the limit state of Near Collapse (NC). As can be seen from Figures 2 and 3, these values do not always guarantee a safe design.

In general, the deformation capacity of structural masonry is influenced not only by the failure mechanism but also by many other factors, such as constituent materials, geometry, pre-compression level, etc. Hence, despite the inherent randomness associated with masonry as a material, the classification of the URM piers based only on their failure mechanisms seems to be the main reason for the observed scatter in the values of the ultimate drift capacity. Unfortunately, due to inhomogeneous experimental data and a lack of reliable mechanical models (as described in the following sections), we are still not able to properly take into account the influence of all factors affecting the deformation capacity of structural masonry. Furthermore, the assignment of a specific failure mechanism to a test is often a fairly subjective matter. Obviously, to get a clearer picture of the problem, in addition to conducting more tests, we need to develop reliable mechanical models to describe the load-deformation behaviour of structural masonry.

Analytical research on the deformation capacity of unreinforced masonry walls

Computational strategies for structural masonry

A substantial amount of theoretical work has been invested in modelling structural masonry. Simple models are based on linear theory of elasticity and its application to structural masonry. Regarding the serviceability limit state, i.e. when investigating the behaviour of masonry subjected to load levels up to 40-50% of the ultimate load, the applicability of the linear theory of elasticity is beyond dispute. However, when approaching higher load levels nonlinear modelling is generally required. Hereby, both geometrical and material nonlinearities must be taken into account. In general, there are three sources of nonlinearity in the in-plane response of URM piers: geometrical nonlinearity due to the evolutionary partialization of cross-sections as cracking spreads within the panel, material nonlinearity in the elastic range and material nonlinearity in the plastic range (Augenti and Parisi 2009a). Hence, a reliable model must be able to consider all the above-mentioned sources of nonlinearity in a proper way. Since only very few closed-form solutions for nonlinear problems are available, numerical solution methods must be applied. Such solutions are usually obtained by means of Finite Element Method (FEM) procedures. However, the main problem when applying FEM is related to the modelling of the material. Since masonry is composed of two components, i.e. masonry units and mortar, and is highly anisotropic and nonlinear, modelling the physical reality is very demanding.

In general, three different approaches are found in literature for modelling the seismic response of URM structures: micro-modelling, macro-modelling and macro-element discretization. In the micro-modelling strategy, the different components, i.e. the units, mortar, and the unit-mortar interface are distinctly represented. In detailed micro-models, masonry units and mortar joints are represented by continuum elements, whereas the masonry unit-mortar interface is represented by discontinuous elements (Figure 4a). Detailed micro-modelling requires much computational effort. This drawback is partially overcome by the simplified micro-models. In the simplified micro-modelling strategy, masonry units are represented by continuum elements whilst the mortar joints and masonry unit-mortar interface are lumped into discontinuous

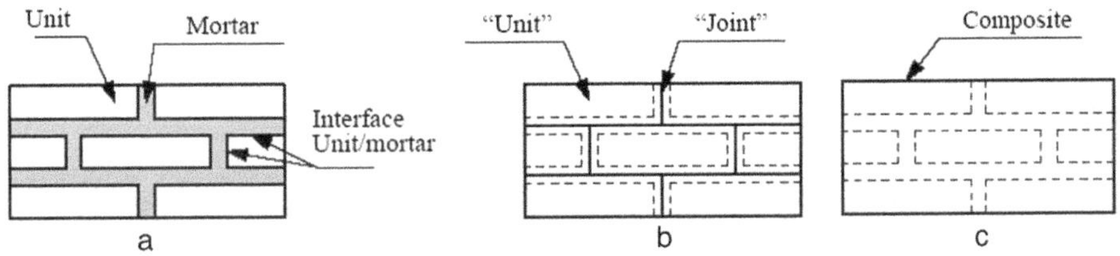

Figure 4 Detailed (a) and simplified (b) micro-modelling; macro-modelling (c) (Lourenco 1996).

elements (Figure 4b). The micro-modelling approaches are suitable for small structural elements with particular interest in strongly heterogeneous states of stress and strain. The primary aim is to closely represent masonry based on knowledge of the properties of each constituent and of the interface (Roca et al. 2010). In the macro-modelling strategy, masonry is treated as a fictitious homogeneous orthotropic continuum with different tensile and compressive strengths as well as different inelastic properties along the material axes (Figure 4c). In particular, FE meshes are simpler since they do not have to accurately describe the internal structure of masonry and the finite elements can have dimensions greater than the single brick units (Roca et al. 2010).

Although significant progress has been made in the field of micro- and macro-modelling strategies, e.g., development of the so-called homogenized modelling, see, e.g., (Lourenco et al. 2007), these approaches are still not suitable for the analysis of whole buildings in everyday engineering practice. This is because a considerable number of material parameters are needed as input for a meaningful analysis using these approaches, and these parameters are usually unavailable. Furthermore, the current micro and macro models have a limited range of validity and also require significant computational resources and high expertise. In addition, due to the great difficulty in the formulation of robust numerical algorithms representing satisfactorily the inelastic behaviour of masonry, micro and macro analyses of masonry structures are often limited to the structural pre-peak regime (Maruccio 2010 and Xu et al. 2012). However, the importance of the post-peak response is evident in order to evaluate the deformation capacity and to assess the structural safety.

As a consequence, several methods based on macro-element discretization have been developed, particularly in Italy. In this approach, each panel in the structure, i.e. piers and spandrels, is modelled by using a single element. Such elements called macro-elements are based on the simplification of both the material behaviour and the stress field within the panel. These elements seem to be the most appropriate for the design and assessment of masonry buildings because of the simplicity of modelling, the straightforward interpretation of the results, particularly in terms of collapse mechanisms, and the accuracy demonstrated in different validations (Lourenco et al. 2009 and Grande et al. 2011). The use of macro-elements for the nonlinear analysis of masonry structures has been introduced in several guidelines, e.g., FEMA 356 (Applied Technology Council ATC 2000) and Eurocode 8 (CEN-EN 1998-3 2005), with particular reference to the use of the pushover analysis method. Given the above, in the following section a review on macro-elements for the seismic analysis of unreinforced masonry piers is presented with emphasis on the ways in which they deal with the deformation capacity. Readers interested in the micro- and macro-modelling approaches can find a recent comprehensive review in (Roca et al. 2010).

Macro-elements for URM piers

One-dimensional macro-elements

The simplest one-dimensional macro-elements are single shear springs which represent experimental resistance envelopes of URM piers with idealized bilinear (linear elastic, perfectly plastic) relationships. In order to determine an idealized bilinear envelope curve, three parameters must be identified: the effective stiffness (K_{eff}), the ultimate shear strength (V_u) and the ultimate displacement capacity (δ_u).

In general, the effective stiffness is a complex parameter and difficult to determine. For practical applications it is usually taken as the elastic stiffness, K_O (see Figure 1), which is calculated based on the elastic beam theory incorporating shear deformation, or as 50% of the elastic stiffness; see, e.g., (Tomazevic 1999), Eurocode 8 and FEMA 356. It should be noted here that, in general, the determination of the effective stiffness is subject to variation and the data obtained from tests exhibits a large scatter. Comparison with experimental results shows that the effective stiffness varies between 40% and 80% of the elastic stiffness and it may be influenced by the pre-compression level. The influence of the pre-compression level on the effective stiffness is a controversial subject. While some experiments have indicated that the ratio between the effective stiffness and the elastic stiffness (K_{eff}/K_O) strongly depends on the pre-compression level, e.g., (Bosiljkov et al. 2005), no such dependencies were found in other experiments; see, e.g., (Magenes et al. 2008). Regarding the ultimate shear strength, it was found by evaluating the results of several tests that the choice of $V_u = 0.9V_{max}$ is appropriate for the ultimate shear strength (Tomazevic 1999). The maximum shear resistance of URM piers, V_{max}, can be predicted with acceptable accuracy using the formulations provided by the codes of practice. Different recommendations could be found in the literature for the ultimate displacement capacity parameter that are in effect based on a statistical analysis of the results of past experiments. Unfortunately, the proposed values are not always readily applicable, because as discussed before the data obtained from tests exhibits a rather large scatter.

Among the methods using bilinear shear springs, the assessment method proposed by (Tomazevic 1978) which is usually known as the POR method and the FEMA 356 method are well-known and extensively used. The POR method is an equivalent static, simplified nonlinear assessment method which assumes that the failure occurs only in the piers without any damage of spandrels. This method, which is historically the first seismic assessment method for structural masonry, is based on the storey mechanism approach. The procedure consists of a separate inter-storey shear-displacement curve for each storey, in which each masonry pier is typically modelled by a linear elastic, perfectly plastic shear spring of limited ductility. The FEMA 356 method also employs nonlinear shear springs to model the force-displacement response of individual piers. The spandrels of URM walls are only considered to affect the boundary conditions of the piers, i.e. fixed ends or cantilever. The force-displacement relationship for each pier is defined based on the governing failure mode, which is taken as the failure mode of least lateral resistance. In conclusion, the macro-elements that are based on bilinear idealization of the experimental resistance envelope are not reliable, particularly regarding the prediction of the ultimate displacement capacity, due to large scatter in available experimental data.

(Magenes and Della Fontana 1998) and (Magenes 2000) proposed an improvement to the POR method based on the so-called equivalent frame idealization. In the proposed method, termed SAM (Simplified Analysis of Masonry buildings), both the spandrels and the piers are modelled as beam-column elements with shear deformation, while their intersections are modelled by means of rigid offsets at the ends of the pier and spandrel elements. To describe the nonlinear response of piers and spandrels, the SAM method employs several plastic hinges that are located by the user to account for the possible failure modes. Typically these plastic hinges are placed at both ends and at the mid-span of the beam-column elements to capture the flexural and the shear failure modes, respectively. This approach can be easily implemented by the conventional commercial programs, e.g., SAP2000 (Pasticier et al. 2008). However, since the properties of the plastic hinges are mainly based on the available experimental data, the model suffers from the lack of reliability of the results, particularly in terms of the displacement capacity.

Figure 5a shows the macro-elements proposed by (Chen et al. 2008) for URM piers and spandrels. The proposed approach is meant to be used in conjunction with FEMA 356. This approach, which provides rotational, shear, and axial springs in series, was first introduced for the analysis of reinforced concrete (RC) shear walls (Kabeyasawa et al. 1982 and James and Kunnath 1994). The elements developed for RC shear walls were improved for the modelling of URM piers by the addition of two shear springs at the top and bottom of the macro-element to account for bed joint sliding deformation in these regions. For modelling spandrels constructed of running bond masonry, these sliding springs are not needed because the interlocking of

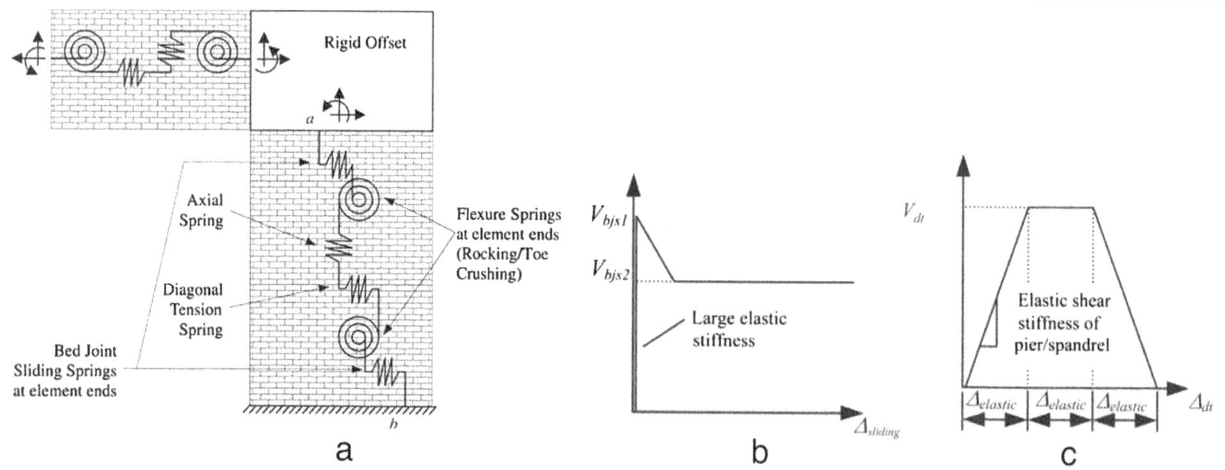

Figure 5 Macro-elements for URM piers and spandrels (a); adopted force-displacement relationships for bed joint sliding (b) and diagonal shear behaviour (c) (Chen et al. 2008).

units prevents sliding along head joints. As shown in Figure 5a, these macro-elements have three degrees of freedom (DOF) at each end and thus can be used within the equivalent frame idealization strategy for the analysis of perforated URM walls.

Regarding the pier element, the axial spring has linear elastic response in compression and no resistance to tensile forces beyond the tensile strength of masonry. The properties of the flexural springs are based on the moment-curvature response of the top and bottom sections of the pier, which are established through the use of a fibre model. The moment-rotation properties of the rotational springs are then obtained by integrating the curvature along the height of the pier. The adopted force-displacement relationships for the shear springs corresponding to the bed joint sliding and diagonal shear behaviour are shown in Figures 5b and c.

Figure 6 provides a comparison between the results of the macro-element simulation and experimentally obtained hysteresis response for the different failure modes. It can be seen that in the case of rocking-flexural and shear sliding failure modes, the simulation results are almost satisfactory, but in the case of diagonal shear failure, the simulation is highly erroneous. The error arises mainly from the adopted force-displacement response for the diagonal shear spring and also from the inability of the element to model the geometrical nonlinearity.

Two-dimensional macro-elements

There are some limitations in the use of one-dimensional macro-elements, namely due to inaccurate simulation of the interaction between piers and spandrels, and due to the weak modelling of the geometrical nonlinearity of the panels (Marques and Lourenco 2011). Two-dimensional macro-elements cannot be applied within the equivalent frame idealization framework and require more computational effort compared to one-dimensional macro-elements. However, they offer a more accurate simulation of the nonlinear response of the masonry piers and spandrels, particularly in the pre-peak regime. This is because unlike one-dimensional elements, these elements are able to simulate the propagation of the tensile cracks along the height of the pier, i.e. geometrical nonlinearity.

The "no-tension multi-fan panel element" was developed by (Braga and Liberatore 1990) based on the idea that the stress field of a masonry panel with free edges

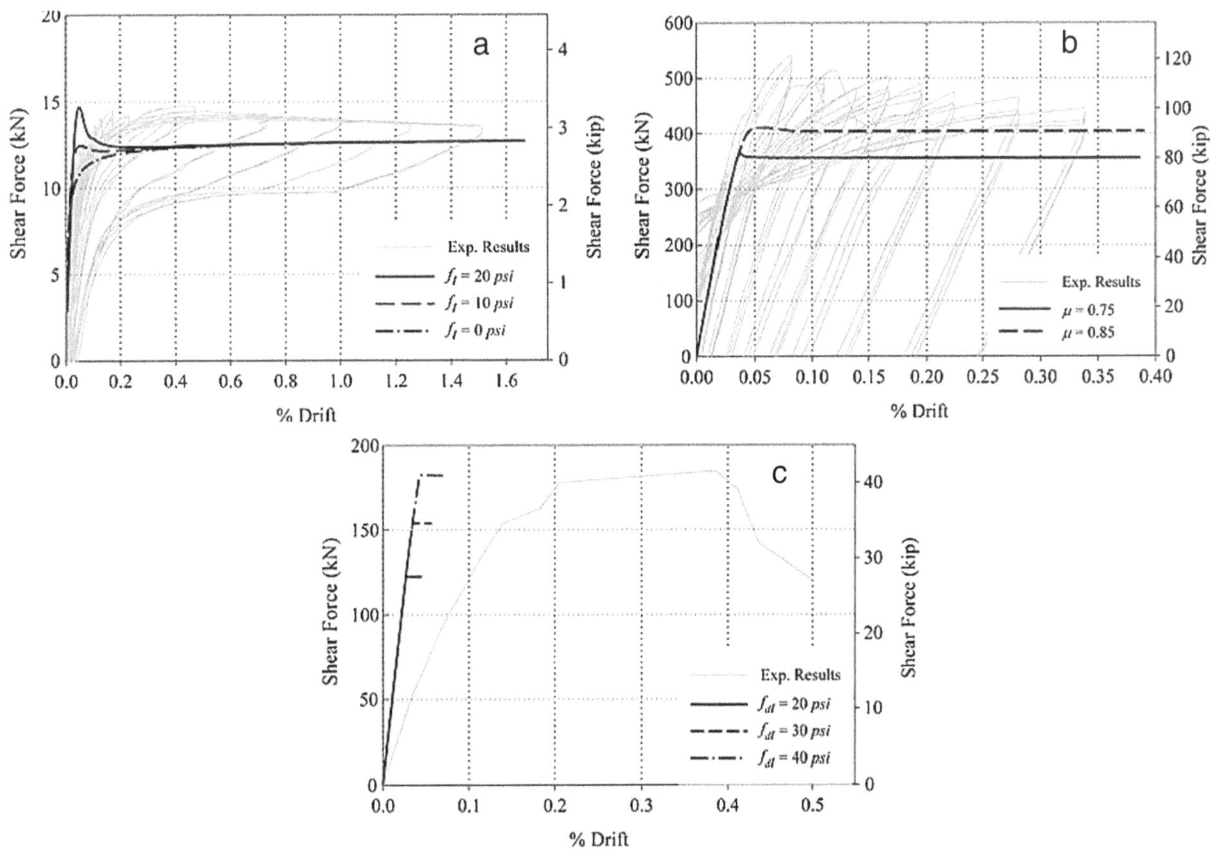

Figure 6 **Comparison between the macro-element simulation and experimental results for flexural (a), bed joint sliding (b) and diagonal shear (c) failure modes (Chen et al. 2008).**

follows a multi-fan pattern (Figure 7a). In addition, it is assumed that the upper and lower faces of the panel are rigid, and that there is no interaction in the circumferential direction between the infinitesimal fans. The material behaviour is assumed linear elastic in compression and non-reacting in tension. There is very good agreement between simulation results obtained using the multi-fan element and experimental results up to a certain level of lateral displacement, but for larger displacements, the accuracy of simulation decreases rapidly because of the adopted elastic constitutive law (Liberatore et al. 1996).

A modification of the multi-fan element is proposed by (Maruccio 2010). The updated multi-fan element, introduces zero-length springs into the multi-fan element to add failure mechanisms in the constitutive law; see Figure 7b. In fact, the behaviour in the elastic stage is defined by a set of radial stress fields in the panel, while the springs are required to define failure mechanisms and the inelastic response. The properties of these springs are based on past component tests. Therefore, as discussed previously in detail, the simulation results could not be considered reliable in terms of displacement capacity. The multi-fan model seems to have a considerable potential for further development.

(Yi et al. 2005) proposed a macro-element, named "Effective pier model", to describe the nonlinear behaviour of an individual URM pier subjected to external forces. The model describes the effective area of the pier by eliminating the flexural tensile cracks and the toe crushing regions; see Figure 8a. Regarding toe crushing, the model assumes that the compressive strength of masonry immediately drops to zero after the ultimate compressive strength of the masonry has been exceeded. Although this assumption is questionable, it greatly simplifies the problem and results in a conservative strength estimate (Yi et al. 2005). However, more sophisticated stress-strain relationships that account for the nonlinear stress-strain behaviour of masonry can be employed if needed; e.g., the macro-element model developed by (Augenti and Parisi 2009b). In order to address the problem of bed joint shear sliding, the model employs the Mohr-Coulomb friction model. Regarding diagonal shear, a smeared crack technique is employed. It is assumed that the effective area of the pier remains continuous even after the development of diagonal cracks. To model the potentially rapid and unstable propagation of these cracks, the effective tangent elastic modulus of masonry is assumed to have a negative value. Since no test data is available for the softening behaviour of URM piers after diagonal cracking, the tangent modulus of URM piers with diagonal shear crack is set equal to $-0.1E$, where E is the initial elastic modulus of masonry (Figure 8b).

According to this model, when a URM pier experiences a reduction of cross section due to either tensile or compressive failure, the remaining part of the pier will typically be inclined at some angle. Hence, after cracking, a portion of the lateral force will be resisted through axial deformation, and the lateral force that causes shear and flexural deformation will be $V - P\tan(\theta)$, where θ is the angle between the central axis of the pier and the vertical line, and V and P are applied lateral and vertical loads, respectively. The proposed model only considers the lateral deformation of the pier caused by flexure and shear:

$$\Delta = \frac{V - P\tan(\theta)}{K} \qquad K = 1/\left[\frac{4\gamma h^3}{EL^3 t} + \frac{h}{GLt}\right]$$

where Δ is the lateral deformation; K is the lateral stiffness of the pier; γ is a coefficient that describes the boundary conditions of the pier (γ is equal to 0.83 for fixed ends and 3.33 for cantilever boundary conditions); E is the elastic modulus of masonry, and G is the shear modulus of masonry, which is taken as $0.4E$. In order to consider the nonlinear compressive stress-strain behaviour of the

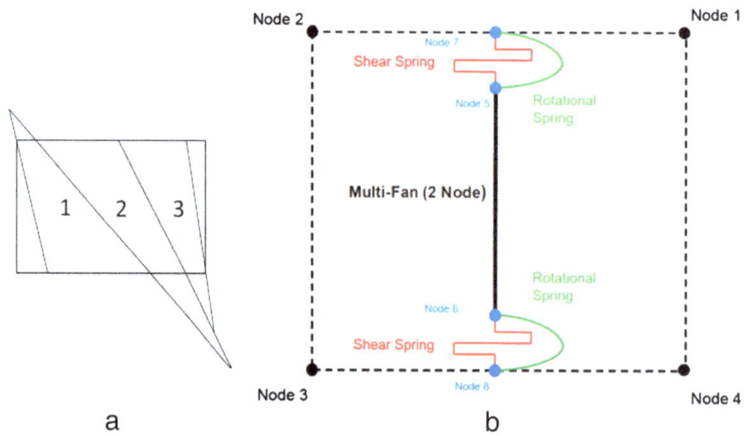

Figure 7 Multi-fan element (a); modified multi-fan element (b) (Maruccio 2010).

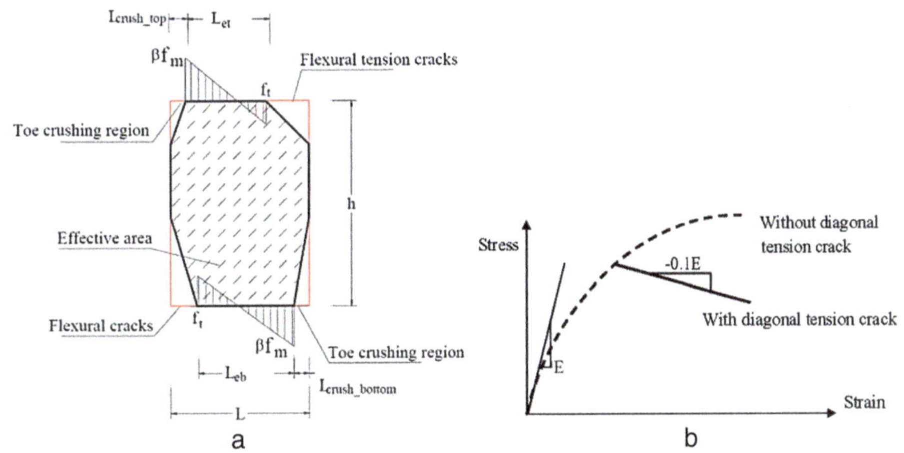

Figure 8 Effective pier model (a); principal compressive stress-strain relationship of masonry before and after diagonal shear cracks (b) (Yi et al. 2005).

masonry, the model uses the relationship proposed by (Naraine and Sinha 1989).

Figure 9 provides a comparison between the results of the effective pier model and the experimentally obtained hysteresis response for the different failure modes. It can be seen that the effective pier model is able to partially predict the behaviour of masonry piers failing in sliding shear or rocking-flexural modes, but in the case of diagonal shear, there is no good agreement between the model and the experimental results (except at the beginning of the

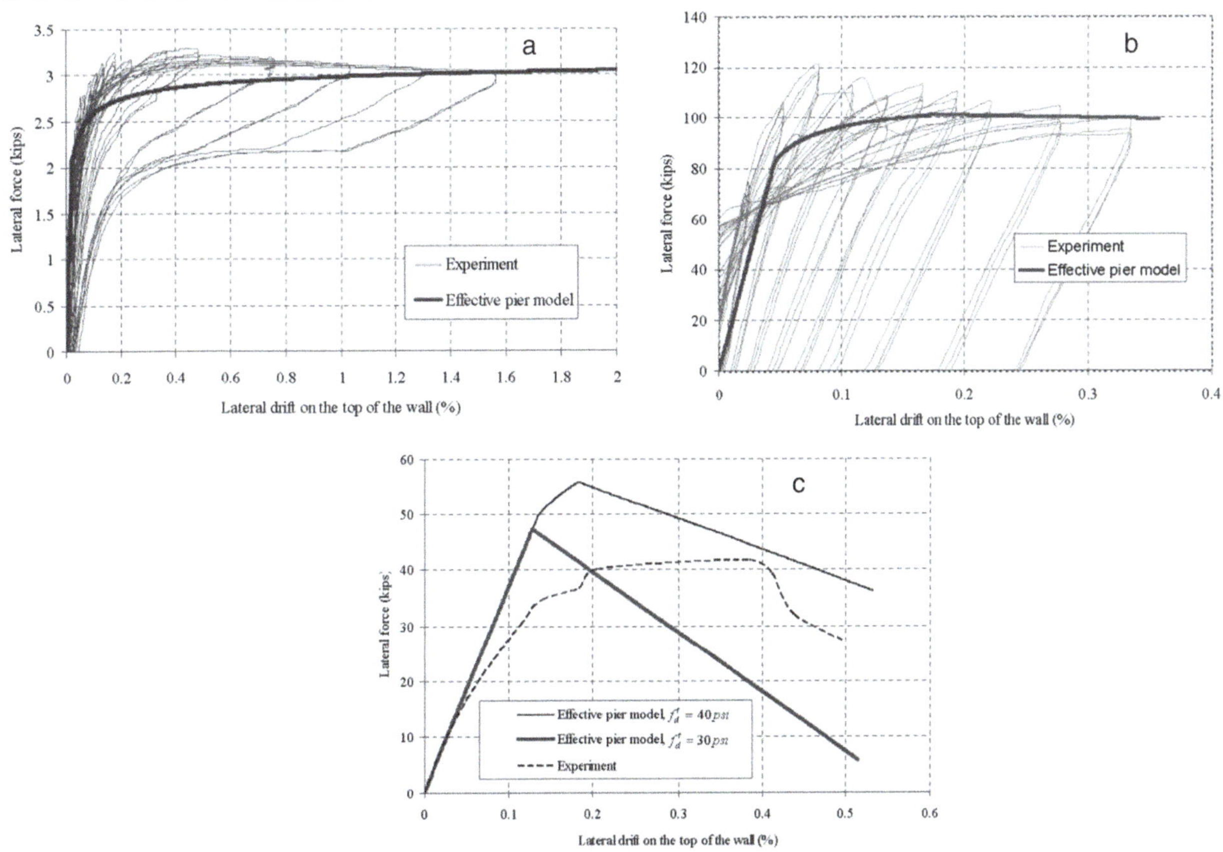

Figure 9 Macro-element simulation of the flexural (a), bed joint sliding (b) and diagonal shear (c) response using Effective pier model (Yi et al. 2005).

response), which could be attributed to the assumed stress field and stress-strain relationship after the development of diagonal shear cracks.

In conclusion, although available macro-elements (specifically two-dimensional ones) provide acceptable results in the case of sliding and rocking-flexural failure modes, they are still not able to simulate the response of unreinforced masonry piers with the diagonal shear failure mode closely enough. In general, there are two main obstacles to the simulation of the in-plane response of URM piers, especially in the post-peak regime: complexity of the stress field after development of the shear cracks and more important, lack of experimental data about the softening of masonry (as a material) under biaxial loading. Furthermore, current macro-elements cannot distinguish between the diagonal cracks passing through the units and those running through bed and head joints, while in the former case, the deformation capacity is limited, but in the latter case there is a considerable displacement capacity.

Conclusions

The ultimate deformation capacity is the most important parameter in the seismic design and evaluation of structures. Our current state of knowledge about the deformation capacity of structural masonry is limited. The available experimental data exhibits too much scatter and it is not possible to identify a rational value for the deformation capacity of masonry structures based only on such experimental data. Furthermore, there are no reliable and practical analytical models for the force-deformation relationship of structural masonry: refined finite element models, besides being too complex for everyday engineering practice, suffer from numerical instabilities in the post-peak regime, and available structural macro-elements are still so far from being considered accurate enough regarding the deformation capacity parameter, especially in the case of the diagonal shear failure mode. It should be mentioned here that there are several other macro-elements that have not been reviewed in this paper because they do not introduce further improvement or different concepts regarding the deformation capacity.

Obviously, to get a clearer picture of the problem, in addition to conducting more tests, we need to develop reliable mechanical models to describe the load-deformation behaviour of structural masonry. This task is being approached within the framework of the current research project. The research project includes several cyclic quasi-static shear tests on full-scale, storey-high masonry walls as well as developing and introducing new sophisticated mechanical models for structural masonry. A novel approach will be developed and utilized for the purpose of applying experimental evidence collected from our own tests performed for the development of reliable mechanical models.

Additional file

Additional file 1: Table S1. Summary of the experimental studies on the deformation capacity of unreinforced masonry walls.

Competing interests
The authors declare that they do not have any competing interests.

Authors' contributions
AHS carried out the review and drafted the manuscript. NM and JS supervised the research. All authors read and approved the final manuscript.

Acknowledgments
Funding from the Swiss National Science Foundation (Grant 200021_131971) is gratefully acknowledged.

Author details
[1]Institute of Structural Engineering, ETH Zurich, CH-8093, Zurich, Switzerland. [2]Institute of Technology in Architecture, ETH Zurich, CH-8093, Zurich, Switzerland.

References

Abrams DP, Shah N (1992) Cyclic load testing of unreinforced masonry walls: report #92-26-10. Advanced Construction Technology Centre, College of Engineering, University of Illinois, Urbana-Champaign

Anthoine A, Magonette G, Magenes G (1994) Shear-compression testing and analysis of brick masonry walls. In: Duma G (ed) Proceedings of the 10th European conference on earthquake engineering. Vienna, Austria, A.A. Balkema, Rotterdam p 1657

Applied Technology Council (ATC) (2000) Prestandard and commentary for the seismic rehabilitation of buildings, FEMA-356. Federal Emergency Management Agency, Washington, D.C

Augenti N, Parisi F (2009a) Non-linear static analysis of masonry structures. In: Proceedings of the 13th Italian national conference on earthquake engineering, paper no. S4-01

Augenti N, Parisi F (2009b) Influence of the constitutive law on the flexural strength of masonry panels. In: Proceedings of the 13th Italian national conference on earthquake engineering, paper no. S4-03

Beyer K, Dazio A (2012) Developing force-deformation characteristics of brick masonry spandrels in historic buildings. In: Proceedings of the 15th world conference on earthquake engineering. Lisbon, Portugal, Paper No. WCEE2012-2352

Bosiljkov V, Page A, Bokan-Bosiljkov V, Zarnic R (2003) Performance based studies of in-plane loaded unreinforced masonry walls. Masonry Int 16(2):39–50

Bosiljkov V, Tomazevic M, Lutman M (2004) Optimization of shape of masonry units and technology of construction for earthquake resistant masonry buildings: research report-part one and two. ZAG Ljubljana, Slovenia

Bosiljkov V, Tomazevic M, Lutman M (2006) Optimization of shape of masonry units and technology of construction for earthquake resistant masonry buildigs: research report-part three. ZAG Ljubljana, Slovenia

Bosiljkov V, Totoev YZ, Nichols JM (2005) Shear modulus and stiffness of brickwork masonry: an experimental perspective. Struct Eng Mech 20(1):21–43

Braga F, Liberatore D (1990) A finite element method for the response of masonry buildings under seismic actions. In: Proceedings of the 5th North American masonry conference, 1st edn., p 201

CEN-EN 998-2 (2003) Specification for mortar for masonry, Part 2. Masonry mortar

CEN-EN 1998-3 (2005) Eurocode 8: design of structures for earthquake resistance, Part 3: strengthening and repair of buildings

Chen SY, Moon FL, Yi T (2008) A macroelement for the nonlinear analysis of in-plane unreinforced masonry piers. Eng Struct 30(8):2242–2252

Da Porto F, Grendene M, Modena C (2009) Estimation of load reduction factors for clay masonry walls. Earthq Eng Struct Dyn 38:1155–1174
Fehling E, Stuerz J, Emami A (2007) Tests results on the behaviour of masonry under static (monotonic and cyclic) in plane lateral loads: technical report of the collective research project ESECMaSE, Deliverable D7.1a. Institute of Structural Engineering, University of Kassel, Germany
Frumento S, Magenes G, Morandi P, Calvi GM (2009) Interpretation of experimental shear tests on clay brick masonry walls and evaluation of q-factors for seismic design. IUSS Press, Pavia
Grande E, Imbimbo M, Sacoo E (2011) A beam finite element for nonlinear analysis of masonry elements with or without Fiber-Reinforced Plastic (FRP) reinforcements. Int J Archit Herit: Conserv, Anal, and Restoration 5(6):693–716
James J, Kunnath SK (1994) Macroelement model for shear wall analysis. In: Proceedings of the first congress on computing in civil engineering. p 1505
Kabeyasawa T, Shioara H, Otani S, Aoyama H (1982) Analysis of the full-scale seven-story reinforced concrete test structure test PSD3: paper presented at the 3rd joint technical coordinate committee meeting
Liberatore D, Braga F, Mancinelli E (1996) Numerical simulation of the experimental behaviour of masonry panels and walls under horizontal loads. In: Proceedings of the 11th world conference on earthquake engineering, Acapulco, Mexico, Paper No. 1408
Lourenco PB (1996) Computational strategies for masonry structures: dissertation. Delft University of Technology, The Netherlands
Lourenco PB, Mendes N, Marques R (2009) Earthquake design and assessment of masonry structures: review and applications. In: Topping BHV, Costa Neves LF, Barros RC (eds) Trends in civil and structural engineering computing. Saxe-Coburg Publications, Stirlingshire, pp 77–101
Lourenco PB, Milani G, Tralli A, Zucchini A (2007) Analysis of masonry structures: review of and recent trends of homogenisation techniques. Can J Civ Eng 34:1443–1457
Magenes G, Morandi P, Penna A (2008) Tests results on the behaviour of masonry under static cyclic in plane lateral loads: technical report of the collective research project ESECMaSE, Deliverable D7.1c. Department of Structural Mechanics, University of Pavia, Italy
Magenes G, Calvi GM (1992) Cyclic behaviour of brick masonry walls. In: Bernal A (ed) Proceedings of the 10th world conference on earthquake engineering., Madrid, Spain, A.A. Balkema, Rotterdam, p 3517
Magenes G, Calvi GM (1997) In-plane seismic response of brick masonry walls. Earthq Eng Struct Dyn 26:1091–1112
Magenes G, Della Fontana A (1998) Simplified non-linear seismic analysis of masonry buildings. In: Proceedings of the British masonry society, 8th edn., p 190
Magenes G (2000) A method for pushover analysis in seismic assessment of masonry building. In: Proceedings of the12th world conference on earthquake engineering., Auckland, New Zealand, New Zealand Society for Earthquake Engineering, Upper Hutt, p 1866
Magenes G, Modena C, Da Porto F, Morandi P (2009) Seismic behaviour and design of new masonry buildings: recent developments and consequent effects. In: Proceedings of the Eurocode 8 perspectives from the Italian standpoint workshop., p 99
Magenes G, Penna A (2011) Seismic design and assessment of masonry buildings in Europe: recent research and code development issues. In: Ingham J, Dhanasekar M, Masia M (eds) Proceedings of the 9th Australasian masonry conference., Queenstown, New Zealand, Australian Masonry Conference, Auckland, p 583
Manzouri T, Shing PB, Amadei B, Schuller M, Atkinson R (1995) Repair and retrofit of unreinforced masonry walls: experimental evaluation and finite element analysis: report CU/SR-95/2. Department of Civil, Environmental and Architectural Engineering, University of Colorado, Boulder
Marques R, Lourenco PB (2011) Possibilities and comparison of structural component models for the seismic assessment of modern unreinforced masonry buildings. Comput Struct 89(21-22):2079–2091
Maruccio C (2010) Numerical analysis of FRP strengthened masonry structures: dissertation. University of Minho, Portugal and Sapienza University of Rome, Italy
Modena C, Da Porto F, Garbin F (2005) Ricerca sperimentale sul comportamento di sistemi per muratura portante in zona sesmica: draft 2005/01. University of Padua, Italy
Naraine K, Sinha S (1989) Behaviour of brick masonry under cyclic compressive loading. J Constr Eng Manag 115(2):1432–1445
Pasticier L, Amadio C, Fragiacomo M (2008) Non-linear seismic analysis and vulnerability evaluation of a masonry building by means of the SAP2000 V. 10 code. Earthq Eng Struct Dyn 37:467–485
Roca P, Cervera M, Gariup G, Pela L (2010) Structural analysis of masonry historical constructions: classical and advanced approaches. Comput Methods Eng 17(3):299–325
Tomazevic M (1978) The computer program POR: institute for testing and research in materials and structures-ZRMK. Ljubljana, Slovenia
Tomazevic M (1999) Earthquake-resistant design of masonry buildings. Imperial College Press, London, U.K
Xu C, Xiangli C, Bin L (2012) Modeling of influence of heterogeneity on mechanical performance of unreinforced masonry shear walls. Construct Build Mater 26(1):90–95
Yi T, Moon FL, Leon RT, Kahn LF (2005) Effective pier model for the nonlinear in-plane analysis of individual URM piers. Masonry Soc J 23(1):21-35

Thermo-mechanical properties of silicon carbide-filled chopped glass fiber-reinforced epoxy composites

Gaurav Agarwal[1], Amar Patnaik[2*] and Rajesh Kumar Sharma[2]

Abstract

The effect of addition of silicon carbide (SiC) filler in different weight percentages on physical properties, mechanical properties, and thermal properties of chopped glass fiber-reinforced epoxy composites has been investigated. Physical and mechanical properties, i.e., hardness, tensile strength, flexural strength, interlaminar shear strength, and impact strength, are determined with the change in filler content to notice the behavior of composite material subjected to loading. Thermo-mechanical properties of the material are measured with the help of a dynamic mechanical analyzer. The result shows that the physical and mechanical properties of SiC-filled glass fiber-reinforced epoxy composites are better than unfilled glass fiber-reinforced epoxy composites. Viscoelastic analysis for different compositions indicate that adding too much SiC content results in degradation in energy absorption capacity of the material and hence overall performance of the composites, whereas adding too much (more than 10 wt.%) SiC content increases the elastic behavior of the composite.

Keywords: Mechanical properties; Dynamic mechanical analyzer; SiC; Chopped glass fiber

Introduction

A test for mechanical properties is a prime area of interest. The mechanical properties of a material are those properties that involve a reaction to an applied load. The mechanical properties of a composite determine the range of usefulness of a material and establish the service life that can be expected. Most structural materials are anisotropic, which means that their mechanical properties vary with orientation (Shalin 1995). The variation in the properties can be due to the change in the microstructure of fiber/filler and matrix reinforcement. The materials can be subjected to many different loading scenarios, and the composite performance is dependent on the loading conditions. The common properties are considered to determine the loading constraints on the durability and service life effect tensile strength, flexural strength, hardness, interlaminar shear strength (ILSS), and impact strength (Rawlings 1999; Reid and Zhou 2000).

It is also necessary to determine the effect of change in temperature on the properties of fiber-reinforced epoxy composites because an increase in temperature results in the decrease in strength and ductility, whereas a decrease in temperature results in the increase in ductility and strength (Nielsen and Landel 1994). Apart from fiber-reinforced polymer composites, the composites made from both fiber/filler reinforcement performed well in many practical situations. Silicon carbide (SiC) filler particles were added to fiber-reinforced epoxy to enhance the mechanical properties of composites. A change in particle size, shape, and percentage content on the mechanical properties of fiber-reinforced polymer composites is an area of keen interest (Koh et al. 1993; Imanaka et al. 2001; Yamamoto et al. 2003). From research, it has been found that the structure and shape of silica particles have significant effects on the mechanical properties such as fatigue resistance, tensile, and fracture properties (Nakamura et al. 1991a, b; Nakamura et al. 1992).

The literature mentioned above gives a brief overview about the effects of addition of fiber/filler to mechanical and thermal properties of SiC-filled glass fiber-reinforced epoxy composites. It has been observed that various

* Correspondence: amar_mech@sify.com
[2]Department of Mechanical Engineering, National Institute of Technology, Hamirpur 177005, India
Full list of author information is available at the end of the article

mechanical properties at optimum fiber loading conditions can be determined with the addition of fiber/filler reinforcement on matrix composites. To the author's knowledge, no work has been done to determine the optimum composition with the combination of fiber/filler and matrix to optimize the performance of the composite material.

Materials and methods

Material required

Chopped E-glass fiber (5 to 10 mm long, 200 G.S.M. density) manufactured by Ciba Geigy and locally supplied by Northern Polymers Ltd., New Delhi, India, is used as a fiber reinforcement. Commercially available SiC powder also known as carborundum of particle size 25 to 60 μm (density 2.6 g/cm^3) obtained from Silcarb Recrystalized Pvt. Ltd., Bangalore, India, is used as a particulate filler. The matrix material consists of epoxy resin LY 556 and room-temperature curing hardener HY951 supplied by Crystal Chemicals, Delhi, India. The composites were fabricated by blending epoxy resin, glass fiber, and SiC filler in a certain weight percentage reinforcement. Five different compositions of composites were prepared by varying the SiC filler reinforcement with fixed weight percentage (wt.%) of chopped glass fiber reinforcement. SiC filler in five different weight percentages (0, 5, 10, 15, and 20 wt.%) are added with fixed 20 wt.% of chopped glass fiber and remaining epoxy so as to notice the effect of SiC reinforcement on physical, mechanical, and thermal properties of chopped glass fiber-reinforced epoxy composites. The composites were prepared by blending certain weight percentage of fiber/filler and epoxy resin in separate containers and then poured in a mold of desired dimensions. Labeling was done with the help of rollers, and suitable weights were applied on top of the mold. Similarly, five different compositions of fiber, filler, and epoxy resin are poured in separate molds by varying five percentages of SiC filler content with that of the epoxy resin keeping the chopped glass fiber weight percentage as constant. The composites are then left for solidification at room temperature for 24 h. After the solidification process, the composites are then removed from the mold and marking is done as per the test standards. Specimens were prepared as per American Society for Testing Materials (ASTM) test standards for tensile, flexural, interlaminar shear strength, impact strength, and hardness tests.

Tests for mechanical properties

The mechanical properties were determined to understand the behavior of the material under different loading conditions. To understand how materials deform or break as a function of applied load, time, temperature, and other conditions, tensile strength, impact strength, flexural strength, interlaminar shear strength, and hardness of the composite specimens were determined.

The capability of the composite material to resist abrasion or indentation is known as hardness. A Rockwell hardness tester is used to measure the hardness of the composite specimen. A spherical indenter (1/16 in or 1.5875 mm diameter) was forced into the surface of the material under conditions of controlled magnitude and rate of loading. The size of indentation is the measure of amount of hardness of the specimen.

Tests for mechanical properties, i.e., tensile test, interlaminar shear strength, and flexural strength, are done on a universal testing machine (UTM) Instron 1195. The tensile test is performed to measure the ductility of the material. Equal and opposite loads are applied at both ends of the rectangular specimen on the UTM Instron 1195 (ASTM 1976). Dimensions of the test specimen are 200 mm long, 11.5 mm wide, and with variable thickness (depending on the composition of the composite material). Flexural strength is the ability of the material to resist bending under the application of load. A three-point bend test was performed on the same UTM Instron 1195 with a span length of 30 mm in between the supports and a crosshead speed of 10 mm/min to measure the amount of failure when subjected to loading. Rectangular test specimen with dimensions 157 mm length, 12.7 mm width, and 6.35 mm thickness has been used for the experiment. Interlaminar shear strength tests are conducted as per ASTM D 2344-84 test standards on UTM Instron 1195 with span length of 50 mm and crosshead speed of 10 mm/min is maintained for testing (ASTM 1984).

Impact tests are done as per ASTM D 256 test standards (ASTM 1997) on plastic/composite specimen digital impact testing machine. Impact test is done to notice the energy consuming capacity of the composite material before fracture. Charpy V-notch impact test samples with dimensions 64 mm × 12.7 mm × 3.2 mm were prepared with a V-notch groove (2.5 mm depth) in the center of the specimen. The specimen was fixed on the impact tester such that the notch was at the opposite face to the striking end of the hammer.

Results and discussion

The physical and mechanical properties describe the behavior of the material to various practical applications. The properties such as hardness describe the physical state of the system. The mechanical property of the material is a measure of the behavior of the material under different loading conditions. Tests were done to notice the effect of addition of SiC filler on the physical and mechanical properties and the optimum fiber/filler loading of glass fiber-reinforced epoxy composites at which specific wear rate in minimum.

Effect of hardness on SiC-filled chopped glass fiber-reinforced epoxy composites

Hardness is a surface property and is a measure of wear resistance on the surface of the composite. The measured hardness values of all the five weight percentages of SiC reinforcement are shown in Figure 1. It can be seen that the hardness increases with the increase in the SiC content at a constant rate. Hardness increases linearly from 0 wt.% SiC content to 15 wt.% SiC content, but further increase in SiC content results in the decrease in the value of hardness. The increase in the value of hardness are attributed to the fact that as the density of composition increases due to filler particles introduced between the fiber and the matrix, the compression value of the specimen decreases and hardness increases. Addition of a small volume fraction of graphite, SiC, and alumina can significantly improve the hardness and wear resistance of the composites. With the addition of SiC filler content, the filler particles (SiC) fill in the gap between the fiber and the matrix and form a more dense structure and hence hardness increases (Mohamed et al. 2009).

Effect of tensile strength on SiC-filled chopped glass fiber-reinforced epoxy composites

The ultimate tensile strength is an engineering value calculated by dividing the maximum load on a material by the initial cross section of the test specimen. Figure 2 shows the graph of tensile strength versus the SiC-filled chopped glass fiber-reinforced epoxy composite. Tensile strength increases from 0 to 10 wt.% SiC content, whereas further increase in SiC content results in the decrease in the value of tensile strength. This is because filler particles act as a barrier in transferring stress from one point to another, and an increase in SiC content beyond 10 wt.% results in the increase of transfer of stresses from one point to another. Also as the fiber/filler content increases, the bonding surface area increases and hence bonding strength decreases. Due to insufficient amount of bonding between three different constituents, the loads may not effectively be transferred from one end to another and hence there is reduction in tensile strength of the composite. Tensile strength of carbon fiber-reinforced polyetheramide improves significantly with the addition in the percentage of fiber/filler reinforcement. The extent of improvement of tensile strength however is not proportional to the amount of carbon fiber reinforcement (Sua et al. 2005; Devendra and Rangaswamy 2012).

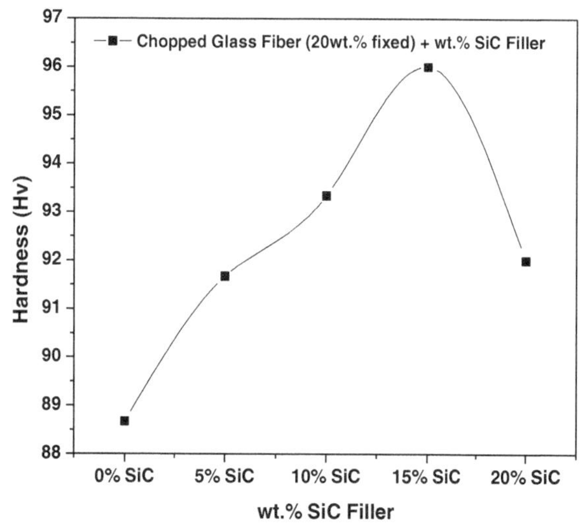

Figure 1 Effect of hardness on SiC-filled chopped glass fiber-reinforced epoxy composites.

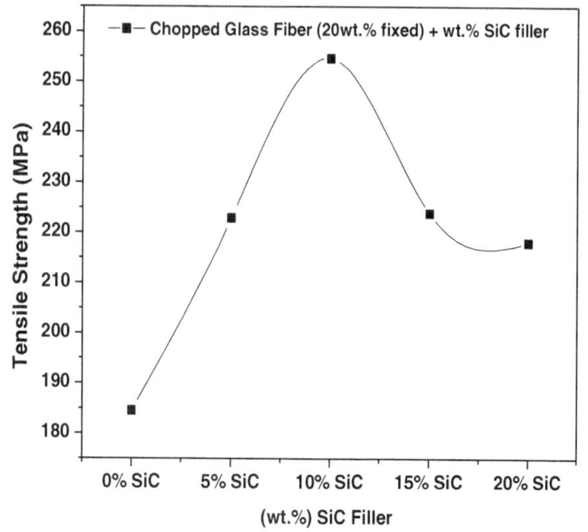

Figure 2 Effect of tensile strength on SiC-filled chopped glass fiber-reinforced epoxy composites.

Effect of flexural strength and interlaminar shear strength on SiC-filled chopped glass fiber-reinforced epoxy composites

Figure 3 shows the flexural strength of SiC-filled glass fiber-reinforced epoxy composites. Flexural strength increases from 55 to 87 MPa with the increase in SiC filler content from 0 to 10 wt.%; an increase in the flexural strength beyond 10 wt.% SiC content results in the decrease in the value of SiC content. When the specimen is placed on two support points and load is applied from the top of the specimen, then the specimen is subjected to bending and the top layer is subjected to compressive loading, whereas the bottom layer is subjected to tensile loading. When the bonding between the fiber/filler and the matrix is increased, a flexural strength increases and strong bonding transfers loads from one end to another resulting in the increase in flexural strength of the specimen, whereas when the percentage of fiber/filler exceeds the required percentage, then the surface area increases,

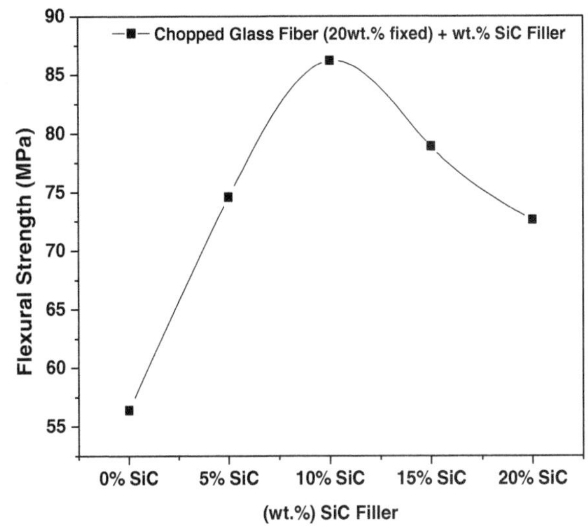

Figure 3 Effect of flexural strength on SiC-filled chopped glass fiber-reinforced epoxy composites.

while the weight percentage of the matrix decreases; therefore, the bonding strength decreases, and loads cannot be effectively transferred from one part to another resulting in the decrease in the flexural strength of the composite. Another reason for the reduction in flexural strength with the addition of SiC filler is that the addition of SiC content beyond 10 wt.% disturbs the matrix continuity and reduction in bonding strength between filler, matrix, and fiber (Kaundal et al. 2012). Interlaminar shear strength is the measure of shear strength between the layers of fibers. Figure 4 shows the ILSS value of SiC-filled glass fiber-reinforced epoxy composites. The ILSS value increases with the increase in the value of SiC content because SiC filler particles stick to the ends of fibers and act as a barrier in the transfer of shear stress from one part to another. A proportional graph is obtained between the weight percentage of SiC filler reinforcement and ILSS value, i.e., ILSS increases in equal proportion with the increase in the value of SiC content.

Effect of impact strength on SiC-filled chopped glass fiber reinforced epoxy composites

The material's resistance to fracture is known as toughness. It is often expressed in terms of the amount of energy a material can absorb before fracture. A ductile material can absorb a considerable amount of energy before fracture, while a brittle material absorbs very little. Figure 5 shows the increase in the value of impact strength with the increase in the percentage of fiber/filler reinforcement. The value of impact strength increases from 0 to 15 wt.% and further decreases for 20 wt.% SiC filler reinforcement. The increase in impact strength is due to the weaker bond, i.e., as the bond strength between the glass fiber, SiC filler, and the epoxy resin reduces, the impact energy absorbing capacity of the composite increases. Also, a large amount of energy is absorbed by the crack initiated along the fiber/filler/matrix interface during debonding. The energy absorbing capability of composites depends on the properties of the constituents. Based on the literature review, it has been found that the benefits of using SiC as reinforcement are improved stiffness, strength, and chemical stability (Kaundal et al. 2012).

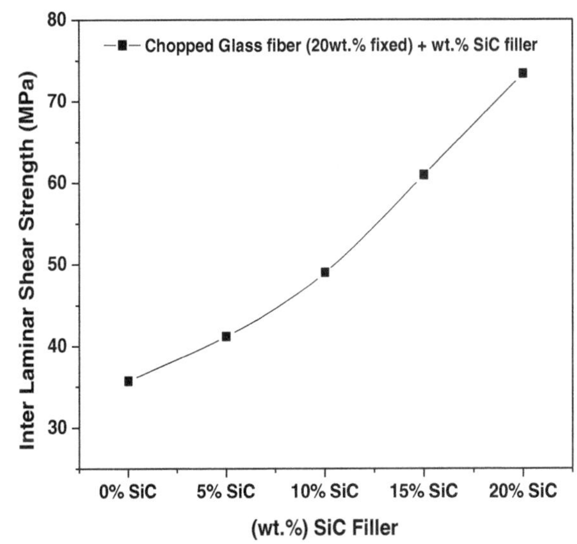

Figure 4 Effect of ILSS on SiC-filled chopped glass fiber-reinforced epoxy composites.

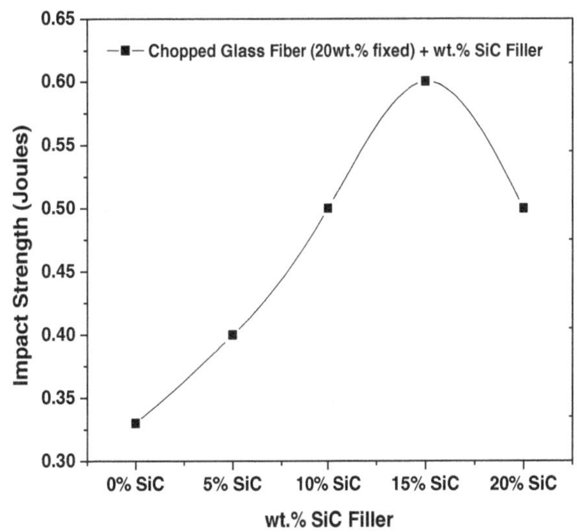

Figure 5 Effect of impact strength on SiC-filled chopped glass fiber-reinforced epoxy composites.

Thermo-mechanical properties of composite (dynamic mechanical analysis)

A common cause of tribological failure of composite parts is the inability to support a given load due to the loss in modulus at frictionally induced elevated temperatures. Dynamic mechanical analysis (DMA) has emerged as one of the most powerful tools available for the study of the behavior of materials. Simply stated, DMA measures the viscoelastic properties of materials. DMA plot provides values of those temperature regions where material properties are very stable with temperature and where rapid changes will occur that would render the material useless (Ward and Sweeney 2004). One important application of DMA is measurement of the glass transition temperature of polymers. Amorphous polymers have different glass transition temperatures, above which the material will have rubbery properties instead of glassy behavior and the stiffness of the material will drop dramatically with an increase in viscosity. At the glass transition, the storage modulus decreases dramatically and the loss modulus reaches a maximum (Agarwal et al. 2013). DMA of chopped glass fiber SiC filler-reinforced epoxy composite have been carried out in this study to investigate the variation of storage modulus (E′), loss modulus (E″) and damping factor (tan delta) i.e.:

$$\tan\delta = \frac{E'}{E''} = \frac{\text{storage modulus}}{\text{loss modulus}} \tag{1}$$

Equation 1 represents that tan δ is the ration of storage modulus to loss modulus, i.e., as the value of loss modulus increases, the value of storage modulus decreases (Agarwal et al. 2013). Similar trend is noticed in Figures 6 and 7. All tests run as per ASTM D4065-94 test standards using a fixed frequency of oscillation of 1 Hz and a sample heating rate of 2°C/min. All tests were initiated from room temperature and goes to 200°C temperature. The mode of stress applied is flexure, and the fixture configuration is a single cantilever beam. Figures 6, 7, and 8 show the most common graphic representation for elastic or storage modulus, the viscous or loss modulus, and tan delta as a function of temperature, respectively.

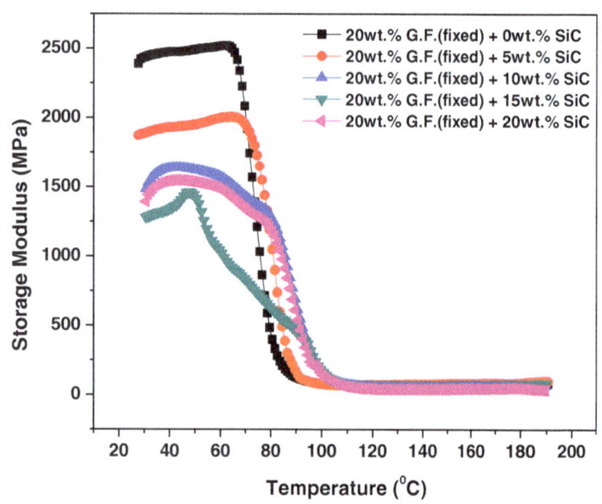

Figure 6 Variation of storage modulus for SiC filled chopped glass fiber-reinforced epoxy composites.

Figure 6 shows a graph of storage modulus versus temperature for SiC-filled chopped glass fiber-reinforced epoxy composite. Storage modulus values for chopped glass fiber SiC filler-reinforced epoxy composites show a linear trend with slight decrease in the value of storage modulus with the change in temperature (room temperature to 70°C). This shows glassy regime in region 1, i.e., from room temperature to 65°C. Region 2 shows a sharp decline in the values of storage modulus with the increase in the values of temperature (65°C to 90°C), i.e., the material changes from glassy to rubbery transition regime in the temperature range of 65°C to 90°C; due to a sharp drop in the value of storage modulus, the material has lost its usefulness as a structural material. Due to amorphous structure in the polymer and the presence of unorganized intermolecular structure, the storage modulus value drops suddenly. Region 3 again is a straight line showing negligible decrease in the values of storage modulus with the increase in the value of temperature. Here, the values of storage modulus nearly tend to zero, which shows a rubbery regime indicating degradation of the moduli above 90°C. In the glassy state, i.e., at low temperatures, stiffness is related to the small displacement of molecules, whereas in the rubbery state at high temperature, molecular chains have considerable flexibility; so that in the undeformed state, they can adopt conformations that lead to maximum entropy. The higher values in region 1 and sharp decrease in the values in region 2 are due to the fact that in region 1, the material is in glassy state in which the contribution of elastic modulus is more than the viscous modulus, whereas in region 2, the material is in glass transition stage in which a change from glass transition state into rubber elastic state takes place. In the glass transition region, i.e., from glassy to the rubbery transition regime region, the value of storage modulus falls during heating to a level of 1,000th to 10,000th of its original value (Jinmin et al. 1994).

Figure 7 clearly shows the value of loss modulus with the increase in the value of temperature. From the values obtained in the graph, we notice that the values of loss modulus rise to a maximum as the storage modulus values are in their most rapid rate of decent. The peak of the loss modulus curve denotes the glass transition temperature (*Tg*). Figure 7 denotes the glass transition temperature as 90°C and desired loss modulus values (minimum values) obtained for 10 to 20 wt.% SiC content. This may be due to the increase in the percentage of fiber/filler content, i.e., as the percentage of fiber/filler content increases, the height of the E″ curve decreases and the peak width spreads across a wider temperature

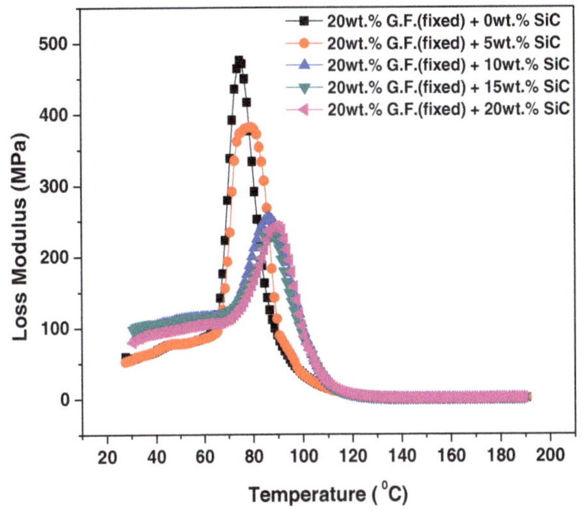

Figure 7 Variation of loss modulus for SiC filled chopped glass fiber-reinforced epoxy composites.

range leading to a decrease in the value of loss modulus (desired value obtained). Further, it has also been observed from Figure 7 that the composition with lower fiber/filler content peak of glass transition temperature shifted towards lower temperature with a sharp peak and higher value of loss modulus.

Figure 8 shows the graph of tan delta versus temperature. The tan delta curve closely resembles the loss modulus curve leading to the glass transition tan delta values well below 0.1. The rapid rise in tan delta values closely resembles the rapid decline in storage modulus values. The peak value of tan delta indicates that the material is nonelastic, whereas the lower value of tan delta indicates that the material is elastic in nature (Agarwal et al. 2013). Region 1 shows that the material is highly elastic in this region, and as the value of tan delta shifts to region 2, the nature of material shifts from elastic to plastic. The peak value of tan delta indicates that any further increase in temperature beyond this changes the material from elastic to plastic zone. Figure 8 shows that as the percentage of fiber/filler increases, the peak value shifts towards higher temperature resulting in wider elastic range of temperature, whereas as the percentage of fiber/filler decreases, the peak value shifts towards the lower temperature range resulting in narrow elastic range.

Cole-Cole plot analysis

Figure 9 shows the analysis carried on SiC-filled chopped glass fiber-reinforced epoxy composites using Cole-Cole analysis. A Cole-Cole plot is used to predict the nature of a composite whether it be homogeneous or heterogeneous (Kumar et al. 2011). A Cole-Cole is plotted for storage modulus and loss modulus where the values of loss modulus (log E″) are plotted on x-axis and the values of storage modulus (log E′) are plotted on y-axis. Cole-Cole plot is used to describe the analysis of molecular architecture. Changes in the architecture of the model samples were readily detected as systematic variations in shape and displacements on Cole-Cole plot. A homogeneous system typically exhibits a semicircular curve, whereas a heterogeneous system typically exhibits an irregular curve (imperfect semicircle); also, the nature of curvature obtained in Cole-Cole curve determines the adhesion between fiber and matrix.

Investigated values of SiC-filled chopped glass fiber-reinforced epoxy composites represent imperfect semicircles and hence are heterogeneous in nature. Also, unfilled glass fiber-reinforced epoxy composites exhibit

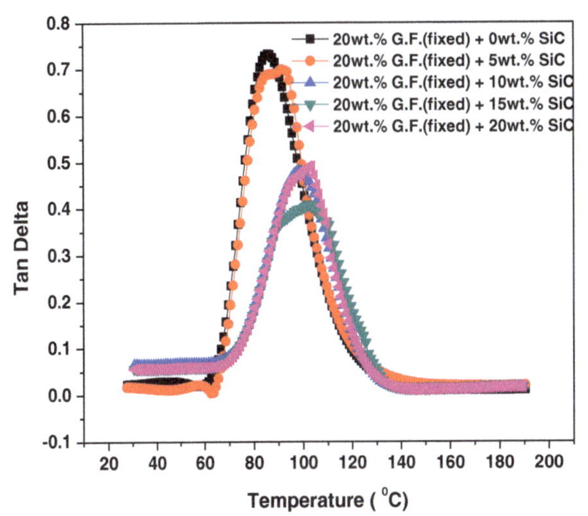

Figure 8 Variation of tan delta for SiC filled chopped glass fiber-reinforced epoxy composites.

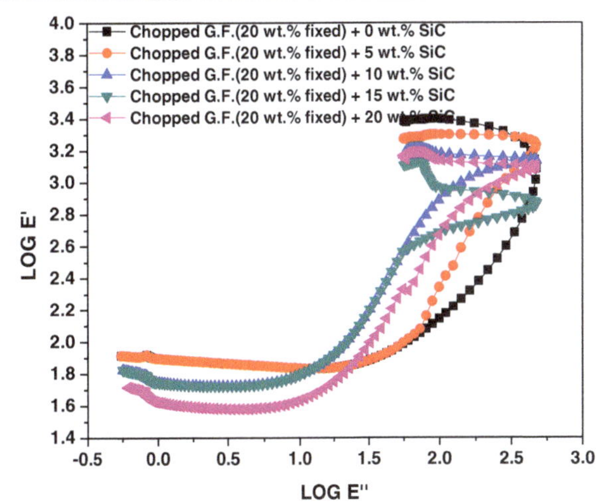

Figure 9 Cole-Cole plot for SiC-filled chopped glass fiber-reinforced epoxy composites.

better interfacial characteristics and some homogeneous nature with respect to SiC-filled glass fiber-reinforced epoxy composites. This is because adding SiC content breaks the matrix continuity, and the nature is shifted more towards heterogeneous tendency.

Conclusions

Experimental observations are carried out on SiC-filled chopped glass fiber-reinforced epoxy matrix composite to notice the effect of physical, mechanical, and thermal properties. Based on the experimental observations, the following conclusions can be drawn:

1. Mechanical properties such as hardness, tensile strength, interlaminar shear strength, flexural strength, and impact strength increase with the increase in SiC filler content up to 10 to 15 wt.%, whereas addition of SiC content beyond 15 wt.% results in the decrease in mechanical property.
2. The value of hardness and impact strength increases up to 15 wt.% filler content; the value of tensile strength and flexural strength increases up to 10 wt.% SiC filler content, whereas an exceptional increase in the value of ILSS is noticed up to 20 wt.% SiC filler content.
3. Storage modulus values for SiC-filled glass fiber-reinforced epoxy composite decreases with the increase in SiC content. Storage modulus values decrease with the increase in the percentage of SiC content, whereas elastic range of the material increases with the increase in SiC content.
4. As the percentage of SiC content increases, the area under the curve spreads across the wider range and the peak value shifts towards the higher temperature region resulting in the increase in elastic range of the composite.
5. SiC-filled chopped glass fiber-reinforced epoxy composites represent imperfect semicircles and hence are heterogeneous in nature. Also, unfilled glass fiber-reinforced epoxy composites exhibit better interfacial characteristics and some homogeneous nature with respect to SiC-filled glass fiber-reinforced epoxy composites. The imperfect circles and heterogeneous nature is due to the uneven distribution of SiC particles inside the composite.
6. From the analysis of the results of mechanical properties and thermal properties, it has been concluded that the optimum properties are obtained for 15 wt.% SiC content in addition to 20 wt.% chopped glass fiber-reinforced epoxy composite.

Competing interests
The authors declare that they have no competing interests.

Authors' contributions
All authors read and approved the final manuscript.

Author details
[1]Department of Mechanical Engineering, S.R.M.S.C.E.T, Bareilly 243122, India. [2]Department of Mechanical Engineering, National Institute of Technology, Hamirpur 177005, India.

References

Agarwal G, Patnaik A, Sharma RK (2013) Parametric optimization of three-body abrasive wear behavior of long and short carbon fibre reinforced epoxy composites. Tribology Material, surface and interfaces 7(3):150–160
ASTM (American Society for Testing and Materials) (1976) ASTM D 3039-3076: tensile properties of fiber–resin composites. American National Standard, West Conshohocken
ASTM (American Society for Testing and Materials) (1984) ASTM D 2344-84: standard test method for apparent inter-laminar shear strength for parallel fiber composites by short beam method. Annual book of ASTM standards. ASTM, West Conshohocken, pp 15–17
ASTM (American Society for Testing and Materials) (1997) Standard D 256–97: standard test methods for determining the pendulum impact resistance of notched specimens of plastic (1999) Annual book of ASTM standards, 8(1). ASTM, West Conshohocken, pp 01–20
Devendra K, Rangaswamy T (2012) Determination of mechanical properties of Al_2O_3, Mg $(OH)_2$ and SiC filled E-glass/epoxy composites. Int J Eng Res Appl 2(5):2028–2033
Imanaka M, Takeuchi Y, Nakamura Y, Nishimura A, Lida T (2001) Fracture toughness of spherical silica-filled epoxy adhesives. Int J Adhesin Adhes 21:389–396
Zhang J, Perez RJ, Wong CR, Lavernia EJ (1994) Effects of secondary phases on the damping behavior of metals, alloys and metal matrix composites. Mater Sci Eng 13(8):325–389
Kaundal R, Patnaik A, Satapathy A (2012) Comparison of the mechanical and thermo-mechanical properties of unfilled and SiC filled short glass polyester composites. Silicon 4:175–188
Koh SW, Kim JK, Mai YW (1993) Fracture toughness and failure mechanisms in silica-filled epoxy resin composites: effects of temperature and loading rate. J Polymer 34(16):3446–3455
Kumar S, Satapathy BK, Patnaik A (2011) Visco-elastic interpretations of erosion performance of short aramid fibre reinforced vinyl ester resin composites. J Mater Sci 46:7489–7500
Mohamed SS, Mahmoud TS, Eimahallawi IS, Khalifa TA (2009) Mechanical and tribological characteristics of epoxy based PMC's reinforced with different ceramic particulates. Eng Res J 121:29–49
Nakamura Y, Yamaguchi M, Kitayama A, Okubo M, Matsumoto T (1991a) Effect of particle size on fracture toughness of epoxy resin filled with angular-shaped silica. J Polymer 32(12):2221–2229
Nakamura Y, Yamaguchi M, Okubo M, Matsumoto T (1991b) Effect of particle size on impact properties of epoxy resin filled with angular shaped silica particles. J Polymer 32(16):2976–2979
Nakamura Y, Yamaguchi M, Okubo M, Matsumoto T (1992) Effects of particle size on mechanical and impact properties of epoxy resin filled with spherical silica J. Appl Polym Sci 45:1281–1289
Nielsen LE, Landel RF (1994) Mechanical properties of polymers and composites, 2nd edition. Marcel Deckker, New York
Rawlings MR (1999) Composite materials: engineering and science. Woodhead Publishing, Cambridge
Reid SR, Zhou G (2000) Impact behavior of fiber reinforced composite materials and structures. Taylor & Francis, London
Shalin RE (1995) Polymer matrix composites. Soviet advanced composites technology series, 4th edition. Springer, Heidelberg
Sua F, Zhang Z, Wang K, Jiang W, Liu W (2005) Tribological and mechanical properties of the composites made of carbon fabrics modified with various methods. Compos Part A 36:1601–1607

Ward IM, Sweeney J (2004) An introduction to the mechanical properties of solid polymers. Wiley, New York

Yamamoto I, Higashihara T, Kobayashi T (2003) Effect of silica-particle characteristics on impact/usual fatigue properties and evaluation of mechanical characteristics of silica-particle epoxy resins. Int J JSME 46(2):145–153

Experimental investigation on the ultimate strength of partially infilled and steel-braced reinforced concrete frames

Shailendra Kumar Damodar Dubey[1*] and Sunil Y Kute[2]

Abstract

Infilled walls are normally considered as nonstructural elements. However, these walls are effective in carrying lateral loads. In this regard, an experimental investigation was planned and conducted to study the effect of braced and partially concrete-infilled, reinforced concrete (RC) frames in comparison to the bare frames. All these frames were tested up to collapse and subjected to only horizontal loads to obtain an effective and possible solution for soft story. In comparison to bare RC frames, partially infilled frames have more lateral load capacity. Central bracing is more effective than that of corner bracing. For the same load, braced and partially infilled frames deflected significantly less than that of the bare frames. Based on experimental observations, a mathematical model has been proposed to calculate theoretical ultimate load for braced and partially infilled RC frames.

Keywords: Central-braced frame; Partially infilled frame; Lateral load; Soft story; Diagonal bracing; RC frame

Introduction

Usually, the infill walls of a building are not considered as elements of the force-resisting system. In many instances, structural walls are assumed to resist racking loads due to wind and earthquake. Such walls are generally known as shear walls and can also carry vertical loads and are designed as an integral part of the structures. Other types of walls are called filler panels, which are used to partition space or to enclose structures. Such panels occupy a rectangular space between two adjacent columns and two successive stories. But these filler panels can have considerable influences on the lateral response of a building. Nowadays, soft story in buildings is common at the parking level as there is absence of infill walls, whereas the stories above are filled with partition walls. Such frames have less capacity to bear lateral loads. Portal frames tested up to collapse have drawn attention of several investigators in the recent past for inherent structural advantage of such frames.

Available literature review shows that attempts to evaluate the strength and stiffness of these frames. Infilled frames investigated (Wood 1958) by conducting several tests on concrete encased steel frames with brick and concrete-infilled panel walls on the resistance of structural frame works against racking loads. (Benjamin and Williams 1958, 1959) had tested many prototypes as well as models of RC frames with plain and reinforced concrete infill walls. The foundations were considered rigid. It was observed that there was no scale effect; i.e., test can be performed on any scale model, and results of the scale models were found to be consistent with the prototype. (Smith 1966) investigated the behavior of square frames and tried to compare the theoretical results with experimental ones. He had derived expressions for diagonal stiffness and suggested that the infill acts as a diagonal strut. A method had been described by (Mallick and Severn 1967) which make use of finite element concept. For the purpose of calculation, the actual frame with infinite degrees of freedom was replaced by a system with finite number of degrees of freedom, and the elements of infill were assumed as rectangular in shape. The approach to analyze infilled frames by (Mainstone 1971; Smolira 1973) was based on the concept of diagonal struts. (Barua and Mallick 1977) had tested a series of mortar-infilled model steel frames with a tightly fitted, homogeneous and elastic infill. Tests were performed by Mali and Saldoga (1981) on

* Correspondence: dubey.dhule@rediffmail.com
[1]Civil Engineering Department SSVPS BSD College of Engineering, Dhule, India
Full list of author information is available at the end of the article

RCC frames with brick as infill. In addition to the racking load, they applied a uniformly distributed vertical load, which caused pre-compression of wall and reported cracking and failure behavior of infilled frames. Their experimental values were close to the theoretical values.

(Liauw and Kwan 1983) had proposed plastic theory for analysis to both single story and multi-story integral infilled frames. The problem of non-integral infilled frame was overcome by introduction of material with strong bond or shear connectors at the frame/infill interface. In 1999 in Kocaeli (Turkey), the complete failure of the first story or the bottom two stories happened due to earthquake. (Dolšek and Fajfar 2001) demonstrated that a soft story mechanism is formed in such structural systems if the intensity of ground motion is above a certain level and collapse will occur if the global ductility of the bare frames is low. (Hashmi and Alok 2008) studied an analytical investigation of seismic performance of masonry-infilled, RC-framed building with an open storey generally termed as soft storey. It was observed that if the storey is partially infilled (in comparison to no infills), it decreases story drift and deformations in the column in the storey and reduces the damage to columns and overall frame. The main objective of (Liao 2010) study was to develop and validate a seismic design methodology for reinforced concrete special moment frames which is able to produce structures with predictable and intended seismic performance. For existing reinforced concrete buildings, (Viswanath et al. 2010) retrofitted to overcome deficiencies to resist seismic loads by using concentric steel bracing systems. A reinforced concrete frame is modeled for finite element sensitivity analysis by (Alam and Dookie 2012) followed by finite element reliability analysis under both static and dynamic load cases. Building damage by earthquake action is a serious problem; in this regard, (Rachana and Mohod 2012) studied seismically deficient structures by carrying out the pushover analysis of frame structures using structural analysis and programming software. Forces increased as per the height of structure: low-rise structures have higher resonant frequency and hence lower-frequency, high-rise structures had been studied. As per (Vijayakumar and Venkatesh Babu 2012), Indian buildings built over the past two decades are seismically deficient because of the lack of awareness regarding the seismic behavior of structures. This paper aims to evaluate a selected existing reinforced concrete building in zone III to conduct pushover analysis. The analysis shows the pushover curves, capacity spectrum, plastic hinges, and performance level of the existing building.

Considering all these factors, high-yield, strength mild-steel bracings with partial infills of concrete, the

Table 1 Description of various frames

Serial number	Frame notation		Description
1	$R1_A$	} R1	Bare RC frame
	$R1_B$		
2	$R2_A$	} R2	Corner top bracing frame
	$R2_B$		
3	$R3_A$	} R3	Central top bracing frame
	$R3_B$		
4	$R4_A$	} R4	Diagonal bracing frame
	$R4_B$		
5	$R5_A$	} R5	Corner top bracing frame filled with concrete
	$R5_B$		
6	$R6_A$	} R6	Central top bracing frame filled with concrete
	$R6_B$		
7	$R7_A$	} R7	Diagonal bracing frame filled with concrete
	$R7_B$		

RC frames were tested under lateral loads to understand the behavior and contribution of infill. In this study, tests are conducted on 14 numbers of different models of bare, braced, and infilled frames as shown in Table 1. For main reinforcement and bracings, tor steel bars and steel square bars are used for frames. The behavior of frames has been studied with respect to the following:

- Bracing system-bare frames and different types of braced RC frames
- Partial infill-steel-braced RC frames without infill and partially infilled with concrete
- Strength and deformation of frames

In the present work, experiments have been performed on models up to failure. Studies have been carried out on single-bay, single-story frames. For each frame, two

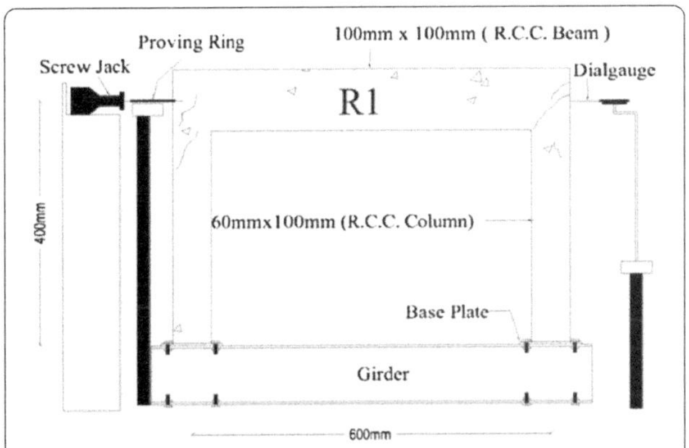

Figure 1 Details of bare frame.

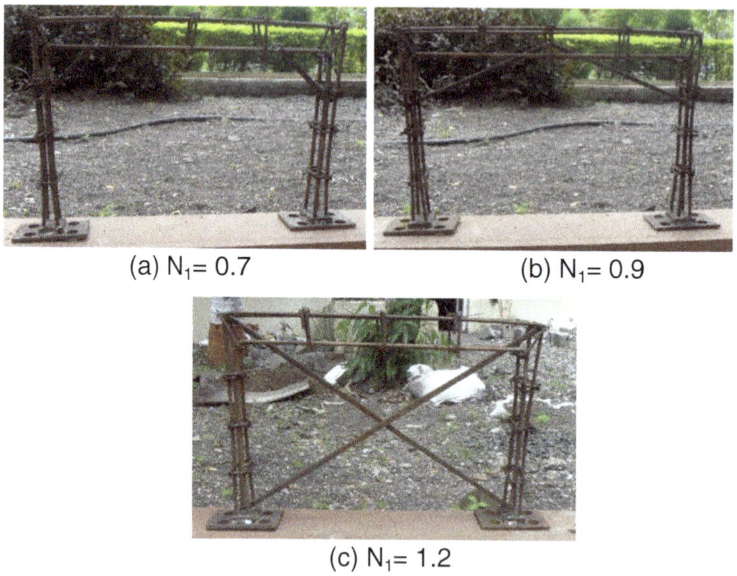

Figure 2 Arrangement of reinforcement and bracing for various frames. (a) Corner bracing, R2; **(b)** central bracing, R3; and **(c)** diagonal bracing, R4.

models were tested, and the average value is considered for experimental loads and deflections.

Methods

Experimental set-up

For the experimental setup, the RC portal frame with a welded base plate was rigidly bolted to the supporting girder. All the experiments were performed on models with single-bay, single-story frames. The two models were tested of each category as mentioned in Table 1. The details of the model's dimensions, proving ring, loading jack, and position of the dial gauge are shown in Figure 1. The different bracing systems such as corner, central and diagonal, and the steel skeleton of reinforcement are shown in Figure 2. The frame consists of two vertical members of height 400 mm connected by a horizontal member at the top with a span of 600 mm. For the bare frames' proving ring capacity of 10 kN and for remaining frames, a hydraulic jack of 500 kN was attached to measure the load. Dial gauge with range of 20 mm was used to measure the horizontal displacement at the beam level. The size of column is 60 mm × 100 mm and for the beam is 100 mm × 100 mm. Side cover plates of 10 mm thick were welded to such frames, and these plates were bolted to the girder by four bolts of 20-mm diameter. The RC frames and infill of concrete were cast by laying the frame on the horizontal surface.

Materials for models and control specimen

The following materials were used for the frame, bracing, and the partial infill:

- Tor steel of 8 mm of weight 3.9 N/m as main reinforcement and 6 mm for ties and stirrups were used for the RC frame.

Table 2 Comparison of ultimate loads for different frames

Frame	Experimental ultimate load (kN)	Contribution of bracings in comparison to bare RC frames (%)	Contribution of infill with similar type of braced RC frames (%)	Theoretical ultimate load (kN)	Percentage change in theoretical load in comparison to experimental load
R1	9.35	-	-	9.72	+3.95
R2	25	167.3	-	24.85	−0.6
R3	30	220.8	-	29.18	−2.73
R4	37.5	301	-	39.37	+4.98
R5	30	-	20	30.75	+2.5
R6	35	-	16.6	37.02	+5.77
R7	85	-	126.6	84.76	−0.28

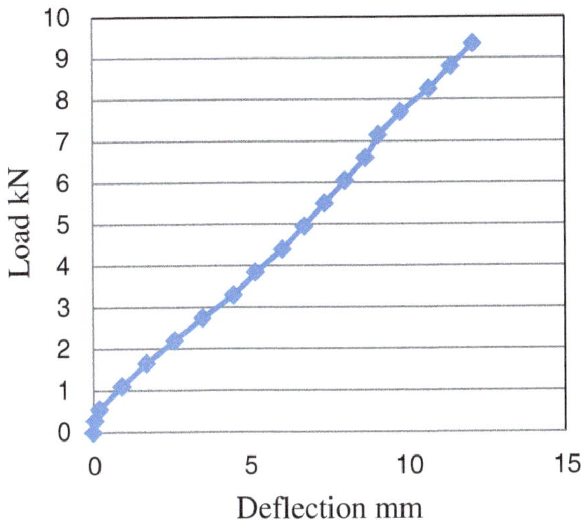

Figure 3 Load vs. deflection graph for bare RC frame R1.

- For bracings, 10-mm square bars of high-yield, strength mild- steel is used.
- The partial infills of concrete with thickness of 50 mm which are made up with cement, river sand, and 12-mm coarse aggregate of 1:1.5:3 ratios were used.

Three samples for the tensile test of 8-mm and 6-mm tor steel and high-yield strength square bars of 10 mm were prepared and tested to obtain the characteristic strength and yield stress of steel. Three cubes of size 150 mm × 150 mm × 150 mm were cast and tested after 28 days to determine the compressive strength of concrete, and slump was also observed for fresh concrete.

Test procedure

While conducting the test on the various frames, bolts were fully tightened ensuring full fixity at the supports. The dial gauge was checked for proper functioning before it was mounted in the position, and the contact between the dial gauge tip and the frame surface was ensured. The jack centerline was aligned along the beam axis.

The horizontal load was applied with the help of a screw/hydraulic jack. The load was applied at a uniform rate. The load and the deflection were recorded at regular intervals for each test setup. The load was applied continuously until it remained constant for a particular time on the proving ring and then moved in a reverse order. This is called as plastic state condition. The collapse load corresponding to this stage was recorded as an ultimate load.

Results

While conducting the experiments, precautions were taken to keep the proving ring at its position as it was trying to lift itself. During load application, attention was paid to crack formation and its propagation in the infill. The direction and progress of cracks at different load levels were recorded. The locations and extent of loss of contact between the frame and infill were noted down. The final collapse modes were photographed for full details. The compressive strength of concrete mix cubes was observed to be 24.2 N/mm^2, tested after 28 days. The value of slump was 50 mm for the fresh concrete. The characteristic strength of 8-mm and 6-mm tor steel is 419 and 350 N/mm^2, respectively. The yield stress of bracing bars was observed to be 450 N/mm^2, and allowable bending tensile strength was taken as 0.66 times yield stress. The increase in lateral load capacity of braced frames R2, R3, and R4 with respect to the bare frame R1 is 167.3%, 220.8%, and 301%, respectively. The contribution of infill is 20%, 16.6%, and 126.6% for R5, R6, and R7 in comparison to the braced frames of similar type without an infill as shown in Table 2. The load deflection curves for R1, R2, R3, and R4 and R5, R6, and

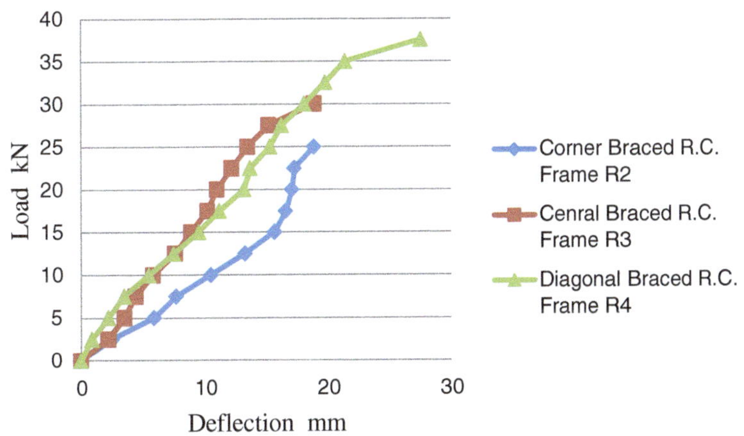

Figure 4 Load vs. deflection graph for braced frames.

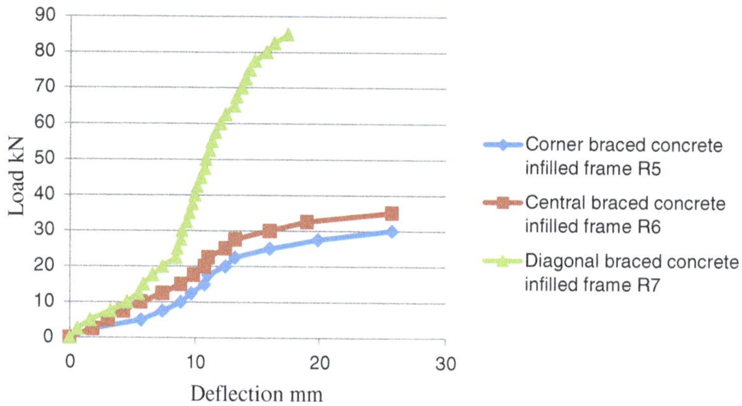

Figure 5 Load vs. deflection graph for braced and concrete-infilled frames.

R7 are shown in Figures 3, 4, 5 respectively. These figures show that central-braced infilled frames are more effective than that of corner-braced and diagonal-braced, partially infilled frames for the deflection criteria. The comparison of experimental and theoretical ultimate loads for various frames is shown in Table 2. Figures 6, 7, 8, 9, 10, 11, 12, illustrate the crack pattern indicated by the red painted lines for all bare, braced, and infilled frames correspondingly.

Proposed analytical method

The analytical methods for bare, braced, and infilled frames are described as follows.

Method 1: proposed methods for bare RC frame

While particularly applying the load on the bare frames, it was observed that although failure mechanism was predominantly the sway mechanism, plastic hinges are formed at base of column and at junction of column and the beam. The idealization of the proposed method is illustrated in Figure 13. To develop a mathematical model certain assumptions were made:

1. The concrete material is homogenous and isotropic.
2. The mode of collapse is a sway mechanism for frames.
3. For frames at joints, initial fixity exists perfectly.
4. The small deformation theory is applicable.

Let virtual displacement, corresponding to sway mode, be represented by 'Δ' at q and r and hinge rotation 'θ' at p, q, r, and s.

The external work done due to ultimate load is as follows:

$$W_E U = W_u \Delta = W_u(H\theta) \tag{1}$$

The internal work done due to hinge rotation is presented as

$$W_I H = 4Mp\theta, \tag{2}$$

where external work done = internal work done is shown as

$$W_u = 4Mp/H \tag{3}$$

Here Mp depends on σ_c. The value of $Mp = [(0.67) * (\sigma_c/\gamma c) * Zp]$.

Figure 6 Bare RC frame.

Figure 7 Corner top bracing RC frame.

Figure 8 Central top bracing RC frame.

Figure 10 Corner top bracing frame filled with concrete.

Method 2: proposed methods for braced and partially infilled RC frame

The proposed analytical model to calculate theoretical ultimate lateral load capacity is on the basis of the experimental observations of crack patterns and formation of plastic hinges for various braced and partial concrete-filled RC frames. It is observed that due to compressive force from diagonal compression band, tensile cracks are developed along tension column for all RC-infilled frames. The idealization of the proposed method is illustrated in Figure 14. To develop a mathematical model certain assumptions were made:

1. Plastic hinges are formed at column base and column beam junction, and mode of collapse is the sway mechanism.
2. For partially infilled frames, perfect initial fit exists.
3. The small deformation theory is applicable.

The crushing strength of concrete infill was taken as 0.67 times the experimental compressive strength. As per assumption, the failure mode is the sway mechanism with plastic hinges at *p*, *q*, *r*, and *s*. *For all infills, strength is considered along the diagonal compression only.*

Figure 9 Diagonal bracing RC frame.

Let virtual displacement, corresponding to the sway mode, be represented by 'Δ' at q and r and hinge rotation 'θ' at p, q, r, and s. External work done due to the ultimate load and internal work done due to hinge rotation are presented as Equations 1 and 2, respectively.

Internal work done due to infill can be presented as follows:

$$W_I I = F_C[(H + H_2)/2]\theta \cos\alpha, \tag{4}$$

where external work done = internal work done is shown as

$$W_u = 4M_p/H + F_C[(H + H_2)/2] \cos\alpha/H \tag{5}$$

and where the value of 'Fc' for various frames are as follows:

1. For braced RC frames

$$F_C = \sigma_{bt} A_{sb} N_1 \tag{6}$$

2. For concrete partial infill

$$F_C = 0.67\sigma_c t_w W_C + \sigma_{bt} A_{sb} N_1. \tag{7}$$

The values of N_1 are shown in Figure 2 and are obtained from bracing present in diagonal compression strut and H_2 = (3/4) H. The value of W_C will be taken as 15 H/B and 20 H/B for R5 and R6, respectively.

Figure 11 Central top bracing frame filled with concrete.

Figure 12 Diagonal bracing frame filled with concrete.

Method 3: proposed methods for diagonally braced and infilled RC frame

On the basis of the experimental observations of crack patterns and formation of plastic hinges for diagonal bracing and concrete-filled RC frames, plastic hinges are formed at column base and column beam junction, and the mode of collapse is the sway mechanism. To develop a mathematical model, certain assumptions were made:

1. The crushing strength of concrete infill was taken as 0.67 times the experimental compressive strength.
2. For infilled frames, perfect initial fit exists.
3. The small deformation theory is applicable.

The idealization of the proposed method is illustrated in Figure 15. As per assumption, the failure mode is a sway mechanism with plastic hinges at p, q, r, and s. *For all infills, strength is considered along the diagonal compression only.*

Let virtual displacement, corresponding to the sway mode, be represented by 'Δ' at q and r and hinge rotation 'θ' at p, q, r, and s.

External work done due to ultimate load and internal work done due to hinge rotation are presented as Equations 1 and 2, respectively.

The internal work done due to infill is as follows:

$$W_I I = F_C\{\Delta_1 - [(C_Y/2)\theta \cos\alpha]\}, \tag{8}$$

where external work done = internal work done as

$$W_u = 4M_p/H + F_C[(H - C_X \tan\alpha)] \cos\alpha/H \tag{9}$$

Where the value of 'Fc' for various diagonally braced frames are as follows:

1. For braced RC frames

$$F_C = \sigma_{bt} A_{sb} N_1 \tag{10}$$

2. For concrete infill

$$F_C = 0.67\sigma_c t_w W_C + \sigma_{bt} A_{sb} N1 \tag{11}$$

The values of N_1 are shown in Figure 2 and are obtained from the bracing present in the diagonal compression strut. The value of $W_C = 2C_X \sin\alpha$. The value of $C_Y = C_X \tan\alpha$, where the value of C_X has been taken as zero for R4 and 50 B/H for R7.

Discussion

The behavior of partially infilled, braced RC frames subjected to racking load was studied with different patterns of steel bracings such as corner, central, and

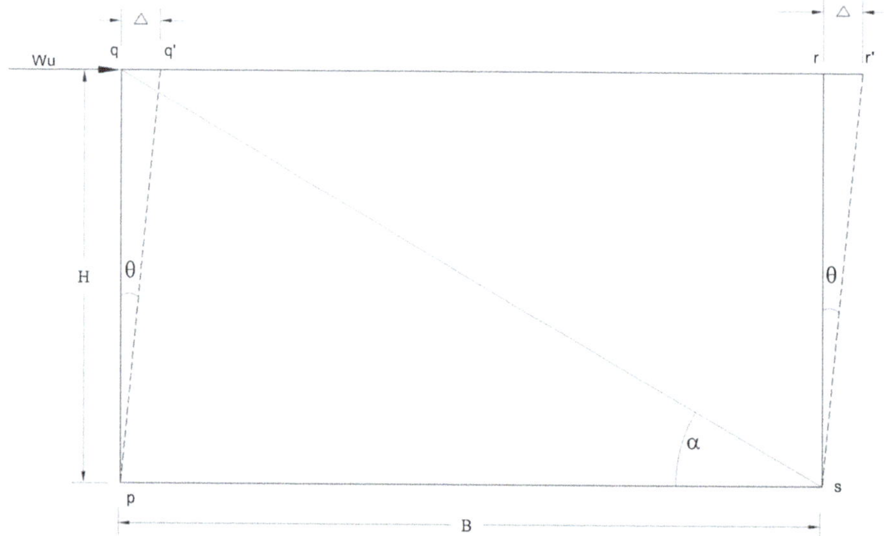

Figure 13 Method 1: idealization for bare RC frame.

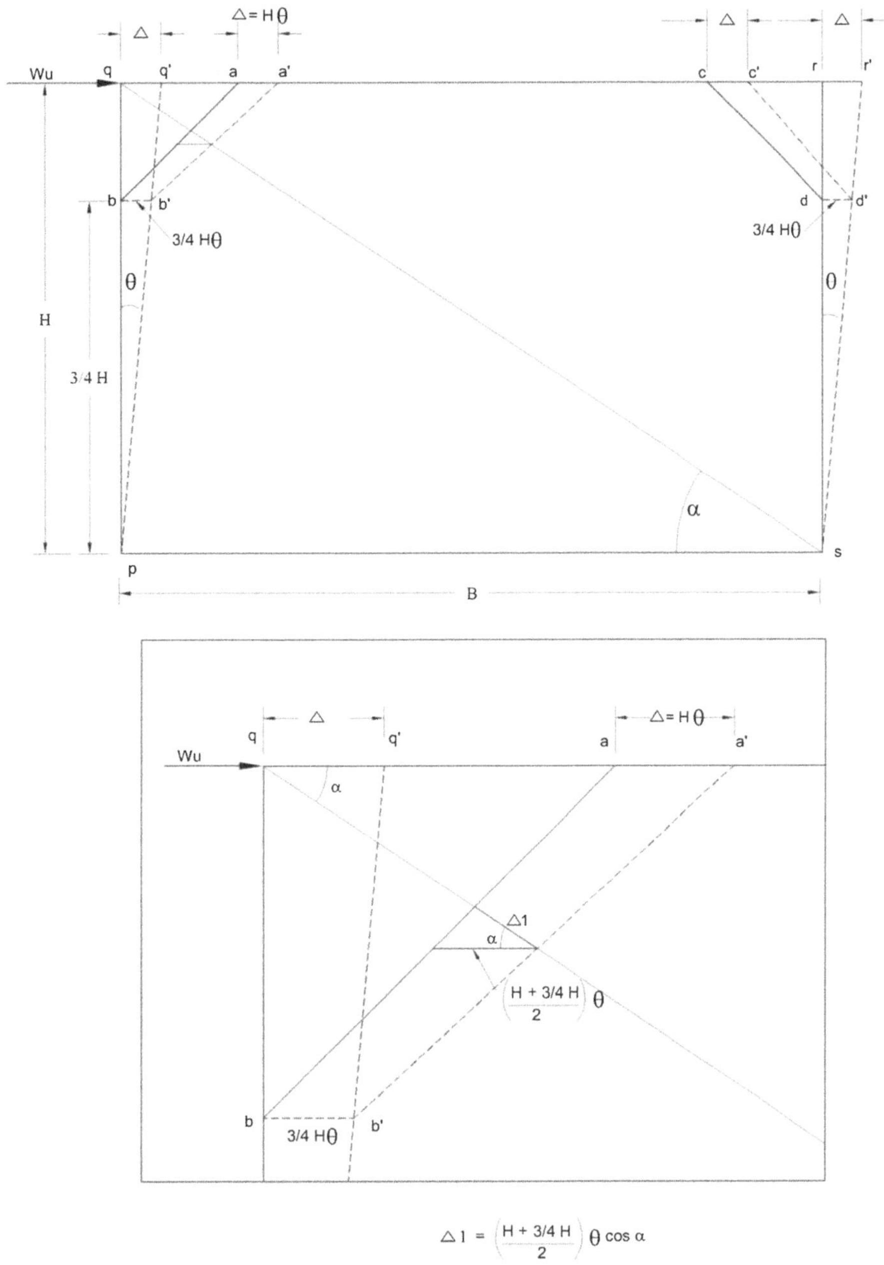

Figure 14 Method 2: idealization for braced and partially infilled RC frame.

diagonal and infilled material such as concrete. It is observed that due to the compressive force along the diagonal compression band, tensile cracks are developed along tension column for all RC-infilled frames. The cracks developed at various places are indicated by the red painted lines on different infills as shown in figures. It can be observed from the photo plates that failure was predominately caused by the sway mechanism. The partial safety factor for concrete 'γc' is taken as 1.5. The ultimate load versus deflection comparison shows a significant increase in the lateral load capacity for braced and partially infilled frames than that of the bare frames. Although no lugs were provided between partial infill and the RC frames, the concrete partial infill contribution was still observed to be 20% more than the braced frames.

Since cracks developed prior to failure, the tensile strength of concrete does not take into account for ultimate strength of infill. Though diagonally braced system shows better results than that of the other two

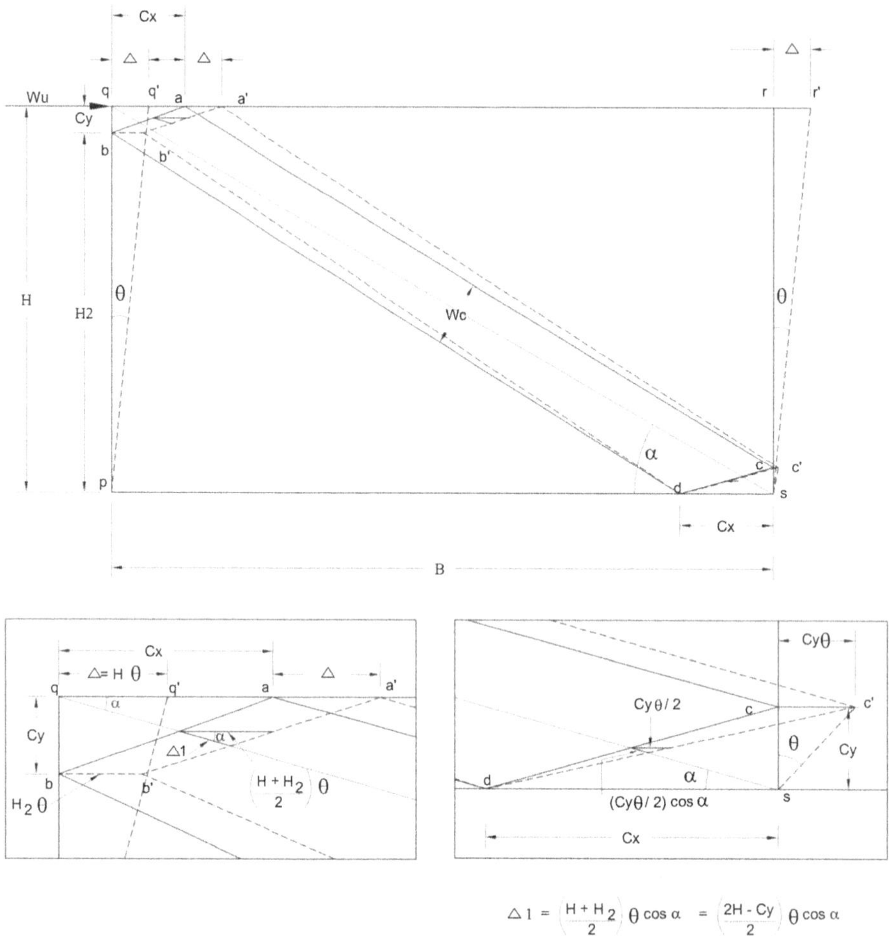

Figure 15 Method 3: idealization for diagonally braced and infilled RC frame.

systems, practically, it is difficult to implement as it would hinder the movement of users around the space and thus central bracing system is more effective for soft story frames.

Shear walls which are used in all four corners of multi-story buildings can be strengthened by using diagonal steel bracing with concrete infill, as its lateral strength contribution is remarkable. It can be further seen that the experimental ultimate loads are fairly close to the theoretical ultimate loads based on the proposed method.

Conclusions

The purpose of this study was to evaluate the strength of braced and partially infilled RC frames then those studied by previous researchers as well as to add to the database of strengthening test results in order to lead to changes or acceptance in design codes and standards. To study the ultimate load two bare, six braced and six partially infilled RC frames were constructed and tested up to collapse.

Based on the results of the investigation, the following findings and conclusions are presented for such frames:

- If the infill is stronger than the frame, the failure mode corresponds to sway mechanism with major tension cracks along the tension column. For braced and partially infilled RC frames, possible plastic hinge locations are at the column beam junction and the bottom of column.
- Based on a comparison with the bare RC frames, braced frames have shown an increase of 167.3%, 220.8%, and 301% in lateral load capacity. It specifies that the contribution of bracings is up to a significant level for the lateral strength for braced frames.
- All braced and partially infilled braced frames have significantly less deflection in comparison to the bare frames.
- Practically, the partially infilled and center-braced system may be a viable solution which may not affect architectural or interior function than that of

corner and diagonal bracing partially infilled system for soft story frames.

- For shear walls, use of diagonal steel bracing with RCC infill will be more effective, as its lateral strength contribution is remarkable.
- Though contribution to the strength of partially infilled frames is 20% in comparison to braced frames, it stiffens the frames considerably.
- The proposed analytical method gives results close to the experimental results and can be used to calculate ultimate collapse load for such type of infilled system.

It is suggested for future study to use 3D models for testing and analyzing by using ANSYS software.

Abbreviations

A_{sb}: Area of square bracing bar (mm^2); B: Width of frame (mm); C_X: Effective width of crack along x direction (mm); C_Y: Effective width of crack along y direction (mm); F_C: Compressive force along diagonal (N); H: Height of frame (mm); M_P: Full plastic moment of the section (N/mm); N_1: Number of bracing bar in compression zone; t_w: Thickness of partial infill (mm); W_C: Width of diagonal compression (mm); W_u: Theoretical ultimate load (N); γc: Partial safety factor for concrete; Zp: Plastic modulus (mm^3); Δ: Virtual lateral displacement (mm); Δ_1: Virtual lateral displacement due to Fc (mm); θ: Virtual hinge rotation (radian); α: angle between compression diagonal and horizontal (radian); σ_{bt}: Allowable bending tensile strength of bracing bars (N/mm^2); σ_c: Ultimate compressive strength of concrete (N/mm^2); σ_y: Yield stress of mild steel bracing bars (N/mm^2).

Competing interests

The authors declare no competing interests.

Authors' contributions

SKDD prepared the experimental model, carried out the experiments, drafted the manuscript, and prepared mathematical models. SYK had given the idea, participated in designing the experimental model, helped in the interpretation of experimental output, and also helped in drafting the manuscript. Both authors read and approved the final manuscript.

Authors' information

SKDD has done his M.Tech. in the field of structural engineering from IIT Mumbai in 1991. He is an associate professor in SSVPS COE Dhule and doing his Ph.D. in the field of structural engineering. He has published papers in the field of structural engineering and building materials. SYK is a Ph.D. and is a professor in K. K. Wagh COE Nasik and has published papers in the field of structural and hydraulics engineering.

Author details

[1]Civil Engineering Department SSVPS BSD College of Engineering, Dhule, India.
[2]Civil Engineering Department, K.K. Wagh College of Engineering, Nasik, India.

References

Alam MI, Dookie K (2012) Effect of constitutive material models on seismic response of two-story reinforced concrete frame. International Journal of Concrete Structures and Materials 6:101–110. doi:10.1007/s40069-012-0010-3

Barua HK, Mallick SK (1977) Behavior of mortar infilled steel frames. Build Environ 12:262–272

Benjamin JR, Williams HA (1958) The behavior of one storey brick shear walls. ASCE J Struct Div 84:1–30

Benjamin JR, Williams HA (1959) Behavior of reinforced concrete shear wall. Trans Am Soc Civil Eng 124:669–708

Dolšek M, Fajfar P (2001) Soft storey effects in uniformly infilled reinforced concrete frames. J Earth Eng 5:1–12

Hashmi AK, Alok M (2008) Damage forecast for masonry infilled reinforced concrete framed buildings subjected to earthquakes in India. Curr Sci 94(1):61–73

Liao W-C (2010) Performance-based plastic design of earthquake resistant reinforced concrete moment frames. University of Michigan, Doctoral thesis

Liauw TC, Kwan KH (1983) Plastic theory of infilled frames with finite interference shears strength. Proc Inst Civil Eng 75(2):707–723

Mainstone RJ (1971) On the stiffness and strength of infilled frames. Proc Inst Civil Eng IV:57–90

Mali SG (1981) Model test on R.C. frames with brick infill with vertical and horizontal load. Indian Concrete J 55:13–21

Mallick DV, Severn RJ (1967) The behavior of infilled frame under static loading. Proc Inst Civil Eng 38:639–656

Rachana RG, Mohod MV (2012) Seismic evaluation of framed structures. Int J Sci Eng Res 3:1–8

Smith BS (1966) Behavior of square infilled frames. J Struct Div ASCE 92(1):381–403

Smolira M (1973) Analysis of infilled shear walls. Proc of Ins Civil Eng 55:895–912

Vijayakumar A, Venkatesh Babu DL (2012) Pushover analysis of existing reinforced concrete framed structures. Eur JSci Res 2:195–202

Viswanath KG, Prakash KB, Desai A (2010) Seismic analysis of steel braced reinforced concrete frames. Int J of Civil Struc Eng 1:114–122

Wood RK (1958) The stability of tall buildings. Proc Ins Civil Eng 11:69–102

Numerical and experimental analyses of an integral bridge

Yasser Khodair[1*] and Sophia Hassiotis[2]

Abstract

In this paper, we present the analysis of a numerical and an experimental study of the Scotch Road integral abutment bridge located in Trenton, NJ, USA. Three-dimensional, nonlinear finite element (FE) model of the full bridge has been developed to study the effect of thermal loading on the bridge substructure. The bridge substructure was fully instrumented. Data analysis was performed to study the effect of several design parameters on axial stress in piles. An analysis of the pile-soil system was performed using the finite difference software LPILE. The maximum displacement of the bridge superstructure obtained from the FE model due to a maximum expected temperature change of ±42°C during the lifetime of the bridge was applied to the substructure model. The effect of bridge skew on the build-up of soil pressure behind the abutment was studied. A significant increase in the soil pressures and axial stresses behind the abutment at the obtuse side versus the acute was observed. The effect of the size of the galvanized steel sleeve on the induced axial stresses in piles was studied. We found that increasing the size of the steel sleeve increases their capacity to resist bending.

Keywords: Integral abutment, Soil-pile interaction, Laterally loaded piles, Full-scale testing, Finite element analysis

Introduction

Integral abutments are continuous bridges, where the deck is rigidly connected to the abutments and approach slabs. This rigid connection allows integral bridges to act as a single unit in resisting thermal and brake loads. The need for jointless bridges evolved from the desire to eliminate the use of expansion joints and bearings. The leak of deck drainage water contaminated with deicing chemicals has caused the deterioration of beam ends, pedestals, and piers. Moreover, expansion joints are expensive to buy, maintain, or replace. Bearings may grind to a halt due to lack of lubrication or progressive corrosion (Wasserman and Walker 1996). However, the impact of thermal loading on the abutment-pile system and the surrounding soil represents a crucial factor in the design of integral bridges and is considered the biggest obstacle in the analysis and design of such bridges. As such, the soil-structure interaction constitutes the focus of the research presented herein, in particular, the development of axial stresses in the piles supporting the abutment as a result of thermal expansions and contractions of the bridge superstructure.

The daily and seasonal temperature changes, the changing temperature conductivity between the different components of the bridge superstructure, and the nonuniformity of the temperatures within the bridge deck in both the horizontal and vertical directions result in cyclic bridge expansions and contractions (Arsoy et al. 1999).

The behavior of pile-soil interaction has been studied using experimental and numerical analyses. The subgrade reaction approach provides the simplest solution for the pile-soil interaction problem. In this approach, the pile is treated as an elastic laterally loaded beam. The soil is idealized as a series of independent springs with constant stiffness, where the lateral load at one point does not affect the lateral load at other points along the depth of the pile (Reese and Matlock 1956; Broms 1964). The spring stiffness, or modulus of subgrade reaction, is defined as the ratio of the soil reaction per unit length of the pile as described in the following equation:

$$p = K_h y \tag{1}$$

* Correspondence: ykhodair@bradley.edu
[1]Department of Civil Engineering and Construction, Bradley University, 1501 West Bradley Avenue, Peoria, IL 61625, USA
Full list of author information is available at the end of the article

where p is the soil resistance per unit length of the pile, K_h the modulus of subgrade reaction, and y the lateral deflection of the pile. However, a drawback of the method is its inability to account for the nonlinearity and continuity of soil.

The concept of a *p-y* curve was first introduced by McCelland and Focht (1958). Brown and Shie (1990, 1991) developed three-dimensional (3D) finite element (FE) models to derive *p-y* curves. In these models, a simple elastic–plastic material model was used for the soil to study undrained static loading in clay soils. A plain strain finite element model that is symmetrical about the centerline of the bridge was developed by Arsoy et al. (1999). Linear stress–strain constitutive model was used to model the abutment. The approach fill and the foundation soil were modeled using hyperbolic material properties. The loads applied on the model represent the loads reflected from the superstructure and the abutment. Faraji et al. (2001) studied the effect of thermal loading on integral bridges using a three-dimensional FE model. The *p-y* method was used to model the nonlinear behavior of the soil behind the abutment and surrounding the piles. The soil pressure distribution on the abutment is typically nonlinear and varies with depth, amount, and mode of wall displacement. A small parametric study was conducted to study the effect of the level of soil compaction on the response of the pile-soil system. Thompson (1977) used a two-dimensional finite element model to obtain *p-y* curves for laterally loaded piles. The soil was modeled as an elastic-hyperbolic material. A one-dimensional finite element model was developed by Desai and Kuppusamy (1980). In their model, the soil was modeled using nonlinear springs and the pile as a beam column element. The soil was defined using the Ramberg-Osgood model. Faruque and Desai (1982) implemented both numeric and geometric nonlinearities in their three-dimensional finite element model. The Drucker-Prager plasticity theory was adopted to model the nonlinear behavior of the soil. Kumar (1992) investigated the behavior of laterally loaded single piles and pile groups using a three-dimensional nonlinear finite element modeling. Rajashree and Sitharam (2001) developed a nonlinear finite element model of batter piles under lateral loading. In their model, the nonlinear soil behavior was modeled using a hyperbolic relation for static load condition and modified hyperbolic relation, including degradation and gap for cyclic load condition. Olson et al. (2009) developed three-dimensional finite element analyses using SAP2000 to evaluate skew and maximum length of the integral bridges. The numerical analyses showed that acceptable bridge length and skew combinations that result in stresses below the yielding stress are affected by the grade of steel used for the piles and by the use of moment-reducing elements.

Other researchers have studied integral bridges using experimental analysis. Jorgensen (1983) instrumented an integral bridge for a period of 1 year. The analysis of the data proved that stresses at the top of the pile were sufficient to initiate a yield stress in the steel. Lawver et al. (2000) instrumented and monitored a bridge near Rochester, Minnesota for several years. It was found that the piles deform in double curvature, with the pile strains on the approach panel sides of the piles which indicates the beginning of yielding. Girton et al. (1991) instrumented two skewed bridges in IA, USA. The authors found that the stress in the piles may reach 60% to 75% of the yield stress, over the 1-year period.

The objectives of this research are to study (1) the maximum displacement that the Scotch Road integral abutment bridge can be subjected to due to thermal loading; (2) the effect of relevant design parameters which include the abutment temperature, displacement of the superstructure during thermal loading, abutment rotation, and soil pressure behind the abutment on the pile axial stresses; (3) the effect of bridge skew on the variation of soil pressure behind the abutment and axial stresses induced on the piles supporting the abutments; and (4) the effect of increasing the amount of sand surrounding the piles on its induced axial stresses. To that end, we (1) instrumented the abutment and piles during construction, (2) developed 3D FE model of the full bridge, a 2D finite difference (FD) method model using LPILE, (3) analyzed the effect of crucial design parameters on pile axial stress, and (4) studied the effect of the size of the galvanized steel sleeve on pile axial stress. Then, we draw our conclusion.

Bridge description

The Scotch Road I-95 integral bridge is located in the Ewing/Hopewell Township in NJ, USA. The bridge is composed of two spans of 45.5-m long continuous, integral abutment bridge as shown in Figure 1. The new bridge will accommodate two traffic lanes, one auxiliary lane, one shoulder, and one walkway in each direction with a total width of 32.91 m and an approximately 15° skew angle. The superstructure of the bridge consists of a composite section including the reinforced concrete bridge deck supported by ten high performance steel girders. An elevation view of the north integral abutment supported by the steel HP piles is shown in Figure 2.

The substructure is composed of reinforced concrete integral stub abutments (0.9 m in width and 2.88 m in depth) supported by 19 HP 360 × 152 steel piles embedded into the abutment walls to ensure full moment transfer as depicted in Figure 2. The relief slab is connected rigidly to the integral abutment from one side and supported by a sleeper slab from the other side. The soil behind the abutment and under the approach slab

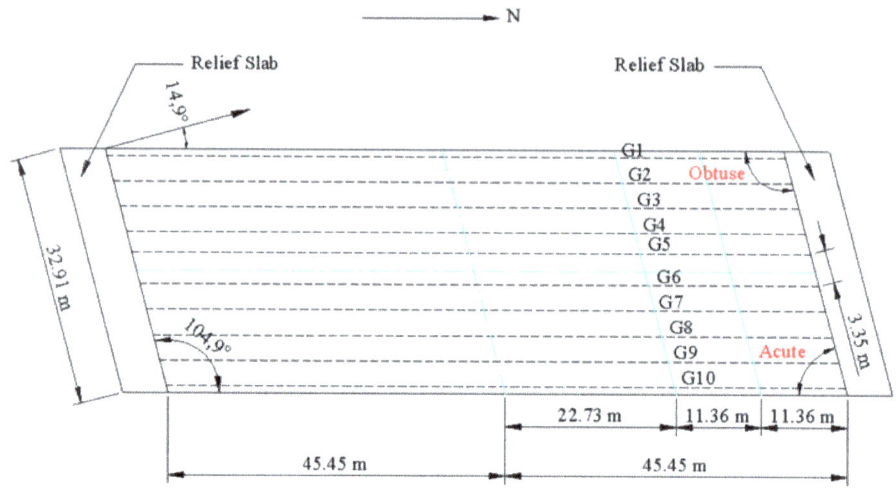

Figure 1 Plan view of the Scotch Road Bridge.

consists of well-compacted porous fill. The piles were first installed into pre-augured holes of 0.76 m in diameter. Then, the concrete was poured to cover up to 60% of the pile length, equivalent to 7.62 m of the total pile length. Next, a corrugated steel sleeve was inserted, supported by the concrete base, and extended to the top of the pile. The gap between the pile and the sleeve was filled with sand to facilitate the movement of the piles as subjected to lateral loads transferred from the superstructure. The piles were oriented such that bending would take place around their weak axes to increase the flexibility of the piles. Compacted crushed stone was used as a backfilling material between the piles, and the steel mesh was used to tie the components of the mechanically stabilized (MSE) wall. The MSE wall is a retaining structure surrounding the piles and the backfill soil between them. It consists of several panels stacked on top of each other and tied laterally using soil-stabilizing steel meshes.

Methods

Experimental approach

The substructure of the Scotch Road I-95 integral abutment bridge was instrumented using six types of measuring devices: (1) 15 electrical resistant strain gages to measure strains along the depth of the piles, where each strain gage is composed of four resistances connected together to form a Wheatstone bridge circuit; (2) two tiltmeters to measure the rotations at the connection between the abutment and the second and fifth girders (G2 and G5); (3) two temperature sensors to measure the temperature of the abutment at the second and fifth girders (G2 and G5); (4) two round displacement transducers that were connected to the four strain gages to

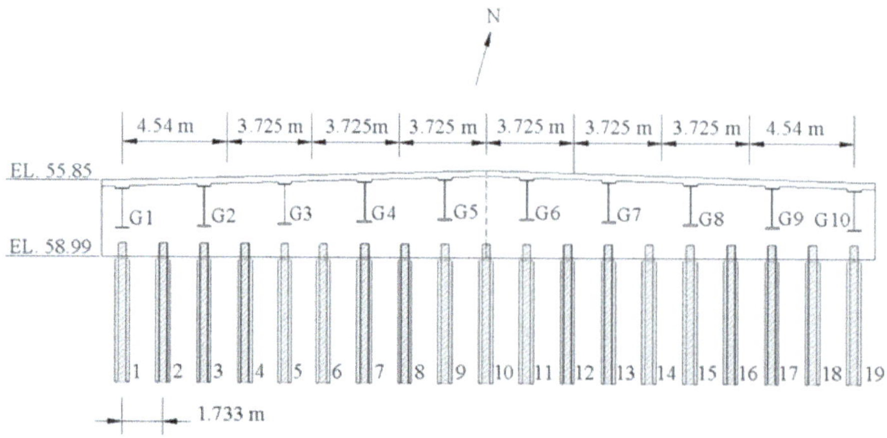

Figure 2 Side view of the Scotch Road Bridge North integral abutment supported by HP steel piles (Khodair and Hassiotis 2013).

measure the displacement at the east and west ends of the north relief slab; (5) eight thermocouples to measure the temperature variations in the bridge deck at the second and ninth girders (G2 and G9); and (6) 18 soil pressure cells to measure the horizontal soil pressure behind the abutment on the galvanized steel sleeves surrounding the piles (Khodair and Hassiotis 2013). The soil pressure cells were manufactured and calibrated in the laboratory before being installed on site. Figures 3 and 4 show the locations of the instrumented piles and north abutment.

Data acquisition was achieved using the CR 5000 measurement and control system manufactured by Campbell Scientific, Inc (UT, USA). The CR 5000 is a standalone system that was connected to a laptop computer. A serial link interface was used to connect the CR5000 system and the computer. A programming code was developed to control the flow of data. The data were gathered every 1 hour at the top of the hour.

Results and discussion

Analysis of field data

The Scotch Road I-95 integral abutment bridge has been instrumented during two construction stages (I and II). The effect of some of the crucial design parameters such as abutment temperature, approach slab displacement, abutment rotation, soil pressure on axial stresses in piles, and the effect of bridge skew on soil pressure is discussed below.

Effect of abutment temperature on pile axial stress

The repetitive increase or decrease in temperatures causes expansion or contraction movements of the superstructure of the bridge during the passive and active pressure cycles. The relation between the axial stresses in piles #3, #6, and #9 and the abutment temperatures at G2 is described in Figure 5. As the abutment temperatures increase during March to August, the magnitude of the axial stresses increase with

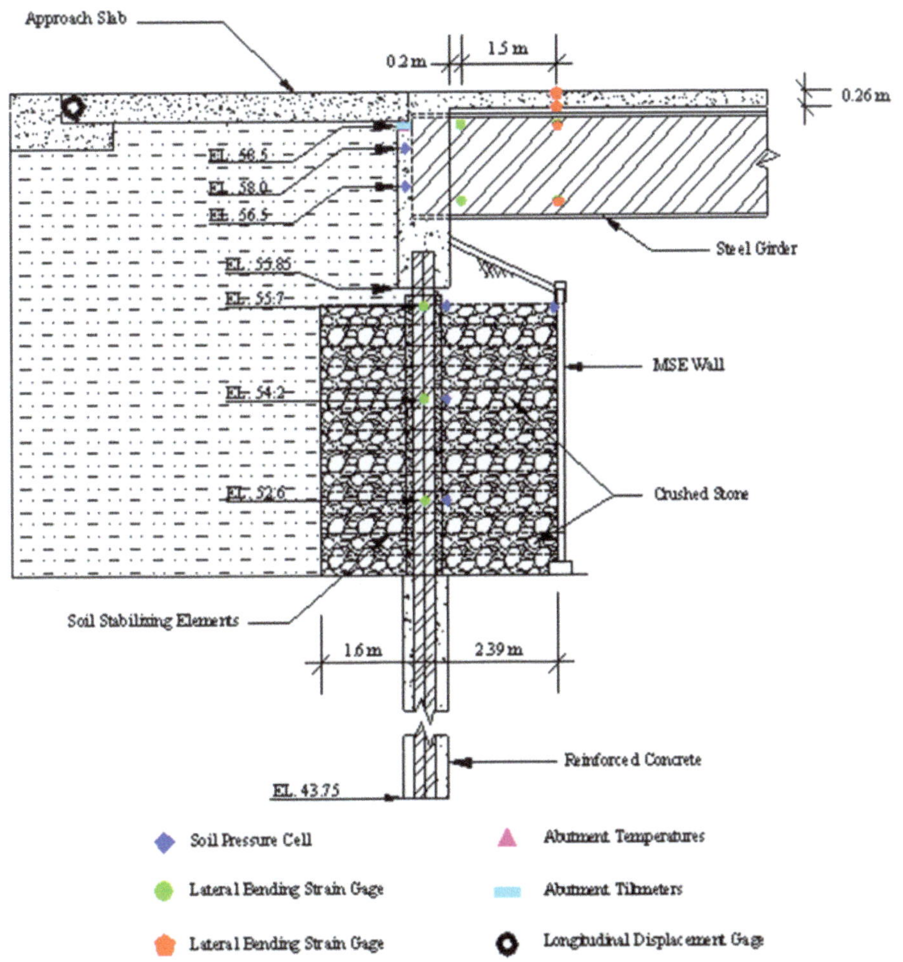

Figure 3 Diagram of the Cross-sectional elevation of the instrumented bridge (Khodair and Hassiotis 2013).

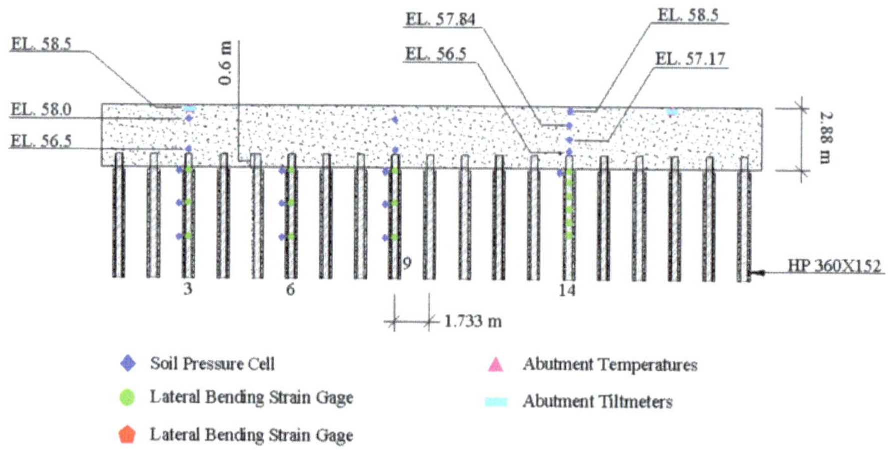

Figure 4 Diagram for the side view of the instrumented bridge (Khodair and Hassiotis 2013).

negative magnitudes during the passive pressure cycle. The negative magnitudes indicate tension stresses at the north face of the pile and compression at the south as shown in Figure 6. To the contrary, as the abutment temperatures decrease during September to February, the magnitude of the axial stresses also increases, with positive magnitudes indicating compression stresses at the north and tension at the south faces, respectively.

Effect of approach slab displacement on piles axial stress

The approach slabs are floored with two sheets of polyethylene acting as bond breakers to alleviate or even eliminate the friction between the approach slab and the soil underneath it. Therefore, the displacements at the end of the approach slabs are approximately the same as those at the top of the abutments. The displacement rings were configured such that negative displacement magnitudes correspond to an increase in the abutment temperatures and an overall expansion of the bridge superstructure during the passive pressure cycle. However, the positive displacement magnitudes correspond to a decrease in abutment temperatures and an overall contraction of the bridge superstructure during the active pressure cycle. Figure 7 shows that as the displacements increase with negative magnitudes, the axial stresses increase with negative magnitudes, indicating tension in the north face of the pile and

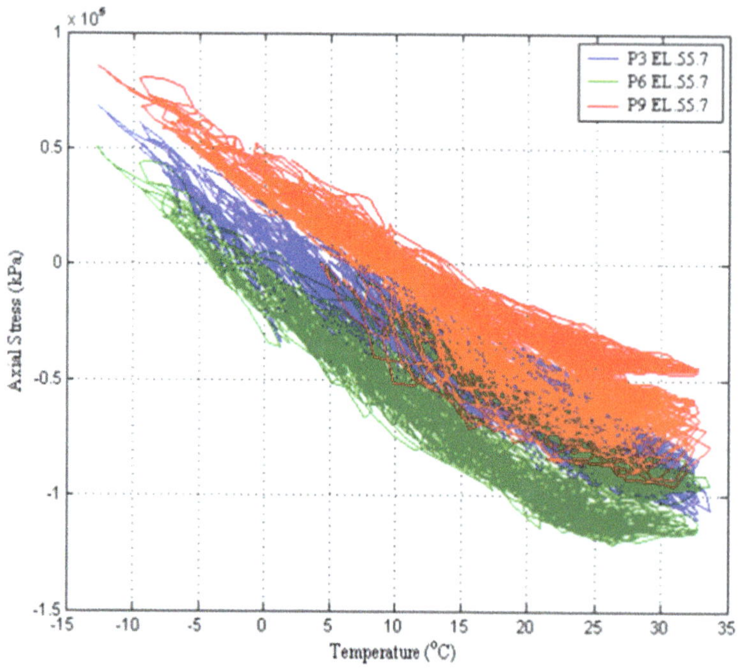

Figure 5 Effect of abutment temperature variations on axial stresses in piles #3, #6, and #9.

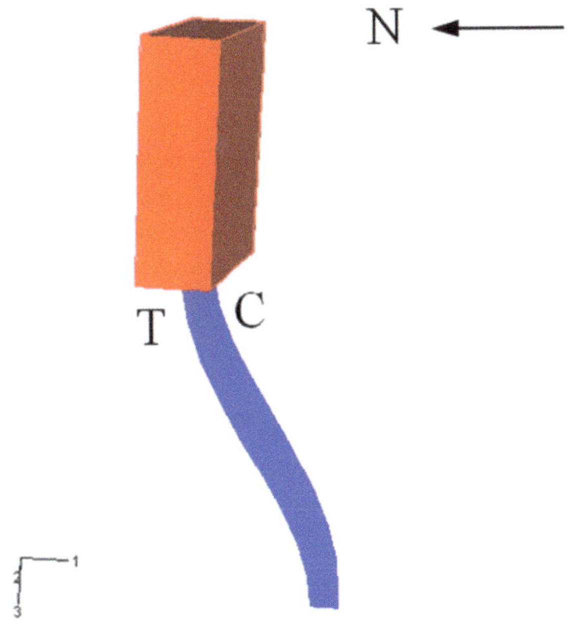

Figure 6 Pile orientation during the passive pressure cycle.

compression on the south as shown in Figure 6. The same trend is observed for the increase in the positive displacements, which are also associated with a corresponding increase in the positive bending stresses.

Effect of abutment rotation on pile axial stress

The tiltmeters used to measure the rotations of the abutment were configured such that rotations towards the retained soil during the passive pressure cycle correspond to the rotations of negative magnitudes, and rotations away from the retained soil during the active pressure cycle correspond to positive magnitudes. Figure 8 shows that as the magnitudes of the measured rotations of the abutment increase with negative magnitudes, the magnitudes of the axial stresses increase with negative magnitudes accordingly. This indicates that the displacements at the end of the approach slabs and rotations at the top of the abutment have the same effect on the axial stresses of the piles, which are increasing the stresses with negative magnitudes during the passive pressure cycles and increasing them with positive magnitudes during the active pressure cycles.

Effect of soil pressure on pile axial stress

As the soil pressures in piles #3, #6, and #9 behind the abutment increase in magnitude during the period from March to August (passive pressure cycle), the bending stresses increase with negative magnitudes. However, as the magnitudes of the soil pressures drop from September to February (active pressure cycle), the bending stresses increase with positive magnitudes as shown in Figure 9.

Effect of bridge skew

The Scotch Road I-95 integral abutment bridge has a 15° skew angle as mentioned before. Figure 10 shows the discrepancy in the magnitudes of the soil pressure at piles #3, #9, and #14. Although, piles #3, #6, and #9 were constructed during stage I of the project, and pile #14 in

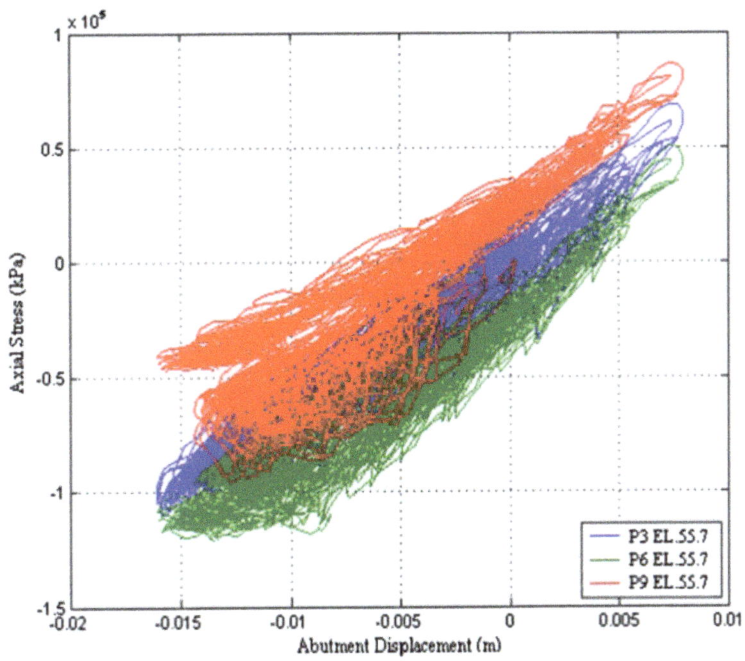

Figure 7 Effect of variation in abutment displacements on axial stresses in piles #3, #6, and #9.

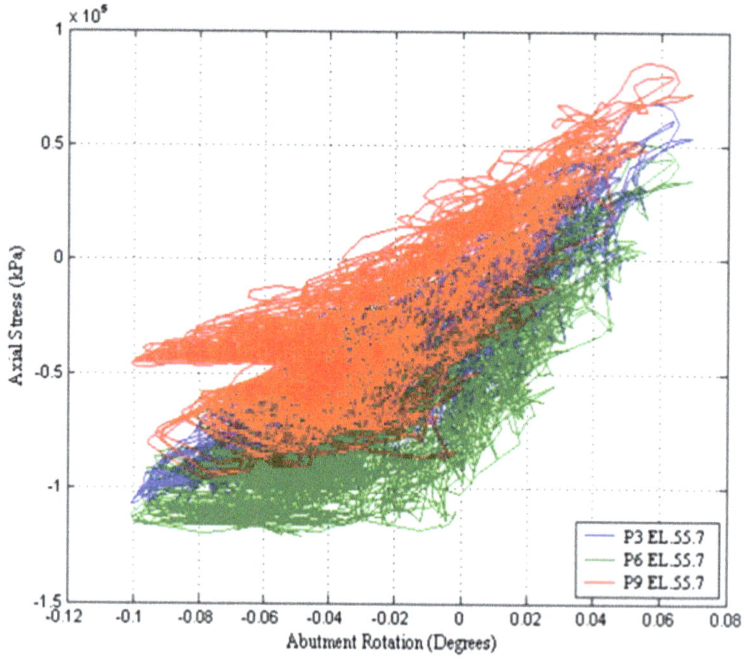

Figure 8 Effect of variation in abutment rotations on axial stresses in piles #3, #6, and #9.

stage II, comparing the soil pressures that accumulated on the abutment for the first few months at each location, it shows that a large difference in magnitudes that cannot be accounted for by the pressure built up due to cyclic loading alone. This difference is primarily attributed to the position of the pressure cells with respect to the skew angle of the bridge. The soil pressure cells in piles #3, #6, and #9 are positioned towards the obtuse side of the bridge, where the level of soil compaction is lower than that of the acute. The reduced level of confinement of the retained soil behind the abutment at the obtuse side of the bridge allows for greater lateral

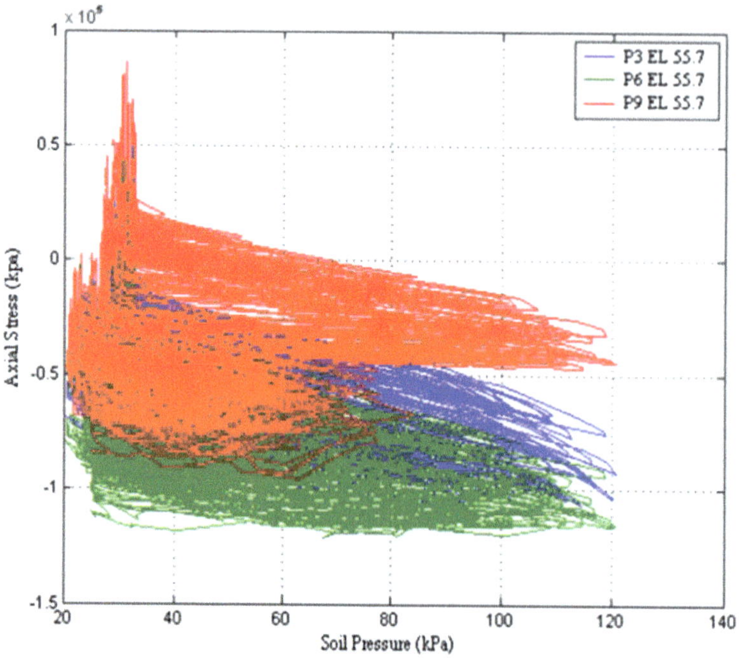

Figure 9 Effect of variation in soil pressures behind abutment on axial stresses in piles #3, #6, and #9.

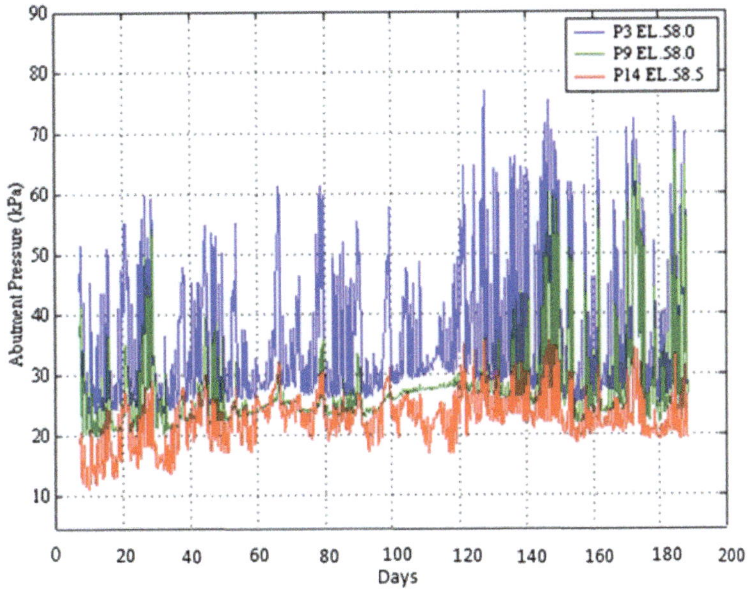

Figure 10 Distribution of abutment soil pressure in piles #3, #9, and #14 with time.

movement of the abutment, and thus higher soil pressure than that accumulated at the acute side. This resulted in higher axial stresses in piles #3, #6, and #9 versus those calculated at pile #14.

Finite element models

Two 3D finite element models were developed to study the effect of thermal loading on the superstructure and substructure of the Scotch Road integral abutment bridge. The FE models were developed using ABAQUS/Cae (ABAQUS Inc., RI, USA), a computer package with capabilities to address problems with soil-structure interaction.

Full bridge model

The geometry was modeled according to the exact dimensions of the superstructure and substructure of the Scotch Road integral abutment bridge, as shown in Figure 11. Three types of elements were used in the full bridge model: (1) the four-nodded, shell elements were used to model the bridge deck, (2) three-nodded beam elements were used to model the I-shaped steel girders, and (3) two-nodded beam elements were used to model the reinforced concrete abutment and piles. The soil behind the abutments and under the approach slabs was modeled using Spring1 elements, which connect from the defined node to the ground. The reinforced concrete material definition for the bridge deck was assumed to be elastic–plastic. The modulus of elasticity for the concrete was considered to be 2.86×10^7 kN/m^2. Grade 70 steel I-beams were used to model the steel beams with a modulus of elasticity of 2.0×10^8. An elastic-perfectly plastic material modeling was adopted for the steel beams. The modulus of elasticity of the steel piles was assumed to be 2.0×10^8 kN/m^2, and their yielding stress to be 344,750 kN/m^2.

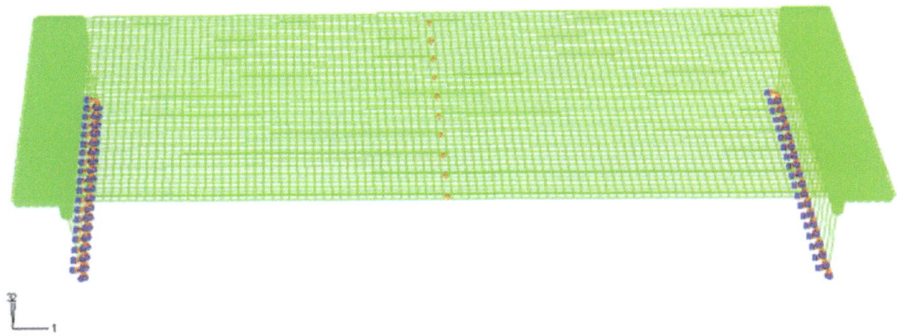

Figure 11 Boundary conditions for full bridge model.

Table 1 Modulus of subgrade reaction for soils behind abutments and under approach slabs

Structure type	Tributary area (m^2)		Modulus of subgrade reaction (kN/m)	
	Interior	Exterior	Interior	Exterior
Abutments	0.84395	0.421975	51,480.95	25,740.48
Approach slabs	0.67	0.335	40,870	20,435

The spring elements used to model the soil were presumed to be linear springs with constant modulus of subgrade reaction. A value of 61,000 kPa/m was used to represent the modulus of subgrade reaction in the model for the soil behind the abutment and below the approach slabs (Bowels 1996). Each interior spring has a tributary area, defined as the centerline-to-centerline distance between each other node, along the longitudinal and transverse directions. However, exterior springs carry half the tributary area defined for the interior. The modulus of subgrade reaction for linear springs is estimated according to the following equation:

$$K_s(\mathrm{kN/m}) = 61,000\ \mathrm{kPa/m} * A_{\mathrm{tributory}} \quad (4.1)$$

Table 1 shows the values of the modulus of subgrade reaction used in the full bridge model for exterior and interior springs for the soil behind the abutments and under the approach slabs. Three boundary conditions were imposed on this model, corresponding to modeling the bridge as a frame-like structure. The embedment of the piles below a depth of 5.182 m into the plain concrete was modeled using a fixed-type connection. A roller support was imposed at the middle of the bridge corresponding to the support provided by the piers for the bridge superstructure, as shown in Figure 11. The composite section between the steel girders and the bridge deck was modeled using multipoint constraints (MPC). MPC-type beam provides a rigid beam between two nodes to constrain the displacement and rotation at the first node to the displacement and rotation at the second node, corresponding to the presence of a rigid beam between the two nodes (Hibbitt, Karlsson & Sorensen, Inc 2009). The temperature variations between the top and bottom flanges of the second girder (G2), along with those between the top and bottom of the bridge deck were applied to the full bridge model for the sake of comparison. However, the field data collection is not sufficient to monitor the thermal behavior of the bridge during the changes in temperature between the different seasons over a long period of time and to determine the movement of the bridge during its predicted lifetime. Therefore, it was necessary to predict the deflections that the bridge undergoes, as subjected to extensive thermal expansions or contractions. A temperature change of ±42°C (AASHTO LRFD 2012) was applied to the bridge deck, assuming an isotropic material behavior. Figure 12 shows the deformed shape of the bridge expansion due to thermal loading, where the maximum displacement of the bridge superstructure was found to be approximately 0.02 m.

Soil-pile interaction FE model

The 3D soil-pile interaction model (Khodair and Hassiotis 2013) was developed to fully simulate the actual behavior

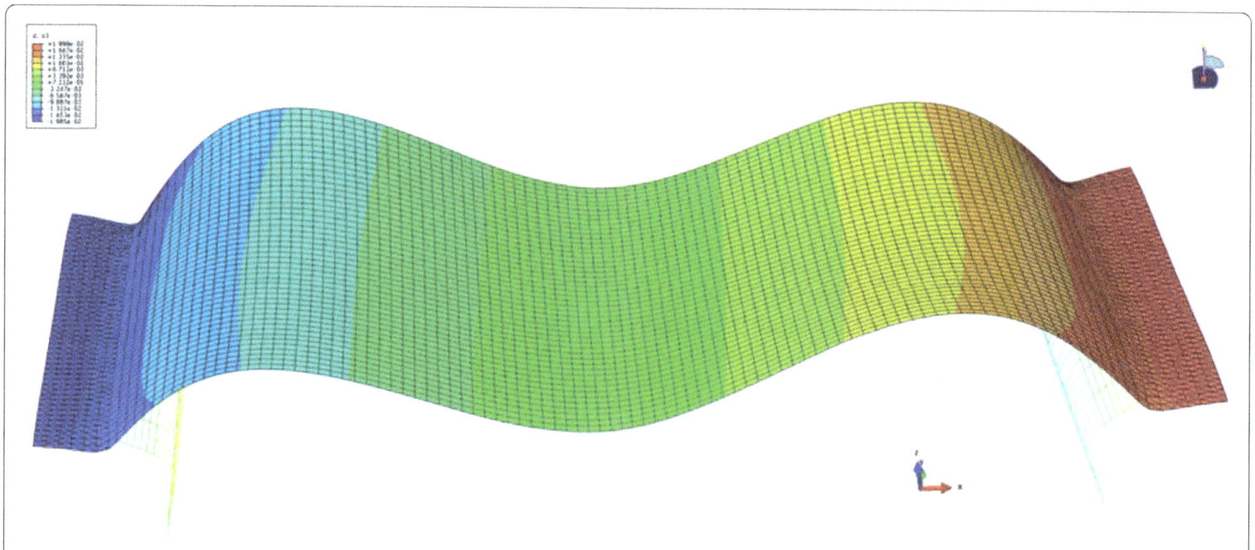

Figure 12 Contour plot of the full bridge under thermal loading. $\Delta T = 42°C$.

of the piles and their interaction with the soil. In this model, the pile and soil were modeled using eight-nodded solid continuum elements. Three boundary conditions were imposed in the FE model: (1) fixed conditions at the bottom of the pile at a depth of 5.182 from the top of the piles to model the embedment of the piles in plain concrete, (2) restraining all degrees of freedom associated with the exterior surface of the sand surrounding the piles to model the confinement of the galvanized steel sleeves by crushed stone backfill, and (3) guided fixation at the top of the pile modeled by tying the nodes at the top surface of the pile to a defined reference point located in the centroid of the cross-section of the pile at its top to simulate the embedment of the piles into the abutments, as shown in Figure 13.

The steel piles were modeled using an elastic-perfectly plastic model. The soil was modeled using a strain hardening model implementing Mohr-Coulomb failure criterion. The soil-pile interaction was simulated by adopting tangential and normal contact behavior in the model. Master and slave surfaces were defined in the model such that the exterior surface of the pile was used to model the master surface, and the interior surface of the sand was used for the slave surface. A friction coefficient of 0.44 (Venkatramiah 2006) and an angle of internal friction ϕ of 30° to 38° (Bowels 1996) were used to define the tangential contact between the two surfaces and the variation in the density (loose to dense) of sand surrounding the pile. A variation in Young's modulus for the sand between 27,000 and 69,000 kN/m^2 was adopted along the depth of the pile in the FE model. The steel piles modulus of elasticity was assumed to be 2×10^8 kN/m^2, and their yielding stress to be 344,750 kN/m^2. The computed dead and live loads of 692 KN were applied as a concentrated load to the reference point of the rigid body along the depth of the pile defined as global (three direction). The self-weight was incorporated into the model using a gravity step, where the density of the materials and gravity acceleration are specified. The dead and live loads were applied prior to the application of the lateral loading to maintain their effect in full during the lateral deflection step.

Parametric study

A limited parametric study was conducted to study the effect of the size of the galvanized steel sleeve and, hence, the amount of sand surrounding the pile on its induced axial stresses (Khodair and Hassiotis 2005). The 3D finite element model of soil-pile interaction (Khodair and Hassiotis 2013) was used to study the effect of the size of the galvanized steel sleeve and, hence, the amount of sand surrounding the pile on its induced axial stresses. The lateral loading that was obtained from the 3D finite element model of the full bridge (0.02 m) which corresponds to the maximum temperature increase/decrease of ±42°C that the bridge could be subjected to during its lifetime was applied to the soil-pile interaction FE model while varying the diameter of the steel sleeve surrounding the pile. The diameter of the steel sleeve was assumed to vary from 0.6 to 2.5 m. The FE model shows that as the diameter of the steel sleeve increases, which indicates an increase in the amount of sand

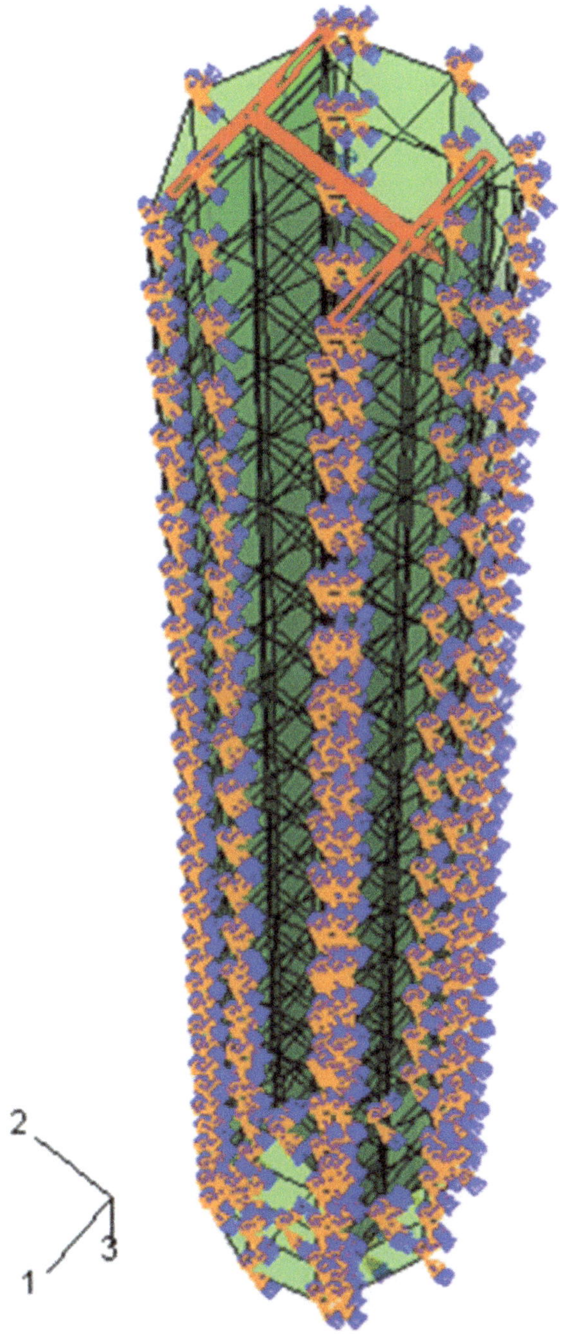

Figure 13 Boundary conditions for single pile embedded in sand with rigid surface at its top (Khodair and Hassiotis, 2013).

surrounding the piles, the magnitudes of the axial stresses decrease accordingly; hence, the gap in the magnitudes of the axial stresses calculated based on the finite element model and those obtained from LPILE (Ensoft 2012) reduced significantly as shown in Figure 14. The reason for the discrepancy in the magnitudes of the axial stresses induced in the piles between LPILE and the soil-pile interaction FE model is attributed to the fact that LPILE analysis is based on a single pile embedded in an extended single layer of sand.

Conclusions

The analysis of the soil-structure interaction in integral abutment bridges under thermal loading has always been a concern. In this paper, experimental and numerical analyses have been conducted to study the effect of thermal loading on the superstructure and substructure of the Scotch Road I-95 integral abutment bridge. The bridge superstructure and substructure were fully instrumented using strain gages, tiltmeters, temperature sensors, round displacement transducers, and thermocouples. The field data were gathered using the CR 5,000 measurement and control system manufactured by Campbell Scientific, Inc. The data were collected every 1 hour at the top of the hour. The data analysis showed excellent correlation between the abutment temperature, approach slab displacement, abutment rotation, soil pressure and the piles axial stress, which indicates that the strain gages are working properly. The data analysis showed that as the temperature increases, the displacements at the end of the approach slab and the rotations at the abutment increase with negative magnitudes, resulting in an overall expansion of the bridge during the passive pressure cycles. This results in an increase in axial stresses with negative magnitudes, indicating tension in the north face of the pile and compression on the south. However, as the temperature decreases, the displacements at the end of the approach slab and the rotations at the abutment increase with positive magnitudes, resulting in an overall contraction of the bridge during the active pressure cycles. This results in an increase in the positive axial stresses, indicating compression in the north face of the pile and tension on the south. The effect of bridge skew on the build-up of soil pressures and the axial stresses in the piles was studied. The data showed that there is an increase in the soil pressure towards the obtuse side of the bridge versus the acute because the reduced level of confinement of the retained soil behind the abutment at the obtuse side results in greater flexibility in the abutment movement, hence higher soil pressures. This larger movement of the abutment at the obtuse side has resulted in higher axial stresses in piles #3, #6, and #9 versus those calculated in pile #14.

A 3D FE model was developed of the Scotch Road I-95 integral abutment bridge (full bridge model) using ABAQUS/Cae. The full bridge FE model consisted of shell elements for the bridge deck, approach slabs, beam elements for the steel girders, abutment, and piles. Spring elements were used to model the soil. The

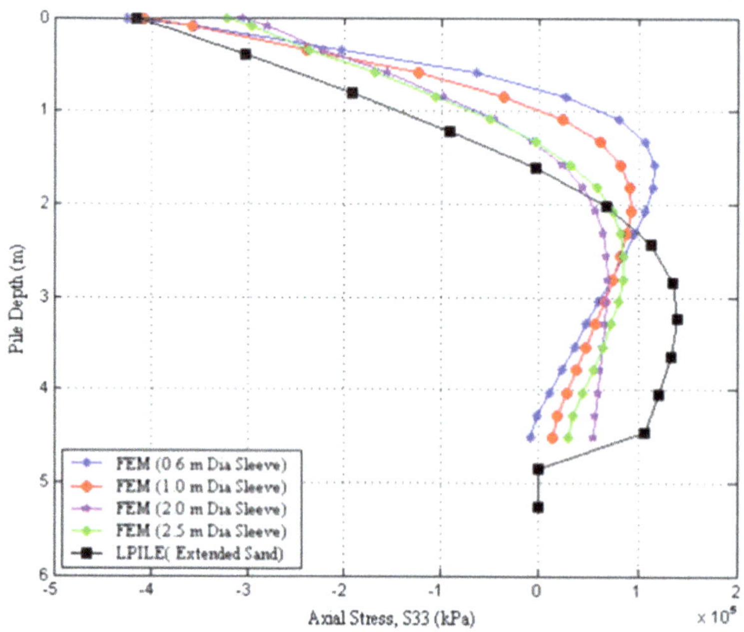

Figure 14 Comparison of the axial stresses. These stresses were obtained from FE and FD solutions at different sand sleeve diameters due to thermal loading. $\delta = 0.02$ m.

composite action between the steel girders and the bridge deck was modeled using MPCs. A temperature change of ±42°C was applied to the bridge deck to determine the maximum displacement that the bridge could be subjected to over its predicted lifetime. The results from the model showed that the maximum displacement of the bridge superstructure is 0.02 m. The 3D soil-pile interaction model consisted of solid continuum elements for both the piles and soil. The contact behavior between the piles and the soil was defined using tangential and normal algorithms in ABAQUS/Cae. Mohr-Coulomb failure criterion was used to model the nonlinear response of the soil. A limited parametric study was conducted to study the effect of the size of the galvanized steel sleeve on the induced axial stresses in piles. The soil-pile interaction model was compared to FD solutions for a single pile embedded in sand under a displacement of 0.02 m obtained from the finite element model of the full bridge while varying the diameter of the steel sleeve surrounding the pile. The results from FE and FD analyses showed that the discrepancy in the magnitudes of the axial stresses and displacements from both analyses was substantially reduced with the increase in the size of the steel sleeve surrounding the pile approaching an extended single layer of sand. This indicates that increasing the size of the steel sleeve and the amount of sand surrounding the piles reduces the induced axial stresses in the piles, thus increases their capacity to resist bending.

Competing interest

The authors declare that they have no competing interests.

Authors' contributions

YK developed the three-dimensional finite element models, analyzed the results, and drafted the manuscript. YK conducted the literature review. Both YK and SH participated in the instrumentation and the analysis of the data collected from the field. Both authors read and approved the final manuscript.

Authors' information

YK holds the position of an Assistant Professor in the Department of Civil Engineering and Construction at Bradley University. SH holds the position of an Associate Professor at the Department of Civil, Environmental, and Ocean Engineering at Stevens Institute of Technology.

Acknowledgments

This work was supported by the New Jersey Department of Transportation (NJDOT).

Author details

[1]Department of Civil Engineering and Construction, Bradley University, 1501 West Bradley Avenue, Peoria, IL 61625, USA. [2]Department of Civil, Environmental and Ocean Engineering, Stevens Institute of Technology, Castle Point on Hudson, Hoboken, NJ 07054, USA.

References

AASHTO LRFD (2012) Bridge design specifications, 6th edn. American Association of State Highway and Transportation Officials, Washington DC

Arsoy S, Barker RM, Duncan JM (1999) The behavior of integral abutment bridges (VTRC 00-CR3). Virginia Transportation Research Council, Charlottesville, VA

Bowels J (1996) Foundation analysis and design, 5th edn. McGraw-Hill, New York

Broms B (1964) Lateral resistance of piles in cohesionless soils. J Soil Mech Found Div, (ASCE) 90(3):136–156

Brown DA, Shie C-F (1990) Three dimensional finite element model of laterally loaded piles. Comput Geotech 10(1):59–79

Brown DA, Shie C-F (1991) Some numerical experiments with a three dimensional finite element model of a laterally loaded pile. Comput Geotech 12(2):149–162

Desai CS, Kuppusamy T (1980) Application of a numerical procedure for laterally loaded structures. In: Numerical methods in offshore piling. Institution of Civil Engineers (ICE), London, pp 93–99

Ensoft (2012) LPILE Plus 6.0 geotechnical analysis program. http://www.ensoftinc.com. Accessed 05 Dec 2012

Faraji S, Ting J, Crovo D, Ernst H (2001) Nonlinear analysis of integral bridges: finite-element model. J Geotech Geoenviron Eng, (ASCE) 27(5):454–461

Faruque MO, Desai CS (1982) 3-D material and geometric non-linear analysis of piles. Paper presented at the second international conference on numerical methods for offshore piling, Austin, TX, 29–30 April 1982

Girton DD, Hawkinson TR, Greimann LF (1991) Validation of design recommendations for integral abutment piles. J Struct Eng 117:2117–2134

Hibbitt, Karlsson & Sorensen, Inc (2009) ABAQUS/Standard 6.9.1 User Manuals. ABAQUS, Inc, RI, USA

Jorgensen JL (1983) Behavior of abutment piles in an integral abutment in response to bridge movements. Transport Res Board 903:72–79

Khodair Y, Hassiotis S (2005) Analysis of soil-pile interaction in integral abutment. Comput Geotech 32:201–209

Khodair Y, Hassiotis S (2013) Rigidity of abutments in integral abutment bridges. Struct Infrastruct Eng 9(2):156–160

Kumar S (1992) Three-dimensional non-linear finite element analysis of laterally loaded piles in clay., Dissertation, Edinburgh University

Lawver A, French C, Shield CK (2000) Field performance of integral abutment bridge. Transport Res Rec 1740:108–117

McCelland B, Focht JA (1958) Soil modulus for laterally loaded piles. Trans Am Soc Civil Eng (ASCE) 123:1049–1063

Olson S, Long J, Hansen J, Renekis D, LaFave J (2009) Modification of IDOT integral abutment design limitations and details. Final Report, FHWA-ICT-09-054. Center of Transportation, Illinois

Rajashree SS, Sitharam TG (2001) Nonlinear finite-element modeling of batter piles under lateral load. J Geotechn Geoenviron Eng 127(7):604–612

Reese LC, Matlock H (1956) Non-dimensional solutions for laterally loaded piles with soil modulus assumed proportional to depth. In: University of Texas (ed) Proceedings of the 8th Texas conference on soil mechanics and foundation engineering, Sp. Pub. 29. Bureau of Engineering Research, University of Texas, Austin

Thompson GR (1977) Application of finite element method to the development of *p-y* curves for saturated clays. Thesis, University of Texas, Austin, Texas

Venkatramiah C (2006) Geotechnical engineering, 3rd edn. New Age International, New Delhi

Wasserman EP, Walker JH (1996) Integral abutments for steel bridges. Highway structures design handbook, vol. II, chap. 5. American Iron and Steel Institute (AISI), Chicago, IL

Effect of epsilon-based record selection on fragility curves of typical irregular steel frames with concrete shear walls in Mashhad city

Hamid Kazemi[1*], Mohsen Ghafory-Ashtiany[2] and Alireza Azarbakht[3]

Abstract

In this paper, the seismic vulnerability of Mashhad city, as the second largest city in Iran, has been investigated using analytical fragility curves. Disaggregation analysis is first performed in order to identify the target epsilon at different hazard levels. The disaggregation results revealed different epsilon values at the first mode period of two representative structures, in the case of 72-, 475-, and 2,475-year return periods. Nonlinear incremental dynamic analyses are then performed for two representative models of a typical steel frame with a concrete shear wall, using independent suites of acceleration time histories that are selected based on the target epsilons. Structural limit states are defined on each incremental dynamic analysis curve, and the corresponding damage measures are estimated. The results show that if ε is neglected in the considered simulations, then the predicted median structural capacities is decreased by around 10%, 15%, and 18%, respectively, for the three abovementioned hazard levels.

Keywords: Record selection; Hazard disaggregation; 3D analysis; Epsilon; Collapse capacity; Strong ground motion; Spectral shape; Vulnerability

Introduction

Mashhad is the second largest city in Iran, and different seismic zonations have shown that the city rest on a seismically active region. One of the main tools for the risk estimation in urban buildings is the seismic vulnerability evaluation in a proper scale for a specific level of seismic hazard. EMS-98, HAZUS, and Risk-UE provide a good and useful library of fragility curves, but they do not really capture the characteristic of Iranian buildings and construction process. Different efforts have been conducted to develop the fragility models for Tehran. Japan International Cooperation Agency (JICA 2000) has developed hybrid fragility curves for Tehran, using available data from major earthquakes. This paper, as part of large studies, is focused on the development of the fragility curve for the most common buildings in the Mashhad city for spectral acceleration (S_a) and epsilon (ε).

* Correspondence: hamid.kazemi@srbiau.ac.ir
[1]PhD candidate, Department of Civil Engineering, Science and Research Branch, Islamic Azad University, Tehran, Iran
Full list of author information is available at the end of the article

In this way, the analytical fragility curves for the existing buildings in Mashhad are one of the present paper goals. The analytical seismic fragility, as an amount of probability of damage, is calculated in this appear based on seismic hazard analysis, selection of typical structures, structural analysis, damage criteria, and probability distribution function.

Based on 2012 municipality census data, the conventional steel frames with concrete shear walls (as the most common types of new buildings) comprise nearly 48% of the residential construction in the Mashhad city during the past decade. Since the seismic behavior of buildings cannot be specified one by one, in order to reduce the number of structural models, it is required that the representative buildings is selected with the aim of being a good sample for a large group of existing buildings by statistical study on desired area. Therefore, the focus in this study is based on two typical five-storey steel frames with cast-in-place concrete shear walls, and for low-rise and high-rise buildings, more studies are needed in the future.

By selecting the representative structural model, the dynamic analysis can be carried out using the appropriate suites of ground motion records. In this way, careful ground motion selection can result in the same reduction in the bias and the variance of structural response as can be gained using advanced intensity measures (IMs). Epsilon (ε), as a spectral shape indicator and a predictor of nonlinear structural responses, is an efficient parameter to reduce the bias in the structural response (Mousavi et al. 2011). The previous studies had shown that epsilon is more effective than finding records with appropriate magnitude and distance values (Baker and Cornell 2006). Therefore, disaggregation of seismic hazard should be carried out to identify the target magnitude, the target distance, and the target epsilon at different hazard levels for ground motion selection (Bazzurro and Cornell 1999).

For damage estimation of structures, employing an appropriate damage assessment method is essential to derive the corresponding fragility curves. Many researchers have used from different criteria, e.g., drift, acceleration, or energy indices (Estekanchi and Arjomandi 2007). However, it should be noted that for the seismic assessment of structures with planar irregularities, a damage measure should be able to reflect three-dimensional (3D) structural response features such as torsion and bidirectional response. The effective damage calculation method is used for irregular building in plan that was defined by Jeong and Elnashai (2005, 2006).

In this paper, first, disaggregation analysis on Mashhad city was performed to identify the target epsilon at different hazard levels. In order to perform nonlinear incremental dynamic analysis by employing OpenSees platform, independent suites of the acceleration time histories were selected based on the target epsilon. Structural limit states (immediate occupancy (IO), life safety (LS), and collapse prevention (CP)) are defined on each incremental dynamic analysis (IDA) curve, and the corresponding damage measure is estimated (FEMA-356). In the last step, the proposed method accounts for the multistorey as well as the asymmetry of the structure using multidirectionality of earthquake motions. The variability of the fragility curves is considered, and the corresponding probability of damage is obtained, in the case of 72-, 475-, and 2,475-year return periods.

Methods

Site seismic hazard

Probabilistic seismic hazard analysis (PSHA) aims to quantify these uncertainties and combine them to produce an explicit description of the distribution of future shaking that may occur at a site. Seismic Hazard Map of Iran (1999) has shown that the Mashhad city is mainly exposed to earthquake, and Figure 1 shows active fault map and level of hazard in the area. There have been three major studies on Mashhad seismic hazard. One was done by IIEES for Holy Shrine site located in downtown and the other one by Hafezi-Moghaddas (2007) for the whole area. Recently, Zolfaghari and Ghafory-Ashtiany (2012) have also carried out an independent PSHA study for the city of Mashhad. PSHA was carried out in a city central region to calculate the expected strong ground motion parameters based on seismicity of the region within the radius of 200 km around the city at the ground level by Zolfaghari and Ghafory-Ashtiany (2012), and S_a is used as IM.

Then, we performed hazard disaggregation analysis for determining target M, R, and ε at probabilities of exceedance of 50%, 10%, and 2% in a 50-year return period, and a sample of the results is shown in Figure 2.

At each of the given hazard levels, the disaggregation results revealed different target epsilon values. For the first mode period and in high hazard level (2,475 years), hazard is dominated by M = 6 to 6.5, R = 0 to 10 km, and $\varepsilon > +2.0$. For relatively high hazard level (475 years), it is also dominated by epsilon values between +1.0 to +2.0, and for low hazard level (72 years), it is limited to −1.0 to 0. Therefore, as shown in Table 1 for each hazard level, independent suites of acceleration histories were selected based on the target epsilon.

Structural selection criteria

Literature surveys (2012 census data) were compiled on the existing typology for residential buildings that has been constructed in the Mashhad city during the last 12 years. It was shown that conventional steel frames with cast-in-place concrete shear walls are most common systems of the residential construction over the past few years. The shear walls in these buildings often can bear walls and the steel frame which is only designed for vertical loads. Lateral loads are transferred by diaphragms to the shear walls, and the steel frame may provide a secondary lateral force resisting system. In this way, 100 existing structures were considered from the database inventory building in a regional study. Since there is low dispersion in the design and construction methods in desired area, it is assumed that the median and the standard deviation values can be suitable criteria for building selection for estimating damage. The median and the standard deviation for structural dimensions of the 100 selected buildings are shown in Table 2. A description of representative model is presented in the 'Overview of mathematical modeling' section.

Based on a statistics study on considering structures, a reasonable sample of buildings has been selected as a representative of the great group of structures. It means that the structural geometric dimensions should be close to the average values. But in this paper, we have selected two models between the upper and lower limits of shear wall dimensions. Shear wall length in Y direction was

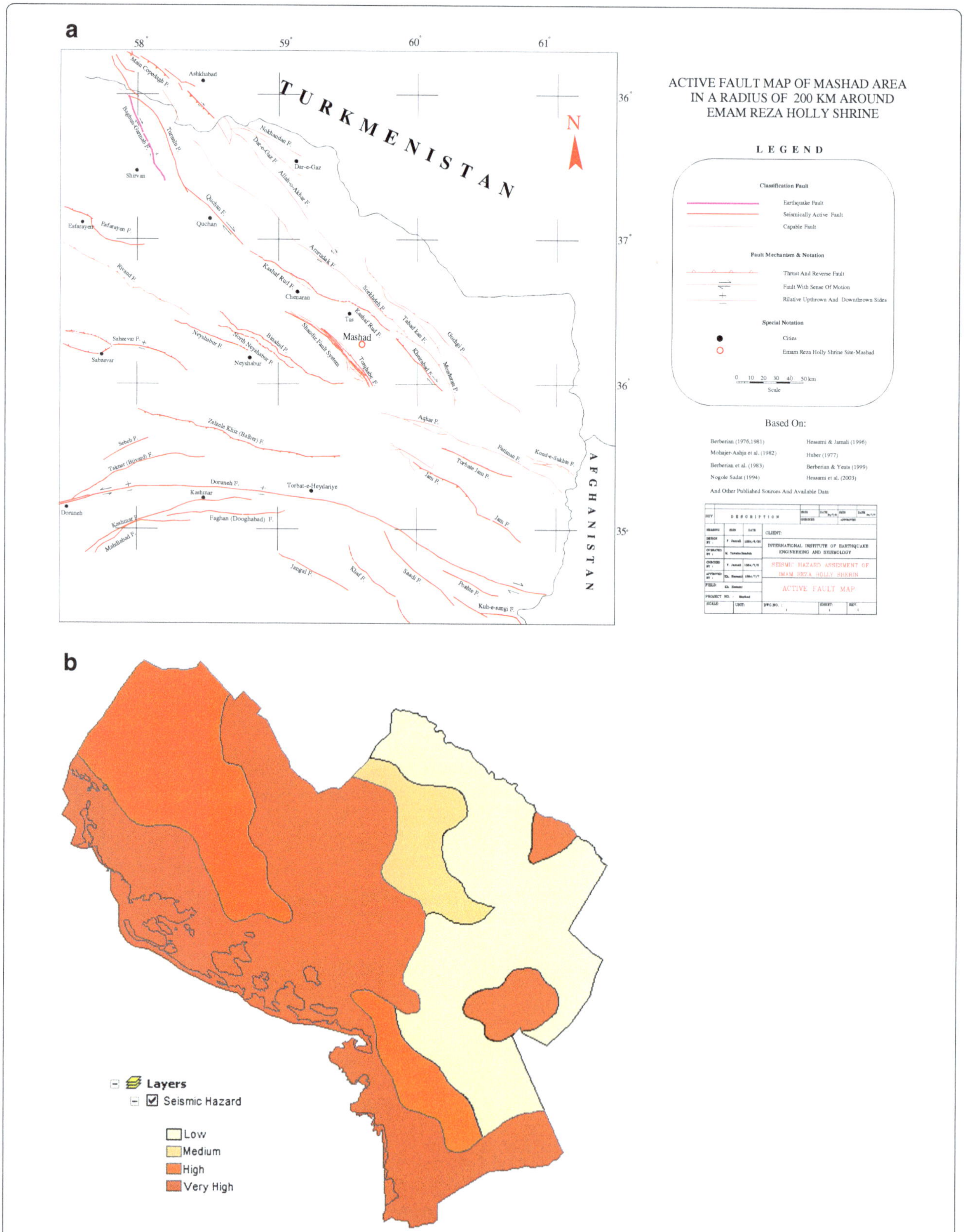

Figure 1 **Regional seismology of the Mashhad city: (a) active faults and (b) seismic hazard map.** Adapted from Hafezi-Moghaddas (2007).

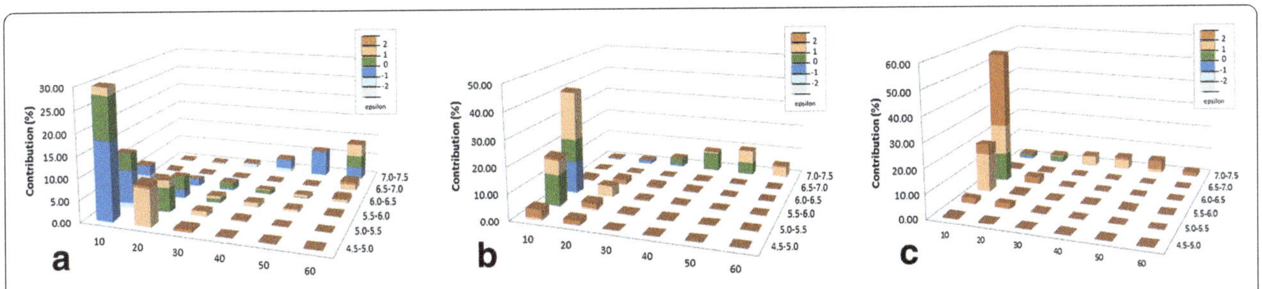

Figure 2 The hazard disaggregation results. **(a)** 72 years, **(b)** 475 years, **(c)** 2,475 years at $T = 0.54$ s.

equal to 7.8 (close to 6.4 + 1.18 according to Table 2) for the first model, and it was equal to 5.2 (approximately 6.4 − 1.18) for the second model.

Two existing five-storey steel frame with shear wall buildings have been selected. The structures were designed according to Iran's seismic code (standard no. 2800, 2nd edition, 1999) and were constructed 7 years ago. The specifications of these structures are shown in Table 3 and Figure 3.

Overview of mathematical modeling

The performance-based earthquake engineering requires structural models to be accurate for frequent and rare ground motions which mostly contribute to damage, financial loss, and collapse risk. To analytically predict the inelastic response of such structural systems under seismic loads, the building structure should be accurately described. Using reliable analytical software and definition of strength of materials, yielding behavior of elements, the effects of confinement in boundary elements, concrete crack or crash, and strain hardening and stiffening deterioration phenomena at large deformations are necessary.

The models were analyzed herein by employing OpenSees software. The masses are lumped at floor levels, whereas the horizontal degrees of freedom are defined. The Rayleigh damping with a specified ratio of $\xi = 0.05$ was assigned at all of the vibration modes, and the effect of nonstructural elements was not considered.

All of the beam end connections within the structure are assumed to be pinned. Therefore, the beams are modeled as elastic elements. These models are built with nonlinear beam column element for columns (batten column with double I section) and shear walls as well as the P-delta effects are taken into account. Fiber elements were used in all of the nonlinear elements, and the spread of plasticity along the elements was considered.

Table 1 Disaggregation results for a central point in Mashhad

	Return period (years)		
	72	**475**	**2,475**
M	4.5 to 5.0	5.5 to 6.0	6.0 to 6.5
R (km)	0 to 10	0 to 10	0 to 10
ε	−1.0 to 0.0	+1.0 to 2.0	Larger than +2.0

Table 2 Median and standard deviation of geometric dimensions for 100 samples

	Floor area (m^2)	Shear wall length		$A_{shear\ wall}/A_{floor}$ [a]	
		X (m)	Y (m)	X	Y
Median	146	8.70	6.40	1.63	1.20
SD	33.05	1.80	1.18	0.4	0.28

[a]Area of cross section for shear walls in plan to floor area multiple to 100%.

Table 3 Specifications of selected building

Code	Number of stories	Dimensions		Height, M	Wall length		W (ton)	T_1 (s)
		X (m)	Y (m)		X (m)	Y (m)		
5SW-1	5	16.70	9.80	16.90	9.0	7.6	827	0.54
5SW-2	5	15.40	9.80	17.10	8.8	5.2	794	1.01

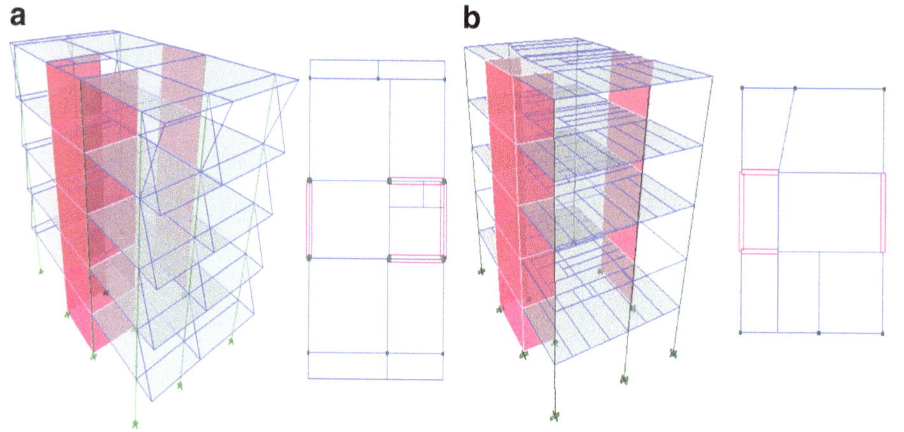

Figure 3 The plan and 3D view configurations of representative structures. (a) 5SW-1 and **(b)** 5SW-2.

The shear walls undergo both shear and flexural deformations. The studies have shown that the shear strength is a function of several parameters such as axial load ratio and horizontal steel ratio. Since the shear strength was enough larger than the flexural strength of the representative structure, the shear failure is definitely an undesirable behavior.

Verification of material properties

In order to calibrate the plastic behavior of batten columns and shear walls, several parameters have been examined in the mathematical model. Therefore, the influential parameters should be validated against the existing experimental test results. Very few tests on entire steel frames with concrete shear wall systems have been performed, especially the ones incorporating details representative of current practice in Iran (Shokrzadeh and Tasnimi 1995). For accurate modeling, the parameters were verified based on existing experimental results by Thomsen and Wallace (1995). The experimental tests are used in order to calibrate the material behavior, which are shown in Figure 4 in which a very good match is observed. There were low differences between the analytical hysteretic loops and the ones obtained experimentally as seen in Figure 4.

It usually use a unique type of reinforcing steel named AIII grade (f_y = 400 MPa), and compressive strength of concrete used for design of the shear walls was 25 MPa in the Mashhad city. The concrete compressive strengths of testing ranged between 19 and 37 MPa, with mean of 26.3 MPa, and similarly for the reinforcement yield stress from 341 to 504 MPa, with mean 445 MPa with yield strain of approximately 0.002 for all specimens.

The parameters of steel02 material for batten columns and for accurate modeling were verified based on existing experimental results obtained by Jafari and Hossaini-Hashemi (2008). The concrete01 for shear walls was used, and the material was modeled as 'unconfined' with peak strength achieved at a strain of −0.002 and minimum post-peak strength achieved at a strain of −0.008. The reinforcement was assumed to have a post-yield modulus equal to 2% of the elastic modulus.

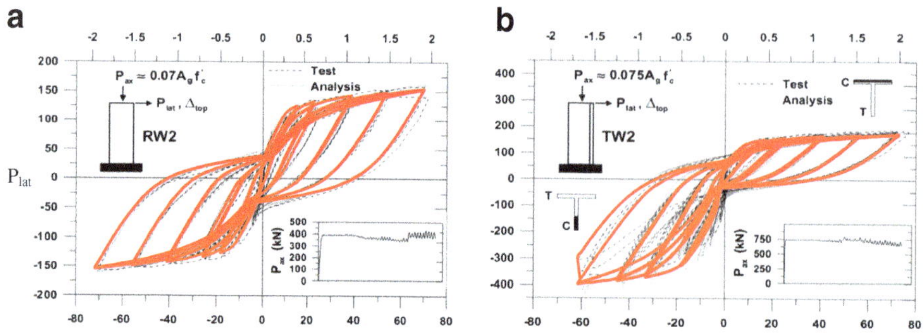

Figure 4 The test and shear wall simulation results: (a) rectangle wall and (b) T-shaped wall.

Ground motion database/selection

Earthquake record database

Earthquake record selection often is considered with the aim of accurately estimating the response of a structure at a specified ground motion intensity, as measured by spectral acceleration at $S_a(T_1)$.

The M6.5 scenario database of SGM records as shown in the 'Appendix' has been used for the time history analysis (Hatefi and Ghafory-Ashtiany 2010). This database belongs to a bin of relatively large magnitudes of 6.0 to 7.6 and moderate distances recorded on II and III soil types. Soil types are classified as type 1, 2, 3, or 4 in accordance with the descriptions defined in 2800 Iran's seismic code and based on geotechnical information. Most of the areas of the Mashhad city are located on II or III soil type. Each record from the database contained two horizontal components for use in the dynamic analysis.

Record selection criteria

When selecting ground motions for dynamic analysis, the efforts should be made to find records with ε-filtration values (as an indicator of spectral shape). It is seen that selecting ground motions based on their epsilon (ε) values is more effective than magnitude (M) and distance (R) (Baker and Cornell 2006). Two studies provide a comparison of the results mentioned here. Zareian (2006) found that a change from $\varepsilon = 0$ to +2.0 caused an approximately 45% to 50% increase in the expected collapse capacity. Haselton and Baker (2006) found that a change consistent with ε causes a 50% shift in the median collapse capacity for some S_a levels. In other works, it has been shown that ε effect scan changes the predicted probability of collapse by 20% to 30% for benchmark project located in the south of downtown Los Angeles (Goulet et al. 2006).

The record selection in the current study is based on the records which have compatible epsilons with the target epsilon of ground motion disaggregation in specified scenarios of different hazard levels in the Mashhad site. The epsilon values of ground motion at T_1 are derived. Epsilon is defined as a measure of the difference between the spectral acceleration of a ground motion record and the mean obtained from an attenuation prediction equation (here, Campbell and Bozorgnia 2008).

The disaggregation results reflected the expectation (−1.0 to 0.0), (+1.0 to +2.0), and (+2.0 to 3.0) for 50%, 10%, and 2% probability of exceedance in 50 years, respectively. Consequently, the database records are classified according to predicted epsilon for analysis of the structure.

Nonlinear dynamic analyses

The nonlinear IDA method involves carrying out a sequence of dynamic analyses in which the intensity of S_a (T_1) is considered incrementally increased until a specific limit state (Vamvatsikos and Cornell 2002). Based on the existing methodologies, failure may be defined by some different methods. First, each of the distinct records was incremented and run until an S_a level was reached at which the maximum interstorey drift ratio (MIDR) grew rapidly implying dynamic instability. Second, the rate of decrease of stiffness with increasing record intensity that exceeds beyond a prescribed MIDR is considered doubtful at 10% (Cornell et al. 2005).

In this work, the time histories for the horizontal displacement were, in the case of the shear walls, reported as the OpenSees output. Both horizontal component of a ground motion are used for IDA analysis, and the results were derived for the ground motion records selected based on the target epsilon. MIDR as the engineering demand parameter is correlated to damage within the structure. Figure 5 shows MIDR from the IDAs for both models using 35 ground motions. In the figure, the IM is the ground motion 5%-damped spectral

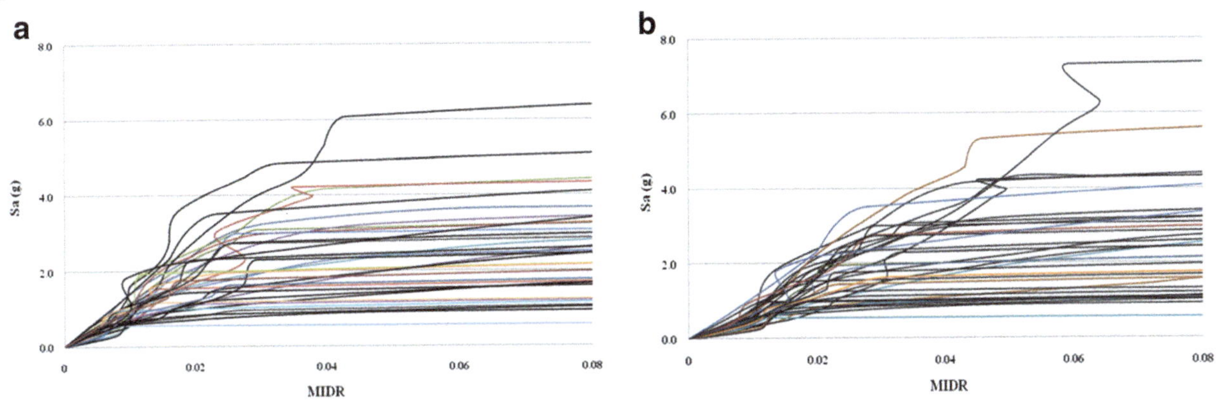

Figure 5 IDA curves for two representative models. (a) 5SW-1 and **(b)** 5SW-2.

acceleration at the first mode period ($S_a(T_1)$) normalized by 'g.' The colored lines are individual IDA curves, and the flat line at the end of the curve represents the collapse region for the particular ground motion (Ibarra and Krawinkler 2004). The different hazard levels can be marked in each figure with horizontal lines, and the intersection of the hazard level line with individual IDA curves leads to results by counting.

Damage index

Except for a few brittle systems and acceleration-sensitive elements, building damage is primarily a function of building relative displacements, rather than force. Hence, successful prediction of earthquake damage to buildings requires reasonably accurate estimation of building drift response in the inelastic range. In this paper, interstorey drift damage indices are selected as one of the most extensively used damage measures.

Structural drift limit states (e.g., IO, LS, or CP) based on FEMA-356 are defined on each IDA curve, and the corresponding capacities are calculated. Furthermore, for the seismic assessment of structures, a damage index should be able to reflect three-dimensional structural behaviors such as torsion and bidirectional response. The procedure advocated in this paper for the damage assessment of structure with planar irregularities is achieved by Jeong and Elnashai (2006). The methodology had been established for one-storey irregular building, while here, we have applied for five-storey irregular structure. Based on the assumption that a critical storey governs the overall damage state of the building, the 3D damage measure of a critical storey is employed as the response variable.

At first, individual shear wall damages are obtained from the transient drift ratio of 3D model for each shear wall. Then, to combine the damage to all frames of a given direction, a weighting mechanism was developed based on effective gravity loads of each frame, such that the importance of each frame to total building damage is a function which that frame resides as well as bearing areas. The local damages were combined using the following equation at each storey in x and y directions separately (Jeong and Elnashai 2006). Finally, the maximum amount of damage index in the critical storey and critical direction is taken (the damage ratio is taken as the maximum over time during seismic loading). The

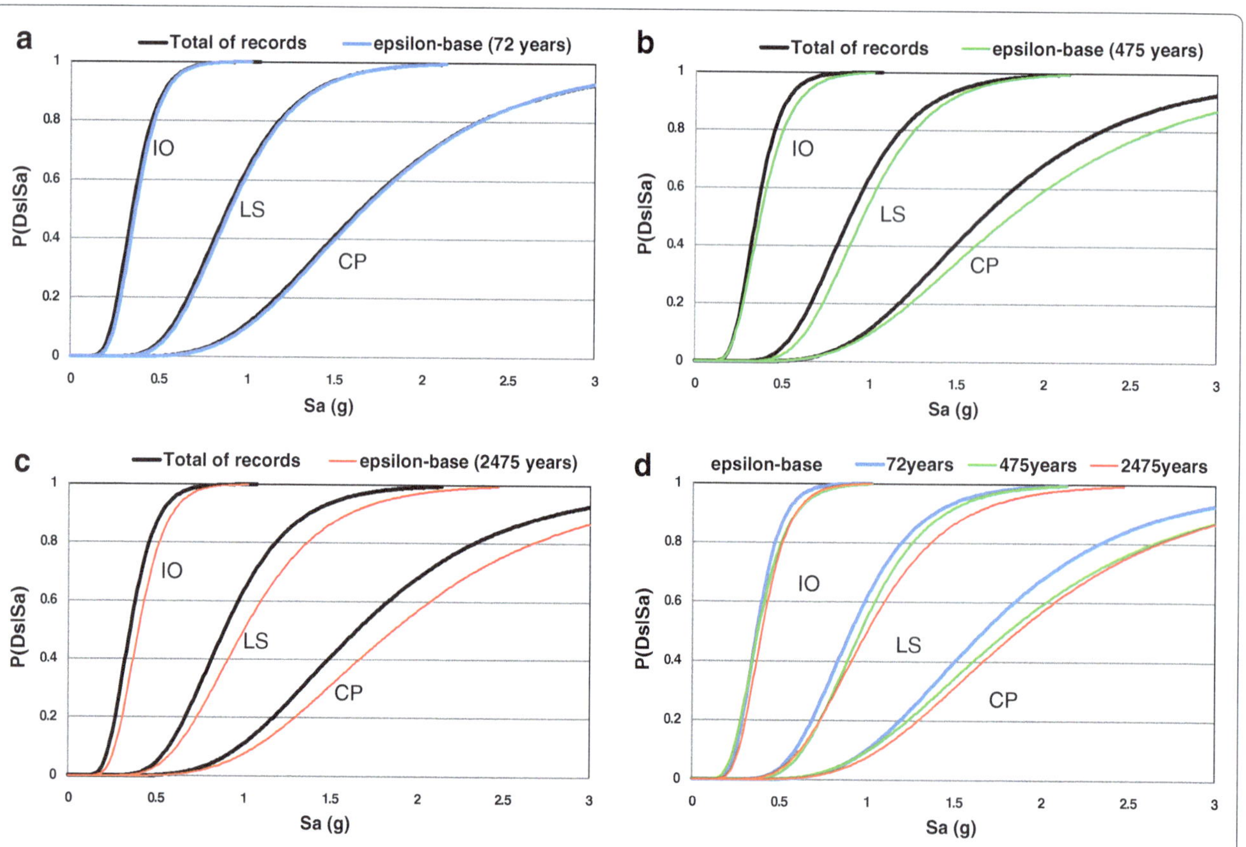

Figure 6 The fragility curves of 5SW-1 model, with and without regard to epsilon. Return periods of **(a)** 72 years, **(b)** 475 years, and **(c)** 2,475 years. **(d)** Final fragility curves using epsilon.

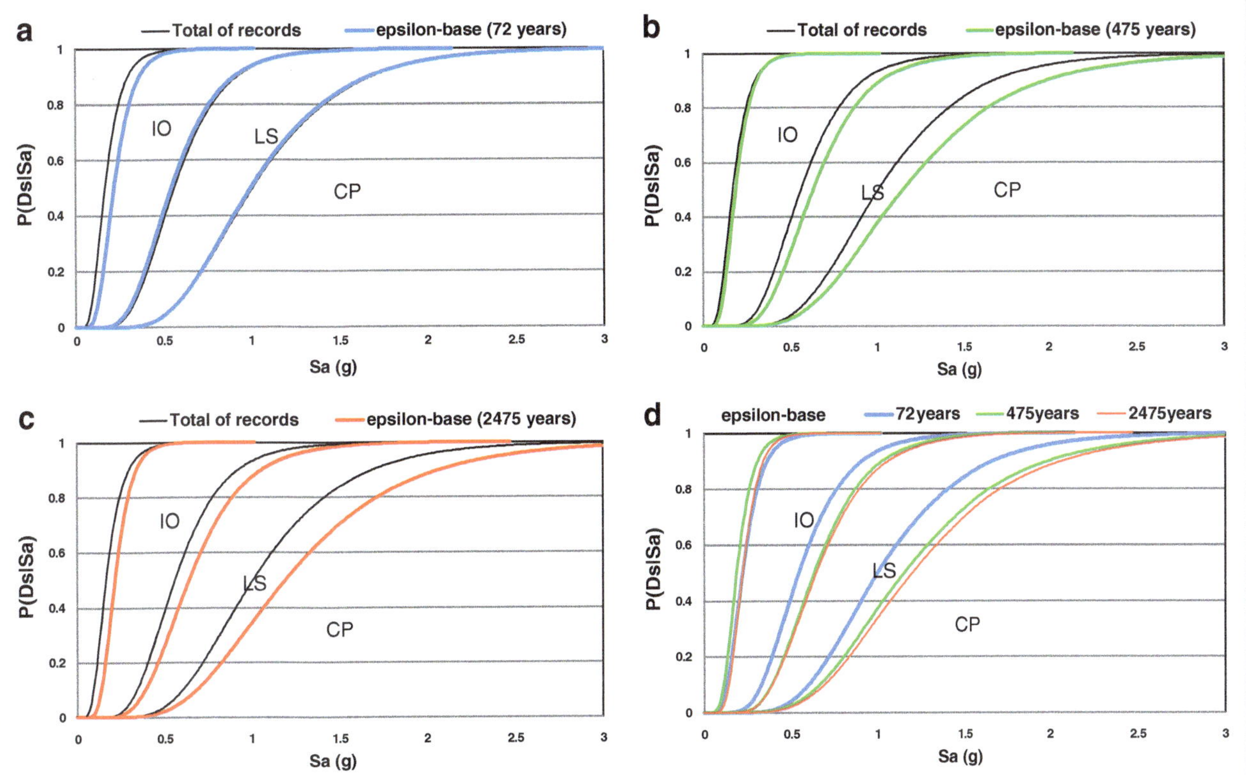

Figure 7 The fragility curves of 5SW-2, with and without regard to epsilon. Return periods of **(a)** 72 years, **(b)** 475 years, and **(c)** 2,475 years. **(d)** Final fragility curves using epsilon.

parameters are described in detail in the related article by Jeong and Elnashai. Mathematically, the damage index for x direction is given as

$$D_{gnx} = \sum_{i=1} D_i \frac{W_{i,\min}}{W_{\text{total}}} + \sum_{i\neq j} \left\{ \frac{W_{CF,ij}}{2 \cdot W_{\text{total}}} \left[\max(D_i, D_j)^2 - D_i \cdot D_j \right] \right\} \tag{1}$$

where D_{gnx} is the x-global damage index in storey n, D_i is the local damage index of shear wall, W_{total} is the total effective weight in storey n, $W_{i,\min}$ is the tributary weight of shear wall i in storey n, and W_{CFij} is the common failure consequence weight between local shear walls i and j.

Fragility curve results

As mentioned previously, the most important element at risk is vulnerability of structures in the city, and the vulnerability of a building can be described using fragility curves. Structural damage fragility curves are described (in this research by values of drift ratio) that define the thresholds of different damage states at a specified

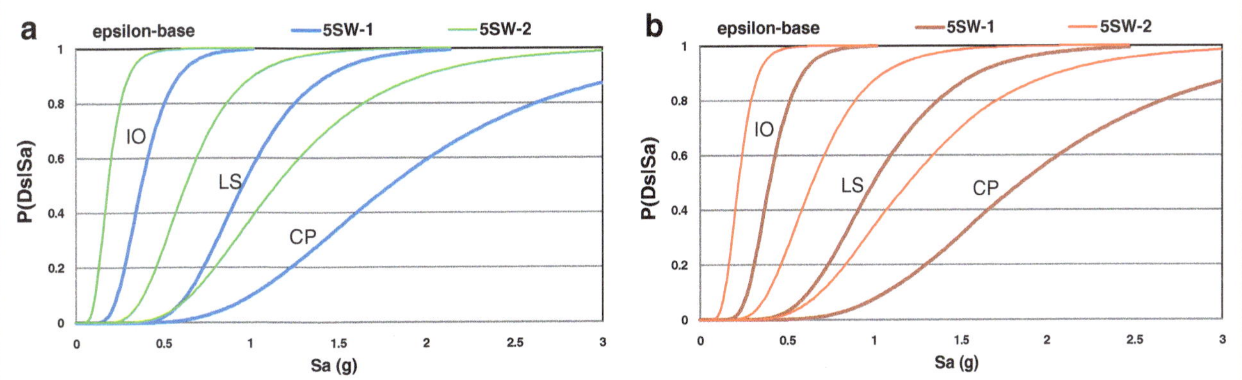

Figure 8 Fragility curves of steel frame with concrete shear walls, with regard to epsilon. Return periods of **(a)** 475 years and **(b)** 2,475 years.

Table 4 Spectral acceleration in the first mode of structure at different levels from PSHA

Model	First mode period (s)	Probability of occurrence		
		50% in 50-year return period (72 years)	10% in 50-year return period (475 years)	2% in 50-year return period (2,475 years)
5SW-1	0.54	0.10 g	0.45 g	0.75 g
5SW-2	1.01	0.07 g	0.16 g	0.28 g

hazard level. Using the strong ground motion indices, S_a, and the damage ratio, fragility curves were constructed and the cumulative probability of occurrence of damage was assumed to be lognormal (Miranda and Aslani 2003). The definition is expressed by the following mathematical formulation:

$$F = P(d > D_i|\mathrm{IM}) \tag{2}$$

where F is the fragility function, P is the probability function, d denotes damage level of structure, D_i presents ith damage states, IM denotes ground motion intensity parameter (S_a), and (i =1 to n) shows different damage states.

Comparison of the results from lognormal fit in Figures 6 and 7 shows higher values of the median damage capacity when record selection based on epsilon is employed. The differences in fragility curves increase with the increase of the epsilon in the record selection as well as hazard level in the site. As shown in Figures 6 and 7, a change of epsilon-based record selection from 72- to 2,475-year return periods reduces the probability of exceedance expected performance levels by less than 20%. Similarly, Figures 6 and 7 show that if ε had been neglected in our simulations, the median predicted structural capacities would be decreased by less than 10%, 15%, and 18% for 72-, 475-, and 2,475-year return periods, respectively. These results demonstrate the importance of ground motion acceleration history selection criteria in accurately predicting building limit states.

Figure 8 shows comparison of fragility curves for two representative structures. According to these curves, we surmise that expected damages for most of the steel frames with concrete shear walls approximately are on these limits.

Table 4 shows the S_a at the first period of the two structural models for three hazard levels in the city of Mashhad (Zolfaghari and Ghafory-Ashtiany 2012). Figure 9 shows the estimated probability of exceedance of expected damage states for the typical buildings.

Conclusions

A procedure for developing fragility curves for spectral acceleration and epsilon to be used for existing five-storey steel frame with cast-in-place concrete shear wall buildings is demonstrated. The methodology employs a proposed technique for calculating the probability of exceedance of three limit states for irregular buildings in plan by Jeong and Elnashai. The distribution of damage indices in each ground motion intensity measure (S_a) are estimated by numerous inelastic incremental dynamic analysis. Unlike alternative methodologies, the proposed procedure allows estimating the fragilities based on weighed combination of interstorey drift demand of the lateral resistance systems, which is an engineering demand parameter closely related to structural damage.

The epsilon-based method was taken into account for earthquake ground motion selection, whereas the disaggregation analysis was implemented on the Mashhad site. The epsilon demands reflected the expectation (−1.0 to 0.0), (+1.0 to +2.0), and (+2.0 to 3.0) for 50%, 10%, and 2% probability of exceedance in 50 years, respectively.

The results show that a change in epsilon-based record selection decreases the probability of exceedance expected performance levels by less than 10%, 15%, and 18% for 72-, 475-, and 2,475-year return periods, respectively. It seems that the regarded typical building (steel frame with concrete shear walls) in high hazard levels is safe for collapse in the site.

Therefore, the proposed method to derive fragility curves is recommended for the probabilistic seismic assessment of typical buildings with significant torsional

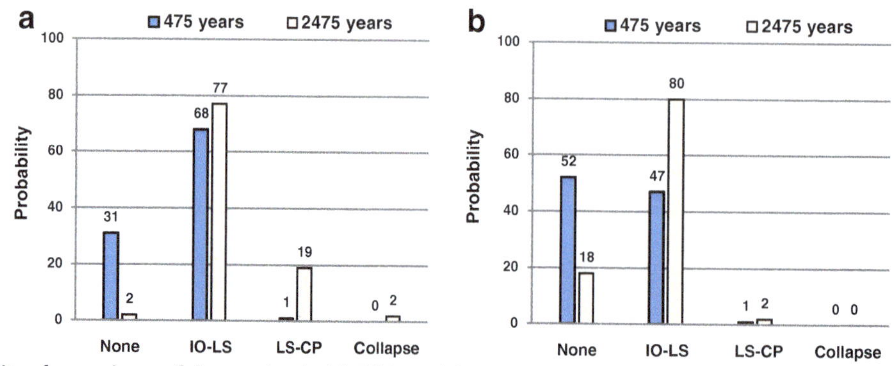

Figure 9 Probability of exceedance of damage levels. (a) 5SW-1 and **(b)** 5SW-2.

and bi-directional responses. Deriving curves leads to more reliable damage assessment in terms of interstorey drifts for typical steel frame with concrete shear walls in Mashhad.

Appendix

The M6.5 scenario database of SGM records is shown in Table 5.

Table 5 M6.5 scenario database of SGM

Number	Event	Year	Station	*M*	PGA	*R*	V_{s30}
1	Bandar Abbas	1975	Bandar Abbas	6.1	0.13	36	337
2	Tabas	1978	Boshruyeh	7.4	0.10	55	564
3	Tabas	1978	Tabas	7.4	0.81	54	645
4	Tularud	1978	Talesh	6.0	0.24	15	539
5	Qaen	1979	Khezri	7.1	0.10	75	701
6	Qaen	1979	Gonabad	7.1	0.10	93	529
7	Golbaft	1981	Golbaft	7.0	0.28	13	365
8	Manjil	1990	Qazvin	7.4	0.27	94	456
9	Manjil	1990	Abhar	7.4	0.21	101	291
10	Manjil	1990	Ab-bar	7.4	0.57	41	291
11	Islamabad	1997	Kariq	6.0	0.57	48	589
12	Avaj	2002	Kaboodar-A	6.5	0.16	62	613
13	Avaj	2002	Razan	6.5	0.20	35	314
14	Firoozabad	2004	Hasan-Keyf	6.3	0.84	42	339
15	Firoozabad	2004	Moalem-Kel	6.3	0.29	99	490
16	Enchehborun	2005	Agh-Gala	6.1	0.12	14	341
17	Zarand	2005	Zarand	6.4	0.32	16	226
18	Erzurum	1983	Meteorologi-Ist	6.6	0.17	35	316
19	Adana	1998	Tarum-Ilce	6.2	0.28	48	263
20	Adana	1998	Meteorologi-Ist	6.2	0.13	65	366
21	Kocaeli	1999	Devlet-Has	7.4	0.14	81	348
22	Kocaeli	1999	Meteorologi-Ist	7.4	0.37	101	282
23	Kocaeli	1999	Marmara-Ara	7.4	0.12	43	701
24	Duzce	1999	Bayindirlik	7.1	0.81	36	294
25	Bingol	2003	Bayindirliklsk	6.3	0.50	12	529
26	Northridge	1994	Beverly Hills	6.7	0.49	13	356
27	Northridge	1994	Can country	6.7	0.48	27	309
28	Friuli	1976	Tolmezzo	6.5	0.34	20	425
29	San Fernando	1971	LA - Hollywood	6.6	0.18	40	316
30	Imperial Valley	1979	Delta	6.5	0.34	34	275
31	Landers	1992	Yermo Fire	7.3	0.24	86	354
32	Loma Prieta	1989	Hollister City Hall	6.9	0.25	47	199
33	New Zealand	2010	Heathcote-V	7.0	0.63	43	-
34	Chi-Chi	1999	Chi-CHY101	7.6	0.40	32	259
35	Chi-Chi	1999	Chi-TCU045	7.6	0.47	76	701

Author details

[1]PhD candidate, Department of Civil Engineering, Science and Research Branch, Islamic Azad University, Tehran, Iran. [2]International Institute of Earthquake Engineering and Seismology (IIEES), No. 26, Arghavan Street, Tehran 19395-3913, Iran. [3]Department of Civil Engineering, Faculty of Engineering, Arak University, P.O. Box 38156-88359, Arak, Iran.

References

Agency FEM (1997a) Earthquake loss estimation methodology: HAZUS. FEMA, Washington DC

Agency FEM (1997b) FEMA 356 NEHRP prestandard and commentary for the seismic rehabilitation of buildings. FEMA, Washington, DC

Baker JW, Cornell CA (2006) Spectral shape, epsilon and record selection. Earthquake Eng Struct Dynam 34(10):1193–1217

Bazzurro P, Cornell CA (1999) Disaggregation of seismic hazard. Bull Seismol Soc Am 89(2):501–520

Cornell A, Zareian F, Krawinkler H, Miranada E (2005) Prediction of probability of collapse. In: Krawinkler H (ed) Van Nuys Hotel building testbed report: exercising seismic performance assessment. vol 4.5. Pacific Earthquake Engineering Research, Berkeley, pp 85–93

Estekanchi H, Arjomandi K (2007) Comparison of damage indexes in nonlinear time history analysis of steel moment frames. Asian J Civ Eng 8(6):629–646

Goulet C, Haselton C, Mitrani-Reiser J, Stewart JP, Taciroglu E, Deierlein G (2006) Evaluation of the seismic performance of a code-conforming reinforced-concrete frame building—part I: ground motion selection and structural collapse simulation. Proceedings of the 8th national conference on earthquake engineering, San Francisco, CA, Paper NCEE-1576

Hafezi-Moghaddas N (2007) Seismic microzonation of the Mashhad city. Technical report. Khorasan-Razavi Housing and Urban Development Organization, Washington, DC

Haselton CB, Baker JW (2006) Ground motion intensity measures for collapse capacity prediction: choice of optimal spectral period and effect of spectral shape. Proceedings of the 8th national conference on earthquake engineering, San Francisco, CA

Hatefi H, Ghafory-Ashtiany M (2010) Nonlinear dynamic analysis based on M6.5 strong ground motion database, M.Sc. Thesis. Department of Civil Engineering, IIEES, Tehran, Iran

Jafari MA, Hossaini-Hashemi B (2008) Experimental investigation on seismic behavior of batten columns. Ph.D. dissertation. Department of Civil and Earthquake Engineering, IIEES, Tehran, Iran

Jeong SH (2005) Analytical damage assessment. Experimental and analytical seismic assessment of buildings with plan irregularities, Ph.D. thesis. University of Illinois, Urbana

Jeong SH, Elnashai AS (2006) New three-dimensional damage index for RC buildings with plan irregularities. J Struct Eng ASCE 132(9):1482–90

JICA (2000) The study on seismic microzoning of the Greater Tehran area in the Islamic Republic of Iran. A report from Center for Earthquake an Environmental Studies of Tehran (CEST) and Japan International Cooperation Agency (JICA), Tehran, Iran

Miranda E, Aslani H (2003) Probabilistic response assessment for building-specific loss estimation. Report PEER 2003/03. Pacific Earthquake Engineering Research (PEER) Center, Richmond, CA

Mousavi M, Ghafory-Ashtiany M, Azarbakht A (2011) A new indicator of elastic spectral shape for the reliable selection of ground motion records. Earthquake Eng Struct Dynam 40(12):1403–1416

OpenSees, Open System for Earthquake Engineering Simulation (1997) Pacific Earthquake Engineering Research Center. University of California, Berkeley, California

Shokrzadeh M, Tasnimi AA (1995) Experimental study on ductility and behavior of shear walls using in R/C conventional Iranian construction under lateral load, M.Sc. Thesis. , Tehran, Iran

Thomsen JH, Wallace W (1995) Displacement-based design of reinforced concrete structural walls: an experimental investigation of walls with rectangular and T-shaped cross-sections. Report no. CU/CEE–95/06. Department of Civil Engineering Clarkson University, Postdam, New York

Vamvatsikos D, Cornell CA (2002) Incremental dynamic analysis. Earthq Eng Struct Dyn 31(3):491–514
Zareian F (2006) Simplified performance-based earthquake engineering, Ph.D. dissertation. Department of Civil and Environmental Engineering, Stanford University, Stanford, CA
Zolfaghari MR, Ghafory-Ashtiany M (2012) PSHA of the Mashhad city. Technical report. , Mashhad, Iran

Permissions

All chapters in this book were first published in IJASE, by Springer; hereby published with permission under the Creative Commons Attribution License or equivalent. Every chapter published in this book has been scrutinized by our experts. Their significance has been extensively debated. The topics covered herein carry significant findings which will fuel the growth of the discipline. They may even be implemented as practical applications or may be referred to as a beginning point for another development.

The contributors of this book come from diverse backgrounds, making this book a truly international effort. This book will bring forth new frontiers with its revolutionizing research information and detailed analysis of the nascent developments around the world.

We would like to thank all the contributing authors for lending their expertise to make the book truly unique. They have played a crucial role in the development of this book. Without their invaluable contributions this book wouldn't have been possible. They have made vital efforts to compile up to date information on the varied aspects of this subject to make this book a valuable addition to the collection of many professionals and students.

This book was conceptualized with the vision of imparting up-to-date information and advanced data in this field. To ensure the same, a matchless editorial board was set up. Every individual on the board went through rigorous rounds of assessment to prove their worth. After which they invested a large part of their time researching and compiling the most relevant data for our readers.

The editorial board has been involved in producing this book since its inception. They have spent rigorous hours researching and exploring the diverse topics which have resulted in the successful publishing of this book. They have passed on their knowledge of decades through this book. To expedite this challenging task, the publisher supported the team at every step. A small team of assistant editors was also appointed to further simplify the editing procedure and attain best results for the readers.

Apart from the editorial board, the designing team has also invested a significant amount of their time in understanding the subject and creating the most relevant covers. They scrutinized every image to scout for the most suitable representation of the subject and create an appropriate cover for the book.

The publishing team has been an ardent support to the editorial, designing and production team. Their endless efforts to recruit the best for this project, has resulted in the accomplishment of this book. They are a veteran in the field of academics and their pool of knowledge is as vast as their experience in printing. Their expertise and guidance has proved useful at every step. Their uncompromising quality standards have made this book an exceptional effort. Their encouragement from time to time has been an inspiration for everyone.

The publisher and the editorial board hope that this book will prove to be a valuable piece of knowledge for researchers, students, practitioners and scholars across the globe.

List of Contributors

Payam Soltani
Department of Mechanical Engineering, Semnan Branch, Islamic Azad University, Semnam, Iran
Department of Mechanical Engineering, Ferdowsi University of Mashhad, Mashhad, Iran

Amir Kassaei
Department of Mechanical Engineering, Semnan Branch, Islamic Azad University, Semnam, Iran

Mohammad Mehdi Taherian
Young Researchers Club, Semnan Branch, Islamic Azad University, Semnam, Iran

Anooshiravan Farshidianfar
Department of Mechanical Engineering, Ferdowsi University of Mashhad, Mashhad, Iran

Hugo Rodrigues
Civil Engineering Department, University of Aveiro, Aveiro, Portugal
Faculty of Natural Sciences, Engineering and Technology, Oporto Lusophone University, Oporto, Portugal
Civil Engineering Department, University of Aveiro, Campus Universitário de Santiago, Aveiro 3810-193, Portugal

Humberto Varum
Civil Engineering Department, University of Aveiro, Aveiro, Portugal

Antonio Arêde
Civil Engineering Department, Faculty of Engineering, University of Porto, Oporto, Portugal

Aníbal G Costa
Civil Engineering Department, University of Aveiro, Aveiro, Portugal

Kishori Lal Verma
Department of Mathematics, Government Post Graduate College, Hamirpur,Himachal Pradesh 177005, India

Emili Bhattacharjee
Department of Civil Engineering, National Institute of Technology, Agartala, Jirania, Barjala 799055, India

Lipika Halder
Department of Civil Engineering, National Institute of Technology, Agartala, Jirania, Barjala 799055, India

Richi Prasad Sharma
Department of Civil Engineering, National Institute of Technology, Agartala, Jirania, Barjala 799055, India

Pramod Kumar Gupta
Department of Civil Engineering, I.I.T. Roorkee, Roorkee, India

A Sofi
VIT University, Vellore, Tamil Nadu 632014, India

K Swathy
VIT University, Vellore, Tamil Nadu 632014, India

G Srija
VIT University, Vellore, Tamil Nadu 632014, India

Hisham M Khater
Housing and Building National Research Centre (HBNRC), 87 El-Tahreer St., Dokki, Giza, P.O. Box 1770, Cairo, Egypt

Heting Qiao
School of Mechanical Engineering, Shenyang University of Technology, Shenyang 110004, People's Republic of China

Haiyan Li
School of Mechanical Engineering, Shenyang University of Technology, Shenyang 110004, People's Republic of China

Shehata E Abdel Raheem
Taibah University, Madinah Munawarh, P.O. 344, Kingdom of Saudi Arabia
Civil Engineering Department, Faculty of Engineering, Assiut University, Assiut 71516, Egypt

Kaza Vijayakumar
Department of Aerospace Engineering, Indian Institute of Science, Bangalore 560012, India

Shehata E Abdel Raheem
Taibah University, Medina, Saudi Arabia
Civil Engineering Department, Faculty of Engineering, Assiut University, Assiut 71516, Egypt

Toshiro Hayashikawa
Graduate School of Engineering, Hokkaido University, Sapporo, Japan

Uday Naik
Department of Civil Engineering, K.K. Wagh Institute of Engineering and Research Centre, Nasik Pin 422003, Maharashtra, India

Sunil Kute
Department of Civil Engineering, K.K. Wagh Institute of Engineering and Research Centre, Nasik Pin 422003, Maharashtra, India

Luis A Montejo
Department of Engineering Science and Materials, University of Puerto Ric at Mayaguez, Mayaguez, 00680, Puerto Rico

Luis E Suarez
Department of Civil Engineering and Surveying, University of Puerto Rico at Mayaguez, Mayaguez, 00680, Puerto Rico

Maziar Ramezani
Centre for Advanced Manufacturing Technology (CAMTEC), School of Engineering, Auckland University of Technology, Auckland 1010 New Zealand

Juan Vilches
Centre for Advanced Manufacturing Technology (CAMTEC), School of Engineering, Auckland University of Technology, Auckland 1010 New Zealand

Thomas Neitzert
Centre for Advanced Manufacturing Technology (CAMTEC), School of Engineering, Auckland University of Technology, Auckland 1010 New Zealand

Ayaho Miyamoto
Graduate School of Science & Engineering, Yamaguchi University, 2-16-1, Tokiwadai, Ube, Yamaguchi 755-8611, Japan

Minoru Motoshita
Kyowa Sekkei CO., LTD. Ushitora, Osaka, Ibaraki 567-0877, Japan

Joan R Casas
Department of Construction Engineering, School of Civil Engineering, Technical University of Catalunya
Campus Nord, c/Jordi Girona 1-3, Modulo C-1, Barcelona 08034, Spain

Satish B Satpal
Department of Mechanical Engineering, Indian Institute of Technology Bombay, Powai, Mumbai 400076, India

Yogesh Khandare
Department of Mechanical Engineering, Indian Institute of Technology Bombay, Powai, Mumbai 400076, India

Anirban Guha
Department of Mechanical Engineering, Indian Institute of Technology Bombay, Powai, Mumbai 400076, India

Sauvik Banerjee
Department of Civil Engineering, Indian Institute of Technology Bombay, Powai, Mumbai 400076, India

Ram Ranjan Sahu
Civil Engineering Department, Indian Institute of Technology Roorkee, Uttarakhand Pin 247667, India
Engineering Research Centre, Tata Motors-Pune, India

Pramod Kumar Gupta
Department of Civil Engineering, Indian Institute of Technology Roorkee, Uttarakhand Pin 247667, India

Morteza Esmaeili
School of Railway Engineering, Iran University of Science and Technology, University Street, Hengam Aveneu, Narmak, Tehran 1684613114, Iran

Jabbar Ali Zakeri
School of Railway Engineering, Iran University of Science and Technology, University Street, Hengam Aveneu, Narmak, Tehran 1684613114, Iran

Parisa Haji Abdulrazagh
School of Civil Engineering, University of Tehran, Tehran 111554563, Iran

Mussa Mahmoudi
Faculty of Civil Engineering, Shahid Rajaee Teacher Training University, Tehran, Iran

Mahdi Zaree
Faculty of Civil Engineering, Shahid Rajaee Teacher Training University, Tehran, Iran

Mallikarjuna D Reddy
Reva Institute of Technology and Management, Bangalore Karnataka 560064, India

Seetharaman Swarnamani
Indian Institute of Technology Madras, Chennai, Tamilnadu 600036, India

Amir Hosein Salmanpour
Institute of Structural Engineering, ETH Zurich, CH-8093, Zurich, Switzerland.

Nebojsa Mojsilovic
Institute of Structural Engineering, ETH Zurich, CH-8093, Zurich, Switzerland

Joseph Schwartz
Institute of Technology in Architecture, ETH Zurich, CH-8093, Zurich, Switzerland

Gaurav Agarwal
Department of Mechanical Engineering, S.R.M.S.C.E.T, Bareilly 243122, India

Amar Patnaik
Department of Mechanical Engineering, National Institute of Technology, Hamirpur 177005, India

Rajesh Kumar Sharma
Department of Mechanical Engineering, National Institute of Technology, Hamirpur 177005, India

Shailendra Kumar Damodar Dubey
Civil Engineering Department SSVPS BSD College of Engineering, Dhule, India

Sunil Y Kute
Civil Engineering Department, K.K. Wagh College of Engineering, Nasik, India

Yasser Khodair
Department of Civil Engineering and Construction, Bradley University, 1501 West Bradley Avenue, Peoria, IL 61625, USA

Sophia Hassiotis
Department of Civil, Environmental and Ocean Engineering, Stevens Institute of Technology, Castle Point on Hudson, Hoboken, NJ 07054, USA

Hamid Kazemi
PhD candidate, Department of Civil Engineering, Science and Research Branch, Islamic Azad University, Tehran, Iran

Mohsen Ghafory-Ashtiany
International Institute of Earthquake Engineering and Seismology (IIEES), No. 26, Arghavan Street,
Tehran 19395-3913, Iran

Alireza Azarbakht
Department of Civil Engineering, Faculty of Engineering, Arak University, P.O. Box 38156-88359, Arak, Iran